Claus W. Heizmann (Ed.)

Novel Calcium-Binding Proteins

Fundamentals and Clinical Implications

With 148 Figures

Springer-Verlag
Berlin Heidelberg New York London
Paris Tokyo Hong Kong Barcelona
Budapest

Professor Dr. CLAUS W. HEIZMANN
Department of Pediatrics
Division of Clinical Chemistry
University of Zürich
Steinwiesstraße 75
CH-8032 Zürich

ISBN 3-540-53277-3 Springer-Verlag Berlin Heidelberg New York
ISBN 0-387-53277-3 Springer-Verlag New York Berlin Heidelberg

Library of Congress Cataloging-in-Publication Data. Novel calcium-binding proteins: fundamentals and clinical implications / Claus W. Heizmann, ed. p. cm. Includes indexes. ISBN 3-540-53277-3 (Berlin: alk. paper). − ISBN 0-387-53277-3 (New York: alk. paper) 1. Calcium-binding proteins. I. Heizmann, Claus W. QP552.C24N68 1991. 591.19′245 − dc20. 91-6326

© Springer-Verlag Berlin Heidelberg 1991
Printed in Germany

31/3145(3011)-543210 − Printed on acid-free paper

TABLE OF CONTENTS

VI

SECTION V   **CALCIUM FLUXES, BINDING, AND METABOLISM**

Takashi Murachi

19 September 1926 to 12 May 1990

The sudden death of Prof. T. Murachi has deprived the scientific community of one of its most gifted biochemists. He had accepted my invitation to contribute a chapter to this book and was in the process of delivering his manuscript entitled: *Calpain and Calpastatin: Structure, Function, and Gene Expression*. On the afternoon of May 12 he returned from a lecture given at the Kansai Medical University and entered his office in order to prepare manuscripts for an upcoming meeting of the Japanese Society of Biochemistry. He was found dead in his office.

Takashi Murachi was born 1926 in Kyoto, Japan, and studied biochemistry at the Kyoto University, where he received his degree (D.Med.Sc). From 1952 to 1957 he was instructor at the Department of Medical Chemistry; from 1957 to 1959 research associate at the Department of Biochemistry (Prof. H. Neurath) at the University of Washington, Seattle, USA. He was appointed Associate Professor (1959-1965) at the Department of Biochemistry, Nagoya City University School of Medicine, Nagoya. Thereafter he was promoted to Professor and Chairman (1966-1973) at the same University. He then moved back to Kyoto University as Professor and Chairman of the Department of Clinical Science and Laboratory Medicine, Faculty of Medicine, and Director of the Central Clinical Laboratory at the Kyoto University Hospital.

Prof. Murachi received many honors; for example, he was a member of the Committee on Biochemistry of the Science Council of Japan; Chairman, Nominating Committee of the International Union of Biochemistry; Foreign Honorary Member of the American Society of Biochemistry and Molecular Biology; Chester Beatty Lecturer at the

14th International Congress of Biochemistry, Prague, Czechoslovakia; and he received the Shimadzu Science Foundation Prize in January 1990.

Prof. Murachi was also President (1987-1989) of the Federation of Asian and Oceanian Biochemists and successfully organized the 5th FAOB Congress held in Seoul, August 1989. As Chairman of the Symposium on Clinical Chemistry I had the honor of introducing Prof. Murachi for his stimulating lecture entitled: The Calpain-Calpastatin System in Health and Disease.

Prof. Murachi had a number of scientific interests (biochemistry, molecular biology, clinical chemistry) and published many key papers on various topics.

He was well recognized among the scientists working in the field of calcium regulation for his outstanding work on the calpain/calpastatin regulatory system. In honor of his work and his excellent contributions to science, I will list some of his more recent publications on the structure and function of calpain.

We mourn the loss of an outstanding scientist, a dear friend and colleague.

C.W. Heizmann

SELECTED RECENT PUBLICATIONS

Ando Y, Miyachi Y, Imamura S, Kannagi R, Murachi T (1988) Purification and characterization of calpains from pig epidermis and their action of epidermal keratin. J Invest Dermatol 90:26-30

Maki M, Bagci H, Hamaguchi K, Ueda M, Murachi T, Hatanaka M (1989) Inhibition of calpain by a synthetic oligopeptide corresponding to an exon of the human calpastatin gene. J Biol Chem 264:18866-18869

Maki M, Tanako E, Osawa T, Ooi T, Murachi T, Hatanaka M (1988) Analysis of structure-function relationship of pig calpastatin by expression of mutated cDNAs in *Escherichia coli*. J Biol Chem 263:10254-10261

Murachi T (1989) Intracellular regulatory system involving calpain and calpastatin. Biochem Int 18:263-294

Murachi T, Murakami T, Ueda M, Fukui I, Hamakubo T, Adachi Y, Hatanaka M (1989) The calpain-calpastatin system in hematopoietic cells. Adv Exp Med Biol 255:445-454

Murachi T, Tanako E, Maki M, Adachi Y, Hatanaka M (1989) Cloning and expression of the genes for calpains and calpastatins. Biochem Soc Symp 55:29-44

Shearer TR, Azuma M, David LL, Yamagata Y, Murachi T (1990) Calpain and calpastatin in rabbit corneal epithelium. Curr Eye Res 9:39-44

Takano E, Park YH, Kitahara A, Yamagata Y, Kannagi R, Murachi T (1988) Distribution of calpains and calpastatin in human blood cells. Biochem Int 16:391-395

Tanako E, Hamakubo T, Kawatani Y, Ueda M, Kannagi R, Murachi T (1989) Multiple forms of calpastatin in pig brain. Biochem Int 19:633-643

Tanako E, Maki M, Mori H, Hatanaka M, Marti T, Titani K, Kannagi R, Ooi T, Murachi T (1988) Pig heart calpastatin: identification of repetitive domain structures and anomalous behavior in polyacrylamide gel electrophoresis. Biochemistry 27:1964-1972

Uemori T, Shimojo T, Asada K, Asano T, Kimizuka F, Kato I, Maki M, Hatanaka M, Murachi T, Hanzawa H et al. (1990) Biochem Biophys Res Commun 166:1485-1493

Table of Contributors

Babitch Joseph A.
Department of Chemistry
Texas Christian University
P.O. Box 32908
Fort Worth, TX 76129
USA

Balasubramanian Doraivajan
Center for Cellular and
Molecular Biology
Uppal Road
Hyderabad 500 007
INDIA

Barraclough Roger
Department of Biochemistry
University of Liverpool
P.O. Box 147
Liverpool L69 3BX
UNITED KINGDOM

Booth Karla S.
Department of Biochemistry
Colorado State University
Fort Collins, CO 80523
USA

Brüggen Josef
Department of Biotechnology
CIBA-GEIGY LTD.
CH-4002 Basel
SWITZERLAND

Buchner Erich
Institut für Genetik und
Mikrobiologie
Universität Würzburg
Röntgenring 11
D-W-8700 Würzburg
GERMANY

Buchner Sigrid
Institut für Genetik und
Mikrobiologie
Universität Würzburg
Röntgenring 11
D-W-8700 Würzburg
GERMANY

Bühler Fritz R.
Department of Medicine and
Research
University Hospital
CH-4031 Basel
SWITZERLAND

Caughey Winston S.
Department of Biochemistry
Colorado State University
Fort Collins, CO 80523
USA

Cerletti Nico
Department of Biotechnology
CIBA-GEIGY LTD.
CH-4002 Basel
SWITZERLAND

Cirino Giuseppe
Department of Experimental
Pharmacology
Via Domenico Montesano 49
I-80131 Naples
ITALY

Cox Jos A.
Department of Biochemistry
Faculty of Science
University of Geneva
CH-1211 Geneva
SWITZERLAND

Crabos Maryse
Department of Medicine
and Research
University Hospital
CH-4031 Basel
SWITZERLAND

Crumpton Michael J.
Cell Surface Biochem. Lab.
The Imperial Cancer
Research Fund
44 Lincoln's Inn Fields
London WC2A 3PX
UNITED KINGDOM

Dumont Jacques
Inst. de Recherche Interdiscipl.
Faculté de Médecine
Université Libre de Bruxelles
808 Route de Lennik
B-1070 Bruxelles
BELGIUM

Edashige Keisuke
Department of Medical Biology
Kochi Medical School
Nankoku-shi, Kochi 781-51
JAPAN

Edwards Helena C.
Cell Surface Biochem. Lab.
The Imperial Cancer
Research Fund
44 Lincoln's Inn Fields
London WC2A 3PX
UNITED KINGDOM

Engelkamp Dieter
Abteilung für Klinische Chemie
Universitäts-Kinderklinik
Steinwiesstrasse 75
CH-8032 Zürich
SWITZERLAND

Erneux Christophe
Inst. de Recherche Interdiscipl.
Faculté de Médecine
Université Libre de Bruxelles
808 Route de Lennik
B-1070 Bruxelles
BELGIUM

Erne Paul
Department of Medicine and
Research
University Hospital
CH-4031 Basel
SWITZERLAND

Faust Claudia
Institut für Genetik und
Mikrobiologie
Universität Würzburg
Röntgenring 11
D-W-8700 Würzburg
GERMANY

2

Ferrus Alberto
Instituto Cajal
CSIC
E-28006 Madrid
SPAIN

Flower R.J.
Department of Biochemical
Pharmacology
St. Bartholomew's Hospital
Medical College
London EC1M 6BQ
UNITED KINGDOM

Galea S.
Department of Immunology
University of Toronto and
Toronto Western Hospital
399 Bathurst Street
Toronto, Ontario M5T 2S8
CANADA

Gerke Volker
Max-Planck-Institute for
Biophysical Chemistry
Department of Biochemistry
Am Fassberg
D-W-3400 Göttingen
GERMANY

Gong Y.
Department of Physiology
University of Manitoba
770 Bannatyne Avenue
Winnipeg R3E OW3
CANADA

Heizmann Claus W.
Abteilung für Klinische Chemie
Universitäts-Kinderklinik
Steinwiesstrasse 75
CH-8032 Zürich
SWITZERLAND

Heimbeck Gertrud
Institut für Genetik und
Mikrobiologie
Universität Würzburg
Röntgenring 11
D-W-8700 Würzburg
GERMANY

Hilt Dana C.
The University of Maryland
School of Medicine
Department of Neurology
22 South Greene Street
Baltimore, MD 21201
USA

Hofbauer Alois
Institut für Genetik und
Mikrobiologie
Universität Würzburg
Röntgenring 11
D-W-8700 Würzburg
GERMANY

Ikeda-Saito Masao
Department of Physiology and
Biophysics
Case Western Reserve University
School of Medicine
Cleveland, OH 44106
USA

Inoue Masayasu
Department of Biochemistry
Kumamoto University
Medical School
Kumamoto 860
JAPAN

Inouye Masayori
Department of Biochemistry
University of Medicine and
Dentistry of New Jersey
675 Hoes Lane
Piscataway, NJ 08854-5635
USA

Inouye Sumiko
Department of Biochemistry
University of Medicine and
Dentistry of New Jersey
675 Hoes Lane
Piscataway, NJ 08854-5635
USA

Jacobowitz David M.
Laboratory of Clinical Science
NIMH, Bldg. 10, Rm. 3D-48
Bethesda, MD 20892
USA

Jongstra Jan
Department of Immunology
University of Toronto and
Toronto Western Hospital
399 Bathurst Street
Toronto, Ontario M5T 2S8
CANADA

Jongstra-Bilen Jenny
Department of Immunology
University of Toronto and
Toronto Western Hospital
399 Bathurst Street
Toronto, Ontario M5T 2S8
CANADA

Kimura Shioko
Laboratory of Molecular
Carcinogenesis
National Cancer Institute
Bethesda, MD 20892
USA

Klein David Peter
Department of Immunology
University of Toronto and
Toronto Western Hospital
399 Bathurst Street
Toronto, Ontario M5T 2S8
CANADA

Klein William H.
Dept. of Biochem. & Mol. Biol.,
Box 117, The Univ. of Texas
M.D. Anderson Canc. Ctr.
1515 Holcombe Boulevard
Houston, TX 77030
USA

Kligman Douglas
University of Miami School
of Medicine
Miami, FL 33103
USA

Krah-Jentgens Imke
Lehrstuhl für Biochemie
Ruhr-Universität Bochum
D-W-4630 Bochum 1
GERMANY

Kretsinger Robert H.
Department of Biology
University of Virginia
Charlottesville, VA 22901
USA

Kuznicki Jacek
Nencki Institute of Experimental
Biology
3 Pasteur St.
PL-02-093 Warsaw
POLAND

Lagasse Eric
Laboratory of Exp. Oncology
Department of Pathology
Howard Hughes Medical Institute
Stanford Univ. School of Med.
Stanford, CA 94305-5428
USA

Lamy Françoise
Inst. de Recherche Interdiscipl.
Faculté de Médecine
Université Libre de Bruxelles
808 Route de Lennik
B-1070 Bruxelles
BELGIUM

Lecocq Raymond
Inst. de Recherche Interdiscipl.
Faculté de Médecine
Université Libre de Bruxelles
808 Route de Lennik
B-1070 Bruxelles
BELGIUM

Lefort Anne
Inst. de Recherche Interdiscipl.
Faculté de Médecine
Université Libre de Bruxelles
808 Route de Lennik
B-1070 Bruxelles
BELGIUM

Libert Frédérick
Inst. de Recherche Interdiscipl.
Faculté de Médecine   -
Université Libre de Bruxelles
808 Route de Lennik
B-1070 Bruxelles
BELGIUM

Lipski Norbert
Institut für Genetik und
Mikrobiologie
Universität Würzburg
Röntgenring 11
D-W-8700 Würzburg
GERMANY

Longbottom David
MRC Human Genetics Unit
Western General Hospital
Crewe Road
Edinburgh EH4 2XU
UNITED KINGDOM

Luan-Rilliet Ying
Department of Biochemistry
Faculty of Science
University of Geneva
CH-1211 Geneva
SWITZERLAND

Masiakowski Piotr
Regeneron Pharmaceuticals, Inc.
777 Old Saw Mill River Road
Tarrytown, NY 10591-6707
USA

Meyers Marian B.
Laboratory of Cellular and
Biochemical Genetics, Mem.
Sloan-Kettering Cancer Ctr.
1275 York Avenue
New York, NY 10021
USA

Moss Stephen E.
Department of Physiology
University College London
Gower Street
London, WC1E 6BT
UNITED KINGDOM

Murphy Leigh C.
Department of Biochemistry and
Molecular Biology
University of Manitoba
770 Bannatyne Avenue
Winnipeg R3E OW3
CANADA

Nakayama Susumu
Department of Biology
University of Virginia
Charlottesville, VA 22901
USA

Noegel Angelika A.
Max-Planck-Institut für
Biochemie
D-W-8033 Martinsried
GERMANY

Pearson William
Department of Biochemistry
University of Virginia
Charlottesville, VA 22901
USA

Pflugfelder Gert
Institut für Genetik und
Mikrobiologie
Universität Würzburg
Röntgenring 11
D-W-8700 Würzburg
GERMANY

Pongs Olaf
Lehrstuhl für Biochemie
Ruhr-Universtät Bochum
D-W-4630 Bochum 1
GERMANY

Quiocho Florante A.
Howard Hughes Medical Inst.
and Dept. of Biochemistry
and Structural Biology,
Baylor College of Medicine
Houston, TX 77030
USA

Reid R.E.
Faculty of Pharmacy
University of Manitoba
770 Bannatyne Avenue
Winnipeg R3E OW3
CANADA

4

Reifegerste Rita
Institut für Genetik und
Mikrobiologie
Universität Würzburg
Röntgenring 11
D-W-8700 Würzburg
GERMANY

Rogers John H.
Physiological Laboratory
University of Cambridge
Downing Street
Cambridge CB2 3EG
UNITED KINGDOM

Rudland Philip S.
Department of Biochemistry
University of Liverpool
P.O. Box 147
Liverpool L69 3BX
UNITED KINGDOM

Sasaki Junzo
Department of Anatomy
Okayama University
Medical School
Okayama 700
JAPAN

Sato Eisuke F.
Department of Medical Biology
Kochi Medical School
Nankoku-shi, Kochi 781-51
JAPAN

Schallreuter Karin U.
Department of Dermatology
University of Hamburg
Martinistrasse 52
D-W-2000 Hamburg 20
GERMANY

Schleicher Michael
Max-Planck-Institut für Biochemie
D-W-8033 Martinsried
GERMANY

Sharma Yogendra
Center for Cellular and
Molecular Biology
Uppal Road
Hyderabad 500 007
INDIA

Shooter Eric M.
Department of Neurobiology
Stanford University
School of Medicine
Stanford, CA 94305
USA

Takagi Takashi
Biological Institute
Faculty of Science
Tohoku University
Sendai 980
JAPAN

Takemasa Tohru
Institute of Biological Sciences
University of Tsukuba
Tsukuba, Ibaraki 305
JAPAN

Tanaka Yoshikazu
Institute for Fundamental
Research Ctr., Suntory LTD.
Shimamoto-cho
Mishima-gun
Osaka 618
JAPAN

Teintze Martin
Depts. of Medicine and Cell
Biology and Anatomy
Cornell Univ. Medical College
1300 York Ave.
New York, NY 10021
USA

Tolbert David
Department of Biology
University of Virginia
Charlottesville, VA 22901
USA

Utsumi Kozo
Department of Medical Biology
Kochi Medical School
Nankoku-shi, Kochi 781-51
JAPAN

van Heyningen Veronica
MRC Human Genetics Unit
Western General Hospital
Crewe Road
Edinburgh EH4 2XU
UNITED KINGDOM

Vyas Meenakshi N.
Howard Hughes Medical Inst.
and Department of Biochem.
and Structural Biology
Baylor College of Medicine
Houston, TX 77030
USA

Vyas Nand K.
Howard Hughes Medical Inst.
and Department of Biochem.
and Structural Biology
Baylor College of Medicine
Houston, TX 77030
USA

Wasserman Robert H.
Department of Physiology
Cornell University
Ithaca, NY 14853-6401
USA

Watanabe Yoshio
Institute of Biological Sciences
University of Tsukuba
Tsukuba, Ibaraki 305
JAPAN

Wessel Gary M.
Dept. of Biochem. & Mol.
Biol. Box 117, The Univ. of
Texas, M.D. Anderson
Canc. Ctr.
1515 Holcombe Boulevard
Houston, TX 77030
USA

Winsky Lois
Laboratory of Clinical Science
NIMH, Bldg. 10, Rm. 3D-48
Bethesda, MD 20892
USA

Witke Walter
Max-Planck-Institut für Biochemie
D-W-8033 Martinsried
GERMANY

Wood John M.
Department of Dermatology
University of Hamburg
Martinistrasse 52
D-W-2000 Hamburg 20
GERMANY

Zinsmaier Konrad
Institut für Genetik und
Mikrobiologie
Universität Würzburg
Röntgenring 11
D-W-8700 Würzburg
GERMANY

Xiang Mengqing
Dept. of Biochem. and Mol. Biol.,
Box 117, The Univ. of
Texas, M.D. Anderson Canc. Ctr.
1515 Holcombe Boulevard
Houston, TX 77030
USA

PREFACE

CALCIUM-BINDING PROTEINS: PAST, PRESENT AND FUTURE

Robert H. Wasserman

The indispensability of the calcium ion in the operation of physiological systems was early shown by the classical studies of Sydney Ringer who, in the latter part of the 19th century, showed that calcium was required for the contractility of the heart, for embryonic development, and cell adhesion. Subsequent studies by Locke, Loewi, Loeb, and Heilbrunn in the early 1900's further emphasized the essentiality of calcium for cell and tissue function. Nerve impulse conduction, muscle contraction, stimulus-secretion coupling, cell division, blood coagulation, and bone and teeth formation are some of the physiological processes that are dependent upon an adequate concentration of calcium in the extracellular fluid. The maintenance of the calcium concentration in the extracellular fluid compartment within the normal range is the function of the calciotropic hormones, i.e., parathyroid hormone, (1,25)-dihydroxyvitamin D, and calcitonin, affecting one or more of the homeo-static regulatory organs: the intestine, kidney, and skeleton.

A significant upturn in the research on the biological role of calcium, as depicted in Figure 1, appears to coincide with the appreciation of calcium as a "second messenger" in the cellular action of certain hormones, growth factors, and other agonists. The "second messenger" function of calcium is suitably illustrated by the agonist stimulation of the phosphoinositide cycle. As shown schematically in Figure 2, the binding of a calcium-mobilizing agonist to its receptor, e.g., a hormone such as insulin on adipose tissue or vasopressin on liver, activates a series of biochemical steps that results in a physiological response. One step in the phosphoinositide cascade is the formation of inositol-1,4,5-tris-phosphate (IP_3) that interacts with a membrane receptor of the endoplasmic reticulum (or a variant thereof), releasing stored calcium. The concentration of free Ca^{2+} transiently increases, activating calcium-dependent enzymes and/or other calcium-dependent processes. The

8

increased influx of Ca^{2+} from the external milieu also contributes to the transient elevation of intracellular free Ca^{2+} concentration. The return of intracellular free Ca^{2+} to basal levels occurs upon extrusion of Ca^{2+} from the cell and the re-sequestration of Ca^{2+} by intracellular organelles.

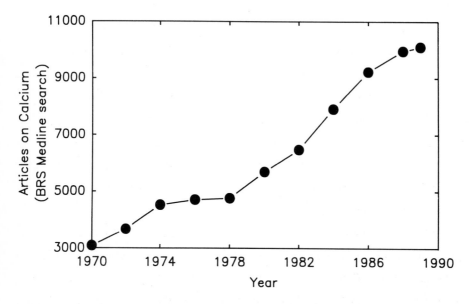

Fig. 1. An estimation of the number of published journal articles on calcium. This information was derived from the BRS Medline data base, using "calcium" as the search word

The intermediary between the elevation of intracellular Ca^{2+} in response to an agonist and the physiological response is some type of calcium-binding protein, whether the protein is a calcium-dependent enzyme or a calcium-dependent modulating protein, such as calmodulin or troponin C. The intracellular calcium-responsive proteins are characterized by relatively high affinity Ca^{2+} binding sites of the order of 10^6 to 10^7 M^{-1}. Within this range of binding affinities, these proteins are nicely poised to bind Ca^{2+} as the intracellular free Ca^{2+} concentration increases from 10^{-7} to 10^{-6} M or greater upon cell activation by agonists. It also turns out that many (or all) of these intracellular Ca^{2+} responsive proteins contain within their sequence the helix-loop-helix configuration (EF-hand) that characterizes the high affinity calcium-binding region, first deduced from the amino acid sequence and the crystal structure of parvalbumin (Kretsinger 1975).

One of the first high affinity calcium-binding proteins uncovered is calbindin-D28K (Wasserman and Taylor 1966). This protein was first found in chick intestine as a vitamin D-induced calcium-binding protein (CaBP), and subsequently identified in a wide variety of tissues (Wasserman and Fullmer 1982). Its apparent affinity constant is about 2×10^6 M^{-1}, binds four Ca^{2+} atoms per molecule, and has an $M_r \approx 30000$. A homologous, immunological-ly cross-reacting protein of about the same molecular size is present in mammalian tissue, prominently in kidney and brain. The mammalian intestinal CaBP, also vitamin D-dependent, is smaller ($M_r \approx 10000$) and binds two atoms of Ca^{2+} per molecule. Epithelial CaBP (intestine, kidney) is suggested to act primarily as diffusional facilitators (Kretsinger et al. 1982), increasing the rate of transfer of Ca^{2+} from the apical region of the cell to the basolateral membrane where the ATP-dependent Ca^{2+} pump is located. In nonepithelial tissues, CaBP might also act to facilitate calcium diffusion, or as a calcium buffer, or activator of some enzyme.

Troponin, identified about the same time as CaBP (Ebashi 1963), and its calcium-binding subunit, troponin C (Greaser and Gergely 1971), are known players in the process of muscle contraction. Calmodulin, identified in 1967 by Cheung and shown to be a calcium-binding protein by Teo and Wang (1971), upon binding Ca^{2+} can activate a number of enzymes, including adenylate cyclase, a cyclic nucleotide phosphodiesterase, protein kinases and calcium ATPase, and participates in cell division and other cellular processes.

Aside from the high affinity calcium-binding proteins that reside within the cell, a large number of extracellular calcium-binding proteins with lower affinities are known. The affinities of these proteins are of the order of 10^3-10^5 M^{-1} and exemplified by osteocalcin of bone origin (Price 1988), vitamin K-dependent plasma proteins (Suttie 1985) and various phosphoproteins, such as milk casein (Farrell and Thompson 1988).

In recent years, investigations designed to identify proteins that bind Ca^{2+}, or the examination of well-known proteins for Ca^{2+}-binding properties, or observations of a serendipitous nature, have added considerably to the list of molecules categorized as calcium-binding proteins. Some of these are described in the present monograph and designated as being novel.

10

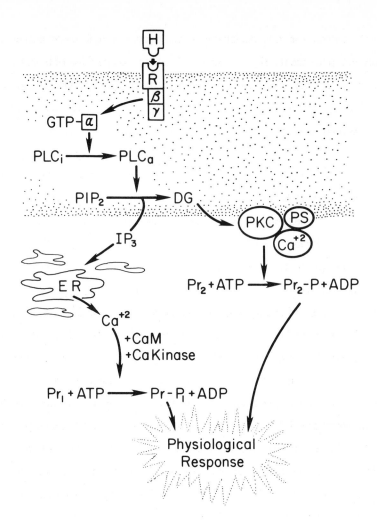

Fig. 2. Formation of second messengers by Ca^{2+}-mobilizing agonists. Hormone (*H*) or other agonists binds to receptor (*R*), which activates the G protein by inducing GTP binding to the alpha subunit. The GTP-alpha subunit stimulates phospholipase C activity which, in turn, hydrolyses PIP_2 to the two second messengers, inositol-1,4,5-trisphosphate (*IP$_3$*) and diacylglycerol (*DG*). IP_3 interacts with an intracellular Ca^{2+} store (endoplasmatic reticulum), releasing Ca^{2+} and elevating the intracellular free Ca^{2+} concentration. Through calmodulin (*CaM*), the Ca^{2+}-dependent protein kinase catalyzes the phosphorylation of appropriate substrates, the latter being responsible for a biological response. Diacylglycerol, in the presence of phosphatidylserine and Ca^{2+}, activates another kinase, protein kinase C (*PKC*), which catalyzes the phosphorylation of proteins and also leads to a biological response. The second messengers, IP_3 and DG, may act synergistically or independently. (PERMISSION GRANTED by "Calcium in Human Biology (ed. B.E.C. Nordin)"

Whatever the reason for this designation, these proteins and those earlier identified emphatically re-emphasize the significance of calcium in the operation of biological systems. Although some of the physiological and biochemical functions of some of the calcium-binding proteins are known, as mentioned for troponin C and calmodulin, we look forward to detailed descriptions of the fundamental roles of these proteins in cellular processes.

In addition to basic research approaches, the possible relationship of calcium-binding proteins to disease processes bears continual scrutiny. Recently, changes in the concentrations of calbindin and its mRNA in specific regions of the brain in relation to aging and neurodegenerative diseases were reported (Iacopino and Christakos 1990). In the aging human brain, significant decreases in calbindin and calbindin mRNA were noted in the cerebellum and other brain sites. In neurodegenerative diseases, there were significant decreases of calbindin and calbindin mRNA in the substantia nigra in Parkinson's disease, in the corpus striatum in Huntington's disease, and in the nucleus basalis in Alzheimer's disease. Oncomodulin, a calcium-binding protein homologous with parvalbumin, was first identified in chemically induced rat liver neoplasms and is present in a number of tumors of human origin (MacManus 1982), as well as in chemically transformed rat fibroblasts (Sommer and Heizmann 1989). Calmodulin (Chafouleas et al. 1981; Zendegui et al. 1984) is known to be present in higher concentration in some transformed cells than in normal cells. The above illustrates the pathological relevance of some calcium-binding proteins and their diagnostic potential. This is an area of investigation that will undoubtedly attract increasing attention in future years.

REFERENCES

Chafouleas JG, Pardue RL, Brinkley BR, Dedman HR and Means AR (1981) Regulation of intracellular levels of calmodulin and tubulin in normal and transformed cells. Proc Natl Acad Sci USA 78:996-1000

Cheung WY (1968) Cyclic-3',5'-nucleotide phosphodiesterase. Pronounced stimulation by snake venom. Biochem Biophys Res Comm 29:478-482

Ebashi S (1963) Third component participating in the superprecipitation of natural actomyosin. Nature 200:1010

Farrel HM, Thompson MP (1988) The caseins of milk as calcium-binding proteins. In: Thompson MP (ed) Calcium-binding proteins, Vol. II, Biological functions, CRC Press, Boca Raton, FL, pp 117-137

Greaser ML, Gergely J (1971) Reconstitution of troponin activity from three protein components. J Biol Chem 246:4226-4233

Iacopino AM, Christakos S (1990) Specific reduction of calcium-binding protein (28-kilodalton calbindin-D) gene expression in ageing and neuro-degenerative disease. Proc Natl Acad Sci USA 87:4078-4089

Kretsinger RH (1975) Hypothesis: calcium modulated proteins contain EF-hands. In: E. Carafoli (ed) Calcium transport in contraction and secretion. North-Holland Amsterdam, pp 469-478

Kretsinger RH, Mann JE, Simmons JB (1982) Model of the facilitated diffusion of calcium by the intestinal calcium binding proteins. In: Norman AW, Schaefer K, Herrath DV, Gregoleit HG (eds) Vitamin D. Chemical, biochemical and clinical endocrinology of calcium metabolism. W. de Gruyter Berlin, pp 233-248

MacManus JP, Whitfield JF, Boynton AL, Durkin JP, Swierenga SH (1982) Oncomodulin, a widely distributed, tumor-specific, calcium-binding protein. Oncodev Biol Med 3:79-90

Price PA (1988) Role of vitamin-K-dependent proteins in bone metabolism. Annual Review of Nutrition 8:565-583

Sommer EW, Heizmann CW (1989) Expression of the tumor-specific and calcium-binding protein oncomodulin during chemical transformation of rat fibroblasts. Cancer Res 49:899-905

Suttie JW (1985) Vitamin K-dependent carboxylase. Annual Review of Biochemistry 54:459-477

Teo TS, Wang JH (1973) Mechanism of activation of a cyclic adenosine-3',5'-monophosphate diesterase from rat heart. J Biol Chem 248:5950-5955

Wasserman RH (1988) Cellular calcium: action of hormones. In: BEC Nordin (ed) Calcium in human biology. Springer, London, pp 383-419

Wasserman RH, Fullmer CS (1982) Vitamin D-induced calcium-binding protein. In: Cheung WY (ed) Calcium and Cell Function. Academic Press, New York, pp 175-216

Wasserman RH, Taylor AN (1966) Vitamin D_3-induced calcium-binding protein in chick intestinal mucosa. Science 152:791-793

Zendegui JG, Zielinski RE, Watterson DM, van Eldik LJ (1984) Biosynthesis of calmodulin in normal and virus transformed chicken embryo fibroblasts. Mol Cell Biol 4:883-889

ACKNOWLEDGMENTS

The editor would like to express his appreciation to the authors for their enthusiasm in participating in this venture. By contributing their expert knowledge and original ideas and datas they helped make this book possible. Many thanks to the authors and publisher. Thanks are also due to my coworker, Frank Neuheiser, for retyping the manuscripts and for bringing uniformity and style to this book.

Section I

Calcium Signaling by Calcium-Binding Proteins

THE EF-HAND, HOMOLOGS AND ANALOGS

Robert H. Kretsinger, David Tolbert, Susumu Nakayama, and William Pearson

INTRODUCTION

Many proteins bind calcium; they have different structures and employ different coordination geometries to achieve a wide range of affinities and selectivities. Kretsinger (1975) proposed that those within the cytosol are calcium modulated, that they are in the magnesium or apoform in the quiescent cell and in the calcium form in the stimulated cell, and that they detect calcium functioning as a second messenger. Further, he proposed that these calcium modulated proteins, in contrast to the vast array of extracytosolic calcium-binding proteins, all contain the EF-hand homolog domain and are all members of one homolog family. The generality of this theory should be questioned. What exactly is a quiescent cell? How do pores and pumps figure in this scheme? Do some proteins lacking EF-hands, such as the annexins, bind messenger calcium? Do some EF-hand proteins have a high enough affinity to bind calcium in quiescent cells? Do some EF-hand proteins function in the extracytosolic environment? Do these generalizations apply to prokaryotic cells? Although we will address some of these questions, our main concern here is to identify and characterize the members of the EF-hand homolog family and to distinguish them from several very similar analogs.

One of the major challenges of molecular evolution is the correct assignment of proteins or protein domains to their proper homolog families. This assignment usually depends upon the correct alignment of amino acid sequences. In some cases this alignment in turn depends upon knowledge of tertiary structure or of functional characteristics. These alignments and assignments are complicated by the evolutionary process. Proteins may diverge to assume different functions and structures. Conversely, proteins from different homolog families may converge toward the same structural solution to a functional selection.

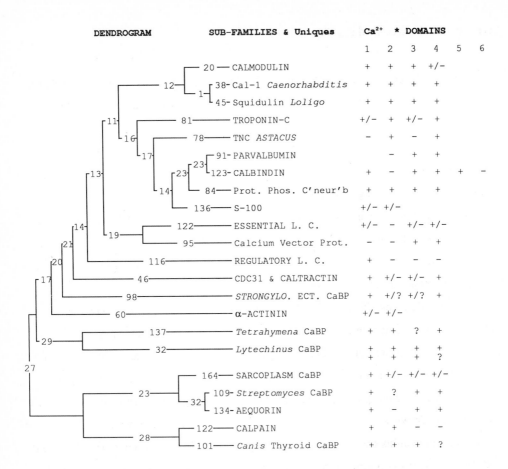

Figure 1. Dendrogram of 22 EF-hand subfamilies. We have identified 14 subfamilies and eight unique proteins, probably the first representives of separate subfamilies; the uniques are indicated in lower case script. The numbers of EF-hand domains are indicated as well as their (inferred) ability to bind calcium. "+/-" indicates that some domains do and some do not bind calcium; "?" indicates uncertainty. The protein from *Lytechinus* has eight domains. Minimum mutation distances are indicated; for proteins having other than four domains distances are normalized to four. Distances to the 14 terminal nodes of subfamilies are weighted averages of all members. The root of the dendrogram has not been determined; the lower five appear most distant from the upper 17 uniques and subfamilies.

Since some proteins judged to be EF-hand homologs have lower similarity scores when compared with some other EF-hand homologs than do some analogs, we will discuss the criteria used to assign proteins to the EF-hand homolog family.

SURVEY OF SUBFAMILIES

Moncrief et al. (1990) evaluated 153 amino acid sequences of EF-hand proteins available in Nov/1988. The second MS in this series on the evolution of the EF-hand homolog family is being prepared. As of May/1990 174 sequences, containing 627 domains, were available. Several of the conclusions of those papers are included in this review; references for all sequences are included in those two papers. Figure 1 provides an overview of fourteen distinct subfamilies and eight unique proteins that are probably the first representatives of additional subfamilies; we will refer to 22 subfamilies. The dendrogram shows the relationships among these subfamilies. The internal numbers indicate inferred minimum mutation distances between internal nodes. The distances to the 22 terminal nodes are weighted to account for all of the members of the subfamily. Most of the subfamilies contain four EF-hand domains. Those with two, three, six, or eight domains have distances normalized to four. One infers, but hardly considers proven, that the branching relationships in this dendrogram reflect the evolutionary pathways of a phenogram and that the indicated lengths reflect rates of evolution. We cannot infer the amino acid sequence of the precursor of all known EF-hand proteins; the root of the dendrogram remains unknown. The dendrogram is drawn with the nodes farthest from the termini on the extreme left.

The calmodulins are the most widely distributed of EF-hand proteins. They are found in all four kingdoms -- fungi, protocists, plants, and animals. Most other EF-hand homologs are restricted to animals, excepting CDC31 & caltractin found in fungi and protocists, α-actinin of slime mold, the essential light chain of myosin found in fungi, and the *Tetrahymena* calcium-binding protein. Several have been found only in invertebrates: the sarcoplasm calcium-binding proteins, squidulin, cal-1, calcium vector protein, *Strongylocentrotus* ectodermal protein, *Lytechinus* calcium-binding protein, and aequorin. The *Streptomyces* calcium-binding protein is the only representative from a prokaryote; whether calcium functions as a second messenger remains to be determined. S-100, calcineurin B, and the thyroid protein have been found only in mammals. A typical mammal would contain EF-hand proteins from at least eleven subfamilies and multiple isotypes from within each family.

The number of EF-hand domains are indicated in figure 1 as well as the observation or inference of calcium binding. Crystal structures are available for parvalbumin (Declerq et al. 1988; Swain et al. 1989; Kumar et al. 1990), intestinal calcium-binding protein (S-100 subfamily) (Szebenyi and Moffat 1986), troponin C (Satyshur et al. 1988; Herzberg and James 1988), and calmodulin (Babu et al. 1988). In addition calcium binding has been determined for calbindin (Leathers et al. 1990), calcineurin B (Klee et al. 1979), essential light chain of myosin [ELC] (Kessler et al. 1980), regulatory light chain of myosin [RLC] (Bagshaw and Kendrick-Jones 1980), calpain (Yoshimura et al. 1983), calcium vector protein (Cox 1986), S-100 (Baudier et al. 1986) and aequorin (Kemple et al. 1990). Based on these precedents calcium binding is inferred for the other proteins. As will be discussed, "?" refers to domains for which there is doubt. Within some subfamilies there are examples of both calcium binding and non-binding within a domain, "+/-".

CANONICAL EF-HAND

Tufty and Kretsinger (1975) proposed a simple mnemonic for identifying the α-helix, calcium binding loop, α-helix of the EF-hand.

```
---α-helix E ------                         ------α-helix F------
                  --calcium-binding loop----
                  1 1 1 1 1 1 1 1 1 2 2 2 2 2 2 2 2 2 2
1 2 3 4 5 6 7 8 9 0 1 2 3 4 5 6 7 8 9 0 1 2 3 4 5 6 7 8 9
E n * * n n * * n X * Y * Z G # I-X * *-Z n * * n n * * n  (1975)
```

The first residue is Glu; "n" refers to hydrophobic residues -- Val, Ile, Leu, Met, Phe, Tyr, Trp -- the inner aspects of the α-helices; X, Y, Z, -X, and -Z refer to calcium coordinating amino acids -- Asp, Asn, Ser, Thr, Glu, Gln -- at the vertices of an octahedron; Gly occurs at a sharp bend and uses dihedral angles restricted to the other 19 amino acids; "I" -- Ile, Leu, Val -- provides a hydrophobic side chain to the core of the molecule. The "#" at position 16, the -Y vertex, can be occupied by any amino acid; here calcium is coordinated with the carbonyl oxygen atom. The "*" indicates that any amino acid is acceptable at those twelve positions. Although this query pattern is found in nearly all of the recognized EF-hand proteins, it does have two shortcomings. It scores some non-homologs as EF-hands and it does not recognize all of the EF-hand domains within some homolog proteins. Before discussing alternate query patterns and search algorithms we summarize several of the more challenging variations and special characteristics of EF-hand domains.

NONCANONICAL DOMAINS

S-100: The first domains of the S-100 proteins have two amino acids inserted into the calcium-binding loops. Instead of coordinating calcium with five side chains and one carbonyl oxygen at vertex -Y (#), the S-100 domain employs four carbonyl oxygen atoms "=O" and only one side chain, Glu at the -Z vertex. The calcium affinity is somewhat reduced (Baudier et al. 1986). Loop positions are numbered from the -Y residue with carbonyl oxygens at 0, -3, -5, and -8. Although the helix-loop-helix motif is recognized, the consensus for the S-100 domain is significantly different from that of the canonical EF-hand. We feel certain that this first domain is a homolog because most of the second domains of the S-100 proteins are canonical domains, guilt by association, an oft repeated theme.

```
E  n  *  *  n  n  *  *  n  X  *  Y  *  Z  G  #  I-X  *  *-Z  n  *  *  n  n  *  *  n  (1975)
                                          =O
*  n  *  *  n  F  *  *  n-8  **E  G  D**  #  L  *  K  *  E  L  K  *  L  n  *  *  E  (S-100)
         =O        =O    =O    =O
```

Deletions: In addition to domain #1 of S-100's many other domains have examples of insertions or deletions. Of 627 unique domains in our data base 146 contain deletions or insertions. Twenty-four occur in domain 1; 78 in 2; 9 in 3; and 35 in 4. In general we find domains 3 and 4 more conserved than domains 1 and 2. Either α-helix E or F might also be altered by an insertion as seen in domain 1 of two isotypes of sarcoplasm calcium-binding protein (SARC) of *Branchiostoma*. Conversely, the α-helix may be normal and the sole effect of the insertion being a different aspect of the helix exposed to solvent. All examples are aligned with the canonical (1975) mnemonic. Domain number, protein abbreviation, and genus of source are indicated to the right of the sequence. The different residues, only, are given for isotypes.

```
E n * * n n * * n X * Y * Z G # I-X * *-Z n * * n n * * n
K  I  K  F  T  F  DFF  L  D  Y  N  K  D  G  S  I  Q  Q  E  D  F  E  E  M  I  K  R  Y  (#1 SARC Branch.)
                   M      H                        D          D          M  T
```

A deletion, as seen in domain 1 of the luciferin binding protein (LBP) of *Renilla*, might simply shorten α-helix F relative to the closely related domain 1 of the aequorins, which do bind calcium.

```
E n * * n n * * n X * Y * Z G # I-X * *-Z n * * n n * * n
K  M  K  T  R  M  K  R  V  D  V  T  G  D  G  F  I  S  R  E  D  .  .  Y  E  L  I  A  V  (#1 LBP Renilla)
```

The *Lytechinus* ectoderm contains a protein with eight tandem EF-hands. The first four are closely related to the last four. They seem to have evolved by a recent gene duplication. Domain 2 has lost two residues when aligned with domain 6, which appears to bind calcium. Again one does not know whether α-helix E is shortened or simply employs two residues from the 1,2 loop, which is already shorter than the 5,6 loop between domains 5 and 6.

```
E n * * n n * * n X * Y * Z G # I-X * *-Z n * * n n * * n
Q  N  .  .  I  I  A  R  L  D  V  N  S  D  G  H  M  Q  F  D  E  F  I  L  Y  M  E  G  S  (#2 CaBP Lyt.)
```

Often such insertions or deletions are accompanied by other changes in sequence; therefore, the loss of calcium binding seems more probable. The *Lytechinus* calcium-binding protein domain 8 provides an example of combined loop insertion and helix deletion.

```
E n * * n n * * n X * Y * Z G # I-X * *-Z n * * n n * * n
E  V  A  K  L  I  K  E  S  SFE  D  D  D  G  Y  I  N  F  N  E  F  V  N  R  F  .  .  .  (#8 CaBP Lyt.)
```

The ectodermal protein, SPEC2C, from *Strongylocentrotus* has a Lys inserted at the beginning of domain 2 as well as Arg at the Gly-15 position and a Pro at 16 whose ϕ angle in the homologs of known crystal structure would preclude Pro.

```
E n * * n n * * n X * Y * Z G # I-X * *-Z n * * n n * * n
EKI  D  E  M  M  S  M  V  D  K  D  G  S  R  P  V  D  F  S  E  I  L  M  K  K  A  L  Q  (#2 SPEC2C Strong.)
```

The fourth domain of the thyroid protein (TPP) of *Canis* has two residues inserted; yet, in common with the two previous examples a full complement of potential calcium binding ligands is present.

```
E n * * n n * * n X * Y * Z G # I-X * *-Z n * * n n * * n
EVL R R F L D N F D SSE K D G Q V T L A E F V A M M T S A (#4 TPP Canis)
```

The third domain of the SPEC2D has a Leu inserted in an otherwise canonical loop.

```
E n * * n n * * n X * Y * Z G # I-X * *-Z n * * n n * * n
K C R K V F K A M D K D D K D S LLS A D E V R Q A V L S F (#3 SPEC2D Strong.)
```

Cysteine: Sulfur is much less electronegative than is oxygen; Cys is not usually considered to be involved in calcium coordination. In addition to the previous examples Cys is found at the X position of the sarcoplasm calcium-binding protein (SARC) of *Penaeus*, at the Y position of protein P11 (~P10~42C) of several mammals, and at the -X position of calpain (light) of *Oryctolagus*. It is not known whether these domains bind calcium.

```
E n * * n n * * n X * Y * Z G # I-X * *-Z n * * n n * * n
E I D D A Y D K L C T E E D K K A G G I N L A R Y Q E L Y (#4 SARC Penaeus)
A V D K I M K D L D Q C R D G K V G F Q S F L S L V A G L (#2 p10 Homo)
                        L           I
K W Q A I Y K Q F D V D R S G T I C S R E L P G A F E A A (#2 CALP Oryct.)
```

Unusual Combinations: Other combinations of potential ligands are so unprecedented as to merit attention; although, singly they might be involved in calcium coordination. Domain 3 of TNC of *Tachypleus* has an unprecedented Gln (Y), Gln (Z). The residues at X, Y, and Z are usually the shorter Ser, Thr, Asp, or Asn.

```
E n * * n n * * n X * Y * Z G # I-X * *-Z n * * n n * * n
E L R E A F R L Y D K Q G Q G F I N V S D L R D I L R A L (#3 TNC Tachy.)
```

Domain 2 of the only recognized EF-protein protein (CMSE) found in a prokaryote, *Streptomyces*, has a Gly (X), Gly (Y). We don't know whether two waters bridge to calcium.

```
E n * * n n * * n X * Y * Z G # I-X * *-Z n * * n n * * n
L F D Y L A K E A G V G S D G S L T E E Q F I R V T E N L (#2 CMSE Strep.)
```

The *Rattus* calmodulin pseudogene 1 has Ile at -X of its domain 4. This vertex frequently has a water molecule coordinating calcium; however, it seems improbable that the hydrophobic Ile can hydrogen bond to water. It is not known whether this pseudogene is expressed.

```
E n * * n n * * n X * Y * Z G # I-X * *-Z n * * n n * * n
E V D G M I R E A D M D G D G Q V I Y E E F V Q M V T A K   (#4 ps-CaM Rattus)
```

Of the nine sarcoplasm calcium-binding proteins (SCBP) of known sequence three do not and two do bind calcium in domain 4. Calcium binding is uncertain for one from *Astacus* and three from *Penaeus* types β, α-A, and α-B. There are both an Ala at X and a Lys at position 15 of *Astacus*. In addition, the three closely related isotypes from *Penaeus* have an infrequent Cys or Thr at X and Lys or Arg at position 15, which is usually Gly.

```
E n * * n n * * n X * Y * Z G # I-X * *-Z n * * n n * * n
E I D D A Y N K L A T D A D K K A G G I S L A R Y Q E L Y   (#4 SCBP Astacus)
E I D D A Y D K L C T E E D K K A G G I N L A R Y Q E L Y   (#4 SCBP Penaeus β)
E I D D A Y D K L T T E D D R K A G G L T L E R Y Q D L Y   (#4 SCBP Penaeus α-A)
E I D D A Y D K L T T E D D R K A G G L T L E R Y Q D L Y   (#4 SCBP Penaeus α-B)
```

Proline: The φ dihedral angle of Pro is restricted to -60°; most EF-hand domains lack Pro. The φ angles observed in the crystal structures of canonical EF-hands are near -60° at positions 11, 19, and 20; Pro is sometimes observed at those positions. Pro cannot be accommodated within an α-helix without introducing at least a 30° bend in the helix. Pro can, though, occur at the first three residues of an α-helix since their nitrogen atoms need not be involved in intra-strand hydrogen bonds.

```
P P P                   P               P P    P P          (Pro OK)
E n * * n n * * n X * Y * Z G # I-X * *-Z n * * n n * * n
E L Q D M I N E V D A D G N G T I D F P E F L T M M A R K   (#2 CAM Homo)
E F R A S F N H F D R D H S G T L G P E E F K A C L I S L   (#1 ACT Gallus)
E F A R I M S L V D P N G Q G T V T F Q S F I D F M T R E   (#2 ACT Gallus)
- - - - - - - - - - - - - - - - - - - - - - - - - - - - -
E I S E M I R E A D I D G D G M V N Y E E F V K M M T P K   (#4 SQUID Loligo)
Y T E I M L R M F D A N N D G K L E L T E L A R L L P V Q   (#4 CALB Gallus)
R L G K R F K K L D L D N S G S L S V E E F M S L P . E L   (#1 CALCIB Bos)
E F K E A F P L F D R D G D Q T V T T K E L G M V M G S L   (#1 ps-CaM Rattus)
  P           P P P P P PPP P P P    P       P    P P   P   (Pro in non-Ca)
```

These "allowed" positions are indicated by "P". In addition Pro is observed at many other positions in domains that would otherwise be considered calcium binding. We do not know whether the inevitable changes from the canonical structure affect calcium binding. At the bottom of the sequence listing we summarize the positions at which Pro's are observed in domains judged not to bind calcium.

Glycine #15: Gly is a special amino acid; its dihedral angles can assume a broad range of conformation forbidden to the other nineteen. The Gly at position 15, though, is not invariant even in those domains demonstrated or strongly inferred to bind calcium. In the canonical domains its ϕ and ψ values are near 90° and 0°. It would be interesting to know the conformations of the following loops:

```
E n * * n n * * n X * Y * Z G # I-X * *-Z n * * n n * * n
E V N D L M N E T D V D G N H Q I E F S E F L A L M S R Q  (#2 CAM Sacch.)
E V D E M M A D G D K N H D S Q I D Y E E W V T M M K F V  (#4 TNC Halocyn.)
E T K A L L K A G D Q D G D D K I G V D E F T N L V K A A  (#4 PARV Latimer.)
E T S N F L A A G D S D G D H K I G V D E F K S M A K M T  (#4 PARV Raja)
D I K K V F G I I D Q D K S D F V E E D E L K L F L Q N F  (#3 PARV Merluc.)
G L K K L M G D L D E N S D Q Q V D F Q E Y A V F L A L I  (#2 S-100β)
V I E H I M E D L D T N A D K Q L S F E E F I M L M A R L  (#2 S-100 MRP-14)
E I A R L M E D L D R N K D Q E V N F Q E Y V T F L G A L  (#2 S-100 2A9)
E I A R L M D D L D R N K D Q E V N F Q E Y V A F L G A L  (#2 S-100 calcyc)
A F Q K L M N N L D S N R D N E V D F Q E Y C V F L S C I  (#2 S-100 42A)
A F Q K V M S N L D S N R D N E V D F Q E Y C V F L S C I  (#2 S-100 PCBP)
E L Q E M I A E A D R N D D N E I D E D E F I R I M K K T  (#4 TRACT Chlamy.)
E F K R R F K N K D T D K S K S I T A E E L G E F F K S T  (#1 SPEC Arbacia)
E F K A C F S H F D K D N D N K L N R L E F S S C L K S I  (#1 ACT Dictyo.)
E W R D L K G R A D I N K D D V V S W E E Y L A M W E K T  (#2 SARC Branch.)
P L P L F F R A V D T N E D N M I S R D E Y G I F F G M L  (#3 SARC Nereis)
P L P L F F R A V D T N E D N M I S R D E Y G I F F N M L  (#3 SARC Periner.)
C F N M I F D V I D T D K D R S I D L N E F I Y A F A A F  (#3 SARC Patino.)
S C R N M V N L M D K D G S A R L G L V E F Q I L W N K I  (#1 CALPh Gallus)
G L A R F F R R L D R D R S R S L D S R E L Q R G L A E L  (#1 TPP Canis)
```

Gly is also observed at all five (X, Y, Z, -X, -Z) calcium-coordinating positions; it is relatively common at -X. In parvalbumin a water molecule coordinates calcium at -X. One assumes water coordination in the otherwise canonical domains. Water can also bridge from the oxygen of the shorter Ser, Thr, Asn, or Asp to calcium as seen in troponin C.

Summary: Of the 627 EF-hand domains available in our UVa database, 355 (57%) are considered to be canonical calcium binders. As noted above, there are 20 examples of S-100 domains and 16 questionable domains. The remaining 236 (38%) probably do not bind calcium.

These various examples of deletions, of uncertain calcium coordination, and of aberrant Cys's, Pro's, and Gly's all make for fascinating chemistry. Their investigations will provide additional insights into the structures and functions of the EF-hands. Of more immediate relevance to this MS they illustrate the difficulty of assigning a simple query pattern when searching for all homologs.

IDENTIFICATION OF EF-HANDS

The initial 1975 query pattern (S-1 of Table 1) codified the characteristics of a domain consisting of a helix, calcium-binding loop, helix as based on the twenty or so sequences available at that time. The sixteen informational positions were scored 0 or 1. A score of "12" indicated a homolog; "10" not; and "11" questionable. At that time scoring for insertions or deletions was not considered. Consistent with recent empirical use, we now suggest a penalty of one for each gap or insertion of any length.

```
                1 1 1 1 1 1 1 1 1 2 2 2 2 2 2 2 2 2 2
1 2 3 4 5 6 7 8 9 0 1 2 3 4 5 6 7 8 9 0 1 2 3 4 5 6 7 8 9
E n * * n n * * n X * Y * Z G # I-X * * *-Z n * * n n * * n
```

The mnemonic is easy to apply; it reflects the known structural and functional characteristics of the canonical calcium-binding domain; and, given its simplicity, it has been remarkably successful in identifying nearly all of the subsequently sequenced EF-hand proteins. As noted, E = Glu; n = Val, Ile, Leu, Met, Phe, Tyr, Trp; X, Y, Z, -X, -Z = Ser, Thr, Asp, Asn, Glu, Gln; G = Gly; I = Ile, Val, Leu; # and * anything. However, it yields many false positives; therefore, we have tried to design more discriminating schemes. We describe three (S-2, S-3, and S-4) and summarize results of their applications.

We used four data bases -- our own (UVa) data base (DB-1) consisting of 174 recognized EF-hand proteins, a subset of the National Biomedical Research Foundation (NBRF) Data Base (release 24, Mar/90), consisting of 6550 protein sequences (DB-2) each assigned to one of over a thousand homolog families (there are 81 EF-hand proteins in family #792), and the total NBRF Data Base consisting of 17730 entries (DB-3). The fourth data base (DB-4) has 46 proteins that may be either analogs or distant homologs. Twenty-two have been suggested in the literature to be similar to EF-hands; 24 additional ones scored high by one or several of our searches. The results of the four searches against the four data bases are summarized in Table 1.

We cite two criteria for the four searches: the number of nonhomologs having a score equal to or greater than that score required to identify 79 of 81 homolog proteins in DB-2 and the number of homolog proteins in DB-2 having scores higher than the highest nonhomolog. For instance, the 1975 search pattern, with penalty of -1.0 for gaps (S-1), identifies 170 of 174 EF-hand homologs proteins and 448 of 627 domains as scoring 12/16 or higher. All of the canonical domains, not surprisingly, score $\geq$ 10. It also identifies 79 of the 81 homologs of DB-2 as having at least one of their multiple EF-hand domains with a score of 12 or higher. However, there are also 406 nonhomologs that also score 12 or higher. The next column under DB-2 indicates that 63 of the 81 EF-hand proteins score higher or equal to the highest scoring non-homolog. In the complete NBRF data base of 17730 proteins 1056 nonhomologs score 12 or higher by S-1. Thirteen of 46 in DB-4 score 12 or higher by S-1. Again we emphasize that 236 of 627 domains do not bind calcium; they frequently score far below 12/16. The sixth domain of *Homo* calbindin scores only 8. The second domain of aequorin scores only 2; one would not even consider it, were it not sandwiched between two canonical domains.

```
E n * * n n * * n X * Y * Z G # I-X * *-Z n * * n n * * n
I T T Y K K N I M A L S D G G K L Y R T D L A L I L C A G (#6 CALB Homo)
K R H K D A V E A F F G G A G M K Y G V E T E W P E Y I E (#2 AEQ Aequorea)
```

Domains with deletions usually score low. Nonetheless, the old faithful S-1 (1975) does quite well and will continue to be used; can one do better?

Search 2 employs a more complex, and highly subjective weighting scheme with a greater emphasis on the calcium-binding loop. Values range from 8 to 70 for the sixteen positions; the gap penalty is -16 per residue, relatively about the same as used in S-1.

```
E  n * * n n * * n X * Y * Z G # I-X * *-Z n * * n n * * n
E  N      N F     C X   Y   Z G   I-X     -Z A      N N      M
E  = E 20,  Q 20
N  = V  8,  L  8,  I  8,  F  8,  M  8,  Y  8,  W  8,  A  8,  C  8
F  = V  8,  L  8,  I  8,  F 16,  M  8,  Y  8,  W  8,  A  8,  C  8
C  = V  8,  L  8,  I  8,  F  8,  M  8,  Y  8,  W  8,  A  8
A  = V  8,  L 16,  I  8,  F 16,  M  8,  Y  8,  W  8,       C  8
M  = V  8,  L  8,  I  8,  F  8,  M  8,  Y  8,  W  8,
X  = D100,  E 70,  S 70,  N 50,  T 50,       F 50
Y  = D 80,  E 60,  S 60,  N 80,  T 40,  Q 60
Z  = D 60,  E 30,  S 60,  N 60,  T 60,  Q 30
G  = D 40.         S 40  N 40,       Q 40,  A 40,  V 40,  H 40,  K 20,  R 40,  G 80
I  = V 80,  L 80,  I 80,  F 40,  M 40,            A 40,  C 60
-X = D 60,  E 60,  S 60,  N 60,  T 60,  Q 40,                          R 40,  G 60
-Z = D 60,  E 80,  S 60,  N 40,  T 40,  Q 40
```

It scores up the lower scoring domains and depresses the higher nonhomologs. It is better than S-1 in that it scores only 71 nonhomologs of DB-2 higher than 524, the score required for 79 of 81 homologs. At score 572, 73 of the 81 are scored higher than any nonhomolog; however, only 413 of the 627 domains of the UVa database are scored. Obviously the main effect of this scheme has been to push down the scores of nonhomologs. Only 272 proteins are scored in the total 17730 protein NBRF database (DB-3). In DB-4 consisting of 46 similars, only 8 score above the 79/81 cutoff. Obviously one could adjust such schemes endlessly trying to optimize preconceived ideas. Searches 3 and 4 involve an algorithm in which both a known homolog and either the calcium-binding loop of S-2 or the entire pattern of S-2 are used as query patterns. Only those proteins having high scores by both criteria are considered, thus eliminating much of the noise associated with searches that employ a single query pattern. Human calmodulin was the homolog used for the search; however, other EF-hand proteins produce similar results. In S-3 only 6 non-homologs appear before the 79th homolog of DB-2; 78 of the homologs score higher than any non-homolog. Of the 174 proteins in DB-1 169 score 180 or higher. Only 16 of the 17730 in DB-3 score 180 or higher; only 6 of the 46 similars (DB-4) pass this cut-off.

Search 4 also uses human calmodulin as the first query; then, it applies the entire S-2 pattern. Seventy-nine homologs of DB-2 score higher than any nonhomolog. Of the 174 in DB-1 167 score 140 or higher; only 5 in DB-3 and only 3 in DB-4 score 140 or higher.

All of these approaches are empirical in that the user makes subjective judgments regarding the chemical or physiological importance of certain positions or of the frequency of occurrence of certain proteins. The 1975 query pattern (S-1) is quite successful in identifying the full length, canonical EF-hand. When this query pattern is modified to emphasize the calcium-binding loop (S-2), it scores all canonical domains higher than any non-calcium binding or questionable domains.

Correspondingly, the specialized S-100 domain query scores those first domains higher than all other EF-hand domains or other proteins of the NBRF data base. It is obviously easier to optimize a query pattern for a subset of this diverse homolog family. A query pattern based on multiple nodal sequences is marginally better at distinguishing all EF-hand containing proteins. Our best query pattern (S-4) is highly empirical. We feel, but certainly cannot prove, that further adjustments of these patterns would improve the resolution only marginally.

Not surprisingly the noncanonical domains are the most difficult to detect. Many are so divergent that we would not accept them as homologs, were they not found in tandem with other more solidly recognized domains.

In the homologs of known structure the domains occur in pairs, with the special exception of the N-terminal domain of three domain parvalbumin in which this unpaired domain contributes to a stable hydrophobic core. An additional criterion would be to demand pairs of domains in any questionable homolog/analog.

Obviously one can continue to refine the query patterns, scoring, and deletion penalties as well as using more domains. However, this approach is inherently flawed and at best can realize only reasonable success. It depends on a chimeric pattern that in some ultimate sense reflects our concepts of chemistry, statistics, or evolution. Further, this approach treats each of the 29 positions independently of the other 28. We know full well from our inability to predict tertiary structure from primary sequence that we must consider multiple interactions between residues.

Table 1. Summary of searches for EF-hand homologs and analogs

	Database-1 UVa	Database-2 NBRF		Database-3 NBRF	Database-4
	174 proteins 627 domains	6550 proteins with 81 EF-hand proteins score for 79 of 81	# before 1st non'	17730 proteins	46 similars
S-1	170 prot 448 dom	406 ≥ 12	63/81	1056 ≥ 12	13
S-2	171 413	71 ≥ 524	73/81	272 ≥ 524	8
S-3	169	6 ≥ 180	78/81	16 ≥ 180	6
S-4	167	0 ≥ 140	79/81	5 ≥ 140	3

See text for descriptions of databases and searches, and discussion of results.

We suggest that our searches for EF-hand homologs would become more discriminating as we include more established homologs in the query patterns. The ultimate and appropriate patterns are the domains that Nature presents us; their multiple interactions have passed Evolution's scrutiny. The strength of this approach is illustrated by searches 3 and 4 in which the actual human calmodulin sequence is used as a preliminary query pattern.

ANALOGS

Based on previously published proposals of analogy or homology and on our searches we have listed 46 proteins that resemble EF-hand(s) in Table 2. A "#" and reference indicates that authors have noted the resemblance to EF-hands; a lack of reference indicates that our searches provided the suggestion of similarity. In addition to the names of the 46 proteins, as they appear in the UVa (#) or NBRF data banks are five scores, that of the Smith, Waterman (1981) algorithm for comparison against human calmodulin and searches 1, 2, 3, and 4 as previously described. For each search those scores higher than that required to identify 79 of the 81 EF-hand proteins in the NBRF data bank are indicated in bold font. Although the rank order of score varies somewhat for the different searches, the general agreement is obvious.

Kretsinger (1987) previously discussed the proteins: sodium channel protein *Rattus* I #201, *Rattus* II #202, *Electrophorus* #203; thrombospondin #301; T4 lysozyme #001; α-lactalbumin *Homo* #051, *Bos* #052: uteroglobulin #151; protein kinase C #351; osteonectin *Mus* #251, *Bos* #252; and D-galactose binding protein #101. For those of known tertiary structure, T4 lysozyme, α-lactalbumin, and D-galactose binding protein, one sees some resemblence to the helix-loop-helix of the EF-hand but large differences that indicate lack of homology.

We focus our attention on the five of the 46 proteins most closely resembling EF-hand proteins by the criteria of our searches. All of those discussed by Kretsinger in 1987 have lower resemblance.

The plasmodium of *Physarum polycephalum* has up to 10^8 nuclei that replicate their DNA in synchrony. The early replication LAV1-2 gene is near an origin of replication. Its 5 kb cDNA has been sequenced by Pallotta et al. (1986); however, no function has been assigned to the LAV1-2 protein. Domain 1, beginning at residue 221, appears to be a canonical calcium-binding EF-hand. Domain 2 also appears to bind calcium; although, it does have the often encountered variation of Asp at position 15, which is usually Gly. Domains 3 and 4 also bind calcium; domain 4 has Lys at 15. Domains 1 and 2 are separated by five residues, 3 and 4 by seven; these separations are in the range normally encountered. Interestingly, there is no linker between pair 1,2 (221-285) and pair 3,4 (286-352); 355 is the C-terminus. This lack of a 1,2 - 3,4 linker, as found in all known four, six, or eight domain homologs, might force the inferred hydrophobic faces of 1,2 and 3,4 into juxtaposition, thereby generating a hydrophobic core with approximate 222 point group symmetry. Although the LAV1-2 protein has not been characterized and there is no evidence of calcium binding, we suggest that calcium functioning as a messenger binds to these four domains thereby regulating a yet to be determined function residing in the first 220 residues.

Sakane et al. (1990) determined the sequence of the cDNA encoding porcine diacylglycerol kinase. They noted the presence of two calcium-binding EF-hands as well as "...two cysteine-rich zinc finger-like sequences similar to those found in protein kinase C".

Table 2. EF-hand analogs or homologs

	SW	S-1	S-2	S-3	S-4
S06939 Hypothetical protein LAV1-2 - Slime mold	128	**13**	**596**	**640**	**384**
A24069 1F8 protein - Trypanosoma cruzi	87	**14**	**612**	**435**	**348**
#491 DGK diacylglycerol kinase -Pig	70	**14**	**588**	**350**	**210**
A23944 Chitin synthase - Yeast	39	**13**	502	58	58
CZCLAM Cellulase A precursor - Clostridium	52	**12**	**552**	**208**	104
DEPGLP Dihydrolipoamide dehydrogenase	73	10	504	**219**	109
MWKW Myosin heavy chain - Caenorhabditis elegans	70	**12**	510	105	105
#451 A25755 Delta crystallin - Chicken	78	10	452	78	78
#202 B25019 Sodium channel protein II - Rat	47	**12**	492	70	70
MNIV2K Nonstructural protein NS2 - Influenza A virus	46	11	**534**	**184**	69
#301 A26155 Thrombospondin - Human	34	10	**560**	136	68
A23498 Alpha-type phaseolin precursor - Kidney bean	43	11	**540**	114	64
QQVZ15 Hypothetical protein B-213 - Vaccinia virus	63	11	418	63	63
A25620 Staphylocoagulase - Staphylococcus aureus	61	**13**	384	93	61
#203 CHEE Sodium channel protein - Electric eel	61	**12**	474	91	61
WMBEH6 UL36 protein - Herpes simplex virus	59	11	486	59	59
#421 HUMPLG2B-1 platelet membrane glycoprotein IIb	58	10	476	58	58
NNBS1 Anthranilate synthase	38	11	492	76	57
LAHU Alpha-lactalbumin precursor - Human	57	10	452	75	57
#001 LZBPT4 Lysozyme - Bacteriophages T4	56	10	482	56	56
#201 A25019 Sodium channel protein I - Rat	55	**12**	460	63	55
#051 LAHU Alpha-lactalbumin - Human	53	10	452	69	53
A23062 Epidermal growth factor receptor - Human	34	**13**	**538**	102	51
LEECA Hemolysin A - Escherichia coli	50	**12**	478	75	50
#151 UGRB Uteroglobin - Rabbit	50	11	356	50	50
#471 A28468 Chromogranin A precursor - Human	47	10	420	47	47
#351 KIBOC Protein kinase C - Bovine	46	10	456	52	46
#472 A24175 Chromogranin A precursor - Bovine	45	9	420	45	45
TWECR Transcription termination factor rho - Escherich	44	10	474	57	44
IDECRK Replication initiation protein - Escherichia co	44	**12**	480	66	44
YRMSCS Monophenol monooxygenase	44	11	414	44	44
BVECAC racC protein - Escherichia coli	42	9	480	42	42
#251 A29227 Osteonectin - Mouse	42	11	474	63	42
QQBE30 Hypothetical BRRF2 protein - Epstein-Barr virus	42	10	476	82	42
#401 HUMNID-1 nidogen -Human	41	10	452	51	41
#511 A26561 Tubulin beta chain - Human	41	11	462	58	41
IMBP5 Immunity region protein 5 - Bacteriophage φ-10	40	11	456	60	40
DJBE16 DNA-directed DNA polymerase - Herpes simplex	40	11	420	40	40
VGXRNH Glycoprotein NCVP5 - Rotavirus	39	11	468	58	39
#402 S03637 Nidogen - Mouse	39	11	462	58	39
#252 A31333 Osteonectin - Bovine	37	11	474	55	37
EFECS Elongation factor EF-Ts - Escherichia coli	37	11	448	55	37
#101 JGECG D-Galactose-binding protein - Escherichia c	37	10	472	51	37
#052 LABO Alpha-lactalbumin - Bovine	37	9	424	42	37
#431 CAATP1 calcium-pumping ATPase - Human	28	9	450	42	28
#411 DZYZSX Development-specific protein S - Myxococcus	28	9	412	28	28

The Smith-Waterman comparison (*SW*) and the four searches are described in the text. Those numbers in *bold font* indicate search scores higher than the minimum value accepted in order to identify 79 of the 81 recognized EF-hand proteins in the NBRF data base. #491 Sakane et al. 1990; #451 Sharma et al. 1989; #201, 202, 203 Babitch and Anthony 1987; #301 Lawler et al. 1982, and Lawyer and Hynes 1986; #421 Poncz et al. 1987; #001 Tufty and Kretsinger 1975; #051, 052 Stuart et al. 1986; #151 Baker 1985; #471 Helman et al. 1988; #351 Coussens et al. 1986; #472 Iacangelo et al. 1986; #251 Engel et al. 1987; #401, 402, Nagayoshi et al. 1989; #511 Fong et al. 1988; #252 Bolander et al. 1988; #101 Vyas et al. 1987; #431 Brandt et al. 1988; #411 Inoue et al. 1983

They found that "...the activity of this (80 kDa) DGK isoform is enhanced by micromolar concentrations of Ca^{2+} in the presence of deoxycholate or sphingosine" and hence "...its action is probably linked with both of the second messengers diacylglycerol and inositol 1,4,5-trisphosphate."

```
E n * * n n * * n X * Y * Z G # I-X * *-Z n * * n n * * n

A L V A D F R K I D T N S N G T L S R K E F R E H F V R L    (#1 LAV1 Physar.)
221
V Q D A L F R Y A D E D E S D D V G F S E Y V H L G L C L    (#2 LAV1 Physar.)
256
V L R I L Y A F A D F D K S G Q L S K E E V Q K V L E D A    (#3 LAV1 Physar.)
286
K F E H Q F S V V D V D D S K S L S Y Q E F V M L V L L M    (#4 LAV1 Physar.)
323
K L E F T F K L Y D T D R N G I L D S S E V D R I I I Q M    (#1 DGK Sus)
111
I L Q E M M K E I D Y D G S G S V S L A E W L R A G A T T    (#2 DGK Sus)
156
R R I E L F K K F D K N E T G K L C Y D E V H S G C L E V    (#12 1F8 Trypan.)
49
E L T V M F D E I D A S G N M L V D E E E L K R A V P K L    (#3 1F8 Trypan.)
131
D P A A L F K E L D K N G T G S V T F D E F A A W A S A V    (#4 1F8 Trypan.)
168
R R I E L F K Q F D T N G T G K L G F R E V L D G C Y S I    (#12 flag. Trypan.)
52
E L T V M F D T M D K D G S L L L E L Q E F R E A L P K L    (#3 flag. Trypan.)
134
D A T T V F N E I D T N G S G V V T F D E F S C W A V T K    (#4 flag. Trypan.)
171
S F Q N I L K N D T I S F E G L I T T E A F R D I V I S L    (#1 chit'syn Sacch.)
890
G F V F V 1 Q T I A T F G T F F T S T Y V L V S I V V S L    (#1 chit'syn Cand.)
538
V T N I N R E A A D V N R D G A I N S S D M T I L K R Y L    (cell'ase A Clost.)
440
E L A E G F R V L S N G Q K T ISI P M K E V S A L M A S V   (#1 EFH5 Trypan.)
47
E V M R V F G Q G E Q T N T E E L S F K D F L S L M M C E    (#2 EFH5 Trypan.)
87
E M R G A F L H Y D K Q K T G F V T K K Q F T E L F A T G    (#3 EFH5 Trypan.)
123
E V E E L L T I A E Q D E T D D K I D Y N R F I N E L I H    (#4 EFH5 Trypan.)
159
```

Gonzalez et al. (1985) reported that an abundant mRNA, 1F8, of *Trypanosoma cruzi* is encoded by a gene of nearly thirty almost identical tandem repeats of 940 base pairs. Although the function of the 1F8 protein remains unknown, Gonzalez noted a "strong homology to troponin". Moncrief et al. (1990) overlooked this publication. Three domains, all of which are inferred to bind calcium, are easily recognized. The last two, which we tentatively label 3 and 4, are separated by eight residues -- a canonical pair of domains. The third domain, labeled 1/2, also binds calcium, albeit with Cys at -X. However, we cannot infer with confidence whether a fragmented domain preceeds or follows it nor whether it is unpaired.

A possible EF-hand domain of chitin synthetase from *Saccharomyces cerevisiae* (Bulawa et al. 1986) scores 13 by S-1 (1975). However, the homologous region, having 7/29 identity, from *Candida albicans* (Au-Young and Robbins 1990) scores only 9. There is no indication of a paired EF-hand. Parsimony suggests these regions are not EF-hand domains.

Beguin et al. (1985) determined the sequence of the DNA encoding the endoglucanase A (cellulase) from *Clostridium thermocellum*. They found little similarity to the cellobiohydrolase I of *Trichodesma reesi* or the β-glucanase of *Bacillus subtilis*. None of these enzymes are known to bind calcium. Although the signature of the calcium-binding loop -- D v N r D G a I N s s D -- is present, we would not consider this cellulase to have an EF-hand homolog. The similarity to α-helix E is weak, there is no second domain to form a pair; and there is no indication of calcium binding.

As this MS was being submitted Lee et al. (1990) identified two EF-hand homolog domains in the amino acid sequence deduced from cDNA encoding a flagellar protein from *Trypanosoma brucei*. We propose a third EF-hand, all three very similar to the three of 1F8.

Campbell kindly sent us his unpublished sequence of a *Trypanosoma brucei* protein, EFH5, as deduced from genomic DNA. We agree with his suggestion of four EF-hand domains. Domains #2 and #3 are inferred to bind calcium; although, there is a Glu at position 15, Gly.

We tentatively add all five proteins, but not chitin synthase and cellulase, to our data base of EF-hand homologs. Note that the flagellar protein and Campbell's protein were not included in data base 4. Three of the 46 proteins in data base 4 quite certainly contain one (diacylglycerol kinase) or two (LAV1-2 and 1F8) pairs of EF-hand domains. Both DGK and 1F8 had already been identified by their discoverers as EF-hand homologs. These three, which might be regarded as controls, did not figure in the design of our searches but did have the three top scores. The recently published flagellar protein (Lee et al. 1990) and EFH5 *Trypanosoma brucei* (communicated by Campbell) are also recognizable as containing pairs of EF-hands by the 1975 mnemonic. We have not yet generated dendrograms incorporating these five homologs into our UVa data base of 174 proteins. We suspect that all five are initial representations of new subfamilies, now a total of 27 (5+22).

Calmodulin seems ubiquitous, possibly occurring in all cells of all eukaryotes, and it activates a score of different target enzymes and structural proteins. It is frequently considered to be the prototypic calcium-modulated protein. Yet, we caution three exceptions to this generalization. First, the functions of two-thirds of these 27 subfamilies remain to be identified. These include cal-1, squidulin, parvalbumin, calbindin, S-100, calcium vector protein, CDC31, SPEC, *Tetrahymena* CaBP, *Lytechinus* CaBP, SARC, *Streptomyces* CaBP, *Canis* thyroid CaBP, LAV1, 1F8, flagellar protein, and *Trypanosoma* protein. Second, the implied targets of the calcium-modulated proteins have for most of these proteins not been identified, in the case of parvalbumin after extensive search. Third, calpain (a -SH protease) and diacylglycerol kinase have evolved by fusion of different homologs with four or with two EF-hands. The N-terminal portions of LAV1 and of the flagellar protein also appear to have evolved from non EF-hand domains.

In a sense the glass is half full. We see a remarkable variety of protein structure and cell physiology based on the simple motif of α-helix, calcium-binding loop, α-helix. More exciting, the data bases are surely half empty. We expect that more discriminating searches and more research will reveal many more EF-hand homolog subfamilies and functions associated with calcium's functioning as a cytosolic messenger.

REFERENCES

Au-Young J, Robbins PW (1990) Isolation of a chitin synthase gene (CHS1) from *candida albicans* by expression in *saccharomyces cerevisiae*. Molec Microbiol 4:197-207

Babitch JA, Anthony FA (1987) Grasping for calcium binding sites in sodium channels with an EF-hand. J Theor Biol 127:451-459

Babu YS, Bugg CE, Cook WJ (1988) Structure of calmodulin refined at 2.2 Å resolution. J Mol Biol 204:191-204

Bagshaw CR, Kendrick-Jones J (1980) Identification of the divalent metal for binding domain of myosin regulatory light chains using spin-labeling techniques. J Mol Biol 140:411-433

Baker ME (1985) Evidence that progesterone binding uteroglobin is similar to myosin alkali light chain. FEBS Lett 189:188-194

Baudier J, Glasser N, Gerard D (1986) Ions binding to S-100 proteins. I. Calcium- and Zinc-binding properties of bovine brain S-100$\alpha\alpha$, S-100a($\alpha\beta$), and S-100b($\beta\beta$) Protein: Zn^{2+} regulates Ca^{2+} binding on S-100b protein. J Biol Chem 261:8192-8203

Beguin P, Cornet P, Aubert JP (1985) Sequence of a cellulase gene of the thermophilic bacterium *Clostridium thermocellum*. J Bacteriol 162:102-105

Bolander ME, Young MF, Fisher LW, Yamada Y, Termine JD (1988) Osteonectin cDNA sequence reveals potential binding regions for calcium and hydroxyapatite and shows homologies with both a basement membrane protein (SPARC) and a serine proteinase inhibitor (Ovomucoid). Proc Natl Acad Sci 85:2919-2923

Brandt P, Zurini M, Neve RL, Rhoads RE, Vanaman TC (1988) A C-terminal, calmodulin-like regulatory domain from the plasma membrane Ca^{2+}-pumping ATPase. Proc Natl Acad Sci 85:2914-2918

Bulawa CE, Slater M, Cabib E, Au-Young J, Sburlati A, Adair WL, Robbins PW (1986) The *S. cerevisiae* structural gene for chitin synthase is not required for chitin synthesis in vitro. Cell 46:213-225

Coussens L, Parker PJ, Rhee L, Yang-Feng TL, Chen E, Waterfield MD, Francke U, Ullrich A (1986) Multiple, distinct forms of bovine and human protein kinase C suggest diversity in cellular signaling pathways. Science 233:859-866

Cox JA (1986) Isolation and characterization of a new M_r 18,000 protein with calcium vector properties in amphioxus muscle and identification of its endogenous target protein. J Biol Chem 261:13173-13178

Declercq J-P, Tinant B, Parello J, Etienne G, Huber R (1988) Crystal structure determination and refinement of pike 4.10 parvalbumin (minor component from *Esox lucius*). J Mol Biol 202:349-353

Engel J, Taylor W, Paulsson M, Sage H, Hogan B (1987) Calcium binding domains and calcium-induced conformational transition of SPARC/BM-40/osteonectin, an extracellular glycoprotein expressed in mineralized and nonmineralized tissues. Biochemistry 26:6958-6965

Fong KC, Babitch JA, Anthony FA (1988) Calcium binding to tubulin. Biochim Biophys Acta 952:13-19

Gonzalez A, Lerner TJ, Huecas M, Sosa-Pineda B, Nogueira N, Lizardi PM (1985) Apparent generation of a segmented mRNA from two separate tandem gene families in *Trypanosoma cruzi*. Nuc Acid Res 13:5789-5804

Helman LJ, Ahn TG, Levine MA, Allison A, Cohen PS, Cooper MJ, Cohn DV, Istael MA (1988) Molecular cloning and primary structure of human chromogranin A (secretory protein I) cDNA. J Biol Chem 263:11559-11563

Herzberg O, James MNG (1988) Refined crystal structure of troponin C from turkey skeletal muscle at 2.0 Å resolution. J Mol Biol 203:761-779

Iacangelo A, Affolter H-U, Eiden LE, Herbert E, Grimes M (1986) Bovine chromogranin A sequence and distribution of its messenger RNA in endorine tissues. Nature 323:82-86

Inoue S, Franceschini T, Inoue M (1983) Structural similarities between the development-specific protein S from a Gram-negative bacterium, *myxococcus xanthus*, and calmodulin. Proc Natl Acad Sci 80:6829-6833

Kemple MD, Lovejoy ML, Ray BD, Prendergast FG, Rao BDN (1990) Mn (II)-EPR measurements of cation binding by aequorin. Eur J Biochem 187:131-135

Kessler D, Eisenlohr LC, Lathwell MJ, Huang J, Taylor HC, Godfrey SD, Spady ML (1980) Physarum myosin light chain binds calcium. Cell Motility 1:63-71

Klee CB, Crouch TH, Krinks MH (1979) Calcineurin a calcium- and calmodulin-binding protein of the nervous system. Proc Natl Acad Sci 76:6270-6273

Kretsinger RH (1975) Hypothesis: calcium modulated proteins contain EF-hands. In: Calcium transport in contraction and secretion. E. Carafoli, F. Clementi, W. Drabikowski and A. Margreth (eds), North-Holland Publishing Co., Amsterdam, pp 469-478

Kretsinger RH (1987) Calcium coordination and the calmodulin fold divergent versus convergent evolution. Cold Spring Harb Symp Quant Biol 52:499-510

Kumar VD, Lee L, Edwards BFP (1990) The refined crystal structure of calcium-liganded carp parvalbumin 4.25 at 1.5 Å resolution. Biochemistry 29:1404-1412

Laroche A, Lemieux G, Pallotta D (1989) The nucleotide sequence of a developmentally regulated cDNA from *physarum polycephalum*. Nuc Acids Res 17:10502-10502

Lawler J, Chao FC, Cohen CM (1982) Evidence for calcium-sensitive structure in platelet Thrombospondin. Isolation and partial characterization of thrombospondin in the presence of calcium. J Biol Chem 257:12257-12265

Lawyer J, Hynes RO (1986) The structure of human thrombospondin, an adhesive glycoprotein with multiple calcium-binding sites and homologies with several different proteins. J Cell Biol 103:1635-1647

Leathers VL, Linse S, Forsén S, Norman AW (1990) Calbindin-D_{28K}, a $1\alpha,25$-dihydroxyvitamin D_3-induced calcium-binding protein, binds five or six Ca^{2+} ions with high affinity. J Biol Chem 265:9838-9841

Lee M G-S, Chen J, Ho AWM, D'Alesandro PA, Vander Ploeg LHT (1990) A putative flagellar Ca^{2+}-binding protein of the flagellum of trypanosomatid protozoan parasites. Nuc Acids Res 18:4252-4252

Moncrief ND, Goodman M, Kretsinger RH (1990) Evolution of EF-hand calcium-modulated proteins I. Relationships based on amino acid sequences. J Mol Evol 30:522-562

Nagayoshi T, Sanborn D, Hickok NJ, Olsen DR, Fazio MJ, Chu M-L, Knowlton R, Mann K, Deutzmann R, Timpl R, Uitto J (1989) Human nidogen: complete amino acid sequence and structural domains deduced from cDNAs, and evidence for polymorphism of the gene. DNA 8:581-594

Pallotta D, Laroche A, Tessier A, Schinnick T, Lemieux G (1986) Molecular cloning of stage specific mRNAs from amoebea. Biochem Cell Biol 64:1294-1302

Pearson WR (1990) Rapid and sensitive sequence comparison with FASTP and FASTA. Meth Enz 183:63-98

Poncz M, Eisman R, Heidenreich R, Silver SM, Vilaire G, Surrey S, Schwartz E, Bennett JS (1987) Structure of the platelet membrane glycoprotein IIb. Homology to the α subunits of the vitronectin and fibronectin membrane receptors. J Biol Chem 262:8476-8482

Sakane F, Yamada K, Kanoh H, Yokoyama C, Tanabe T (1990) Porcine diacylglycerol kinase sequence has zinc finger and EF-hand motifs. Nature 344:345-348

Satyshur KA, Rao ST, Pyzalska D, Drendel W, Greaser, M, Sundaralingam M (1988) Refined structure of chicken skeletal muscle troponin C in the two-calcium state at 2 Å resolution. J Biol Chem 263:1628-1647

Sharma Y, Rao ChM, Narasu ML, Rao SC, Somasundaram T, Gopalakrishna A, Balasubramanian D (1989) Calcium ion binding to δ- and to β-crystallins. The presence of the "EF-hand motif" in δ-crystallin that aids in calcium ion binding. J Biol Chem 264:12794-12799

Smith TF, Waterman MS (1981) Identification of common molecular subsequences. J Mol Biol 147:195-197

Stuart DI, Acharya KR, Walker NPC, Smith SG, Lewis M, Phillips DC (1986) α-Lactalbumin posesses a novel calcium binding loop. Nature 324:84-87

Swain A, Amma S, Kretsinger RH (1989) Restrained least square refinement of native (calcium) and cadmium-substituted carp parvalbumin using X-ray crystallographic data to 1.6 Å resolution. J Biol Chem 264:16620-16628

Szebenyi DME, Moffat K (1986) The refined structure of vitamin D-dependent calcium-binding protein from bovine intestine. Molecular details, ion binding, and implications for the structure of other calcium-binding proteins. J Biol Chem 261:8761-8777

Tufty RM, Kretsinger RH (1975) Troponin and parvalbumin calcium-binding regions predicted in myosin light chain and T4 lysozyme. Science 187:167-169

Vyas NK, Vyas MN, Quiocho FA (1987) A novel calcium-binding site in the galactose-binding protein of bacterial transport and chemotaxis. Nature 327:635-638

Yoshimura N, Kikuchi T, Sasaki T, Kitahara A, Hatanaka M, Murachi T (1983) Two distinct Ca^{2+} proteins (calpain I and calpain II) purified concurrently by the same method from rat kidney. J Biol Chem 258:8883-8889

CALCIUM-BINDING PROTEINS OF THE EF-HAND-TYPE AND OF THE ANNEXIN FAMILY: A SURVEY

Claus W. Heizmann

INTRODUCTION

Calcium as an ubiquitous second messenger regulates many cellular functions including contraction, secretion, cell growth, differentiation, neuronal excitability, and metabolism. Disturbance of calcium homeostasis has been implicated (directly or indirectly) in several degenerative disorders of the central nervous system (e.g., Alzheimer's and Parkinson's diseases, acute and chronic epilepsy) and others (e.g., hypertension, malignant hyperthermia, ischemia, renal and heart failures, or rheumatoid arthritis).

An altered calcium homeostasis has also been reported in some tumor cells. The ability of cultured neoplastic cells, in contrast to normal cells, to proliferate in media with low concentrations of calcium has been proposed as an indication for malignancy (for review see Whitfield 1990). At present, the different mechanisms by which intracellular calcium is regulated in normal and transformed cells is not well understood. Some cancer cells, however, have been reported to have elevated levels of cytosolic calcium (Whitfield 1990; MacManus et al. 1982; Tsuruo et al. 1984; Banyard and Tellam 1985) which may be in part responsible for a permanent activation of DNA synthesis and/or increased motility/invasiveness.

Calcium is believed to control this variety of cellular processes with a high degree of spacial and temporal precision and therefore mechanisms must exist to control this ion in a localized fashion. The calcium signal is transmitted into the intracellular response in part via families of calcium-binding proteins which are thought to be involved in the regulation of many enzymes and cellular activities, in calcium translocation and buffering, i.e., *the EF-*

hand protein family, or *the annexin family* suggested to be involved in phospholipase regulation, cytoskeletal organization, membrane trafficking, cell growth and transformation.

THE EF-HAND PROTEIN FAMILY

Calmodulin (Klee and Vanaman 1982; Means et al. 1988), troponin-C (Herzberg and James 1988; Satyshur et al. 1988), parvalbumin (Heizmann 1984; Heizmann and Braun 1990 ; Heizmann and Hunziker 1990a,b), oncomodulin, belonging to the ß-lineage of parvalbumins (MacManus and Brewer 1987; Banville and Boie 1989; Furter et al. 1989; Mes-Masson et al. 1989; Chalifour et al. 1989; Huber et al. 1990) and calbindins (Szebenyi and Moffat 1986; Perret et al. 1988; Krisinger et al. 1988; Huang and Christakos 1988; Hunziker and Schrickel 1988; Parmentier et al. 1989; Heizmann and Hunziker, 1990a,b) are the most prominent members of this family, whose properties have been reviewed extensively.

This book will focus on *The novel calcium-binding proteins*, most of which were discovered in recent years. Members of this family are listed in Table 1 and are discussed in more detail in the individual chapters of this volume.

Calcium-binding proteins whose functions are known, such as calmodulin, troponin-C, myosin light chains, calpain, or calcineurin are far outnumbered by those whose roles are unknown. In contrast to the multifunctional calcium-receptor calmodulin which is present in all eukaryotic cells, most other members of this family are expressed in a cell-type-specific manner, several of them in the central nervous system.

All of these proteins exhibit a number of common structural features. The general structural principle, the EF-hand, derived from the crystal structure of the calcium-binding carp parvalbumin (pI = 4.25) is thought to be characteristic of this protein family (Kretsinger 1980; Kretsinger et al. 1988; Strynadka and James 1989; Heizmann and Hunziker 1990b). A refined structural analysis of carp parvalbumin has been reported recently at 1.5 Å resolution, showing that although the overall features were very similar, there were significant differences in the amino-terminal region and in the number of oxygen atoms involved in the binding of calcium in the CD and EF sites (Kumar et al. 1990). Besides carp parvalbumin, crystal structures have been reported for *Opsanus tau* parvalbumin (Kahn et al. 1985), pike parvalbumin (Padilla et al. 1988) and rat parvalbumin (McPhalen et al. 1990).

These proteins contain repeat domains which bind calcium selectively with high affinity. Each of these domains consists of a loop of 12-14 amino acids which contribute 6-8 oxygen atoms to the coordination of the calcium ion. The loop is flanked by two α-helices which, in a helix-loop-helix arrangement, is known as EF-hand, designated after the E and F helix of parvalbumin. The essence of this hypothesis is that calcium, acting as a second messenger, interacts with proteins which contain EF-hand structures. This principle received support due to the analysis of the crystal structures of three further members of the EF-hand protein family, the intestinal calcium-binding protein, now named calbindin D-10K (Szebenyi and Moffat 1986), troponin-C (Sundaralingam et al. 1985; Herzberg and James 1988; Satyshur et al. 1988; calmodulin (Babu et al. 1985; Kretsinger et al. 1986), oncomodulin (Przybylska et al. 1988) and a sarcoplasmic calcium-binding protein from the sandworm *Nereus diversicolor* (Ealick et al. 1990).

These proteins display the same loop arrangement of amino acids around the calcium ion, and the α-helices are also in a similar arrangement. However, calmodulin and troponin-C differ from parvalbumin (which is a globular protein) by their central α-helix of 25 amino acids which connect the two pairs of calcium-binding domains.

Putative EF-hand regions can be predicted from the amino acid sequence or the cDNA sequence of a given protein and more than 170 members of the family have been discovered (Kretsinger et al. chapter 1 in this Vol.).

Genes encoding calcium-binding proteins of the EF-type such as calmodulin, parvalbumin, oncomodulin, myosin light chains, Spec proteins, calbindins and calretinin have been analyzed from various species (for reviews see Means et al. 1988; Heizmann and Hunziker 1990a, b; and chapters of this Vol.). Many structural features, including exon length and intron positions, support the view that EF-hand proteins have evolved from a common ancestral single EF-hand motif. During evolution multiple reiterations of this primordial gene gave rise to genes coding for proteins with multiple EF-hands.

Although only a few genes of this family have been analyzed, a family tree is already emerging based on the numbers and positions of the introns. One branch of this tree is that of the S-100 proteins. The analyzed genes of this group all contain an intron in the linker region separating the two calcium-binding domains. The second family branch contains those proteins with four EF-hand domains (calmodulin, myosin alkali light chains, sea urchin Spec-proteins) as well as parvalbumin and oncomodulin with three EF-hands. These genes have five introns (except for parvalbumin and oncomodulin which lack the first two introns due to the absence of the first EF-hand domain), of which four are of the same class and at identical positions. Only the intron in the second to the last EF-hand has a variable class and position. This suggests that this intron was introduced after the divergence of the respective genes.

The calpain gene also contains four EF-hands and five exons, none of which is, however, at positions identical to those of the other families, suggesting different origins.

The third family branch contains the six EF-hand proteins calbindin D-28k and calretinin. The calbindin gene is interrupted by ten introns.

Interestingly, in the S-100 protein family the introns separate functional domains, whereas in the two other families the introns are mostly located within the EF-hand domains.

Despite their structural relationship, little similarity is observed in the promoter regions, which might be expected since the genes show an unique developmental and tissue-specific expression pattern. An interesting sequence has been found in the parvalbumin promoter. Besides the putative transcription regulatory elements, such as a TATA box and a CAAT box, a region was located at position -81 to -112 from the transcription start site which showed 79% homology to the myosin light chain 3F promoter. Both genes are active in the same muscle fiber type (IIB), suggesting that this promoter sequence may be important for the tissue-specific regulation (Berchtold et al. 1987; Berchtold 1989).

The promoter for the rat oncomodulin gene is different from the rat and human parvalbumin. The rat oncomodulin gene contains sequences with high similarities to retroviral long terminal repeats (LTRs), which are most likely responsible for the tumor specific expression of oncomodulin in rat tumors (Banville and Boie 1989; Furter et al. 1989). The concentration of oncomodulin in several human tumor cell lines was found to be at least 100 times lower than in rat tumors when measured by immunological methods (Huber et al. 1990). These results are sustained by recent data indicating that only the rat oncomodulin gene, but not the human and mouse genes, may be under the control of this strong LTR promoter element. Therefore oncomodulin may not be as suitable a marker in human diagnostics as previously suggested.

So far only a few EF-hand protein genes have been mapped with respect to chromosome localization. The parvalbumin gene has been assigned to human chromosome 22. Genetic aberrations of chromosome 22 have been reported in some human cancers and in patients with DiGeorge and Cat Eye syndromes, indicating that the parvalbumin gene may be used as a genetic marker for chromosome 22 associated human diseases (Berchtold et al. 1987).

Table 1. Novel calcium-binding proteins of the EF-type

PROTEINS	Discussed in the following chapters of this Vol.:
S-100 protein family; p11; calcyclin (2Ag); MRP-8 (p8; calgranulin A; L1 light chain); MRP-14[a] (p14; CFAg; calgranulin B; L1 heavy chain); S-100-like proteins; p9Ka; others.	4-12
Calretinin	13 and 14
LSP1	15
α-Actinin (nonmuscle)	16
Crystallins[a]	18
p24 (calciphosine)	19
Sorcin[a]	20
Calcium-binding proteins in invertebrates, fungi, and bacteria	21-28
	SELECTED REFERENCES
Calcineurin	(Klee and Cohen 1988)
Diacylglycerol kinase	(Sakane et al. 1990)
Caltractin	(Huang et al. 1988a,b)
Osteonectin (SPARC; BM-40)	(McVey et al. 1988)
pMP 41	(Kageyama et al. 1989)
Photosystem II calcium-binding proteins	(Webber and Gray 1989)

[a]calcium-binding proteins which were found to be phosphorylated.

The human MRP- and MRP-14 genes are both located on chromosome 1 (Brüggen et al. 1988). The human S-100β, a member of the same subfamily, was found on chromo-some 21 (Allore et al. 1988). It was suggested that some of the impaired brain functions observed in trisomy 21 might be a consequence of the additional copy of this gene since S-100β is expressed in the CNS.

Some additional examples of the chromosomal localization of genes coding for human calcium-binding proteins are listed in Table 2.

Table 2. Chromosomal assignment

Protein	Human chromosome	Reference
Parvalbumin	22	Berchtold et al. 1987
Calbindin D-28K	8	Varghese et al. 1988
Calretinin	16	Parmentier et al. 1989
S-100ß	21	Allore et al. 1988
Calcyclin	1	Ferrari et al. 1987
Myosin alkali light chains (MLC-1 and MLC-3)	2	Seidel et al. 1988
Myosin light chain (MLC-1 emb/A)	17	Seharaseyon et al. 1990
MRP-8 and MRP-16	1	Brüggen et al. 1988
Sorcin	7	van der Bliek et al. 1988
Sorcin-related gene	4	van der Bliek et al. 1988
Calmodulin pseudogene	17	Sen-Gupta et al. 1989

There are also proteins which contain sequences with some resemblance to EF-hand domains or to other parts of the calcium-binding proteins. One example is dystrophin, the protein product of the Duchenne/Becker muscular dystrophy gene. The cystein-rich domain contains two EF-hand loop structures but otherwise it is different from the EF-hand consensus sequence (Monaco 1989). Further examples of proteins with putative EF-hand domains are human lysozyme (Kuroki et al. 1989), bovine glial fibrillary acid protein (Yang et al. 1988) and the sodium channel (Babitch and Anthony, 1988 and chapter 29 in this Vol.).

THE ANNEXIN FAMILY

This family includes a number of structurally related proteins that bind to certain phospholipids in a calcium-dependent manner (Crompton et al. 1988; Haigler et al. 1989; Moss and Crumpton 1990). Several names have been proposed for this family, including lipocortins, calpactins, calelectrin, proteins I and II, chromobindin, calcimedin, endonexin, synhibin, but recently the generic name annexin was suggested (for nomenclature see Crumpton and Dedman 1990, and chapter 31 in this Vol.).

The structural feature which defines this protein family is a repetitive conserved sequence of 70 amino acids. It is assumed that this structural feature is responsible for calcium binding and for interaction with phospholipids. The nonconserved amino-terminal domains may confer a different biological function to each of the members of this family. This domain is the site of phosphorylation by tyrosine kinases, protein kinase C, and other kinases (Johnsson et al. 1988; Abdel-Ghany et al. 1989; Ando et al. 1989; chapter 31-33 in this Vol.). Therefore it was suggested that these proteins may play an important role in intracellular signal transduction.

Although these proteins lack sequences conforming to the typical EF-hand structure characteristic for proteins listed in Table 1, they nevertheless show some structural resemblance. For example, second structure predictions based on the sequence of lipocortin I have suggested that the calcium binding may be achieved through an alternative helix-loop-helix structure in which most of the coordinated ligands are provided by main chain carbonyls. Alignment of the second repeat of human calpactin I with the second EF-hand (which is a mutant EF-hand) of a bacterial calcium-binding protein from *Streptomyces erythraeus* (Moss and Crumpton, 1990) revealed almost 50% sequence homology. It was suggested that these mutant EF-hands in both proteins may still bind calcium and this may suggest a structural relationship of members of the lipocortin/calpactin family with the EF-hand calcium-binding proteins. This observation also indicates that these distinct families of proteins may share common ancestry.

Members of both protein families have also been shown to interact with one another. For example, p11, which is a member of the S-100 proteins containing EF-hand structures, is associated with the calcium- and phospholipid-binding protein, annexin II (p36). This interaction received much attention since p36 represents a major cellular substrate for the tyrosine kinase encoded by the *src* oncogene (see chapter 7 in this Vol.).

CONCLUSION

This book summarizes some of our present knowledge of the protein and gene structures, the tissue distribution, and localization of several novel calcium-binding proteins characterized by the structural principle, the EF-hand, and of another protein family, members of which bind phospholipids in a calcium-dependent manner. Information about the structural and biochemical data of these proteins is accumulating rapidly, hopefully leading to new conclusions about their biological roles in health and disease in the near future.

ACKNOWLEDGMENT

Part of this work was funded by the Swiss National Science Foundation (31-9409.88).

REFERENCES

Abdel-Ghany M, Kole HK, Abon el Saad M, Racker E (1989) Stimulation of phosphorylation of lipocortin at threonine residues by epidermal growth factor (EGF) and the EGF receptor: addition of protein kinase P with polylysine inhibits this effect. Proc Natl Acad Sci USA 86:6072-6078

Allore R, O'Hanlon D, Price R, Neilson K, Willard HF, Cox DR, Marks A, Dunn RJ (1988) Gene encoding the ß subunit of S-100 protein is on chromosome 21: implications for Down's syndrome. Science 239:1311-1313

Ando Y, Imamura S, Hong Y-M, Owada MK, Kakunaga T, Kannagi R (1989) Enhancement of calcium sensitivity of lipocortin I in phospholipid binding induced by limited proteolysis and phosphorylation at the amino terminus as analyzed by phospholipid affinity column chromatography. J Biol Chem 264:6948-6955

Babitch JA, Anthony FA (1987) Grasping for calcium binding sites in sodium channels with an EF-hand. J Theor Biol 127:451-459

Babu YS, Sack JS, Greenhough TJ, Bugg CE, Means AR, Cook WJ (1985) Three-dimensional structure of calmodulin. Nature 316:37-40

Banville D, Boie Y (1989) Retroviral long terminal repeat is the promotor of the gene encoding the tumor-associated calcium-binding protein oncomodulin in the rat. J Mol Biol 207:481-490

Banyard MRC, Tellam RL (1985) The free cytoplasmic calcium concentration of tumorigenic and non-tumorigenic human somatic cell hybrids. Br J Cancer 51:761-766

Berchtold MW (1989) Parvalbumin genes from human and rat are identical in intron/exon organization and contain highly homologous regulatory elements and coding sequences. J Mol Biol 210:417-427

Berchtold MW, Epstein P, Beaudet AL, Payne ME, Heizmann CW, Means AR (1987) Structural organization and chromosomal assignment of the parvalbumin gene. J Biol Chem 262:8696-8701

Brüggen J, Tarcsay L, Cerletti N, Odink K, Rutishauser M, Holländer G, Sorg C (1988) The molecular nature of the cystic fibrosis antigen. Nature 331:570

Chalifour LE, Gomes ML, Mes-Masson AM (1989) Microinjection of metallothionine-oncomodulin DNA into fertilized mouse embryos is correlated with fetal lethality. Oncogene 4:1241-1246

Crompton MR, Moss SE, Crumpton MJ (1988) Diversity in the lipocortin/calpactin family. Cell 55:1-3

Crumpton MJ and Dedman JR (1990) Protein terminology tangle. Nature 345:212

Ealick SE, Babu YS, Cox JA (1990) Three-dimensional structure of a sarcoplasmic calcium-binding protein from the sandworm *Nereus diversicolor*. 7th Int Symp Calcium-binding Proteins in Health and Disease. Banff, Alberta, p 147 (Abstract)

Ferrari SB, Calabretta B, de Riel JK, Battini R, Ghezzo F, Lauret E, Griffin C, Emanuel BS, Gurrieri F, Baserga R (1987) Structural and functional analysis of a growth-regulated gene, the human calcyclin. J Biol Chem 262:8325-8332

Furter C, Heizmann CW, Berchtold MW (1989) Isolation and analysis of a rat genomic clone containing a long terminal repeat with high similarity to the oncomodulin mRNA leader sequence. J Biol Chem 264:18276-18279

Haigler HT, Fitch JM, Jones JM, Schlaepfer DD (1989) Two lipocortin-like proteins, endonexin II and anchorin CII, may be alternate splices of the same gene. TIBS 14:48-50

Heizmann CW (1984) Parvalbumin, an intracellular calcium-binding protein: distribution, properties and possible roles in mammalian cells. Experientia 40:910-921

Heizmann CW, Braun K (1990) Calcium-binding proteins: molecular and functional aspects. In: LJ Anghileri (ed) The role of calcium in biological systems, Vol. V, CRC Press, Boca Raton, FL, pp 21-66

Heizmann CW, Hunziker W (1990a) Intracellular calcium-binding molecules. In: Bronner F (ed) Intracellular calcium regulation. Wiley-Liss Inc, New York, pp 211-248

Heizmann CW, Hunziker W (1990b) Intracellular calcium-binding proteins: more sites than insights. TIBS (in press)

Herzberg O, James NG (1988) Refined crystal structure of troponin C from turkey skeletal muscle at 2.0 Å resolution. J Mol Biol 203:761-779

Huang B, Watterson DM, Lee VD, Schibler MJ (1988a) Purification and characterization of a basal body-associated Ca-binding protein. J Cell Biol 107:121-131

Huang B, Mengerson A, Lee VD (1988b) Molecular cloning of cDNA for caltractin, a basal body-associated Ca-binding protein: homology in its protein sequence with calmodulin and yeast CDC31 gene product. J Cell Biol 107:133-140

Huang YC, Christakos S (1988) Modulation of rat calbindin D-28K gene expression by 1,25α-dihydroxy vitamin D and dietary alterations. Mol Endocrin 2:928-935

Huber S, Leuthold M, Sommer EW, Heizmann CW (1990) Human tumor cell lines express low levels of oncomodulin. Biochem Biophys Res Commun 169:905-909

Hunziker W, Schrickel S (1988) Rat brain calbindin D-28K: six domain structure and extensive amino acid sequence homology with chick calbindin D-28K. Mol Endocrinol 2:465-473

Johnsson N, Johnsson K, Weber K (1988) A discontinuous epitope on p36, the major substrate of *src* tyrosine-protein-kinase, brings the phosphorylation site into the neighborhood of a consensus sequence for Ca-/lipid-binding proteins. FEBS Lett 236:201-204

Kahn R, Fourme R, Bosshard R, Chiadmi M, Risler JL, Dideberg O, Wery JP (1985) Crystal structure study of *Opsanus tau* parvalbumin by multiwavelength anomalous diffraction. FEBS Lett 179:133-137

Klee CB, Cohen P (1988) The calmodulin-regulator protein phosphatase. In: Cohen P, Klee CB (eds) Calmodulin molecular aspects of cellular regulation, Vol. V, Elsevier, Amsterdam, pp 225-247

Klee CB, Vanaman TC (1982) Calmodulin. In: Anfinsen CB, Edsall JT, Richards FM (eds) Calmodulin, molecular aspects of cellular regulation, Vol. V. Elsevier, Amsterdam, pp 225-247

Kretsinger RH (1980) Structure and evolution of calcium-modulated proteins. CRC Crit Rev Biochem 8:119-174

Kretsinger RH, Rudnick SE, Weissman C (1986) Crystal structure of calmodulin. J Inorg Biochem 2:289-302

Kretsinger RH, Moncrief ND, Goodman M, Czelusniak J (1988) Homology of calcium-modulated proteins, their evolutionary and functional relationships. In: Morad M, Naylor WG, Kazda S, Schramm M (eds) The calcium channel, structure, function and implication. Springer, Heidelberg, pp 16-34

Krisinger J, Darwish H, Maeda N, DeLuca HI (1988) Structure and nucleotide sequence of rat intestinal vitamin D-dependent calcium binding protein gene. Proc Natl Acad Sci USA 85:8988-8992

Kumar VD, Lee L, Edwards BFP (1990) Refined crystal structure of calcium-liganded carp parvalbumin 4.25 at 1.5 Å resolution. Biochem 29:1404-1412

Kuroki R, Taniyama Y, Seko C, Nakamura H, Kikuchi M, Ikehara M (1989) Design and creation of a calcium-binding site in human lysozyme to enhance structural stability. Proc Natl Acad Sci USA 86:6903-6907

Kageyama H, Shimizu M, Tokunaga K, Hiwasa T, Sakiyama S (1989) A partial cDNA for a novel protein which has a typical EF-hand structure. Biochim Biophys Acta 1008:255-257

MacManus JP, Brewer LM (1987) Isolation, localization, and properties of the oncodevelopmental calcium-binding protein oncomodulin. Meth Enzymol 139:156-168

MacManus JP, Boynton AL, Whitfield JF (1982) The role of calcium in the control of cell reproduction. In: Anghileri LJ, Tuffet-Anghileri AM (eds) The role of calcium in biological systems, Vol. III, CRC Press, Boca Raton, FL, pp 147-164

McPhalen CA, Sielecki AR, Santarsiero BD, Swensen L, Campbell AP, James MNG (1990) Rat parvalbumin: 3-dimensional X-ray structure. 7th Int Symp Calcium-binding Proteins in Health and Disease, Banff, Alberta, p 146 (Abstract)

McVey JH, Nomura S, Kelly P, Mason IJ, Hogan BLM (1988) Characterization of the mouse SPARC/osteonectin gene. Intron/exon organization and an unusual promoter region. J Biol Chem 263:11111-11116

Means AR, Putkey JA, Epstein P (1988) Organization and evolution of genes for calmodulin and other calcium-binding proteins. In: Cohen P, Klee CB (eds) Calmodulin, molecular aspects of cellular regulation, Vol. V, Elsevier, Amsterdam, pp 17-33

Mes-Masson AM, Masson S, Banville D, Chalifour L (1989) Expression of oncomodulin does not lead to the transformation or immortalization of mammalian cells in vitro. J Cell Sci 94:517-525

Monaco AD (1989) Dystrophin, the protein product of the Duchenne/Becker muscular dystrophy gene. TIBS 14:412-415

Moss SE, Crumpton MJ (1990) The lipocortins and the EF-hand proteins: Ca-binding site and evolution. TIBS 15:11-12

Padilla A, Cavé A, Parello J (1988) Two-dimensional ^{1}H nuclear magnetic resonance study of pike pI 5.0 parvalbumin (*Esox lucius*). Sequential resonance assignments and folding of the polypeptide chain. J Mol Biol 204:995-1017

Parmentier M, DeVijeder JJM, Muir E, Szpirer C, Islam MQ, Geurts vanKessel A, Lawson DEM, Vassart G (1989) The human calbindin 27-kDa gene: structural organization of the 5' and 3' regions, chromosomal assignment and restriction fragment length polymorphism. Genomics 4:309-319

Perret C, Lomri N, Gouhier N, Auffray C, Thomasset M (1988) The rat vitamin D-dependent calcium-binding protein (9-kDa CaBP) gene-complete nucleotide sequence and structural organization. Eur J Biochem 172:43-51

Przybylska M, Ahmed FR, Birnbaum GI, Rose DR (1988) Crystallization and preliminary crystallographic data for oncomodulin. J Mol Biol 199:393-394

Sakane F, Yamada K, Kanoh H, Yokoyama C, Tanabe T (1990) Porcine diacylglycerol kinase sequence has zinc finger and EF-hand motifs. Nature 344:345-348

Satyshur KA, Rao ST, Pyzalska D, Drendel W, Greaser M, Sundaralingam M (1988) Refined structure of chicken skeletal muscle troponin C in the two-calcium state at 2 Å resolution. J Biol Chem 263:1628-1647

Sen-Gupta B, Detera-Wadleigh SD, McBride OW, Friedberg F (1989) A calmodulin pseudogene on human chromosome 17. Nucl Acid Res 17:2868

Seharaseyon J, Bober E, Hsieh C-L, Fodor WL, Francke U, Arnold H-H, Vanin EF (1990) Human embryonic/atrial myosin alkali chain gene: characterization, sequence, and chromosomal localization. Genomics 7:289-293

Seidel U, Bober E, Winter B, Lenz S, Lohse P, Goedde HW, Grzeschik KM, Arnold H-H (1988) Alkali myosin light chains in man are encoded by a multigene family that includes the adult skeletal muscle, the embryonic or atrial, and nonsarcomeric isoforms. Gene 66:135-146

Strynadka NCJ, James MNG (1989) Crystal structures of the helix-loop-helix calcium-binding proteins. Ann Rev Biochem 58:951-998

Sundaralingam M, Bergstrom R, Strasburg G, Rao T, Roychowdhury P (1985) Molecular structure of troponin-C from chicken skeletal muscle at 3 Angstrom resolution. Science 227:945-948

Szebenyi DME, Moffat K (1986) The refined structure of vitamin D-dependent calcium-binding protein from bovine intestine. Molecular details, ion binding, and implication for the structure of other calcium-binding proteins. J Biol Chem 261:8761-8777

Tsuruo T, Iida H, Kawabata H, Tsukagoshi S, Sakurai Y (1984) High calcium content of pleiotropic drug-resistant P388 and K562 leukemia and Chinese hamster ovary cells. Cancer Res 44:5095-5099

van der Bliek AM, Baas F, van der Velde-Koerts T, Biedler JL, Meyers MB, Ozol RF, Hamilton TC, Joenji H, Borst P (1988) Genes amplified and overexpressed in human multidrug-resistant cell lines. Cancer Res 48:5927-5923

Varghese S, Lee S, Huang Y-C, Christakos S (1988) Analysis of rat vitamin D-dependent calbindin D-28K gene expression. J Biol Chem 263:9776-9784

Webber AN, Gray JC (1989) Detection of calcium-binding by photosystem II polypeptides immobilized onto nitrocellulose membrane. FEBS Lett 249:79-82

Whitfield JF (1990) Calcium, cell cycles and cancer. CRC Press Inc., Boca Raton, Florida

Yang ZW, Kong CF, Babitch JA (1988) Characterization and location of divalent cation binding sites in bovine glial fibrillary acidic protein. Biochemistry 27:7045-7050

STAINS-ALL IS A DYE THAT PROBES THE CONFORMATIONAL FEATURES OF CALCIUM BINDING PROTEINS

Yogendra Sharma and Doraivajan Balasubramanian

INTRODUCTION

The field of calcium-binding proteins (CaBP) is still growing rapidly with increasing discoveries of these proteins in a variety of tissues. Significant methodological progress has been made in identifying calcium-binding proteins (CaBP) and in investigating their physicochemical and biochemical properties. The usual methods for the detection and identification of calcium-binding proteins used are ultrafiltration, [45]Ca overlay in gel electrophoresis and subsequent autoradiography, Chelex competitive calcium binding assay, and use of lanthanum luminescence, particularly of terbium, as calcium mimic. Analysis of calcium binding properties implies various methods such as equilibrium dialysis or flow dialysis, equilibrium gel filtration, spectroscopic procedures, and others (for a review, see Schachtele and Marme 1988). Considering the general interest in calcium and CaBP, we review here the less described but widely used technique of staining using the cationic dye called Stains-all.

THE DYE "STAINS-ALL"

A rapid and convenient way of distinguishing calcium-binding proteins from others that has become available recently is the use of the metachromatic cationic carbocyanine dye Stains-all   (1-ethyl-2-{3-(1-ethyl-naphthol[1,2-d]thiazoline-2-ylidine)-2-methylpropenyl}-naphtho[1,2-d]thiazolium bromide) (Figure 1). This dye stains several CaBP blue or purple, and other proteins red or pink. This has led to the suggestion that this dye might be able to identify potential CaBP. Earlier, this dye had been used as a sensitive stain in gel electrophoresis as well as in histochemical staining for various macromolecules. For example, it stains RNA bluish purple, DNA blue and proteins pink or red (Dahlberg et al. 1969) phosphoproteins, mucopolysaccharides (Green et al. 1973; Green and Pastewka 1974a,b), and hyaluronate blue (Badar et al. 1972). This dye has been used to characterize phosphoproteins in neurofilaments (Ksiezak and Yen 1987) and phosphovitin from yolk granules (Wallace and Morgan 1986a) by staining characteristics described by Wallace and Morgan (1986b). The applicability and methodology for staining various protein fractions of erythrocyte membrane in gel electrophoresis was described by King and Morrison (1976). Because of its characteristic for staining almost every type of macromolecules, it was called Stains-all by Dahlberg et al. (1969).

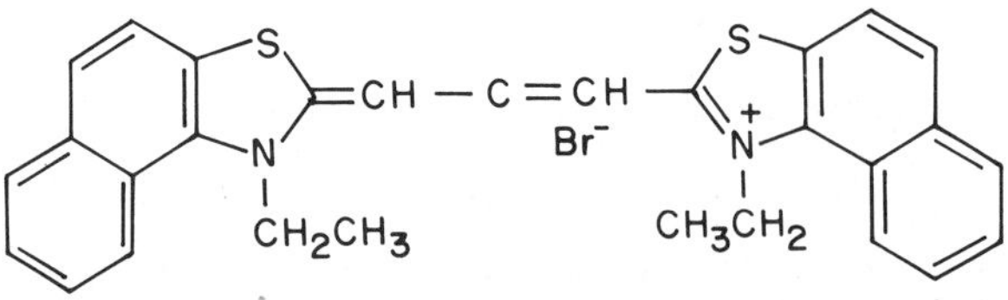

Fig. 1. Structural formula of the dye Stains-all

Five different types of complex states and their corresponding approximate spectral band maxima have been characterized for the dye, namely, the α band with an absorption maximum at 570 nm, the ß band at 535 nm, the γ band at 500-510 nm, the J band at 600-650 nm, a hybrid or mixture ßα band at 550 nm, and the S band at 470 nm. The occurrence of various bands with conditions of dye-dye interaction and dye-macromolecule interactions of Stains-all in solutions has been analyzed and discussed by Kay et al. (1964a,b) and Bean et al. (1965). The monomeric form of the free dye, which is found in organic solvents, as well as in aqueous solutions at low concentrations and high temperatures, has an absorption maximum at 575 nm and corresponds to the α-state (Kay et al. 1964a,b; Bean et al. 1965). The J state occurring around 630 nm, and thereby generating a blue color, characterizes an aggregate form of the dye, arising as a result of the interaction of Stains-all with several classes of macromolecules, and is dependent on the number of anionic sites as well as their spacing and on the macromolecular conformation (Jelley 1936). Other forms of the dye are not very well characterized.

The potentiality and versatility of Stains-all staining of calcium-binding proteins was first pointed out by Jones et al. (1979) and by Jones and Cala (1981); the dye was later shown to combine with several calcium-binding proteins, including calmodulin, troponin C, and calsequestrin, to yield spectrally distinct complexes (Campbell et al. 1983). The problems associated with the self-aggregation of the dye in aqueous media have been minimized by the use of 30% aqueous ethylene glycol as the solvent medium, wherein it has been shown to display a high degree of spectral sensitivity to variations in microenvironment (Caday and Steiner 1985). This feature enables Stains-all to probe microstructural features of the proteins to which it binds.

METHODS FOR STAINS-ALL ASSAY OF CALCIUM-BINDING PROTEINS
STAINING OF *CaBP* IN *SDS*-PAGE

The original staining procedure described by King and Morrison (1976) was modified slightly for CaBP by Campbell et al. (1983). Briefly, gels are fixed in 25% isopropanol overnight and washed exhaustively with the same medium so as to remove SDS. This is important since even traces of the anionic surfactant SDS cause precipitation or decolorization of the dye. The staining of the gels was then performed in the dark (so as not to bleach the dye) for at least 48 h with 0.0025% Stains-all, 25% isopropanol, 7.5% formamide and 30 mM Tris-base, pH 8.8. Ervasti et al. (1989) have used 10% isopropanol. Destaining is not required and gels can be stored in water, and if required, can be restained in Coomassie blue. Stains-all treatment of the gels shows metachromatically blue bands corresponding to CaBP. Photographs can be taken using an orange-red filter, so as to emphasize the blue-stained bands and suppress the red-stained ones. Maruyama and Nonomura (1984) have found that the use of methanol in place of isopropanol remarkably reduces the time necessary for the fixation, washing of SDS, and staining.

This method has been used frequently in the identification of several novel CaBP from a variety of tissues, as for example, calsequestrin-like protein from sea urchin (Oberdorf et al. 1988), its identification in plant cells (Krause et al. 1989) and in smooth muscle endoplasmic reticulum (Wuytack et al. 1987), a plasma lipoprotein and calcium-binding protein from sarcoplasmic reticulum (Hofmann et al. 1989). Several CaBP, e.g., Ca^{2+}-ATPase (Campbell et.al. 1983), calcium-binding crystallins (Sharma et al. 1989), are not stained blue for variety of reasons (Maruyama and Nonomura 1984). Therefore, in addition to Stains-all staining for gels, other methods such as ^{45}Ca autoradiography and if purified protein is available then spectroscopy of the dye-protein complex should also be performed to detect all CaBP.

SPECTROSCOPIC STUDIES

The best method used for absorption and CD spectral monitoring of Stains-all binding has been described by Caday and Steiner (1985). This dye tends to self-aggregate in water and produce the J and γ bands. Since it is these bands that we monitor when the dye binds to proteins, it is important that there is no artifact due to dye aggregation. Caday and Steiner (1985) have provided the reproducible conditions, namely, micromolar dye concentration and the use of ethylene glycol, to obviate this problem. The Stains-all solutions were prepared by dissolving the dye in ethylene glycol to about 0.5 mM and the actual concentration of the dye determined based on a molar extinction coefficient value at 578 nm of 1.13×10^5 in this solvent. In order to make complexes, proteins were dissolved in 2 mM MOPS, pH 7.2, containing 30% ethylene glycol, to which an aliquot of the stock solution of the dye was added and the mixture was incubated for about 30-60 min in the dark with occasional shaking. Under these conditions, reproducible and artifact-free results are obtained. Molecular ellipticities are expressed in terms of the dye molarity, using the extinction coefficient of the dye mentioned above.

STRUCTURAL AND CONFORMATIONAL DEPENDENT BINDING OF STAINS-ALL

The interaction of Stains-all with various CaBP has been studied using absorption and circular dichroism spectroscopy, and it was found that different calcium-binding proteins interact differently with the dye, and yield different spectral bands (Caday and Steiner 1985; Caday et al. 1986; Moore 1988; Sharma et al. 1989b). Of the various bands, namely, α, ß, αß, γ, S, and J bands that the dye displays, the J and the γ bands are the diagnostic ones in this context. Binding of Stains-all with CaBP induces either the J band (600-650 nm) or the γ band (500-520 nm), or both in absorption and in CD. This dye thus appears to enable us to distinguish CaBP from others, and also to probe possible structural differences among CaBP themselves.

CaBP THAT INDUCE BOTH THE J AND γ BANDS UPON STAINS-ALL BINDING

At high dye:protein molar ratio, calmodulin and troponin C complex with the dye similarly and yield the J band. As dye:protein ratio is decreased, the J band is lost, yielding the γ band at 500-510 nm. At a dye:protein ratio of about 10, both the J and γ bands can be seen in the complex (Caday et al. 1986) (Figure 2). Protein S-100 from rat and bovine (Moore 1988) and activation segment of pig procarboxypeptidase A (Vilanova et al. 1988) also induces the J band, which is replaced by the γ band upon the addition of calcium.

CaBP THAT INDUCE ONLY THE J BAND

In case of parvalbumin (Caday et al. 1986) and ß-crystallin (Sharma et al. 1989), the J band is retained at all stoichiometries, but is lost upon the addition of calcium. In these cases, the γ band is not activated (Figure 3). The added calcium simply displaces the dye from the proteins and the spectrum of the free, monomeric Stains-all (the α and the ß bands) results.

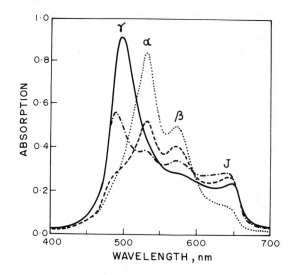

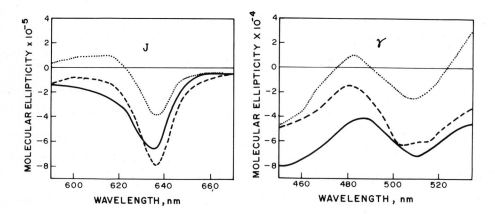

Fig. 2. Absorption and circular-dichroism spectra of the complexes of calmodulin and Stains-all showing both the J and γ bands

CaBP WHICH GENERATE ONLY THE γ BAND

δ-Crystallin, the highly α-helical protein of avian eye lenses (Sharma et al. 1989a) is so far the only CaBP known to generate the γ band (Figure 4) upon Stains-all binding. This band disappears upon the addition of calcium, yielding the α and ß bands of the monomeric dye. This is particularly interesting since, of all the crystallins, δ-crystallin is the one that has a helix-loop helix motif for calcium binding.

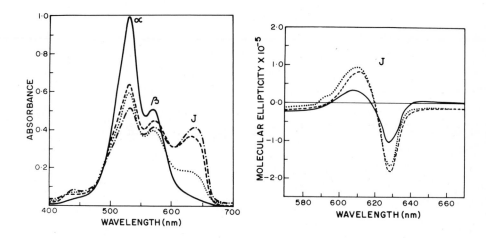

Fig. 3. Absorption and circular-dichroism spectra of the complexes of Stains-all with ß-crystallin showing the J band only

CONCLUSIONS

Caday et al. (1986) earlier argued that it might be the structural difference that is responsible for the difference in the metachromatic behavior of calmodulin, troponin C, and parvalbumin towards Stains all. They proposed that the J band is induced in Stains-all when it binds to the common structural element, namely the globular lobe of calmodulin, and that the γ band arises when the dye is bound to the strand connecting the lobes, i.e., the long helical stretch. We have argued in our paper (Sharma et al. 1989b) that such an assignment of the spectral bands of the bound dye to the conformational status of the substrate raises the possibility of whether a simple water-soluble helical peptide chain that binds Stains-all would activate the γ band. We note in the connection that polyglutamic acid below pH 4, under which conditions it is folded completely in the right-handed α-helical conformation, indeed does so. On the other hand, above pH 4, when it unfolds into a randomly coiled form, it generates neither the γ nor the J band, but the ßα band centered around 550 nm.

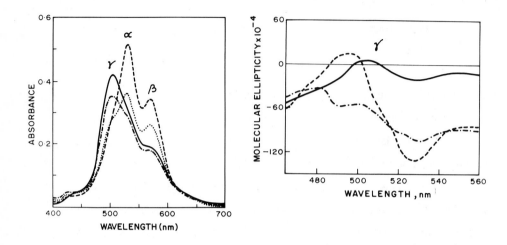

Fig. 4. Absorption and circular-dichroism spectra of the complexes of Stains-all and δ-crystallin showing the γ band only

Polyaspartic acid, which does not adopt any ordered conformation but is a random coil (Fasman 1967), does not generate the γ or the J band in the dye (Bean et al. 1965). Additionally, δ-crystallin, predominantly a α-helical protein, binds to Stains-all and generates only the γ band, whereas ß-crystallin, which adopts a Greek key motif bilobal globular structure induces only the J band in the dye. For the present purpose, therefore, this molecule may be thought of as having the calcium-binding globules which generate the J Band but no helical stretch that generates the γ band.

These observations prompt us to suggest that the absorption spectrum of the bound dye Stains-all is governed by a set of "conformational selection rules," that are summarized below.

1. The J band occurs when the dye is bound to anionic sites that are present in the globular or compact conformations of the protein. In proteins where the binding region occurs elsewhere in the structure, the J band will not be activated and the protein will not stain blue.

2. The γ band of the dye occurs when the dye is bound in a long and exposed helical region of the polymer. Calcium-binding proteins that do not contain binding sites in such helical stretches may not induce the γ band.

3. It may not even be necessary that the protein or the macromolecule be a calcium-binding substrate; it may suffice that it contain anionic charges which enable the binding of the cationic dye Stains-all. It may then generate the J band or the γ band, depending on (1) or (2) above.

4. A substrate that does not have anionic sites properly disposed so as to enable it to bind Stains-all will not induce any color change in the dye.

These rules clarify the question of why some CaBP do not stain the dye Stains-all blue while some others do. The appearance of blue color is due to the J state of the dye. In order to test the validity of the above conformational selection rule, we performed the spectroscopy of the stains-all complex with various fragments of calmodulin and found that C-terminal 120-148 long calmodulin peptide which possesses globular structure showed only J band.

This technique not only help to differentiate various CaBP among themselves but also can be apply to study various systems related to CaBP, modified CaBP, CaBP interaction with other proteins. Caday and Steiner (1986) have used spectroscopic variation in calmodulin-melittin complexes for the study of calmodulin binding domain of target enzyme. Attempts are now being made to apply this method on the study of modified calcium binding crystallins (Sharma et al. 1990).

ACKNOWLEDGMENT

We are grateful to Dr.Gopalakrishna of the laboratory of Chemical Pharmacology, National Heart, Lung and Blood Institute, NIH, Bethesda, Maryland, USA, for a gift of the chemically synthesized calmodulin fragment (120-148) for our use.

REFERENCES

Badar JP, Ray DA, Steck TL (1972) Electrophoretic determination of hyaluronate produced by cells in culture. Biochim Biophys Acta 264:73-84

Beans RC, Shepherd WC, Kay RE, Walwick ER (1965) Spectral changes in a cationic dye to interaction with macromolecules. III. Stoichiometry and mechanism of the complexing reaction. J Phys Chem 69:4368-4379

Caday CG, Steiner RF (1985) The interaction of calmodulin with the carbocyanine dye (Stains-all). J Biol Chem 260:5985-5990

Caday CG, Lambooy PK, Steiner RF (1986) The interaction of calcium binding proteins with the carbocyanine dye Stains-all. Biopolymers 25:1579-1595

Caday CG, Steiner RF (1986) The interaction of calmodulin with melittin. Biochem Biophys Res Comm 135:419-425

Campbell KP, Maclennan DH, Jorgensen AO (1983) Staining of the calcium binding proteins, calsequestrin, calmodulin, troponin C and S-100 with the cationic dye "Stains-all". J Biol Chem 258:11267-11273

Dahlberg AE, Dandman CW, Peacock AC (1969) Electrophoretic characterization of bacterial polyribosomes in agarose-acrylamide composite gels. J Mol Biol 41:139-147

Ervasti JM, Claessens MT, Mickelson JR, Louis CF (1989) Altered transverse tubule dihydropyridine receptor binding in malignant hyperthermia. J Biol Chem 264:2711-2717

Fasman GD (1967) Poly α-amino acids. Marcel Dekker, New York

Green MR, Pastewka JV (1974a) Simultaneous differential staining by a cationic carbocyanine dye of nucleic acids, proteins and conjugated proteins. I. Phosphoproteins. J Histochem Cytochem 22:767-773

Green MR, Pastewka JV (1974b) Simultaneous differential staining by a cationic carbocyanine dye of nucleic acids, proteins and conjugated proteins. II. Carbohydrate and sulfated carbohydrate-containing proteins. J Histochem Cytochem 22:774-781

Green MR, Pastewka JV (1975) Identification of sialic acid rich proteins on polyacrylamide gel. Anal Biochem 65:66-72

Jelley EE (1936) Spectral absorption and fluorescence of dyes in the molecular state. Nature 138:1009-1010

Jones LR, Besch HR Jr, Fleming JW, McConnaughley MM, Watanabe AM (1979) Separation of vesicles of cardiac sarcoplasmic reticulum: comparative biochemical analysis of component activities. J Biol Chem 254:530-539

Jones LR, Cala SE (1981) Biochemical evidence for functional heterogeneity of cardiac sarcoplasmic reticulum vesicles. J Biol Chem 256:11809-11818

Hofman SL, Brown MS, Lee E, Pathak RK, Anderson RG, Goldstein JL (1989) Purification of a sarcoplasmic reticulum protein that binds calcium and plasma lipoproteins. J Biol Chem 264:8260-8270

Kay RE, Walwick ER, Gifford CK (1964) Spectral changes in a cationic dye to interaction with macromolecules. I. Behavior of dye alone in solution and the effect of added macromolecules. J Phys Chem 68:1896-1906

Kay RE, Walwick ER, Gifford CK (1964) Spectral changes in a cationic dye due to interaction with macromolecules. II. Effects of environment and macromolecule structure. J. Phys Chem 68:1907-1916

King LE, Morrison M (1976) The visualization of human erythrocyte membrane proteins and glycoproteins and SDS-polyacrylamide gels employing a single staining procedure. Anal Biochem 71:223-230

Krause KH, Chou M, Thomas MA, Sjolund RD, Campbell KP (1989) Plant cells containing calsequestrin. J Biol Chem 264:4269-4272

Ksiezak-Reding H, Yen SH (1987) Phosphatase and carbocyanine dye binding define different types of phosphate groups in mammalian neurofilaments. J Neurosci 3554-3560

Maruyama K, Nonomura Y (1984) High molecular weight protein in the microsome of seallop striated muscle. J Biochem (Tokyo) 96:859-870

Moore BW (1988) Conformational and hydrophobic properties of rat and bovine S-100 proteins. Neurochem Res 13:539-545

Oberdorf JA, Lebeche D, Head JF, Kaminer B (1988) Identification of a calsequestrin like protein from sea urchin eggs. J Biol Chem 263:6806-6809

Schachtele CN, Marme D (1988) Methods of assay of calcium binding proteins. In: Calcium-binding proteins, Vol I, (Thompson MP ed.), CRC Press Inc., Florida, pp 83-96

Sharma Y, Rao CM, Narasu ML, Rao SC, Somasundaram T, Krishna AG, Balasubramanian D (1989a) Calcium ion binding to ß- and to δ-crystallins. The presence of EF-hand motif in δ-crystallin that aids to calcium binding. J Biol Chem 264:12794-12799

Sharma Y, Rao CM, Rao SC, Krishna AG, Somasundaram T, Balasubramanian D (1989b) Binding site confirmation dictates the color of the dye Stains-all. A study of the binding of this dye to the eye lens proteins crystallins. J Biol Chem 264:20923-20927

Vilanova M, Vendrell J, Cuchillo CM, Ailes FX (1988) Analysis of the conformation and ligand binding properties of the activation segment of pig procarboxypeptidase A. Biochem J 251:901-905

Wallace RA, Morgan JP (1986a) Chromatographic resolution of chicken phosphovitin. Multiple macromolecular species in a classic vitellogenin-derived phosphoprotein. Biochem J 240:871-878

Wuytack F, Raeymaekers L, Verbist J, Jones LR, Casteels R (1987) Smooth-muscle endoplasmic reticulum contains a cardiac-like form of calsequestrin. Biochem Biophys Acta 899:151-158

Section II

EF-Hand Calcium-Binding Proteins:
the S-100 Protein Family

THE S-100 PROTEIN FAMILY:
A BIOCHEMICAL AND FUNCTIONAL OVERVIEW

Dana C. Hilt and Douglas Kligman

INTRODUCTION

The intracellular Ca^{2+} concentration is closely regulated in eukaryotic cells. The concentration of Ca^{2+} transiently increases from a basal level of approximately 0.1 μM as a result of both flux into the cell of extracellular Ca^{2+} or release of intracellular Ca^{2+} stores. This increase in intracellular Ca^{2+} concentration by 10- to 100-fold functions as an important second messenger signal which may be transduced by a variety of pathways to produce a Ca^{2+}-mediated physiological response in the cell. The signal of increased intracellular Ca^{2+} needs to be transduced to a wide variety of cellular proteins or effector molecules in various subcellular compartments. These subcellular compartments have differing Ca^{2+} concentrations. The targets of the signal of increased Ca^{2+} differ dependent upon the cell type. The S-100 proteins are expressed in a cell-type specific fashion and, therefore, are excellent candidates to function as cell-type specific transducers of the signal of increased intracellular Ca^{2+}. Cell-cycle progression, differentiation and a number of specific cellular physiological responses have been demonstrated to be dependent upon changes in the intracellular or subcellular compartment Ca^{2+} concentration. The S-100 proteins may function as cell-type specific and cell-cycle specific mediators of the Ca^{2+} signal.

The S-100 proteins are a group of small acidic Ca^{2+}-binding proteins that are expressed in a cell-type-dependent fashion (for review see Kligman and Hilt 1988). After the initial description of S-100 isolated from brain it was determined that this protein was able to bind calcium which induces a conformational change in the protein. These proteins have sequence similarity with calmodulin and other calcium-binding proteins (Moore 1965; Calissano et al. 1969; Dannies and Levine 1971; Isobe et al. 1977; Isobe and Okuyama 1978 and 1981). Calmodulin functions in a wide range of cell types to transduce the signal of increased intracellular Ca^{2+} and is expressed in virtually all cell types. In contrast, the distribution of particular S-100 proteins appears to be restricted to specific cell types. Cell-type specific responses to increased intracellular Ca^{2+} could either be due to the cell-type specific expression of calmodulin-binding proteins or the expression of cell-type specific Ca^{2+}-binding proteins. These facts, coupled with the relative ubiquity of calmodulin, suggested that the S-100 proteins may have specific cellular functions dependent upon the phenotype of the cell which expresses them and that this function would be dependent upon Ca^{2+} but would be different from that of calmodulin (Kligman and Hilt 1988). The S-100 proteins may, therefore, be involved in transducing the signal of an increase in intracellular calcium in a cell-type specific fashion.

This chapter will outline the physical, biochemical and biological characteristics of the S-100 proteins, briefly discuss members of the S-100 protein family, review the proposed mechanisms of action of the S-100 proteins, point out some of the unanswered questions that must be addressed, and suggest future research directions to be pursued before we can better understand the functions of this interesting family of proteins.

THE S-100 PROTEINS: A HISTORICAL AND EVOLUTIONARY PERSPECTIVE

The first description of S-100 protein was made by Moore who initially characterized a group of abundant low molecular weight (10-12 kDa) acidic proteins that were enriched in the nervous system (Moore 1965). These proteins were present in high abundance and comprised approximately 0.5% of the soluble proteins from bovine brain. Their name derived from the unusual property of solubility in 100% ammonium sulfate. The S-100 fraction as isolated from brain was further characterized by Isobe and coworkers who demonstrated that there were, in fact, multiple protein species in this fraction (Isobe et al. 1977). Two different S-100 subunits were described, α and ß. Three S-100 dimeric isoforms were isolated and characterized; S-100ao, S-100α and S-100ß which differ in their subunit composition (αα, αß, or ßß dimers respectively). Each S-100 monomer protein contained two Ca^{2+}-binding regions known as EF-hands (Persechini et al. 1988; Kretsinger 1980; Tufty and Kretsinger 1975). The S-100 proteins were originally thought to be specific for the nervous system though that is now known to not be the case, as a variety of S-100 proteins are expressed in other tissues. In the nervous system S-100ß is restricted to expression in glial cells while S-100α is expressed at much lower levels in some neuronal cells, as well as non-nervous tissue (Hyden and McEwen 1966; Sviridov et al. 1972; Tabuchi and Kirsch 1975; Matus and Mughal 1975; Cocchia 1981; Haglid et al. 1976).

Over the past few years a number of other proteins closely related to S-100α and/or S-100ß have been isolated from tissues other than the nervous system (Table 1). This "S-100 Protein Family" has expanded with the description of the new members. To date, at least ten proteins have been described in various cells or tissues and will be briefly reviewed in this chapter. These proteins have closely related amino acid compositions, amino acid sequences, and Ca^{2+}-binding sites. These similarities allow one to classify these proteins as related members of a S-100 protein family. However, these structural similarities have made the generation of specific antisera difficult due to the amino acid sequence conservation. Immunological localization of a specific S-100 protein is, therefore, complicated by the cross reactivity of antisera raised against even a completely purified S-100 protein species due to this extensive sequence homology. Ideally, the generation of monospecific reagents utilizing either DNA sequence to generate specific in situ hybridization probes or specific antisera raised against peptides unique to a particular S-100 protein species will be necessary to unambiguously localize the proteins to particular cell types. It also will be necessary to determine if a particular cell type expresses more than one S-100 protein.

The amino acid sequences of a number of S-100 proteins have been conserved in a wide variety of organisms ranging from protozoa to man (Michetti and Cocchia 1982; Cocchia et al. 1985; Moore 1984). This strong conservation argues for an important (and conserved) biological role for the S-100 proteins. An example of this conservation is the remarkable conservation of the S-100ß amino acid sequence in a variety of mammalian species. The amino acid sequence of S-100ß purified from bovine, rat, murine, human, or porcine sources differs by less than 5% (Moore 1984).

Table 1. Properties of the S-100 proteins

Name	Cell type/source	Ca²⁺-binding	Context described	Extracellular/secreted (E)	Cell-cycle regulated or differentiation induced	Inhibition of kinase activity
S-100β	glial cells, other cells	+	brain-specific protein	E	+	+
S-100α	brain	+	brain-specific protein	ND	+	+
p11/calpactin	many cells	-	regulatory subunit of tyrosine kinase substrate	ND	ND	+
42C	PC12 cells	ND	NGF induced mRNA species	ND	+	ND
calcyclin	fibroblasts, other cells	ND	serum-stimulated mRNA G_1 phase cell-cycle-specific	ND	+	ND
PRA	mammary carcinoma, many other cells	ND	co-purifies with prolactin receptor	ND	ND	ND
18A2	fibroblasts, other cells	ND	serum-stimulated mRNA S phase cell-cycle-specific	ND	+	ND
p9Ka	breast myoepithelial cells	ND	induced during differentiation of cuboidal cells to myoepithelial cells	ND	+	ND
pEL68	mammary carcinoma, many other cells	ND	induced in transformed cell lines	ND	+	ND
mts1	tumor cells	ND	increased expression in tumor cells with metastatic potential	ND	+	ND
42A	PC12 cells	ND	NGF induced mRNA species	ND	+	ND
S-100L	lung and other cells	+	expressed in lung cells	ND	ND	ND
CF Ag/MRP-8	myeloid cells	+	expressed in cystic fibrosis and rheumatoid arthritis	E	+	+
MRP-14	myeloid cells	+	expressed in rheumatoid arthritis	E	+	ND
ICaBP	intestinal cells	+	vitamin D_3-induced Ca^{2+}-binding protein	ND	ND	ND

ND= not done

The varied properties of the S-100 proteins are tabulated. The same species of S-100 protein has been described in different cells under different circumstances. p11 and 42C are the same protein species as are calcyclin and PRA. 18A2, p9Ka, pEL68, mts1, and 42A are the same protein

The evident amino acid sequence homology, conservation of calcium-binding sites and similarities of gene organization suggest that this group of proteins evolved from a common ancestral EF-hand gene (Persechini et al. 1988; Kretsinger 1980; Tufty and Kretsinger 1975). The ancestral gene likely contained an EF-hand-like structure capable of binding Ca^{2+}. This primordial gene could have been the subject of gene duplication and subsequent evolution to the forms found in the various EF-hand protein families. In all genes which encode EF-hand-containing proteins that have been studied to date, the EF-hands are on separate exons. The genomic organization of at least three genes which encode S-100 proteins have been examined. The calcyclin, p9Ka, and S-100ß genes all have two separate exons which encode EF-hands (Ferrari et al. 1987; Barraclough et al. 1987; Kuwano et al. 1984; Hilt and Kligman unpubl. observ.). This gene structure is in contrast to the calmodulin-related protein genes which have four separate exons encoding the four EF-hands found in this group of proteins. These observations strengthen the notion that the generation of diversity in the EF-hand protein families is an example of isolation of a biochemical functional motif (the EF-hand) on an exon with subsequent exon shuffling. Interestingly, both the calcyclin and S-100ß genes have the unusual feature of having an intron in the 5' untranslated region (Ferrari et al. 1987; Kligman unpubl. observ.). The significance of this observation is not known.

GENERAL FEATURES OF S-100 PROTEIN FAMILY MEMBERS

All members of the S-100 protein family characterized to date have significant physico-chemical properties in common. S-100 proteins contain two EF-hands (Figures 1A and 2) which function as calcium binding domains. When calcium is bound to these proteins a number of physicochemical properties of the proteins are altered including changes in the absorption, circular dichroism spectra, and α-helix and ß-sheet structure (Calissano et al. 1969; Dannies and Levine 1971; Baudier and Gerard 1983; Moore 1988). These findings indicate significant conformational changes in the protein. The binding of calcium to the EF-hand region(s) and the subsequent induced conformational change has been shown to expose hydrophobic domains of the protein to solvent (Figure 1B). This Ca^{2+}-dependent conformational change may facilitate the interaction of the S-100 protein with a hydrophobic region of a secondary effector or S-100-binding protein.

S100 PROTEIN DOMAINS

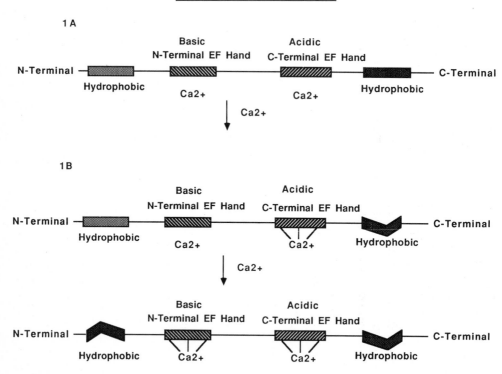

Fig. 1A,B. S-100 protein domains. **A** All S-100 proteins have a common domain structure. Hydrophobic regions are located in both N- and C-termini. The N-terminal EF-hand is located in a region rich in basic amino acids, and the C-terminal EF-hand is located in a region rich in acidic amino acids. These differences may in part explain the differing Ca^{2+} affinities (see text). **B** Ca^{2+} initally binds to the high affinity C-terminal EF-hand and induces a conformational change in the C-terminal hydrophobic domain with exposure of the hydrophobic domain to solvent. At higher Ca^{2+} concentrations the N-terminal EF-hand binds Ca^{2+} inducing a conformational change in the N-terminal EF-hand exposing it to solvent. This demonstrates the heterobifunctional nature of S-100 proteins. The different conformations may have different functions

The biological functions of the S-100 proteins are less clear. No specific enzymatic property has been ascribed to any of the proteins to date. They likely exert their biological effects by binding to or interacting with secondary effector proteins in a Ca^{2+}-dependent fashion. This mode of protein-protein interaction and modulation of the activity of the secondary effector protein is similar to that seen with calmodulin.

N-terminal EF-hand

```
                10         20         30         40         50         60
s-100β     SELEKAVVAL IDVFHQYSGR EGDKHKLKKS ELKELINNEL SHFLEEIKEQ EVVDKVMETL
s-100α    GSELETAMETL INVFHAHSGK EGDKYKLSKK ELKELIQTEL SGFLDAQKDA DAVDKVMKEL
calcyclin ACPLDQAIGLL VAIFHKYSGR EGDKHTLSKK ELKELIQKEL TIGSKLQ    DAEIARLMEDL
p11       PSQMEHAMETM MLTFHRFA   GDKDHLTKE  DLRVIMEREF PGFLENQKDP LAVDKIMKDL
18A2      ARPLEEALDVI VSTFHKYSGK EGDKFKLNKT ELKELLTREL PSFLFKRTDE AAFQKVMSNL
CF Ag     LTELEKALNSI IDVYHKYSLI KGN FHAVYR DDLKKLLETEC PQYIRKKGAD VWF KEL
s-100L    SSPLEQALAVM VATFHKYSGQ EGDKFKLSKG EMKELLHKEL PSFVGEKVDE EGLKKIMGDL
MRP14   TCKMSQLERNIETI INTFHQYSVK LGHPDTLNQGEF KELVRKDL  QNFLKKENK  NEKVIEHIME
ICaBP     AS L KSPEEM KSIFQKYAAK EGDPNQLSKE ELKLLIQSEF PNLLKASSTL DNLFEELDKND
                 *         *        * *  *         *
```

C-terminal EF-hand

```
                70         80         90
s-100β    DSDGDGECDF QEFMAFVAMI TTACHEFFEH E
s-100α    DEDGDGEVDF QEYVVLVAAL TVACNNFFWE NS
calcyclin DRNKDQEVNF QEYVTFLGAL ALIYNEALKG
p11       DQCRDGKVGF QSFISLVAGL TIACNDYFVV NMKQKGKK
18A2      DSNRDNEVDF QEYCVFLSCI AMMCNEFFEG CPDKEPRKK
CF Ag     DINTDGAVNF QEFLILVIKM AWQPTKKAMK KATKSS
s-100L    DENSDQQVDF QEYAVFLALI TIMCNDFFQG SPARS
MRP14     DLDTNADKQLSFEEFIMLMARL TWASHEKMHE GDEGPGHHH KPGLGEGTP
ICaBP     DGEVS YEEFVEFFKKL SQ
          * * * * *  *
```

Fig. 2. The S-100 protein family. The amino acid sequences of the nine distinct S-100 protein species have been aligned using the *single letter* amino acid designation. *Gaps* have been introduced to maximize alignment. Amino acid residues where at least three of the S-100 proteins are homologous are printed in *bold letters* or are *underlined (single or double)*. The N-terminal and C-terminal-EF hands are indicated by *bars*. The amino acids which coordinate Ca^{2+} are shown by *asterisks*. The regions of maximum divergence of the S-100 proteins are in a 'hinge' region between the two EF-hands (amino acid residues 40-60) and the C-terminal region. Even in regions where three S-100 proteins do not share the same amino acid there are conservative amino acid substitutions

Calmodulin also undergoes both conformational changes and exposure of hydrophobic domains to solvent when calcium is bound. The hydrophobic domains so exposed interact with the hydrophobic domains of secondary proteins. These observations suggest that the S-100 proteins exert their effects in a manner similar to that of calmodulin but interact with a different set of secondary effector proteins than does calmodulin.

As other proteins related to S-100ß or S-100α have been described over the past few years it has become evident that this protein family has members which are expressed in a variety of tissues and cell types other than the nervous system. Many of these proteins are differentially expressed with cell-growth stage or degree of cellular differentiation. At least nine S-100 proteins have been identified (Figure 2). The structural properties of S-100-related proteins will first be described then the individual members will be briefly reviewed.

THE S-100-RELATED PROTEINS: STRUCTURAL REQUIREMENTS

All of the S-100-related proteins are low molecular weight (approximately 10-12 kDa) acidic proteins. The most striking conserved feature of these proteins is the presence of two EF-hands per monomer of protein which function as the Ca^{2+}-binding sites (Figure 1 and 2). EF-hand was a term given to the two α-helices (E and F) and the intervening Ca^{2+}-binding loop that was first deduced from the crystal structure of parvalbumin (Tufty and Kretsinger 1975). The Ca^{2+}-binding loop of an EF-hand contains amino acids with side chain oxygens (Ser, Thr, Asp, Asn, Glu, or Gln) which serve to coordinate the bound Ca^{2+} ion. The Ca^{2+} binds to the EF-hand with approximately $K_d=1-10$ μM affinity, although the exact Ca^{2+}-binding affinity of an EF-hand may vary significantly from protein to protein. Binding of Ca^{2+} to the EF-hand induces a conformational change in the proteins and thus transduces the signal of Ca^{2+} binding (Baudier et Gerard 1983).

Other Ca^{2+}-binding proteins such as the calmodulin protein family and troponin C protein family contain four EF-hands while the parvalbumin protein family contains three EF-hands per monomer molecule. The S-100 protein family is the only EF-hand-containing protein family with two EF-hands. The C-terminal EF-hand in the S-100-related proteins contains 12 amino acids and is similar in sequence to that found in the calmodulin protein family. However, the N-terminal EF-hand in each of the S-100-related proteins contains 14 amino acids instead of 12. This variation produces a significant decrease in the calcium affinity of the N-terminal EF-hand when compared to the C-terminal EF-hand. The calcium-binding affinity of the C-terminal EF-hand is approximately K_d=20-50 µM while the Ca^{2+}-binding affinity of the N-terminal EF-hand is K_d=200-500 µM (Kretsinger 1980). Interestingly, the 14 amino acid EF-hand is conserved in all S-100 proteins. This observation argues for an important and conserved function in spite of its low Ca^{2+}-binding affinity.

The crystal structure of only one S-100 protein has been solved: intestinal Ca^{2+}-binding protein (ICaBP) (Szebenyi et al. 1981). In the crystal state the N-terminal EF-hand is much less accessible to solvent than is the C-terminal EF-hand. A solvent H_2O molecule likely functions as a Ca^{2+}-coordinating ligand in the C-terminal EF-hand but the N-terminal EF-hand does not have space for solvent H_2O molecules as it is buried inside the protein molecule. The crystal structure of the C-terminal 12 amino acid EF-hand is very similar to the parvalbumin EF-hands but the N-terminal 14 amino acid EF-hand is very different.

The presence of two binding sites for Ca^{2+} with such different affinities and the presence of only two EF-hands distinguishes S-100 proteins from calmodulin-related proteins which have four EF-hands. Additionally, members of these two groups of proteins lack substantial sequence similarity outside of the C-terminal EF-hand. The lack of amino acid sequence similarity coupled with the variant EF-hand structure allows one to assign an EF-hand protein to either the S-100 or calmodulin-related families.

All the S-100 proteins have amino acid sequence similarity and secondary-structure similarity in specific domains or regions of the proteins (Figures 1A and 2). There are conserved N- and C-terminal hydrophobic amino acid domains. The variant 14 amino acid containing EF-hand is located in a conserved basic domain near the N-terminal of the protein while the 12 amino acid EF-hand is located in a conserved acidic domain in the C-terminal region. The decreased Ca^{2+} affinity of the 14 amino acid EF-hand may in part be attributable to both its variant number of amino acids and its location in an area rich in basic amino acids (Figures 1A and 2).

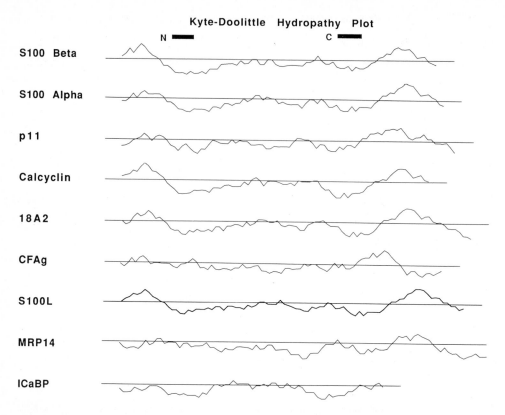

Fig. 3. Kyte-Doolittle hydropathy analysis of S-100 proteins. Unique S-100 species have been subjected to Kyte-Doolittle hydropathy analysis. The N-terminus is on the *left* and the C-terminus is on the *right* of the figure. Areas of relative hydrophobicity are indicated where the tracing goes *above the horizontal* while areas of relative hydrophilicity are *below the horizontal line*. The N- and C-terminal EF-hands are indicated by *bars*. The S-100 proteins have characteristic N- and C-terminal hydrophobic regions immediately flanked by EF-hands in regions of relative hydrophilicity. A plateau which is neither hydrophobic nor hydrophilic separates the two EF-hands. This is a characteristic 'signature' of the S-100 proteins.

Comparison of the S-100 proteins by Kyte-Doolittle hydropathy plot analysis shows that there is a characteristic or fingerprint structure of an S-100 protein (Figure 3). There is a hydrophobic mountain in both the N-terminal and C-terminal regions separated by a neutral plateau that is neither hydrophobic nor hydrophilic. This analysis demonstrates that the overall hydrophobicity and hydrophilicity of the S-100 proteins are conserved even in regions of amino acid sequence divergence. These discrete biochemical or structural domains may represent functional domains.

The S-100 proteins have significant amino acid sequence similarity (Figure 2). Even in areas where the amino acid sequence is not completely conserved there are frequent conservative substitutions which preserve the charge and hydrophobicity/hydrophilicity of the protein. The amino acid homology exists in areas outside the EF-hands and suggests that these regions have important and conserved functions.

However, there are two areas of significant amino acid sequence divergence. These are in the C-terminal region and a 'hinge' region that separates the two EF-hands (Figure 2). These two unique regions may confer biochemical and/or functional specificity to a particular S-100 protein by interacting with a particular effector protein. Each of the EF-hand regions could then be envisioned as a functional domain which binds Ca^{2+} which leads to a conformational change exposing the hydrophobic regions to solvent. These hydrophobic regions in concert with the divergent domains as noted above would then interact in a specific manner with a secondary or effector protein.

S-100α AND S-100ß

The S-100 proteins were originally isolated from bovine brain as a mixture of low molecular weight acidic proteins (Moore 1965; Dannies and Levine 1971). They were subsequently demonstrated to contain both α- and ß-subunits in homo- and heterodimeric forms (Isobe et al. 1977). S-100α exists as a homodimer of the α-subunit while S-100ß was demonstrated to be a homodimer of the ß-subunit (Isobe et al. 1977; Isobe and Okuyama 1978 and 1981). The primary amino acid sequence of both the α- and ß-subunits were determined and shown to share greater than 50% amino acid sequence homology (Figure 2).

There are some significant differences between S-100α and S-100ß. S-100ß contains more histidines than does S-100α. Zn^{2+} binds with high affinity (K_d=0.1 to 1 μM) to these histidine residues without blocking the binding of Ca^{2+} to the proteins. The binding of Zn^{2+} induces a conformational change in the hydrophobic C-terminal domain of S-100ß and increases the Ca^{2+} affinity of the N-terminal EF-hand (Baudier and Gerard 1983; Baudier et al. 1985; Mani et al. 1982; Baudier et al. 1982a; Mani and Kay 1983; Baudier et al. 1983).

It is not known whether the binding of Zn^{2+} plays a role in the normal physiological function of S-100ß.

Another difference between S-100α and S-100ß is the presence of only one cysteine residue in S-100α while S-100ß has two cysteine residues. S-100ß homodimers could, therefore, exist in any one of five isomeric forms while S-100α homodimers exist in only one form. Similarly, there could be two different heterodimeric forms containing one α- and ß-subunit (Figure 4). The varying isomers could have different physicochemical or biological properties. For example, only the disulfide homodimeric form of S-100ß has neurite extending activity (Kligman and Marshak 1985).

ISOMERIC FORMS OF S100 BETA AND ALPHA

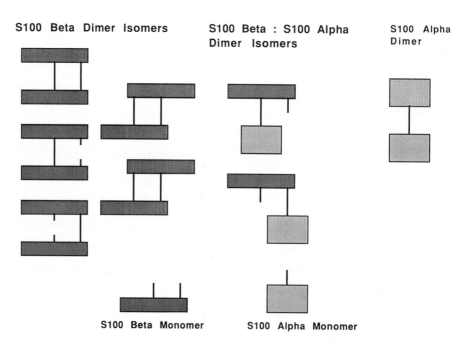

Fig. 4. Possible S-100ß and S-100α dimers. S-100ß contains two cysteine residues at amino acid positions 68 and 84. S-100α contains one cysteine at amino acid position 84. The possible homo- and heterodimers are shown. The different forms may have different biochemical properties or biological activities

S-100ß is found in all glial cells in the nervous system including astrocytes, microglial cells, ependymal cells, oligodendrocytes and Schwann cells (Hyden and McEwen

1966; Sviridov et al. 1972; Tabuchi and Kirsch 1975; Matus and Mughal 1975; Cocchia 1981; Haglid et al. 1976). It is either not expressed in neurons or expressed in a minority of neurons at low levels (Hyden and McEwen 1966; Sviridov et al. 1972; Tabuchi and Kirsch 1975; Matus and Mughal 1975; Cocchia 1981; Haglid et al. 1976). Expression of S-100ß in a variety of cell lines derived from neural tissues has been described (Benda et al. 1968; Pfeiffer et al. 1970; Ishikawa et al. 1983; Gaynor et al. 1980; Stavrou et al. 1980; Kahn et al. 1982). These cell lines provide important model systems in which to study its possible functions and regulation of its expression. Although S-100ß is expressed at high levels by glial cells in the nervous system it is not uniquely expressed in the nervous system. S-100ß also has been purified to homogeneity from adipocytes (Hidaka et al. 1983; Kato et al. 1988) and its expression by immunochemical techniques has been reported in a wide variety of cells including chondrocytes, pituitary cells, melanocytes, melanomas, Langerhans cell of the skin, and cells of the lymphoreticular system including a subpopulation of CD8+ T lymphocytes (Kanamori et al. 1982; De Panfilis et al. 1988; Sansoni et al. 1988).

S-100α is found in the central and peripheral nervous systems where it is predominantly expressed in neurons (Isobe et al. 1984; Kimura et al. 1984). It also is expressed in renal tubular cells, melanocytes and muscle cells (Kato et al. 1986). Skeletal muscle type I fibers (slow twitch fibers) and cardiac muscle fibers express S-100α (Haimoto and Kato 1987, 1988). Skeletal muscle type II fibers (fast twitch fibers) express S-100α at a much lower level. The function of S-100α in muscle physiology has not been determined but Donato and coworkers have proposed that S-100α stimulates the Ca^{2+}-induced release of Ca^{2+} from the sarcoplasmic reticulum and stimulates adenylate cyclase activity (Fano et al. 1989a,b). These observations indicate that S-100α may play a role in the transduction of the signal of increased intracellular Ca^{2+} in a muscle fiber type-specific fashion. Both type I and type II skeletal muscle fibers derive from a common mesodermal precursor cell. The fiber-type-specific expression of S-100α is, therefore, an example of expression of a S-100 protein in one daughter cell population (type I fibers) but not another (type II fibers).

Until recently there were no definite biological function(s) that could be convincingly demonstrated for either S-100α or S-100ß. Recently, however, the homodimeric disulfide bonded form of S-100ß was purified from bovine brain as a protein which had significant neurite extension activity in both chick embryonic cortical neuronal and murine neuroblastoma assays (Kligman 1982; Kligman and Marshak 1985; Kligman and Hsieh 1987). Kligman and coworkers have named this protein Neurite Extension Factor (NEF). The specificity of S-100ß in inducing neurite extension was demonstrated by the fact that both S-100α and calmodulin both have no activity in the assay. The reduced, or monomeric form, of S-100ß similarly has no activity in inducing neurite extension of either dissociated chick embryonic cortical neurons or murine neuroblastoma cells in culture. The identity of NEF with the disulfide bonded homodimeric form of S-100ß was further confirmed by the fact that recombinant expressed S-100ß has neurite extending activity (van Eldik et al. 1988). van Eldik and coworkers have performed a series of experiments that demonstrate that the formation of the S-100ß homodimer species that has neurite extending activity was dependent upon the presence of the two cysteine residues (at amino acid positions 68 and 84) (Figure 2). They also have demonstrated a neurotrophic (neuronal survival) as well as neurite extending activity for S-100ß (van Eldik et al. 1988; Winningham-Major et al. 1989).

The particular neuronal subpopulations which respond to NEF have not been completely elucidated. Chick embryonic optic tectal neurons do not respond to NEF while chick embryonic cortical neurons do respond (Kligman pers. observ.). Rat embryonic cholinergic basal forebrain neurons do not respond to S-100ß but do respond to Nerve Growth Factor. S-100ß but not other nervous system growth factors supports the growth of rat embryonic serotonergic neurons (Azmitia et al. 1990) while locus coeruleus neuron survival is apparently not enhanced by NEF (Hilt and Clark pers. observ.). These results demonstrate that there are specific cell types or subpopulations of neurons that respond to S-100ß/NEF. Complete elucidation of the responding neuronal populations remains to be performed.

The mechanism by which the dimeric form of S-100ß induces neurite outgrowth and/or neuronal survival is not yet known. Is the protein internalized by the target cell? Are there cell surface receptors which transduce the S-100ß signal? A number of S-100-binding proteins have been reported (Zimmer and van Eldik 1989; Baudier and Cole 1988; Donato et al. 1989b; Patel et al. 1983; Qi and Kuo 1984; Hagiwara et al. 1988; Fujii et al. 1990; Baudier et al. 1989). Are one or more of these proteins targets for S-100ß and essential for producing the biological effect(s)? Preliminary experiments have demonstrated that murine neuroblastoma cells (Neuro 2A cells) have specific saturable binding sites for S-100ß/NEF (Muller and Kligman pers. observ.).

Several other biochemical activities have been suggested for S-100α and/or S-100ß. Best documented among these is the ability of S-100α and S-100ß to promote the Ca^{2+}-dependent and pH-dependent dissociation of microtubules and the inhibition of microtubule assembly by blocking the nucleation and elongation processes (Baudier et al. 1982b; Endo and Hidaka 1983; Donato 1988; Donato et al. 1989a). These effects of S-100 require the presence of microtubule associated proteins (MAPs) in addition to tubulin. S-100 has been shown to bind to MAP- and cytoskeletal-associated proteins in a Ca^{2+}-dependent manner. These include tau, MAP2, calpactin, caldesmon, and phosphomyristin C (Fujii et al. 1990; Baudier et al. 1989). S-100ß will form mixed disulfide complexes with tau in a Ca^{2+}-dependent fashion (Baudier et al. 1989).

S-100ß specifically inhibits the phosphorylation of a number of protein kinase C substrates including tau, growth associated protein 43 (GAP43), and phosphomyristin C (PMC) (Baudier et al. 1989; Patel and Kligman 1987; Hornbeck et al. 1989). GAP43 is a neuronal protein specifically expressed during regeneration (Muller et al. 1985). PMC is a prominent protein kinase C substrate that is rapidly phosphorylated in a wide variety of cell types in response to growth factors or other effectors which stimulate phosphoinositide (PI) turnover. In the nervous system S-100ß and PMC appear to be co-expressed primarily in glial cells (Ouimet et al. 1990). These findings suggest that S-100ß may be involved in modulating Ca^{2+}/PI-mediated signal transduction in glial cells in the nervous system.

The gene for S-100ß has been assigned to human chromosome 21q22 (Allore et al. 1988). This finding led some investigators to speculate on a possible causal relationship between the defects of the nervous system of patients with Down syndrome (trisomy 21) and S-100ß. It was suggested that triplication of chromosome 21 in Down syndrome may result in an increased gene dosage of the S-100ß gene and subsequent over-expression of S-100ß protein. The possibility that S-100ß may, indeed, be involved in the pathogenesis of Down syndrome was strengthened when it was determined that some patients with Down syndrome have only a partial chromosome 21 triplication by virtue of a partial triplication-translocation event. The critical region of chromosome 21 which is sufficient to cause Down Syndrome is 21q22.2-22.3. The S-100ß gene has been mapped at high resolution to this region (Duncan et al. 1989).

S-100ß protein expression is increased in both Down syndrome and Alzheimer's disease brains (Griffin et al. 1989). The increased expression was observed in both astrocytes and microglia; cells which normally express S-100ß. Interestingly, the neuropathology of both disorders is identical; a variety of abnormally phosphorylated MAP proteins (tau and MAP2 among others) form insoluble complexes called paired helical filaments (PHFs). These PHFs are the primary constituent of neurofibrillary tangles. Senile plaques are amorphous extracellular structures which are composed of amyloid proteins and other proteins possibly including cytoskeletal-associated proteins. These are the two pathological hallmarks of these disorders. S-100ß modulates the phosphorylation of at least one protein (tau) involved in these pathological processes and is overexpressed in both conditions suggesting that it may play a physiological role in these disorders.

Although the etiological role of S-100ß in either Down syndrome or Alzheimer's disease is not yet proven, S-100ß transgenic mice have been recently shown to have significant behavioral/cognitive abnormalities and may serve as one experimental model for Down syndrome and/or Alzheimer's disease (Reeves et al. 1990). This finding also suggests that the increased level of S-100ß expression observed in Down syndrome and Alzheimer's disease is not simply a reactive process but that specific over-expression of S-100ß in a transgenic mouse may produce significant physiological abnormalities.

Hippocampal long-term potentiation (LTP) is a experimental paradigm for long-term memory (Zalutsky and Nicoll 1990). LTP is a Ca^{2+}-dependent process and involves protein kinase C-mediated protein phosphorylation (Malinow et al. 1989). Two proteins which are rapidly phosphorylated in animals undergoing LTP are GAP43 and PMC, both of which interact with S-100ß (Lewis and Teyler 1986). Anti-S-100 antiserum prevents the development of LTP when infused before the tetanic stimulation necessary to induce LTP. This result suggests that S-100 may be involved in the establishment of LTP and in memory in vivo especially when considered in the context of the previously described findings suggesting that S-100ß is abnormally expressed in Down syndrome and Alzheimer's disease and is capable of altering protein kinase C-mediated protein phosphorylation. These two groups of experiments indicate that a fruitful area of future research will be in exploring the role of S-100 proteins in memory processes and neurodegenerative diseases such as Down syndrome and Alzheimer's disease.

S-100ß gene expression is regulated by a number of mechanisms. Steady state S-100ß mRNA levels in C6 rat glioma cells are increased in response to cAMP and, to a lesser degree, by glucocorticoids, but are decreased by phorbol esters (Hilt et al. 1989). cAMP induces morphological and biochemical changes in C6 glioma cells similar to those seen in differentiation of glial cells in vitro; cell growth rate decreases and the cells extend fine processes. Synthesis of S-100ß protein increases with the degree of C6 glioma cell culture density. Confluent cell cultures synthesize more S-100ß than do logarithmically growing cells but synthesis of S-100ß rapidly falls as cell cultures reach a post-confluent stage. A cell-cell contact effect in initially increasing S-100ß synthesis in these cells has been postulated (Gysin et al. 1980; Zimmer and van Eldik 1988). These experiments suggest that quiescent C6 glioma cells synthesize less S-100ß. Other workers have noted that expression of S-100ß is increased during the G_1 phase of the cell cycle (Fan 1982). The control of S-100ß expression by various signal transduction pathways and in various growth conditions suggests that it is not simply constitutively expressed and may have important physiological functions. Marks and coworkers have demonstrated that S-100ß mRNA and protein levels are specifically and rapidly decreased in stationary-phase C6 glioma cells by microtubule depolymerizing agents such as vinblastine and colchicine (Marks and Labourdette 1977; Dunn et al. 1987). Inactive congeners of colchicine did not have this effect. Whether or not the decreases in S-100ß mRNA were transcriptional or post-transcriptional (such as is seen with both α- and ß-tubulin) is not known. However, these experiments suggest that there is a feedback system involving the degree of tubulin polymerization and the level of S-100ß expression. This may be of physiological importance given the ability of S-100ß to promote tubulin dissembly in a Ca^{2+}-dependent manner.

Elucidation of the many mechanisms controlling S-100ß gene expression is at an early stage. Multiple signal transduction pathways appear to be involved as does a complex cytoskeletal feedback regulation of S-100ß expression.

S-100 PROTEINS AS MODULATORS OF MEMBRANE-CYTOSKELETAL INTERACTIONS: p11 OR CALPACTIN LIGHT CHAIN

p11, or calpactin light chain, was described as a S-100 protein which inhibits the phosphorylation of a 36 kDa protein known as p36 or calpactin heavy chain (Gerke and Weber 1984; Glenney 1986; Klee 1988). Calpactin is a major intracellular substrate for tyrosine kinase activity (e.g., pp60[src] and EGF-receptor kinase). Calpactin is composed of a p11 dimer and two p36 subunits. The p11 subunit binds to the phosphorylation site on p36 and blocks its phosphorylation (Glenney and Tack 1985). Calpactin can interact with phospholipids and actin, hence the name calpactin. This interaction of calpactin with actin and membrane phospholipids is dependent upon Ca^{2+} and p11. Apparently, the interaction, although dependent upon Ca^{2+} does not involve the actual binding of Ca^{2+} to the p11 subunit (Klee 1988). Instead, the p36 subunit binds Ca^{2+} at low affinities ($K_d=1x10^{-4}$ M) even though it does not contain any EF-hand motif. The binding of Ca^{2+} induces a conformational change which facilitates the p11-p36 interaction. The p11 thus blocks the access of tyrosine kinase with its target phosphorylation site in the N-terminal region of p36 (Glenney and Tack 1985). The location or number of Ca^{2+} binding sites in p36 is not known. It is, however, known that phospholipids increase the Ca^{2+} affinity of calpactin and induce Ca^{2+} binding (Glenney 1986). The fact that the p11 subunit does not bind Ca^{2+} is an exception to the general observation that S-100 proteins bind Ca^{2+} to EF-hand regions.

Calpactin may be involved in Ca^{2+}-dependent processes such as exocytosis, subcellular cytoskeletal organization and interactions of the cytoskeleton with other subcellular compartments, actin bundling or processing, and coupling of the cytoskeleton to the plasma membrane (Klee 1988). Calpactin can promote the fusion of adrenal chromaffin granules in vitro at physiological Ca^{2+} concentrations (μM) (Drust and Creutz 1988). Calpactin has a wide tissue distribution including the nervous system. These facts suggest that calpactin may have a functional physiological role in neurotransmitter release.

Calpactin has been shown to be identical to lipocortin, a phospholipase A2 inhibiting protein with potent anti-inflammatory activity (Wallner et al. 1986; Davidson et al. 1987). The anti-inflammatory effects of calpactin/lipocortin are thought to be due to inhibition of phospholipase A2 which generates arachidonic acid by hydrolysis of phospholipids.

Arachidonic acid is the synthetic precursor to a group of potent mediators of inflammation, prostaglandins, and leukotrienes. Glucocorticoids block inflammation and have been shown to induce expression of calpactin/lipocortin.

S-100 PROTEINS ASSOCIATED WITH CELL-CYCLE PROGRESSION OR TRANSFORMATION: CALCYCLIN, 18A2, pEL68, AND mts1

A growing number of S-100 proteins have been characterized when isolating mRNA species or proteins whose expression is increased dependent upon the state of cellular growth, transformation, or differentiation (Figures 5 and 6). As noted above, S-100ß was shown to be expressed during the G_1 phase of the cell cycle (Fan 1982). Calcyclin was the first S-100 protein specifically identified as being related to the state of cellular proliferation.

Calcyclin was initially identified by Baserga and coworkers as a cDNA clone encoding an mRNA species whose level is induced when quiescent/growth arrested human fibroblasts are stimulated to progress from the G_0 to the G_1 phase of the cell cycle (Ferrari et al. 1987; Calabretta et al. 1986; Ghezzo et al. 1988). Serum depletion induces growth arrest in these cells and they remain in the G_0 or stationary phase of the cell cycle. When serum, platelet derived growth factor (PDGF), or epidermal growth factor (EGF) but not insulin is added to the culture medium the cells enter the G_1 phase of the cell cycle and calcyclin is maximally expressed. The cell cycle-specific expression of calcyclin has led to the proposal that it may be one of the proteins involved in cell-cycle progression (Calabretta et al. 1986; Ghezzo et al. 1988) (Figure 5). The increased expression of calcyclin is regulated at the level of gene transcription. DNA regions conferring serum and PDGF responsivity have been located in the transcriptional control region of the calcyclin gene (Ghezzo et al. 1988). A negative regulatory region also has been mapped demonstrating that the gene is under complex transcriptional control. The protein encoded by the calcyclin mRNA has been purified to homogeneity and is expressed in high levels in murine skeletal and cardiac muscle, lung, kidney, spleen, and is particularly enriched in murine smooth muscle (Kuznicki et al. 1989).

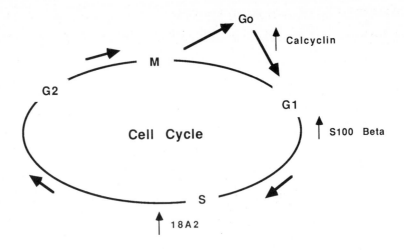

Fig. 5. Cell-cycle specific expression of S-100 proteins. Calcyclin expression is induced when cells pass from the G_0 to G_1 phase of the cell cycle. S-100β expression is increased during the G1 phase of the cell cycle. 18A2 mRNA levels are elevated during the S phase of the cell cycle. The cell-cycle specific expression of other S-100 proteins has not been investigated

The possibility that S-100 proteins are involved in cell-cycle progression was reinforced by the description of another putative protein, 18A2, encoded by an mRNA species induced by serum stimulation of growth-arrested fibroblasts (Jackson-Grusby 1987). Are calcyclin and 18A2 the same protein? Calcyclin and 18A2 appear to be unique proteins as evidenced by the following observations. In contrast to calcyclin, the putative 18A2 protein is expressed during the S phase of the cell cycle not the G_1 phase. The increased expression of the 18A2 gene is dependent upon protein synthesis after serum stimulation while the increase in calcyclin is not dependent upon *de novo* protein synthesis. Genomic Southern blots show that calcyclin and 18A2 are encoded by unique genes. Calcyclin and 18A2, therefore, appear to be different gene products. These observations strongly suggest that two very similar Ca^{2+}-binding proteins may have different functions in the same cell type, be expressed in different phases of the cell cycle, and that cell-cycle progression may depend upon the sequential expression of calcyclin (G_1 phase) and then 18A2 (S phase).

A number of proteins which influence the progression of the eukaryotic cell cycle have been described over the past few years (for review see Lee and Nurse 1988).

S100 Protein Expression and Differentiation

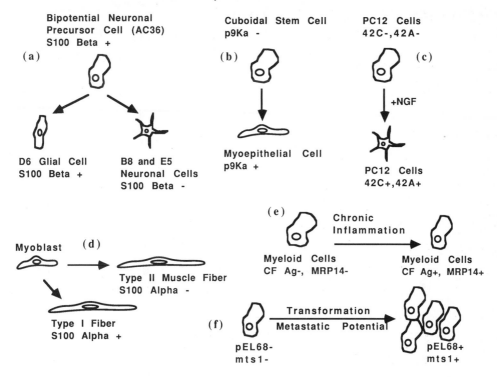

Fig. 6a-e. S-100 protein expression and differentiation. The levels of S-100 proteins are regulated in a variety of cells during differentiation. **a** In the RT4 rat neural tumor cell family, a bipotential precursor cell (AC36) may differentiate into either glial (D6) or neuronal (B8 and E5) daughter cells. The stem cell and glial cell lines are S-100ß+ and the neuronal cell lines are S-100ß-. **b** Mammary cuboidal stem cells are p9Ka-. When this cell differentiates into a myoepithelial-like cell it synthesizes p9Ka. **c** When PC12 rat pheochromocytoma cells are differentiated with NGF the synthesis of 42C and 42A is induced. **d** Type I (slow twitch) muscle fibers express S-100α while Type II (fast twitch) muscle fibers do not. **e** Normal tissue macrophages do not express CF Ag or MRP-14. However, during chronic inflammtion (or induction of HL60 promyelocytic cell differentiation with retinoic acid or dimethylsulfoxide) synthesis of CF Ag and MRP-14 is increased. **f** pEL68 and mts1 expression are induced in tumor cell lines which are transformed or have metastatic potential.

Many of these proteins are either protein kinases, kinase substrates, or proteins which modulate protein kinase pathways. If the S-100 proteins exert their effect by inhibition of specific cellular protein kinases it is tempting to speculate that calcyclin and 18A2 also may function in this manner and that there may be other S-100 proteins involved in cell-cycle progression. These proteins may function to transduce the signal of Ca^{2+} to modulate one or more of the kinase activities during cell-cycle progression.

Prolactin associated receptor protein (PRA) was isolated as a S-100 protein which copurified with the prolactin receptor isolated from T-47D cells (a human breast cancer cell line) (Murphy et al. 1988). The PRA mRNA has 37 additional nucleotides in the 5' untranslated region when compared to the calcyclin mRNA but is otherwise identical to calcyclin. This led to the suggestion that a different transcription start site is utilized in epithelial cells which express PRA (e.g., breast cancer cells) versus fibroblasts which express calcyclin (Murphy et al. 1988). Interestingly, PRA is not expressed in a breast cancer cell line which does express the prolactin receptor and binds prolactin but which does not respond to prolactin morphologically or biochemically. This is compatible with PRA having an essential role in transducing the signal of prolactin receptor activation in a Ca^{2+}-dependent manner.

pEL68 was described as an mRNA species whose expression is induced in transformed murine cells (Goto et al. 1988). Its sequence, mRNA size, and genomic Southern Blot are virtually identical to those of 18A2 suggesting that they are the same protein. Similarly, the protein encoded by the mts1 gene is virtually identical to 18A2 (Ebralidze et al. 1989). The mts1 protein is differentially expressed in tumor cell subclones with metastatic potential. Many genes whose overexpression is correlated with the metastatic phenotype are amplified and/or rearranged (e.g., oncogenes). The mts1 gene overexpression observed in clonal cell lines capable of metastasis is not due to mts1 gene amplification, rearrangement, nor variant gene methylation but induced expression of a single copy of the mts1 gene.

The calcyclin subfamily of S-100 proteins are important in that they are all increased in cell types with some degree of transformed phenotype. These observations would strongly suggest that S-100 proteins may have an important role in tumor cell biology and furthermore that constitutive overexpression of certain S-100 proteins may be sufficient to transform cells.

S-100 PROTEINS ASSOCIATED WITH CELLULAR DIFFERENTIATION

During cellular differentiation either during normal development or in response to specific differentiating agents in vitro, S-100 proteins may be differentially expressed (Figure 5 and 6). Differential expression of S-100ß in the RT4 rat neural tumor cell system is one example of this phenomenon. S-100ß is expressed in a bipotential precursor cell line (AC36) which can differentiate to either glial (D6) or neuronal daughter cell lines (E5 or B8) (Freeman et al. 1989). The neuronal daughter cell lines extinguish S-100ß expression while the glial daughter cell line continues to express S-100ß. This suggests both positive and negative controls of S-100ß. Another S-100 protein expressed in a differentiation-specific fashion is S-100α. S-100α is expressed in only type I skeletal muscle fibers and cardiac muscle fibers but not type II skeletal muscle fibers (Haimoto and Kato 1987, 1988).

p9Ka is a S-100 protein whose expression is induced nearly 15-fold during the differentiation of mammary cuboidal epithelial stem cells (Rama 25 cells) to elongated myoepithelial cells in cell culture (Barraclough 1982, 1987). p9Ka is virtually identical to 18A2 and is very likely to be the rat homolog of 18A2.

Nerve growth factor (NGF) treatment of PC12 cells (a rat pheochromocytoma cell line) induces a neuronal differentiation response in the cells. There are morphological changes including neurite extension and cessation of cell division (they go into the G_0 phase of the cell cycle) and biochemical changes such as induction of choline acetyltransferase enzyme activity. NGF also induces the synthesis of two mRNA species which encode putative S-100 proteins with a high degree of homology to p11 and 18A2 (Masiakowski and Shooter 1988). These species are named 42C and 42A respectively. 42C differs by only five amino acids from p11, and 42A differs by only four amino acids from 18A2. It is likely, therefore, that these are the rat homologs of p11 and 18A2.

It is notable that NGF induces two separate S-100 mRNA species upon cellular differentiation of a single cell type (PC12). During the cellular differentiation response to NGF in PC12 cells, 42C (the homolog of p11) may modulate calpactin-mediated cytoskeletal/ membrane interactions necessary for growth cone development or neurotransmitter vesicle fusion with the cellular membrane. In PC12 cells 42A (the homolog of 18A2) is expressed upon cessation of cellular division, while in 3T3 cells 18A2 is expressed upon induction of cell growth during the S phase of the cell cycle. Thus, the synthesis of a specific S-100 protein may be either increased or decreased upon induction of growth dependent upon the particular cell type. These observations suggest that there are complex controls on p11 and 18A2 expression which may be regulated in a cell-type specific manner. In a larger sense, these accumulated data strongly point to the involvement of S-100 proteins in both cell-cycle progression and cellular differentiation processes.

Interestingly, many of the cellular differentiation paradigms in which differential expression of S-100 proteins have been described involve significant and particular changes in cellular morphology. The generation of a differentiated glial phenotype from a bipotential precursor cell, neuronal differentiation of PC12 cells upon NGF stimulation, and differentiation of myoepithelial cells from cuboidal precursor cells all involve generation of a specialized cell type with a complex elongated morphology from a compact less complex precursor cell type. If the S-100 proteins function to modulate cytoskeletal-associated proteins (such as MAPs) and their interaction with other intracellular constituents or plasma membranes, their expression may be crucial mediators of these complex morphological responses.

S-100L is a low molecular weight acidic S-100 protein purified from bovine lung (Glenney et al. 1989). It shares significant sequence homology with other S-100 proteins (Figure 2), has the requisite two EF-hands, and exposes hydrophobic domains upon Ca^{2+} binding. S-100L is expressed in high levels in bovine lung and kidney but at low levels in brain and intestine. The S-100L cDNA was cloned from a bovine kidney cell line (MDBK cells). These cells express both p11 and S-100L but each is targeted to different subcellular compartments. The presence of the two S-100 proteins in different subcellular compartments suggests that they have differing physiological roles. Co-expression of p11 and S-100L is yet another example of expression of multiple S-100 proteins in a particular cell type.

S-100 PROTEINS DESCRIBED IN ASSOCIATION WITH HUMAN DISEASE: CYSTIC FIBROSIS ANTIGEN, *MRP*-8 AND *MRP*-14

Two recently reported members of the S-100 protein family have been described in association with particular human disease states. The cystic fibrosis antigen (CF Ag) is a S-100 protein which is present at high levels in the serum of individuals with cystic fibrosis (Dorin et al. 1987). CF Ag is exclusively expressed in normal and leukemic granulocytes and cells of the myeloid lineage (Dorin et al. 1987). The CF Ag gene has been mapped to chromosome 1 and is not the defective gene in individuals with cystic fibrosis which maps to chromosome 7 and encodes a chloride channel protein. During retinoic acid and dimethyl-sulfoxide induction of differentiation of the promyelocytic cell line, HL60, expression of the CF Ag increases.

MRP-8 and MRP-14 are two Ca^{2+}-binding S-100 proteins (8 kDa and 14 kDa respectively) expressed by cells of myeloid origin and present at elevated levels in patients with rheumatoid arthritis (Odink et al. 1987). Circulating granulocytes, monocytes, and macrophages but not normal tissue macrophages express these two proteins. However, in chronic inflammatory conditions tissue macrophages express both MRP-8 and MRP-14. These two S-100 proteins were originally isolated from synovial fluid as part of a complex composed of macrophage migration inhibition factor (MIF) and MRP-8 and MRP-14. Even though MRP-8 and MRP-14 bind to MIF they do not apparently alter its activity. MRP-8 is very likely identical to the CF Ag as the last 15 amino acids of MRP-8 differ from the CF Ag as a result of a single base change in the cDNA sequence which may have occurred due to a sequencing error. We will, therefore, consider them to be the same protein. MRP-14 is a unique S-100 protein that has less sequence similarity to the other S-100 proteins and is larger (14 kDa) than the other S-100 proteins (Figure 2).

CF Ag/MRP-8 and MRP-14 expression is linked to a number of human diseases characterized by chronic inflammation including rheumatoid arthritis and cystic fibrosis. Induction of CF Ag and MRP-14 expression also has been observed in keratinocytes in chronic inflammatory dermatoses (Kelly et al. 1989; Wilkinson et al. 1988). It has been suggested that these proteins may function as inflammatory cytokines in cutaneous lesions (Wilkinson et al. 1988).

The CF Ag also was isolated as part of a protein complex from human spleen cell nuclei which inhibits the protein kinase activity of casein kinase I and II but not cAMP-dependent protein kinase, protein kinase C, or insulin receptor tyrosine kinase activities. Casein kinases phosphorylate a wide spectrum of proteins including topoisomerase II and RNA polymerase I and II (all enzymes involved in transcriptional regulation) (Murao et al. 1988). The kinase-inhibitory protein complex was differentially distributed in monocytes as compared to granulocytes. In monocytes the major location of the CF Ag-containing complex was nuclear while in granulocytes the complex is primarily cytoplasmic

(Murao et al. 1988). The kinase-inhibitory properties of CF Ag coupled with its nuclear location in monocytes (which are the precursors to tissue macrophages) suggest that it may play an important role in modulation of macrophage gene expression or biological function.

Recently, antibacterial and antifungal activity of a human leukocyte protein L1 was described (Steinbakk et al. 1990). L1 at low concentrations inhibited the growth of bacterial and fungal cultures and at higher concentrations L1 was bactericidal and fungicidal. L1 is a 36.5 kDa complex composed of one $L1_L$ and two $L1_H$ chains which are identical to CF Ag/MRP-8 and MRP-14. The name calprotectin was proposed for the L1 complex due to its antimicrobial activity. The antimicrobial effects of L1 were partially dependent upon Ca^{2+}. These results suggest that a secreted or extracellular form of a S-100 protein may have a physiological role in controlling microbial infections. Whether or not this is a direct effect of L1 or whether it is biologically important is not yet known.

INTESTINAL CA^{2+}-BINDING PROTEIN

Intestinal Ca^{2+}-binding protein (ICaBP) was originally described as a 7 kDa protein whose synthesis is induced by Vitamin D3 (Szebenyi et al. 1981). This is yet another example of the inducible expression of an S-100 protein. Although ICaBP contains the two requisite EF-hands and other characteristics of an S-100 protein it is smaller than the other S-100 proteins and has significantly less sequence similarity to the other S-100 proteins (Figure 2). Additionally, ICaBP lacks the signature N-terminal and C-terminal hydrophobic regions when analyzed by Kyte-Doolittle hydropathy plots (Figure 3).

TRANSDUCTION OF THE CA^{2+} SIGNAL BY S-100 PROTEINS

If the S-100 proteins have no specific enzymatic properties, how do they exert their varied effects? It is likely that the S-100 proteins exert their effects by modulating the activity of specific secondary or effector proteins (Figure 7). This model predicts that at basal intracellular Ca^{2+} levels the S-100 proteins do not associate with the effector protein. When the intracellular Ca^{2+} concentration increases (due to either flux of Ca^{2+} into the cell or release of intracellular Ca^{2+} stores) the S-100 protein binds Ca^{2+} which induces a conformational change. The conformational change exposes hydrophobic domains of the protein to solvent. These hydrophobic domains may then interact with similar hydrophobic domains on the effector protein and modulate its effects by blocking protein phosphorylation. The specificity of a particular S-100: effector protein interaction must reside with the domains which are unique to individual S-100 proteins. The cycle is completed when the intracellular Ca^{2+} level returns to its basal state and the S-100 protein dissociates from the effector protein.

Ca2+-Dependent Modulation of S100 Protein: Effector Protein Interaction

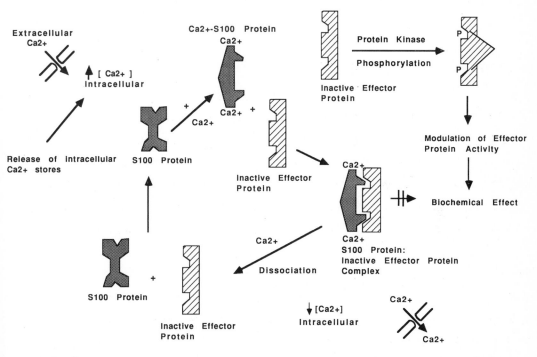

Fig. 7. Ca^{2+}-dependent modulation of S-100 protein: effector protein interaction. S-100 proteins bind Ca^{2+} upon increase in free intracellular Ca^{2+} levels due to Ca^{2+} flux into the cell or release of intracellular Ca^{2+} stores. The S-100 protein undergoes a conformational change and binds to an effector protein blocking its phosphor-ylation. The modulation of effector protein phosphorylation either activates or inactivates the effector protein which may have different activites in the phosphorylated versus nonphosphorylated states. The cycle is completed when free intracellular Ca^{2+} levels decrease and the S-100-effector protein complex dissociates

There are a number of examples of S-100 protein: effector protein interactions in which the S-100 protein inhibits the phosphorylation of an effector protein. S-100ß and S-100α block the phosphorylation of tau, GAP43, and PMC. p11 inhibits the phosphorylation of calpactin and CF Ag blocks the phosphorylation of casein kinase substrates.

This model allows a number of predictions to be made. First, the affinity constant (Ka) of a particular S-100 protein for Ca^{2+} must be greater than the basal intracellular Ca^{2+} concentration. This is essential if the S-100 proteins are to transduce the signal of Ca^{2+} as they can not be saturated with Ca^{2+} in the basal state. Second, the transient intracellular increases in Ca^{2+} must be sufficient to allow the binding of Ca^{2+} to at least one EF-hand. The Ca^{2+} affinity of the C-terminal EF-hand of S-100ß (K_d=20-50 μM) and the transient intracellular Ca^{2+} concentration (10-100 μM) fulfill these requirements. Third, a number of S-100 proteins which have not yet had biochemical properties attributed to them may exert their effects by binding to kinase substrates and thus inhibiting phosphorylation by their respective kinases. This prediction will require additional research on these members of the S-100 protein family.

The interaction of S-100 proteins with effector proteins is reversible and dependent upon Ca^{2+} ions. As the intracellular Ca^{2+} concentration decreases the interaction of the S-100 protein with an effector protein weakens and the S-100: effector protein complex dissociates.

The N-terminal EF-hand is degenerate and has been shown in S-100ß and S-100α to be a poor Ca^{2+}-binding site (Calissano et al. 1969; Dannies and Levine 1971). The affinity of the C-terminal EF-hand for Ca^{2+} is greater than that of the N terminal EF-hand (K_d=20-50 μM versus K_d=200-500 μM) (Kretsinger 1980). These dissociation constants are calculated from in vitro studies. The intracellular binding of Ca^{2+} to these proteins and the effective affinity at the subcellular level may depend upon the in vivo microenvironment or subcellular compartment, and thus may be lower. These affinity constants also suggest that the S-100 proteins may be activated only in subcellular compartments where the Ca^{2+} concentration reaches a relatively high level.

Why are two EF-hands with such disparate calcium affinities conserved in all these proteins? What is the physiological function of the conserved N-terminal EF-hand which has such a low Ca^{2+}-binding affinity? The S-100 proteins could be viewed as heterobifunctional molecules; at lower Ca^{2+} concentrations the C-terminal EF-hand binds Ca^{2+}, changing the conformation of the C-terminal hydrophobic region and allowing interaction with an effector protein (Figures 1B and 7). The N-terminal EF-hand may bind Ca^{2+} only when the intracellular Ca^{2+} concentration reaches a higher level (i.e., when the cell is depolarized) (Figure 1B). At these higher Ca^{2+} concentrations, a similar conformational change could expose the N-terminal hydrophobic region and allow its interaction with either a different region of the same effector protein or another effector protein. The S-100 protein could have specific functions for each of these Ca^{2+} bound and conformational states.

Additional modes of control might involve modulation of Ca^{2+} affinity by either Zn^{2+} (which increases Ca^{2+} affinity) or K^+ (which decreases Ca^{2+} affinity) (Baudier and Gerard 1983). Such control mechanisms would provide fine-tuned regulation in situations such as calcium-induced exocytosis. These modes of control also suggest that when a cell is depolarized (and the intracellular K^+ decreases) the effective affinity of an S-100 protein for Ca^{2+} is increased. Alternatively, the low Ca^{2+}-affinity N-terminal EF-hand may function to modulate the protein's properties when it has been secreted. In the extracellular space (with its high Ca^{2+} concentration) the C-terminal EF-hand will have Ca^{2+} bound constantly while the lower affinity N-terminal EF-hand may have Ca^{2+} variably bound depending upon a number of factors. This may be a mechanism to specify separate conformations and functions to extracellular versus intracellular S-100 proteins.

The S-100 proteins are expressed in a cell-type specific fashion. It has been suggested that all cells which transduce the Ca^{2+} signal may have a requirement for a S-100 protein (Glenney et al. 1989). It is possible that a particular cell type may express more than one S-100 protein. There are a number of examples of expression of multiple S-100 proteins in the same cell type. These include expression of 42C and 42A in PC12 cells induced to differentiate with NGF; sequential expression of calcyclin and 18A2 in fibroblasts; expression of S-100L and calpactin in bovine kidney cells; and expression of CF Ag and MRP-14 in myeloid cell lines. These observations demonstrate that a single cell type may express more than one S-100 protein each of which likely binds to a different effector protein(s) in one or more subcellular compartments. This may be a mechanism by which the Ca^{2+} signal is transduced to different effector proteins in a cell-type specific fashion. In the example of expression of calpactin and S-100L in bovine kidney cells it was observed that the subcellular distribution of the S-100 proteins differ suggesting that they regulate different subcellular processes (Glenney et al. 1989). Similarly, the subcellular distribution of S-100ß and calmodulin during C6 glioma cell differentiation differ with the same implication (Zimmer and van Eldik 1988).

S-100 proteins and calmodulin may interact with the same effector protein. S-100ß binds to and inhibits the phosphorylation of both GAP43 and PMC. Calmodulin interacts with both of these proteins. In the basal (low Ca^{2+}) intracellular state, calmodulin binds to GAP43. This interaction is not dependent upon Ca^{2+}-calmodulin interaction. When the Ca^{2+} concentration increases, calmodulin binds Ca^{2+} and dissociates from GAP43. The interaction of S-100ß with GAP43 is, however, dependent upon Ca^{2+}. S-100ß will only bind to GAP43 and inhibit its phosphorylation in the presence of Ca^{2+}. This is an example of how the same target protein may be differentially modulated by both calmodulin and a S-100 protein dependent upon the Ca^{2+} concentration. Although calmodulin binding domains of calmodulin-modulated proteins have been characterized, the binding domains of both the S-100 proteins and the effector proteins which interact with them have not yet been characterized.

THE ROLE OF S-100 PROTEINS IN CELL-CYCLE PROGRESSION AND DIFFERENTIATION

Many of the S-100 proteins described to date are expressed in either a particular phase of the cell cycle, linked to differentiation events, or differentially expressed in cells types that are transformed. Some of these observations are diagrammatically shown in Figure 6. These events include cell-type differentiation and transformation in a number of model systems. S-100ß is expressed in a bipotential cell line and its expression is extinguished in neuronal daughter cell lines and it is continued to be expressed in glial cell lines. p9Ka expression is induced upon differentiation to a myoepithelial cell type. 42C and 42A are induced by NGF which induces differentiation of PC12 cells. mts1 is expressed in subclones of mammary cancer cell lines with metastatic potential. CF Ag/MRP-8 and MRP-14 are expressed at high constitutive levels in leukemic leukocytes. Similarly, the expression of pEL68 was increased in murine cell lines transformed by either oncogenes or chemical carcinogens. Increased expression of S-100 has been described in a number of tumors including melanoma, Schwannoma, and glioma (Kahn et al. 1982). Calcyclin, 18A2 and S-100ß have been shown to be expressed in particular phases of the cell cycle. Expression of S-100 proteins appears to be coupled to both cell-cycle and lineage events.

The S-100 proteins whose expression are related to cell-cycle progression (calcyclin, 18A2, and S-100ß) must have short half lives. If the observed cell-cycle specific expression of these proteins is biologically important the protein levels must vary significantly during the cell cycle in contrast to the vast majority of cellular proteins which show no such variation. The cell-cycle specific regulation of intracellular S-100 protein levels could be at the level of mRNA stability or cell-cycle specific proteolysis of the proteins. Alternatively, expression of these S-100 protein genes could be activated during a restricted phase of the cell cycle. At the present time, the mRNA or protein half lives of these S-100 proteins are not known.

The various demonstrated and proposed effects of the S-100 proteins in cell-cycle-progression, cell-type differentiation, signal transduction, and morphological differentiation may, in fact, all be mediated by the same basic cellular processes. The ability of S-100 proteins to inhibit the phosphorylation of various cytoskeletal-associated proteins may be the key in controlling all of these processes.

SECRETION OF S-100 PROTEINS AND POSSIBLE EXTRACELLULAR ROLES

At least three S-100 proteins have been demonstrated to be secreted and may have extracellular effects. S-100ß is secreted by clonal glioma cell lines and primary rat cortical astrocytes (Shashoua et al. 1984). The synthesis and release of S-100ß from C6 glioma cell lines is under complex control. The secreted levels of S-100ß are highest in subconfluent or confluent cell cultures but decrease once the cells become post-confluent. The intracellular levels of S-100ß, however, are higher in the postconfluent cultures than in any other growth condition (Gysin et al. 1980; Zimmer and van Eldik 1988). These data suggest that the intracellular and secreted levels are differentially regulated. Adrenocorticotropic hormone (ACTH) and isoproterenol stimulate the release of S-100ß (Suzuki et al. 1987).

If S-100ß functions in the nervous system directly on neurons as a neurite extension factor and/or neurotrophic factor it must be secreted from glial cells in vivo which synthesize the protein to exert its effect on neurons. S-100ß can be measured in cerebrospinal fluid and is elevated in a wide spectrum of neurological disorders including cerebral hemorrhage, stroke, multiple sclerosis, and central nervous system tumors (Sindic et al. 1982; Persson et al. 1987). Cerebrospinal fluid S-100ß levels have not yet been investigated in Down syndrome and Alzheimer's Disease. It would be particularly interesting if these studies were done as brain S-100ß levels are increased in these disorders. Determination of S-100ß levels in cerebrospinal fluid may potentially have some diagnostic utility in a number of diseases of the nervous system.

CF Ag/MRP-8 and MRP-14 also are secreted (Dorin et al. 1987; Odink et al. 1987). They are present in both serum and synovial fluid. It is not clear if the synthesis of CF Ag/MRP-8 and MRP-14 is increased in cystic fibrosis and/or rheumatoid arthritis or that the release of these proteins is increased due to enhanced secretion or lysis of myeloid cells. The presence of other S-100 proteins in the extracellular space has not been closely examined, but it is likely that other S-100 proteins are secreted.

The possible extracellular functions of S-100 proteins may relate to the different affinities of the EF-hands. Ca^{2+} may bind to the N-terminal EF-hand only when the protein is exposed to the high extracellular Ca^{2+} environment. In such a case, the S-100 protein would have Ca^{2+} bound to both low affinity (N-terminal) and high affinity (C-terminal) EF-hands. The presence of two Ca^{2+} ions bound will induce a different conformation in the S-100 protein than will the binding of just one Ca^{2+}. This may be a mechanism for generating distinct extracellular versus intracellular roles for S-100 proteins.

How are the S-100 protein secreted? None of the S-100 proteins have a classic leader peptide which would function to direct the protein to the secretory pathway. The N-terminal EF-hand and the binding of Ca^{2+} could be involved in targeting the proteins for secretion. They all have a hydrophobic domain in amino terminal region which may function as an internal leader peptide (Figure 1A). The exposure of this hydrophobic domain to solvent is dependent upon the binding of Ca^{2+} to the N-terminal EF-hand. The binding of Ca^{2+} to this

EF-hand, the subsequent conformational change and exposure of the hydrophobic domain, and the targeting of the protein to a membrane bound form destined for secretion may be a novel Ca^{2+}-sensitive release pathway. This hypothesis is directly testable with clones expressing recombinant S-100 proteins and mutagenesis experiments.

CONCLUSIONS, UNANSWERED QUESTIONS AND FUTURE RESEARCH

In summary, the S-100 proteins are candidates for cell-type specific and cell-cycle specific Ca^{2+}-binding proteins which transduce the signal of increased intracellular Ca^{2+} in a specific fashion. They likely exert their effects by regulation of the phosphorylation of specific effector proteins. These effector proteins are cytoskeletal-associated proteins which have roles in signal transduction, cell motility and cytokinesis, cellular differentiation, cell-cycle progression, and inflammation.

Additional members of the S-100 protein family will undoubtedly be described. They will likely be expressed in both a cell-type specific manner and differentially during the cell cycle. It may be possible that every cell expresses one or more S-100 proteins, perhaps in different subcellular compartments or in different phases of the cell cycle. The key to understanding the functions of the S-100 proteins likely resides with investigation of the effector proteins whose activities they modulate. Many of these proteins are cytoskeletal-associated proteins with functions that are not yet well understood.

The availability of molecular clones for the S-100 proteins and their effector proteins will facilitate expression of the proteins. Structure-function studies and mutagenesis experiments will assist in clarifying the interaction of these two groups of proteins. These experiments will determine the regions necessary for binding to and activating/inhibiting the effector proteins that regulate biological activities such as neurite extension, cytoskeletal stability, cytokinesis, and signal transduction. Description of the domains of S-100 proteins and effector proteins that interact with each other will allow comparison to similar domains on calmodulin and calmodulin-binding proteins. Expression of recombinant S-100 proteins and mutagenesis experiments will facilitate the description of S-100 protein domains necessary for biochemical or biological functions such as the neurite extending activity of

S-100ß or the ability of p11 to block p36 phosphorylation. Expression and mutagenesis studies also will be able to address the question of why the S-100 proteins retain the low affinity (N-terminal) EF-hand?

Overexpression of the S-100 proteins related to cell-cycle progression may produce cells with an altered cellular phenotype, perhaps deregulated growth or transformation. These experiments may provide valuable insights into the contribution of Ca^{2+} signal transduction to altered cellular growth, transformation and malignancy. Further investigation of the contribution of S-100 proteins to cell-cycle progression may reveal an interaction or involvement with one or more of the multiple kinase pathways recently described as regulating eukaryotic cell-cycle progression. Transgenic animal studies may produce useful animal models of S-100-related diseases such as the overexpression of S-100ß as a possible model system for Down syndrome or Alzheimer's disease.

The availability of molecular clones for S-100 proteins will make possible studies designed to determine the mechanism(s) of their secretion and possible extracellular functions. Are the extracellular effects (e.g., neurite extension) of the S-100 proteins dependent upon their interaction with specific cellular receptor molecules or are they internalized by the target cell and thus exert their effect(s) from an intracellular compartment? Are the extracellular effects via a pathway other than modification of protein phosphorylation?

Which cells express more than one S-100 protein? Are they localized to separate subcellular compartments? Will specific cellular signals promote the subcellular relocalization of one S-100 protein but not another? Are the effector proteins also differentially distributed at the subcellular level?

Future research will hopefully answer some of these questions. As we learn more about the S-100 proteins and their associated effector proteins the specific mechanism of action and functions of this group of proteins will become more clear.

REFERENCES

Allore R, O'Hanlon D, Price R, Neilson K, Willard HF, Cox DR, Marks A, Dunn RJ (1988) Gene encoding the b subunit of S-100 protein is on chromosome 21: implications for Down syndrome. Science 239:1311-1313

Azmitia EC, Dolan K, Whitaker-Azmitia PM (1990) S-100b but not NGF, EGF, insulin or calmodulin is a CNS serotonergic growth factor. Brain Res 516:354-356

Barraclough R, Dawson KJ, Rudland PS (1982) Control of protein synthesis in cuboidal epithelial cells in culture. Eur J Biochem 129:335-341

Barraclough R, Savin J, Dube SK, Rudland PS (1987) Molecular cloning and sequence of the gene p9Ka. A cultured myoepithelial cell protein with strong homology to S-100, a calcium-binding protein. J Mol Biol 198:13-20

Baudier J, Holtzhurer C, Gerard D (1982a) Zinc-dependent affinity chromatography of the S-100b protein on phenyl-Sepharose. FEBS Lett 148:231-234

Baudier J, Briving C, Deinum J, Haglid K, Sorskog L, Wallin M (1982b) Effect of S-100 proteins and calmodulin on Ca^{2+}-induced disassembly of brain microtubule proteins in vitro. FEBS Lett 147:165-167

Baudier J, Gerard D (1983) Ions binding to S-100 proteins: structural changes induced by calcium and zinc on S-100α and S-100ß proteins. Biochem 22:3360-3369

Baudier J, Haglid K, Haiech J, Gerard D (1983) Zinc binding to human brain calcium binding proteins, calmodulin and S-100ß protein. Biochem Biophys Res Comm 114:1138-1146

Baudier J, Labourette G, Gerard D (1985) Rat brain S-100ß protein: purification, characterization and ion binding properties. A comparison with bovine S-100ß protein. J Neurochem 44:76-84

Baudier J, Cole RD (1988) Reinvestigation of the sulfhydryl reactivity in bovine brain S-100ß protein and the microtubule-associated tau proteins. Ca^{2+} stimulates disulfide cross-linking between the S-100ß ß-subunit and the microtubule-associated tau protein. Biochem 27:2728-2736

Baudier J, Bronner C, Kligman D, Cole RD (1989) Protein kinase C substrates from bovine brain. J Biol Chem 264:1824-1828

Benda P, Lightbody J, Sato G, Levine L, Sweet W (1968) Differentiated rat glial cell strain in tissue culture. Science 161:370

Calabretta B, Battini R, Kaczmarek L, de Riel JK, Baserga R (1986) Molecular cloning of the cDNA for a growth factor-inducible gene with strong homology to S-100, a calcium-binding protein. J Biol Chem 261:12628-12632

Calissano P, Moore BW, Friesen A (1969) Effect of calcium ion on S-100, a protein of the nervous system. Biochem 8:4318-4326

Cocchia D (1981) Immunocytochemical localization of S-100 protein in the brain of adult rat. Cell Tiss Res 214:529-540

Cocchia D, Michetti F, Raffioni S, Donato R (1985) Immunocytochemical localization of S-100 protein in the cilia of cell types of different species. In: Alia E (ed) Contractile proteins in muscle and non-muscle systems: biochemistry, physiology, and pathology. Praeger Scientific Pub, New York, p108-113

Dannies PS, Levine L (1971) Structural properties of bovine brain S-100 protein. J Biol Chem 246:6276-6283

Davidson FF, Dennis EA, Powell M, Glenney JR (1987) Inhibition of phospholipase A2 by 'lipocortins' and calpactins. J Biol Chem 262:1698-1705

De Panfilis G, Rowden G, Manara GC, Ferrari C, Torresani C, Sansoni P (1988) The S-100b protein in normal peripheral blood is uniquely present within a discrete suppressor-T-cell compartment. Cell Immunol 114:398-404

Donato R (1988) Calcium-independent, pH-regulated effects of S-100 proteins on assembly-disassembly of brain microtubule protein in vitro. J Biol Chem 263:106-110

Donato R, Giambanco I, Aisa MC (1989a) Molecular interaction of S-100 proteins with microtubule proteins in vitro. J Neurochem 53:566-571

Donato R, Giambanco I, Aisa MC, Ceccarelli P (1989b) Identification of S-100 proteins and S-100-binding proteins in a detergent-resistant EDTA/KCl-extractable fraction from bovine brain membranes. FEBS Lett 247:31-35

Dorin JR, Novak M, Hill RE, Brock D, Secher DS, van Heyningen (1987) A clue to the basic defect in cystic fibrosis from cloning the CF antigen gene. Nature 326:614-617

Drust DS, Creutz CE (1988) Aggregation of chromaffin granules by calpactin at micromolar levels of calcium. Nature 331:88-91

Duncan A, Higgins J, Dunn RJ, Allore R, Marks A (1989) Refined sublocalization of the human gene encoding the ß subunit of the S-100 protein (S-100ß) and confirmation of a subtle t(9;21) translocation using in situ hybridization. Cytogenet Cell Genet 50:234-235

Dunn R, Landry C, O'Hanlon D, Dunn J, Allore R, Brown I, Marks A (1987) Reduction in S-100 protein ß subunit mRNA in C6 rat glioma cells following treatment with anti-microtubular drugs. J Biol Chem 262:3562-3566

Ebralidze A, Tulchinsky E, Grigorian M, Afanasyeva A, Viacheslav S, Revazova E, Lukanidin E (1989) Isolation and characterization of a gene specifically expressed in different metastatic cells and whose deduced gene product has a high degree of homology to a Ca^{2+}-binding protein family. Genes Devel 3:1086-1093

Endo T, Hidaka H (1983) Effect of S-100 protein on microtubule assembly- disassembly. FEBS Lett 161:235-238

Fan K (1982) S-100 protein synthesis in cultured glioma cell is G_1-phase of cell cycle dependent. Brain Res 237:498-503

Fano G, Angelella P, Mariggio D, Aisa MC, Giambanco I, Donato R (1989a) S-100$\alpha\alpha$ protein stimulates the basal (Mg^{2+}-activated) adenylate cyclase activity associated with skeletal muscle membranes. FEBS Lett 248:9-12

Fano G, Marsili V, Angelella P, Aisa MC, Giambanco I, Donato R (1989b) S-100$\alpha\alpha$ protein stimulates Ca^{2+}-induced release from isolated sarcoplasmic reticulum vesicles. FEBS Lett 255:381-384

Ferrari S, Calabretta B, de Riel JK, Battini R, Ghezzo F, Lauret E, Griffin C, Emanuel BS, Gurrieri F, Baserga R (1987) Structural and functional analysis of a growth-regulated gene, the human calcyclin. J Biol Chem 262:8325-8332

Freeman MR, Beckmann SL, Sueoka N (1989) Regulation of the S-100 protein and GFAP genes is mediated by two common mechanisms in RT4 neuro-glial cell lines. Exp Cell Res 182:370-383

Fujii T, Machino K, Andoh H, Satoh T, Konodo Y (1990) Calcium-dependent control of caldesmon-actin interaction by S-100 protein. J Biochem 107:133-137

Gaynor R, Irie R, Morton D, Herschman HR (1980) S-100 protein is present in cultured human malignant melanomas. Nature 286:400-401

Gerke V, Weber K (1984) Identity of p36K phosphorylated upon Rous sarcoma virus transformation with a protein purified from brush borders; calcium-dependent binding to nonerythroid spectrin and F-actin. EMBO J 3:227-233

Ghezzo F, Lauret E, Ferrari S, Baserga R (1988) Growth factor regulation of the promoter for calcyclin, a growth-regulated gene. J Biol Chem 263:4758-4763

Glenney J (1986) Phospholipid-dependent Ca^{2+} binding by the 36-kDa tyrosine kinase substrate (calpactin) and its 33-kDa core. J Biol Chem 261:7247-7252

Glenney JR, Tack BF (1985) Amino-terminal sequence of p36 and associated p10: identification of the site of tyrosine phosphorylation and homology with S-100. Proc Natl Acad Sci 82:7884-7888

Glenney JR, Kindy MS, Zokas L (1989) Isolation of a new member of the S-100 protein family: amino acid sequence, tissue, and subcellular distribution. J Cell Biol 108:569-578

Goto K, Endo H, Fujiyoshi T (1988) Cloning of the sequences expressed abundantly in established cell lines: identification of a cDNA clone highly homologous to S-100, a calcium binding protein. J Biochem 103:48-53

Griffin WS, Stanley LC, Ling C, White L, MacLeod V, Perrot LJ, White CL, Araoz C (1989) Brain interleukin 1 and S-100 immunoreactivity are elevated in Down syndrome and Alzheimer disease. Proc Natl Acad Sci 86:7611-7615

Gysin R, Moore B, Proffitt RT, Deuel TF, Caldwell K, Glaser L (1980) Regulation of the synthesis of S-100 protein in rat glial cells. J Biol Chem 255:1515-1520

Hagiwara M, Ochiai M, Owada K, Tanaka T, Hidaka H (1988) Modulation of tyrosine phosphorylation of p36 and other substrates by the S-100 protein. J Biol Chem 263:6438-6441

Haglid K, Hamberger A, Hansson H, Hyden H, Persson L, Ronnbach L (1976) Cellular and subcellular distribution of the S-100 protein in rabbit and rat central nervous system. J Neurosci Res 2:175-192

Haimoto H, Kato K (1987) S-100αα protein, a calcium-binding protein, is localized in the slow-twitch muscle fiber. J Neurochem 48:917-923

Haimoto H, Kato K (1988) S-100αα protein in cardiac muscle. Eur J Biochem 171:409-415

Hidaka H, Endo T, Kawamoto S, Yamada E, Umekawa H, Tanabe K, Hara K (1983) Purification and characterization of adipose tissue S-100b protein. J Biol Chem 258:2705-2709

Hilt DC, Joshi J, Kligman D (1989) Coordinate regulation of S-100ß and glial fibrillary acidic protein (GFAP) gene expression by cAMP and glucocorticoids. Evidence of glial-specific gene regulation? In: Advances in gene technology: Molecular Neurobiology and Neuropharmacology Proceedings of the 1989 Miami Biotechnology Winter Symposium, ICSU Short Reports 9:133, IRL Press, Oxford, UK

Hornbeck P, Nakabayashi H, Fowlkes BJ, Paul WE, Kligman D (1989) A major myristylated substrate of protein kinase C and protein kinase itself are differentially regulated during murine B- and T-lymphocyte development and activation. Mol Cell Biol 9:3727-3735

Hyden H, McEwen B (1966) A glial protein specific for the nervous system. Proc Natl Acad Sci 55:354-358

Ishikawa H, Nogami H, Shirasawa N (1983) Novel clonal strains from adult rat anterior pituitary producing S-100 protein. Nature 303:711-713

Isobe T, Nakajima T, Okuyama T (1977) Reinvestigation of extremely acidic proteins in bovine brain. Biochim Biophys Acta 494:222-232

Isobe T, Okuyama T (1978) The amino-acid sequence of S-100 protein (PAP I-b) and its relationship to calcium-binding proteins. Eur J Biochem 89:379-388

Isobe T, Okuyama T (1981) The amino-acid sequence of the a-subunit in bovine brain S-100 protein. Eur J Biochem 116:79-86

Isobe T, Takahasi K, Okuyama T (1984) S-100α protein is present in neurons of the central and peripheral nervous system. J Neurochem 43:1494-1496

Jackson-Grusby LL, Swiergiel J, Linzer D (1987) A growth-related mRNA in cultured mouse cells encodes a placental calcium-binding protein. Nuc Acid Res 15:6677-6690

Kahn HJ, Marks A, Thom H, Baumal R (1982) Role of antibody to S-100 protein in diagnostic pathology. Amer J Clin Path 79:341-347

Kanamori M, Endo T, Shirakawa S, Sakurai M, Hidaka H (1982) S-100 antigen in human lymphocytes. Biochem Biophys Res Comm 108:1447-1453

Kato K, Kimura S, Haimoto H, Suzuki F (1986) S-100αα protein: distribution in muscle tissues of various animals and purification from human pectoral muscle. J Neurochem 46:1555-1560

Kato K, Suzuki F, Ogasawara N (1988) Induction of S-100 protein in 3T3-L1 cells during differentiation to adipocytes and its liberating by lipolytic hormones. Eur J Biochem 177:461-466

Kelly SE, Jones DB, Fleming S (1989) Calgranulin expression in inflammatory dermatoses. J Path 159:17-21

Kimura S, Kato K, Semba R, Isobe T (1984) Regional distribution of S-100αα, S-100αß and S-100ßß in the bovine central nervous system determined with a sensitive enzyme immunoassay system. Neurochem Int 6:513-518

Klee CB (1988) Ca²⁺-dependent phospholipid- (and membrane-) binding proteins. Biochem 27:6645-6653

Kligman D (1982) Isolation of a protein from bovine brain which promotes neurite extension from chick embryo cerebral cortex neurons in defined medium. Brain Res 250:93-100

Kligman D, Marshak D (1985) Purification and characterization of a neurite extension factor from bovine brain. Proc Natl Acad Sci 82:7136-7139

Kligman D, Hsieh LS (1987) Neurite extension factor induces rapid morphological differentiation of mouse neuroblastoma cells in defined medium. Dev Brain Res 33:296-300

Kligman D, Hilt DC (1988) The S-100 protein family. Trends Bio Sci 13:437-443

Kretsinger R (1980) Structure and evolution of calcium modulated proteins. CRC Crit Rev Biochem 8:119-174

Kuwano R, Usui H, Maeda T, Fukui T, Yamanari N, Ohtsuka E, Ikehara M, Takahasi Y (1984) Molecular cloning and the complete nucleotide sequence of cDNA to mRNA for S-100 protein of rat brain. Nuc Acid Res 17:7455-7465

Kuznicki J, Filipek A, Heimann P, Kacmarek L, Kaminska B (1989) Tissue specific distribution of calcyclin-10.5 kDa Ca²⁺-binding protein. FEBS Lett 254:141-144

Lee M, Nurse P (1988) Cell cycle control genes in fission yeast and mammalian cells. Trends Genet 4:287-290

Lewis D, Teyler TJ (1986) Anti-S-100 serum blocks LTP in the hippocampal slice. Brain Res 383:159-164

Malinow R, Schulman H, Tsien RW (1989) Inhibition of postsynaptic PKC or CaMKII blocks induction but not expression of LTP. Science 245:862-866

Mani RS, Kay CM (1983) Circular dichroism studies on the zinc-induced conformational changes in S-100a and S-100ß proteins. FEBS Lett 163:282-286

Mani RS, Boyes BE, Kay CM (1982) Physicochemical and optical studies on calcium- and potassium-induced conformational changes in bovine S-100b protein. Biochem 22:2607-2612

Marks A, Labourdette G (1977) Succinyl concanavalin A stimulates and microtubular drugs inhibit the synthesis of a brain-specific protein in rat glial cells. Proc Natl Acad Sci 74:3855-3858

Masiakowski P, Shooter EM (1988) Nerve growth factor induces the genes for two proteins related to a family of calcium-binding proteins in PC12 cells. Proc Natl Acad Sci 85:1277-1281

Matus A, Mughal S (1975) Immunohistochemical localization of S-100 protein in brain. Nature 258:746-748

Michetti F, Cocchia D (1982) S-100-like immunoreactivity in a planarian. Cell Tiss Res 223:575-582

Moore B (1965) A soluble protein characteristic of the nervous system.
Biochem Biophys Res Comm 19:739-744

Moore B (1984) The S-100 protein. In: Marangos PJ, Campbell IC, and Cohen RM (eds) Neuronal and glial proteins. Academic Press Pub, San Diego, pp137-167

Moore B (1988) Conformational and hydrophobic properties of rat and bovine S-100 proteins.
Neurochem Res 13:539-545

Muller HW, Harter PG, Hangen DH, Shooter EM (1985) A specific 37,00-dalton protein that accumulates in regenerating but not nonregenerating mammalian nerves. Science 228:499-501

Murao S, Collart FR, Huberman E (1988) A protein containing the cystic fibrosis antigen is an inhibitor of protein kinases. J Biol Chem 264:8356-8360

Murphy LC, Murphy LJ, Tsuyuki D, Duckworth ML, Shiu R (1988) Cloning and characterization of a cDNA encoding a highly conserved, putative calcium binding protein, identified by an anti-prolactin receptor antiserum. J Biol Chem 263:2397-2401

Odink K, Cerletti N, Brüggen J, Clerc RG, Tarcsay L, Zwadlo G, Gerhards G, Schlegel R, Sorg C (1987) Two calcium-binding proteins in infiltrate macrophages of rheumatoid arthritis. Nature 330:80-82

Ouimet CC, Wang J, Walaas IW, Albert K, Greengard P (1990) Localization of the MARCKS (87kDa) protein, a major specific substrate for protein kinase C, in rat brain. J Neurosci 10:1683-1698

Patel J, Marangos PJ, Heydorn WE, Chang E, Verma A, Jacobowitz D (1983) S-100 mediated inhibition of brain protein phosphorylation. J Neurochem 41:1040-1045

Patel J, Kligman D (1987) Purification and characterization of an M_r 87,000 protein kinase C substrate from rat brain. J Biol Chem 262:16686-16691

Persechini A, Moncrief ND, Kretsinger RH (1988) The EF-hand family of calcium-modulated proteins.
Trends Neuro Sci 12:462-467

Persson L, Hardemark HG, Gustafsson J, Rundstrom G, Mendel I, Esscher T, Pahlman S (1987) S-100 protein and neuron-specific enolase in cerebrospinal fluid and serum: markers of cell damage in human central nervous system. Stroke 18:911-918

Pfeiffer SE, Herschman HR, Lightbody J, Sato G (1970) Synthesis by a clonal line of rat glial cells of a protein unique to the nervous system. J Cell Phys 75:329-340

Qi D, Kuo JF (1984) S-100 modulates Ca^{2+}-independent phosphorylation of an endogenous protein (Mr=19K) in brain. J Neurochem 43:256-260

Reeves R, Crowley M, Hilt DC, Moran T, Gerhart J (1990) Transgenic mice expressing elevated levels of the S-100ß protein display behavioral abnormalities and learning deficits. Mouse Molecular Genetics Meeting, Cold Spring Harbor

Sansoni F, Rowden G, Manara GC, De Panfilis G (1988) One half of the CD11b+ human peripheral blood lymphocytes coexpresses the S-100 protein. J Exp Immunol 72:357-361

Shashoua VE, Hesse GW, Moore BW (1984) Proteins of the brain extracellular fluid: evidence for release of S-100 protein. J Neurochem 42:1536-1541

Sindic C, Chalon MP, Cambiaso CL, Laterre EC, Masson PL (1982) Assessment of damage to the central nervous system by determination of S-100 protein in the cerebrospinal fluid.
J Neurol Neurosurg Psy 45:1130-1135

Stavrou D, Rieske E, Anzil AP, Haglid KG, Isenberg G (1980) Definition of a cell clone with astroglial characteristics derived from a chemically induced rabbit brain glioma. J Neurol Sci 45:287-301

Steinbakk M, Andresen C, Lingaas E, Dale I, Brandtzaeg P, Fagerhol MK (1990) Antimicrobial actions of calcium binding leucocyte L1 protein, calprotectin. The Lancet 336:763-765

Suzuki F, Kato K, Kato T, Ogasawara N (1987) S-100 protein in clonal astroglioma cells is released by adrenocorticotropic hormone and corticotropin-like intermediate-lobe peptide. J Neurochem 49:1557-1563

Sviridov SM, Korochkin LJ, Lvanov VN (1972) Immunohistochemical studies on S-100 protein during postnatal ontogenesis of the brain of two strains of rats. J Neurochem 19:713-718

Szebenyi DM, Obendorf SK, Moffat K (1981) Structure of vitamin D3-dependent calcium-binding protein from bovine intestine. Nature 294:327-332

Tabuchi K, Kirsch WM (1975) Immunohistochemical localization of S-100 protein in neurons and glia of hamster cerebellum. Brain Res 92:175-180

Tufty RM, Kretsinger RH (1975) Troponin and parvalbumin calcium binding regions predicted in myosin light chain and T4 lysozyme. Science 187:167-169

van Eldik LJ, Staecker JL, Winningham-Major F (1988) Synthesis and expression of a gene coding for the calcium-modulated protein S-100ß and designed for cassette-based, site-directed mutagenesis. J Biol Chem 263:7830-7837

Wallner BP, Mattaliano RJ, Hession C, Cate RL, Tizard R, Sinclair LK, Foeller C, Chow EP, Browning JL, Ramachandran KL, Pepinsky RB (1986) Cloning and expression of human lipocortin, a phospholipase A2 inhibitor with potential anti-inflammatory activity. Nature 320:77-81

Wilkinson MM, Busuttil A, Hayward C, Brock DJ, Dorin JR, van Heyningen V (1988) Expression pattern of two related cystic fibrosis associated calcium-binding proteins in normal and abnormal tissues. J Cell Sci 91:221-230

Winningham-Major F, Staecker JL, Barger SW, Coats S, van Eldik LJ (1989) Neurite extension and neuronal survival activities of recombinant S-100ß proteins that differ in the content and position of cysteine residues. J Cell Biol 109:3063-3071

Zalutsky RA, Nicoll RA (1990) Comparison of two forms of long-term potentiation in single hippocampal neurons. Science 248:1619-1624

Zimmer DB, van Eldik LJ (1988) Levels and distribution of the calcium-modulated proteins S-100 and calmodulin in rat C6 glioma cells. J Neurochem 50:572-579

Zimmer DB, van Eldik LJ (1989) Analysis of the calcium-modulated proteins, S-100 and calmodulin, and their target proteins during C6 glioma cell differentiation. J Cell Biol 108:141-151

p9Ka, A CALCIUM-ION-BINDING PROTEIN OF CULTURED MYOEPITHELIAL CELLS

Roger Barraclough and Philip S. Rudland

INTRODUCTION

Eukaryotic cells contain small, potential calcium-ion-binding proteins that consist of polypeptide chains which contain two EF-hand-like calcium-ion-binding regions. Such proteins are present in a range of normal tissues and in cultured cells (Kligman and Hilt 1988), and are related to S-100 protein (Van Eldick et al. 1982). Some small calcium-ion-binding proteins, such as S-100 protein itself (Moore and McGregor 1965), the vitamin D-dependent calcium-ion binding protein of the small intestine (Fullmer and Wasserman 1981), p10 of pig intestine (Gerke and Weber 1985) and of fibroblasts (Erikson et al. 1984), macrophage migration inhibitory factor-related proteins (MRP) 8 and 14 (Burmeister et al. 1986; Odink et al. 1987), have been isolated and their amino acid sequences either fully or partially determined. The amino acid sequence of other members of the S-100 family of proteins, such as calcyclin (Hirschhorn et al. 1984; Calabretta et al. 1986; Ferrari et al. 1987), have been predicted from the open reading frames of their messenger RNA (mRNA) molecules isolated from certain cultured cell lines.

A small potential calcium-ion-binding protein, p9Ka, is induced when cultured rat mammary epithelial cells convert, in culture, to elongated myoepithelial-like cells (Barraclough et al. 1982). p9Ka is also designated by the names of independently isolated, cloned complementary DNA (cDNA) molecules, which encode a protein with the same derived amino acid sequence as p9Ka (Barraclough et al. 1987b), namely L8A2 (Jackson-Grusby et al. 1987), 42A (Masiakowski and Shooter 1988), pEL98 (Goto et al. 1988) and mts (Ebralidze et al. 1989). This present chapter reviews some recent work on p9Ka, its mRNA, and its gene in rat mammary cells.

p9Ka IN MAMMARY CELLS

Epithelial cell lines derived from benign rat mammary tumors, such as the cell line Rama 25 (Bennett et al. 1978), yield elongated cells (Warburton et al. 1982) with a frequency of 1-3% (Bennett et al. 1978). One cell line derived from these elongated cells is called Rama 29 (Bennett et al. 1978). The conversion process is not specific for one cell line, but has been found in cultures of mouse mammary tumors (Sanford et al. 1961; Dexter et al. 1978; Hager et al. 1981) and in cultures of normal rat mammary glands (Ormerod and Rudland 1985). From a comparison of the ultrastructure and immunocytochemical staining characteristics of histological sections of rat mammary glands (Ormerod and Rudland 1982; Warburton et al. 1982), and of their primary cultures (Warburton et al. 1985), the elongated cells derived from cultures of the cell lines such as Rama 25 are related to myoepithelial cells.

The conversion process of the epithelial cells to the myoepithelial-like cells is associated with reproducible changes in the pattern of intracellular polypeptides separated by two-dimensional polyacrylamide gel electrophoresis (Barraclough et al. 1982; 1984a). Some 6% of the polypeptides thus fractionated show a change when the myoepithelial-like cells and the parental epithelial cells are compared (Figure 1). Some of the major changes are in the cytoskeletal, Triton X-100-insoluble matrix of the cell (Barraclough et al. 1982). However, one relatively abundant polypeptide, which is not isolated with this matrix, is present in the myoepithelial-like cells, but is virtually absent in the parental epithelial cells. This polypeptide is called p9Ka (Figure 1), and the formation of myoepithelial-like cells from the epithelial cells is associated with at least a 16-fold increase in its intracellular level (Barraclough et al. 1984a). p9Ka has been identified initially by its apparent molecular weight (9 ± 0.52 kDa; data from 13 determinations) and isoelectric point (5.5 ± 0.27; data from 26 determinations) (Barraclough et al. 1984a).

BINDING OF CALCIUM IONS BY p9Ka IN VITRO

p9Ka has been substantially purified from the myoepithelial-like cell line, Rama 29, using reversed-phase HPLC, and it behaves as a relatively hydrophobic protein (Barraclough et al.

submitted). The resulting preparations consist of 50-80% p9Ka, as judged by densitometric scanning of the pattern of the stained bands arising from electrophoresis of such preparations on one-dimensional, sodium dodecylsulfate polyacrylamide gels. The p9Ka in these preparations binds calcium ions in vitro. This binding can be demonstrated by subjecting the preparations of p9Ka to either one-dimensional sodium dodecylsulfate polyacrylamide gel electrophoresis or to two-dimensional polyacrylamide gel electrophoresis, and transferring the fractionated polypeptides to nitrocellulose filters. Upon incubation of the filters with $^{45}CaCl_2$, (Maruyama et al. 1984), radioactivity binds only to the band or spot corresponding exactly to p9Ka (Barraclough et al. submitted). In low salt conditions p9Ka binds calcium ions with an apparent K_d of 76 ± 14 µM, and the results suggest that a model in which p9Ka contains a single high-affinity site best fits all the data.

The apparent affinity of p9Ka for calcium ions is less than for the regulatory calcium-ion-binding protein, calmodulin (3-20 µM) (Crouch and Klee 1980), a calcium storage protein, parvalbumin (0.1 µM) (Pechere et al. 1977), and a calcium-ion transport protein, the vitamin D-dependent intestinal calcium-ion-binding protein (2 µM) (Van Eldik et al. 1982). With regard to closely related, small calcium-ion-binding proteins, p9Ka has a slightly lower affinity for calcium than has calcyclin (3 µM) (Kuznicki and Filipek 1987) and S-100 protein (1-15 µM) (Baudier et al. 1986), but p9Ka has a higher affinity than p10, the regulatory subunit of calpactin (Gerke and Weber 1985). The affinity of p9Ka for calcium ions is close to that of the phospholipid-interacting proteins, the large subunit of calpactin (Shadle et al. 1985) and membrane-bound proteins such as clathrin (Mooibroek et al. 1987).

The binding of calcium by p9Ka is strongly antagonized by the presence of potassium ions in the medium. In this regard, p9Ka resembles S-100 protein (Baudier et al. 1986). Since the binding of calcium ions by these proteins will be reduced in the presence of physiological concentrations of K^+, it is possible that p9Ka binds calcium in a subcellular compartment in which calcium ions are elevated above normal levels, or where the potassium ion concentration is low. This idea has been investigated by experiments which examined the distribution of p9Ka in subcellular fractions of the myoepithelial-like cells. When proteins extracted from subcellular fractions, under certain conditions, are subjected to two-dimensional polyacrylamide gel electrophoresis, p9Ka is associated with the 700 g(av),

108

12000 g(av), and 100000 g(av) membrane/cytoskeletal microsomal pellet fractions, as well as with the soluble supernatant fraction (Figure 2).

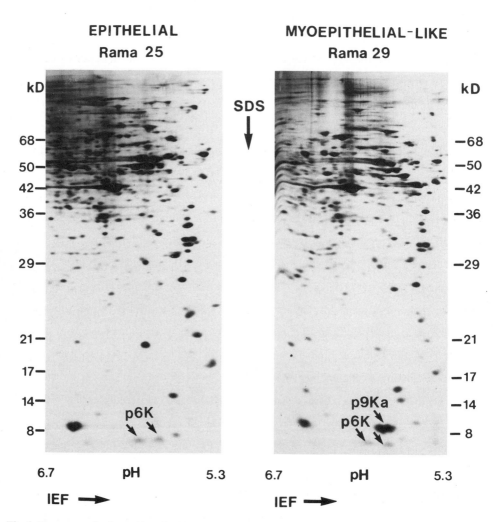

Fig. 1. The pattern of polypeptides of epithelial and myoepithelial-like cell lines following two-dimensional gel electrophoresis. Extracts of epithelial cells, Rama 25 (*left panel*) or myoepithelial-like cells, Rama 29 (*right panel*) radioactively labeled overnight with [^{35}S]methionine, were subjected to isoelectric focusing in the first dimension, and SDS gel electrophoresis in the second dimension. The autoradiographies of the dried-down gels are shown. The *arrow* labeled *p9Ka* points to the radioactive spot corresponding to the polypeptide p9Ka. The *arrows* labeled *p6K* point to the two isoelectric focusing variants of the polypeptide p6K referred to in the text

However, in some experiments, the protein spot corresponding to p9Ka, which is usually elongated in the isoelectric focusing direction (Figure 1), splits into two separate, but closely spaced, isoelectric focusing variants (Figure 2) that are present in about equal amounts in total extracts from the myoepithelial-like cells (Figure 2, Panel A). The more-acidic variant predominates in the membrane/cytoskeletal/microsomal fractions (Figure 2, Panels B-D), whilst the more basic variant predominates in the soluble fraction (Figure 2, Panel E). These results suggest that a proportion of the p9Ka in these cells may be modified in some as-yet unidentified manner, and that the modified form has a subcellular location different to that of the unmodified form.

p9Ka IS AN EF-HAND-CONTAINING CALCIUM-ION-BINDING PROTEIN

A cloned complementary DNA (cDNA) corresponding to 400 nucleotides of the 3' end of p9Ka mRNA has been isolated from a cDNA library constructed from poly(A)-containing RNA of the myoepithelial-like cell line, Rama 29 (Barraclough et al. 1984b). This cloned cDNA hybridizes specifically to a mRNA molecule that is translated by the reticulocyte lysate cell-free protein synthesizing system into a polypeptide product that comigrates exactly with p9Ka on two-dimensional polyacrylamide gel electrophoresis (Barraclough et al. 1984b).

The cloned cDNA corresponding to p9Ka mRNA was used to screen a genomic library containing fragments of the DNA from normal rat liver (Barraclough et al. 1987b). An 18-kilobase (kb) fragment of the normal rat DNA which contains the gene for p9Ka and its flanking regions has been isolated from the recombinant library. The sequence of 2700 nucleotides of this gene region has been determined and the region corresponding to p9Ka mRNA has been delineated.

The part of the gene sequence which corresponds to p9Ka mRNA contains only one open reading frame of sufficient length to encode a polypeptide of molecular mass at least 9 kDa. This open reading frame contains a potential coding region of 100 amino acids, excluding the initiating methionine residue, and ends with two termination codons (Barraclough et al. 1987b). The potential amino acid sequence of this open reading frame is shown in Figure 3.

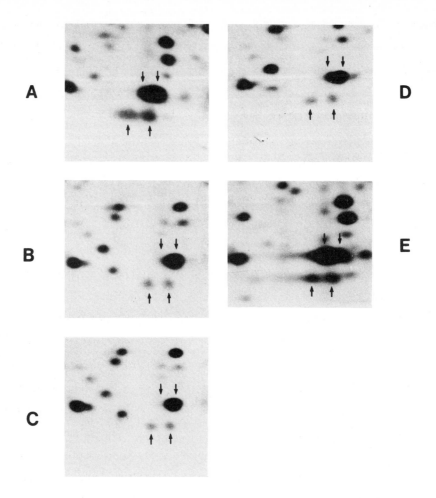

Fig. 2A-E. The subcellular distribution of p9Ka. Myoepithelial-like cells, Rama 29, were radioactively-labeled overnight with [^{35}S]methionine and the cells were homogenized in an aqueous buffer. The extract was subjected to differential centrifugation and the pellets from each centrifugation step were dissolved in electrophoresis sample buffer. Samples of the homogenate and supernatant fractions were freeze-dried and dissolved in electrophoresis sample buffer. Samples of the homogenate (**A**), pellet fractions resulting from centrifugation at 770 g(av) (**B**), 12000 g(av) (**C**), 100000 g(av) (**D**) and the supernatant resulting from centrifugation at 100000 g(av) (**E**) were subjected to isoelectric focusing in the first dimension and SDS gel electrophoresis in the second dimension. The autoradiographies of the relevant portions of the two-dimensional gels are shown. The *arrows point down* to the acidic (*right*) and basic (*left*) isoelectric focusing variants of p9Ka, and *point up* to the two isoelectric focusing variants of p6K

A search of computerized data bases of protein sequences shows limited homology to both the alpha (43%) and beta (43%) chains of bovine S-100 protein (Isobe and Okuyama 1978; 1981) and to rat S-100 protein (42%) (Kuwano et al. 1984). Homology is also observed between the nucleic acid sequences; a cloned cDNA corresponding to a portion of rat S-100 protein mRNA (Kuwano et al. 1984) has an overall homology of 66% with the corresponding region of the p9Ka gene.

The protein p9Ka also shows weak homology (34%) with the bovine vitamin D-dependent, intestinal calcium-ion-binding protein (Fullmer and Wasserman 1981) (Figure 3). This protein has two potential calcium-ion binding sites, one of which conforms to the EF-hand structure of known calcium-ion-binding proteins (Moews and Kretsinger 1975; Szebenyi et al. 1981). When only those amino acid residues thought to be involved in calcium-ion binding by the two potential sites of the bovine, intestinal protein (Szebenyi et al., 1981) are compared with those in the equivalent position in p9Ka, 11 out of 14 of these residues are identical in the two polypeptides (Barraclough et al. 1987b). p9Ka contains two potential calcium-ion binding loops, one of which lies between residues 19 and 32 and the second between residues 62 and 73 (Figure 3). The C-terminal loop (residues 62-73) corresponds to an almost perfect EF-hand sequence, with five residues containing carboxylic acid derivatives in their side chains (aspartate 62, asparagine 64, aspartate 66, aspartate 70, glutamate 73 of the p9Ka amino acid sequence) in the exact positions of the loop region that are thought to be important in calcium-ion binding by the vitamin D-dependent, intestinal, calcium-ion-binding protein (Szebenyi et al. 1981; 1986). In p9Ka, this potential binding site contains asparagine at position 67 instead of the smaller glycine residue. This latter residue is thought to play a role in maintaining the EF-hand structure in both S-100 protein and the intestinal calcium-ion binding protein (Van Eldik et al. 1982). However, this particular substitution, whilst not preventing calcium-ion binding by p9Ka, may account for the slightly reduced affinity for calcium by p9Ka, in comparison with the intestinal calcium-ion-binding protein.

The N-terminal EF-hand loop of p9Ka is less likely to bind calcium ions than the C-terminal structure, and this observation is supported by the experimentally determined calcium-ion-binding data that suggest a single high-affinity calcium-ion-binding site. However, it is possible that the N-terminal site binds calcium, but at a lower affinity than the

C-terminal site, and that the binding to the former site was not detected in our calcium-ion-binding studies, which were carried out using the method of Hummel and Dreyer (1962), as originally pointed out by Shadle et al. (1985). The closely related protein, calcyclin, which also contains two potential calcium-ion binding sites, possesses a C-terminal site which displays detectable binding of calcium ions, whereas binding by the low affinity N-terminal site is not detectable using the same procedure (Kuzniki and Filipek 1987).

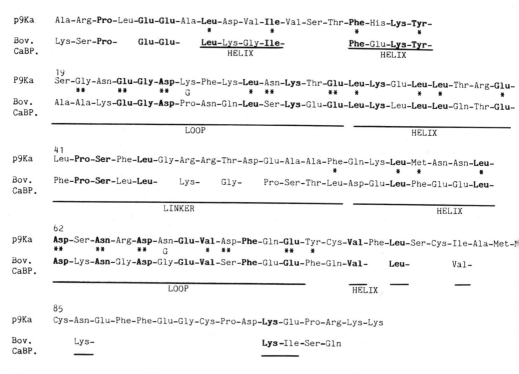

Fig. 3. The potential amino acid sequence of p9Ka and its relationship to the bovine intestinal calcium-ion-binding protein. Amino acids are shown using the three-letter code and are arranged according to the maximum homology using the algorithm of Dayhoff (1978). Amino acids that are invariant between p9Ka and the intestinal calcium-ion-binding protein (*Bov. CaBP*) are shown in *bold type*. The *stars* between the sequences show the preferred side chain for the EF-hand structure as indicated by Szebenyi et al. (1981). *Asterisk* (*) indicates an amino acid with a hydrophobic side chain; *double asterisk* (**) indicates an amino acid with an oxygen-containing side chain; **G** indicates the glycine residue which is replaced by asparagine in p9Ka. The helix-loop-helix arrangement of the calcium-ion-binding sites is indicated by underscoring

There is at present little knowledge of the roles for the small calcium-ion-binding proteins, other than for the intestinal protein. However, evidence suggests that the mRNA's for at least two of these proteins, calcyclin (Hirschhorn et al. 1984) and p9Ka (18A2, 42A) are induced when some cells are stimulated to grow by serum (Calabretta et al. 1986; Jackson-Grusby et al. 1987) or to differentiate with nerve growth factor (Masiakowski and Shooter 1988). In order to understand the induction of p9Ka the mechanisms of control of synthesis of p9Ka have been studied in the rat mammary cell lines.

STRUCTURE OF THE RAT p9Ka GENE REGION

The entire p9Ka gene sequence contains two introns of 1172 base-pairs (bp) and 675 bp (Figure 4) which exhibit 5' and 3' boundaries that conform to consensus intron/exon boundary sequences (Mount 1982). These introns divide the p9Ka gene into three coding regions of 37 bp, 156 bp and 295 bp. The rat p9Ka gene therefore contains a short 5' exon corresponding to part of the 5' untranslated region of the mRNA. The arrangement of the rat p9Ka gene is almost identical to that reported elsewhere for the human genes for calcyclin (Ferrari et al. 1987) (Figure 4), MRP 8 and MRP 14 (Lagasse and Clerc 1988).

The cloned fragment of normal rat DNA, 18 kb in length, which contains the gene for p9Ka (Barraclough et al. 1987b), also contains the gene for an additional protein of molecular mass 6 kDa (p6K) (Barraclough et al. 1988). This 6 kDa protein consists of two well-separated isoelectric focusing variants, visible as separate spots when subjected to two-dimensional polyacrylamide gel electrophoresis (Figure 1); however, it is thought that both these protein spots are the product of a single gene which gives rise to a single mRNA of size 0.8 kb (Barraclough et al. 1988). When the 18-kb genomic fragment containing the genes for p9Ka and p6K is digested with restriction enzyme EcoR1, the genes for p9Ka and p6K are on separate fragments of DNA and do not overlap one another (Barraclough et al. 1987b). At present, the identity of the 6 kDa protein is unknown; however, since related eukaryotic genes are often clustered together at the same location in the chromosomes, it is possible that p6K is also a small calcium-ion-binding protein, either novel or previously described.

EXPRESSION OF p9Ka mRNA IN NORMAL RODENT TISSUES

Experiments in different laboratories have sought to establish the tissue distribution of the mRNA that corresponds to that of p9Ka in rats and mice. These experiments employing cloned cDNA molecules, 18A2 (Jackson-Grusby et al. 1987), mts (Ebralidze et al. 1989) or the cDNA to p9Ka mRNA (Barraclough et al. 1984b), have apparently failed to show a consistent pattern of expression. In the mouse, mRNA corresponding to mts has been detected in spleen, bone marrow, thymus, lymphocytes (Ebralidze et al. 1989), and mRNA corresponding to 18A2 has been detected in uterus, placenta, and kidney with very low levels in the thymus and testes (Jackson-Grusby et al. 1987). In the rat, the mRNA for p9Ka has been detected in the spleen, and low levels have been detected in the mammary gland, and uterus with variable levels being found in liver (Barraclough et al. 1984b). These variable levels of p9Ka reported by different laboratories may be a consequence of its expression in some blood cells which may contaminate normal tissues.

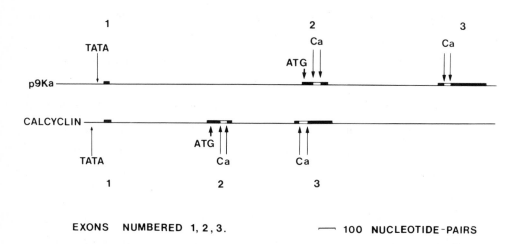

Fig. 4. The arrangement of the genes for p9Ka and calcyclin. The genes for these two calcium-ion-binding proteins are shown diagrammatically drawn to the same scale. The regions corresponding to mRNA (exons) are shown as *black boxes*. Nonexon DNA, including the intervening sequences, is shown as a *line*. The potential calcium-ion-binding regions (*Ca*) are shown as *white boxes* within the exon regions. The positions of the initiating methionine residues are indicated (*ATG*), and the locations of the TATA boxes (*TATA*) are shown. (Data from Barraclough et al. 1987b and Ferrari et al. 1987)

In order to clarify the results with p9Ka with regard to the mammary gland, immunocytochemical experiments are being carried out. Although p9Ka is not a particularly antigenic protein, a mouse antiserum has been raised to the small amounts of p9Ka that are available so far, and to synthetic peptides derived from the p9Ka amino acid sequence. Preliminary immunocytochemical experiments suggest that these antisera bind to the myoepithelial cells and to smooth muscle in histological sections of rat mammary glands (Haynes 1988).

EXPRESSION OF p9Ka AND ITS m*RNA* IN CELL LINES

The increased level of p9Ka in cultured mammary myoepithelial-like cells which occurs during their formation from the parental epithelial cells provides an opportunity to study the regulation of synthesis of this particular small calcium-ion-binding protein under controlled conditions in vitro. The 15-fold increase in the level of p9Ka in the myoepithelial cells over that in the epithelial cells is accompanied by a similar increase in the level of translatable p9Ka mRNA as measured in vitro in the reticulocyte lysate system of cell-free protein synthesis (Barraclough et al. 1982).

In order to find out whether p9Ka mRNA levels are similarly increased, the expression of the mRNA molecules for p9Ka has been compared in the myoepithelial-like cells and in the parental epithelial cells using the technique of Northern hybridization (Barraclough et al. 1984b). p9Ka mRNA is detectable in the epithelial cells, but at a very much reduced level when compared to that in the myoepithelial-like cells (Figure 5A).

Quantification of these levels using "dot-blot" hybridization techniques reveals that there is at least tenfold more p9Ka mRNA in the myoepithelial-like cells than in the epithelial cells (Barraclough et al. 1984b). In contrast, the mRNA for p6K is present at the same level in both the myoepithelial-like and in the parental epithelial cell lines (Figure 5B). Thus the controls which lead to differential synthesis of p9Ka mRNA are probably specific for the p9K gene and do not act on the closely located p6K gene.

SPECIFIC MECHANISMS REGULATE THE EXPRESSION OF THE p9Ka GENE IN CULTURED MYOEPITHELIAL-LIKE CELLS IN VITRO

The conversion of the epithelial cells to the myoepithelial-like cells in culture is associated with changes in the synthesis of a range of markers of myoepithelial cells in vivo, in addition to p9Ka. These markers were originally detected using immunocytochemical staining techniques. Thus, the myoepithelial cells are normally stained by antisera to vimentin, actin, myosin (Warburton et al. 1982; Dulbecco 1982), keratin monoclonal antibody LP34 (Taylor-Papadimitriou et al. 1983; Warburton et al. 1987) and by the lectins Griffonia simplicifolia-1 and pokeweed mitogen (Hughes 1988). The basement membrane which is closely associated with, and possibly synthesized at least in part, by the myoepithelial cells stains with antisera to laminin, type IV collagen and Thy-1 antigen (Warburton et al. 1982; Dulbecco 1982; Monaghan et al. 1983; Rudland et al. 1982). We have examined the synthesis of some of these potential markers of myoepithelial cells in the cultured myoepithelial-like cell lines. The protein markers p9Ka, laminin, type-IV collagen, and Thy-1 antigen are respectively 16-fold, 3.7-fold, 3.5-fold, and 17-fold more abundant in the myoepithelial-like cell line, Rama 29, than in the parental cuboidal, epithelial cells, Rama 25 (Barraclough et al. 1984a; Rudland et al. 1982; Rudland et al. 1986; Warburton et al. 1986). The relative levels of mRNA for the marker proteins in the epithelial and myoepithelial-like cells have been determined by hybridizing poly(A)-containing RNA from the two cell lines to radioactive, cloned cDNAs corresponding to the mRNAs for p9Ka (Barraclough et al. 1984b), laminin, Type IV-collagen[1] and Thy-1 antigen[2]. The relative levels of the mRNA's have been compared to the relative levels of the proteins themselves in the same cell lines (Table 1).

---

[1] Cloned cDNA's corresponding to laminin and type IV collagen were kindly supplied by Dr. Lorraine J. Gudas and Sho-Ya Wang of the Dana Faber Cancer Institute, Boston, MA.

[2] A cloned cDNA corresponding to Thy-1 mRNA was kindly supplied by Drs M.J. Clark, A.N. Barclay and A.F. Williams of the M.R.C. Cellular Immunology Unit, Oxford, U.K.

The 17-fold increase in the amount of Thy-1 in the elongated myoepithelial cells can be completely accounted for by an equivalent increase in Thy-1 mRNA, as measured by hybridization to Thy-1 cDNA (Barraclough et al. 1987a), suggesting that Thy-1 antigen levels are regulated by the amount of Thy-1 mRNA.

The 16-fold increase in p9Ka accumulation in the elongated, myoepithelial-like cells (Barraclough et al. 1984a) is accompanied by a tenfold increase in p9Ka mRNA (Barraclough et al. 1984b), suggesting that in this case translational control may play a minor facilitating role in the increase in p9Ka, but that, like Thy-1 antigen, the level of the protein is largely regulated by the level of its mRNA.

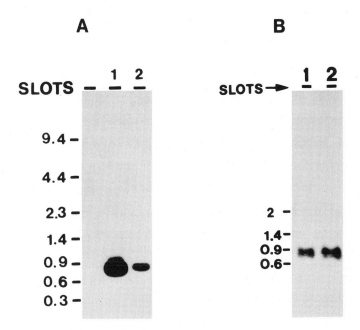

Fig. 5A,B. The expression of mRNA molecules for p9Ka and p6K in epithelial and myoepithelial-like cell lines. Poly(A)-containing RNA was isolated from the myoepithelial-like cell line, Rama 29 (*lane 1*) and from the epithelial cell line, Rama 25 (*lane 2*), and fractionated by agarose gel electrophoresis under denaturing conditions. The RNA was transferred to nitrocellulose filters and incubated with either a 400 bp cloned cDNA corresponding to p9Ka mRNA (**A**), or to a fragment of rat genomic DNA corresponding to the gene for p6K (**B**), which had been radioactively-labeled with ^{32}P. The pattern of radioactive bands is shown after autoradiography. Molecular sizes are shown in kilobases (kb). The *line* across the radioactive bands in **B** arises from the cutting of the filter prior to hybridization

In contrast, in the same cell lines, the 3.7-fold increase in laminin (Rudland et al. 1986) and the 3.5-fold increase in type-IV collagen (Warburton et al. 1986) are accompanied by only a very small increase in their respective mRNAs (Table 1). The increase in type-IV collagen has been shown to arise from a reduced degradation of the protein when the epithelial cells convert to the elongated myoepithelial-like cells (Warburton et al. 1986).The combination of these results strongly suggests that in the same cells, the steady-state levels of different myoepithelial cell marker proteins can arise by an altered control at different molecular levels, including those at transcriptional, post-transcriptional processing or post-translational steps. Thus, each gene studied has its own particular pattern of regulation in these cells.

MOLECULAR CONTROL OF p9Ka PRODUCTION IN MAMMARY CELLS

Changes in the level of cytoplasmic mRNA may arise from altered initiation of transcription, from altered processing of the primary transcripts in the nucleus, or from alterations in the stability of mRNA. The former mechanism of regulation can often be distinguished from the latter two mechanisms by measuring the rate of completion of nascent RNA transcripts by isolated cell nuclei (run-off transcription assays in vitro). However, in our experiments so far, both p9Ka and Thy-1 RNA transcripts can be detected at only a very low level in nuclei isolated from the elongated myoepithelial-like cells and from the epithelial cells. Under the same conditions, RNA transcripts corresponding to the mRNA for actin, and to the mRNA for the protein p6K are produced in abundance, in the nuclei from both cell lines (Barraclough et al. 1987a). These results suggest that the run-off transcription assay in vitro in isolated nuclei is not a true representation of transcription within the myoepithelial-like cell for these proteins studied. One explanation of these results is that putative factors necessary for the transcription of the mRNAs for p9Ka and Thy-1 antigen, but not necessary for the transcription of the mRNAs for actin and p6K, are lost during the isolation of the nuclei from the elongated, myoepithelial-like mammary cells. However, at present, the identity of such factors is unknown.

Table 1. Relative levels of specific marker proteins and their mRNAs in rat mammary cell lines

| Marker protein | Relative level of mRNA | |
	Epithelial cell line (Rama 25)	Myoepithelial-like cell line (Rama 29)
Thy-1 antigen		
Protein ratio[a]	1	17
mRNA ratio[b]	1	16 ± 2
p9Ka		
Protein ratio[c]	1	16
mRNA ratio[d]	1	10 ± 3
Laminin		
Protein ratio[e]	1	4
mRNA ratio[b]	1	1.5 ± 0.2

Poly(A)-containing RNA from cell lines was bound to nitrocellulose filters and the relative levels of specific mRNAs were found by quantitative hybridization to ^{32}P-labeled cloned cDNAs. The relative level of mRNA is the ratio of the level in the myoepithelial-like cells expressed relative to that in the epithelial cells $\pm$ S.D.

[a]Rudland et al. 1982.
[b]Barraclough et al. 1987a.
[c]Barraclough et al. 1984a.
[d]Barraclough et al. 1984b.
[e]Rudland et al. 1986.

The involvement of specific factors in the transcription of eukaryotic genes can be inferred by examining the nucleotide sequences of the DNA, near to the point where the transcription of the mRNA begins, for the presence of particular DNA sequences (sequence motifs). Two such sequence motifs are the so-called TATA (Breathnach and Chambon 1981) and CAAT (Benoist et al. 1980) boxes located, respectively, 30 and 70-90 nucleotides upstream of the start-site of the mRNA. The gene for p9Ka contains the sequence TATAAA 32 nucleotides upstream of the start-site of transcription of p9Ka mRNA (Figure 4), but there is not a clearly defined motif that resembles the CAAT sequence. The absence of a CAAT-like sequence has been reported for a number of eukaryotic genes including the gene for calcyclin (Ferrari et al. 1987). Many genes also contain GC-rich regions which are the binding sites for the transcription factor SP1 (Kadonaga et al. 1986). However, although the potential promoter region of the p9Ka gene (that region adjacent to the start site of mRNA transcription) contains some GC-rich regions, the sequences do not correspond to previously-described SP1 binding sites (Kadonaga et al. 1986). Thus the p9Ka gene may contain a novel promoter that regulates its expression between the epithelial and myoepithelial-like cells in vitro and possibly in vivo. Similarly the Thy-1 gene, which is also expressed in myoepithelial-like, but not in epithelial cells, lacks a CAAT sequence and contains only a single GC rich region (Giguere et al. 1985). A study of these promoter regions of the p9Ka gene to try to establish those regions necessary for the transcription of p9Ka mRNA in the myoepithelial-like cells is underway.

ACKNOWLEDGMENTS

The authors thank Karin Dawson, Rosemary Kimbell, Janet Savin, and Fiona Gibbs for expert technical assistance during the course of the work described, and Julie McGreavey for secretarial assistance. Financial support from the Medical Research Council, the Cancer and Polio Research Fund and the Ludwig Institute for Cancer Research is gratefully acknowledged.

REFERENCES

Barraclough R, Dawson KJ, Rudland PS (1982) Control of protein synthesis in cuboidal rat mammary epithelial cells in culture: changes in gene expression accompanies the formation of elongated cells. Eur J Biochem 129:335-341

Barraclough R, Dawson KJ, Rudland PS (1984a) Elongated cells derived from rat mammary cuboidal epithelial cell lines resemble cultured mesenchymal cells in their pattern of protein synthesis. Biochem Biophys Res Comm 120:351-358

Barraclough R, Kimbell R, Rudland PS (1984b) Increased abundance of a normal cell mRNA sequence accompanies the conversion of rat mammary cuboidal epithelial cells to elongated myoepithelial-like cells in culture. Nucleic Acids Res 12:8097-8114

Barraclough R, Kimbell R, Rudland PS (1987a) Differential control of mRNA levels for Thy-1 antigen and laminin in rat mammary epithelial and myoepithelial-like cells. J Cell Physiol 131:393-401

Barraclough R, Savin J, Dube SK, Rudland PS (1987b) Molecular cloning and sequence of the gene for p9Ka, a cultured myoepithelial cell protein with strong homology to S-100, a calcium-binding protein. J Mol Biol 198:13-20

Barraclough R, Kimbell R, Rudland PS (1988) The identification of a normal rat gene located close to the gene for the potential myoepithelial cell calcium binding protein, p9Ka. J Biol Chem 263:14597-14600

Baudier J, Glasser N, Gerard D (1986) Ions binding to S-100 proteins 1, calcium and zinc-binding properties of bovine brain S-100$\alpha\alpha$, S-100a ($\alpha\beta$), and S-100b ($\beta\beta$) protein: Zn^{2+} regulates Ca^{2+} binding on S-100b protein binding on S-100b protein. J Biol Chem 261:8192-8203

Bennett DC, Peachey LA, Durbin H, Rudland PS (1978) A possible mammary stem cell line. Cell 15:283-298

Benoist C, O'Hare K, Breathnach R, Chambon P (1980) The ovalbumin gene-sequence of putative control regions. Nucleic Acids Res 8:127-142

Breathnach R, Chambon P (1981) Organization and expression of eucaryotic split genes coding for proteins. Annual Rev Biochem 50:349-383

Burmeister GL, Tarcsay L, Sarg C (1986) Generation and characterization of a monoclonal antibody (IC5) to human migration inhibitor factor (MIF). Immunobiology 171:461-474

Calabretta B, Battini R, Kaczmarek L, deRiel JK, Baserga R (1986) Molecular cloning of the cDNA for a growth factor-inducible gene with strong homology to S-100, a calcium-binding protein. J Biol Chem 261:12628-12632

Crouch TH, Klee CL (1980) Positive cooperative binding of calcium to bovine brain calmodulin. Biochemistry 19:3692-3698

Dayhoff MO (1978) Atlas of protein sequence and structure, Vol 5 suppl 3. National Biomedical Research Foundation, Silver Spring MD

Dexter DL, Kowalski HM, Blazer BA, Fligiel S, Vogel R, Heppner GH (1978) Heterogeneity of tumor cells from a single mouse mammary tumor. Cancer Res 38:3174-3181

Dulbecco R (1982) Immunological markers in the study of development and oncogenesis in the rat mammary gland. J Cell Physiol Suppl 2:19-22

Ebralidze A, Tulchinsky E, Grigorian M, Afanayeva A, Senin V, Revazova E, Lukanidin E (1989) Isolation and characterization of a gene specifically expressed in different metastatic cells and whose deduced gene product has a high degree of homology to a Ca^{2+}-binding protein family. Genes and development 3:1086-1093

Erikson E, Tomasiewicz HG, Erikson RL (1984) Biochemical characterization of a 34-kilodalton normal cellular substrate of pp60 v-src and an associated 6-kilodalton protein. Mol Cell Biol 4:77-85

Ferrari S, Calabretta B, deRiel JK, Battini R, Ghezzo F, Lauret E, Griffin C, Emanuel BS, Gurrieri F, Baserga R (1987) Structural and functional analysis of a growth-regulated gene, the human calyclin. J Biol Chem 262:8325-8332

Fullmer CS, Wasserman RH (1981) The amino acid sequence of bovine intestinal calcium-binding protein. J Biol Chem 256:5669-5674

Gerke V, Weber K (1985) The regulatory chain in the p36-Kd substrate complex of viral tyrosine-specific protein kinases is related in sequence to the S-100 protein of glial cells. EMBO J 4:2917-2920

Giguere V, Isobe K-I, Grosveld F (1985) Structure of the murine Thy-1 gene. EMBO J 4:2017-2024

Goto K, Endo H, Fujiyoshi T (1988) Cloning of the sequences expressed abundantly in established cell lines: identification of a cDNA clone highly homologous to S-100, a calcium binding protein. J Biochem 103:48-53

Hager JC, Fligiel S, Stanley W, Richards AM, Heppner GH (1981) Characterization of a variant producing tumor cell line from a heterogenous strain Balb/cf C3H mouse mammary tumor. Cancer Res 42:1293-1300

Haynes GA (1988) Studies on a possible myoepithelial cell marker protein. Ph.D. Dissertation, University of London, England

Hirschhorn RR, Aller P, Yuan Z, Gibson CW, Baserga R (1984) Cell-cycle-specific cDNAs from mammalian cells temperature sensitive for growth. Proc Natl Acad Sci (USA) 81:6004-6008

Hughes CM (1988) Lectin staining of the rat mammary gland. M. Phil. Dissertation, University of London, England

Hummel JP, Dreyer WJ (1962) Measurement of protein binding phenomena by gel filtration. Biochim Biophys Acta 63:530-532

Isobe T, Okuyama T (1978) The amino acid sequence of S-100 protein (PAP 1-b protein) and its relation to the calcium-binding proteins. Eur J Biochem 89:379-388

Isobe T, Okuyama T (1981) The amino acid sequence of the α-subunit of bovine brain S-100a protein. Eur J Biochem 116:79-86

Jackson-Grusby LL, Swiergiel J, Linzer DIH (1987) A growth-related mRNA in cultured mouse cells encodes a placental calcium binding protein. Nucleic Acids Res 15:6677-6690

Kadonaga JT, Jones KA, Tjian R (1986) Promoter-specific activation of RNA polymerase II transcription by Spl. Trends in Biochemical Science 11:20-23

Kligman D, Hilt DC (1988) The S-100 protein family. Trends in Biochemical Science 13:437-443

Kuwano R, Usui H, Maeda T, Fukui T, Yamanari N, Ohlsuka E, Ikehara M, Takahashi Y (1984) Molecular cloning and the complete nucleotide sequence of cDNA to mRNA for S-100 protein of rat brain. Nucl Acids Res 12:7455-7465

Kuznicki J, Filipek A (1987) Purification and properties of a novel Ca^{2+}-binding protein (10.5kDa) from Ehrlich-ascites-tumor cells. Biochem J 247:663-667

Lagasse E, Clerc RG (1988) Cloning and expression of two human genes encoding calcium-binding proteins that are regulated during myeloid differentiation. Mol Cell Biol 8:2403-2410

Maruyama K, Mikawa T, Ebashi S (1984) Detection of calcium binding proteins by ^{45}Ca autoradiography on nitrocellulose membrane after sodium dodecyl sulphate gel electrophoresis. J Biochem (Tokyo) 95:511-519

Masiakowski P, Shooter EM (1988) Nerve growth factor induces the genes for two proteins related to a family of calcium-binding proteins in PC12 cells. Proc Natl Acad Sci (USA) 85:1277-1281

Moews PC, Kretsinger RH (1975) Refinement of the structure of carp muscle calcium-binding parvalbumin by model building and difference Fourier analysis. J Mol Biol 91:201-228

Monaghan P, Warburton MJ, Perusinghe N, Rudland PS (1983) Topographical arrangement of basement membrane proteins in lactating rat mammary gland: comparison of the distribution of type IV collagen, laminin, fibronectin and Thy-1 at the ultrastructural level. Proc Natl Acad Sci (USA) 80:3344-3348

Mooibroek MJ, Michiel DF, Wang JH (1987) Clathrin light chains are calcium binding proteins. J Biol Chem 262 :25-28

Moore BW, McGregor D (1965) Chromatographic and electrophoretic fractionation of soluble proteins of brain and liver. J Biol Chem 240:1647-1653

Mount SM (1982) A catalogue of splice junction sequences. Nucleic Acids Res 10:459-472

Odink K, Cerletti N, Brüggen J, Clerc RG, Tarcsay L, Zwadlo G, Gerhards G, Schlegel R, Sorg C (1987) Two calcium-binding proteins in infiltrate macrophages of rheumatoid arthritis. Nature (London) 330:80-82

Ormerod EJ, Rudland PS (1982) Mammary gland morphogenesis in vitro: formation of branched tubules in collagen gels by a cloned rat mammary cell line. Dev Biol 91:360-375

Ormerod EJ, Rudland PS (1985) Isolation and differentiation of cloned epithelial cell lines from normal rat mammary glands. In Vitro 21:143-153

Pechere J-F (1977) In: Wasserman RH, Corradino R, Carafoli E, Kretsinger RH, MacLennan D, Siegel F (eds) Calcium binding proteins and calcium function. Elsevier, Amsterdam, p 213

Rudland PS, Warburton MJ, Monaghan P, Ritter MA (1982) Thy-1 antigen on normal and neoplastic rat mammary tissue: changes in location and amount of antigen during differentiation of cultured stem cells. JNCI 68:799-811

Rudland PS, Paterson FC, Monaghan P, Twiston-Davies AC, Warburton MJ (1986) Isolation and properties of rat cell lines morphologically intermediate between cultured mammary epithelial and myoepithelial-like cells. Dev Biol 113:388-405

Sanford KK, Dunn TB, Westfall BB, Covalesky AB, Dupre LT, Earle WR (1961) Sarcomatous change and maintenance of differentiation in long-term cultures of mouse mammary carcinoma. J Natl Cancer Inst 26:1139-1161

Shadle PJ, Gerke V, Weber K (1985) Three Ca^{2+}-binding proteins from porcine liver and intestine differ immunologically and physicochemically and are distinct in Ca^{2+} affinities. J Biol Chem 260:16354-16360

Szebenyi DME, Moffat K (1986) The refined structure of vitamin D-dependent calcium-binding protein from bovine intestine. J Biol Chem 261:8761-8777

Szebenyi DME, Obendorf SK, Moffat K (1981) Structure of vitamin D-dependent calcium-binding protein from bovine intestine. Nature (London) 294:327-332

Taylor-Papadimitriou J, Lane EB, Chang SE (1983) Cell lineages and interactions in neoplastic expression in the human breast. In: Rich MA, Hager JC and Furmanski P (eds) Understanding breast cancer, clinical and laboratory concepts. Marcel Dekker Inc, New York pp 215-246

van Eldik LJ, Zendegui JG, Marshak R, Watterson DM (1982) Calcium-binding proteins and the molecular basis of calcium action. Int Rev Cytol 77:1-61

Warburton MJ, Ferns SA, Hughes CM, Sear CHJ, Rudland PS (1987) Generation of cell types with myoepithelial and mesenchymal phenotypes during the conversion of rat mammary tumor epithelial stem cells into elongated cells. JNCI 78:1191-1201

Warburton MJ, Ferns SA, Hughes CM, Rudland PS (1985) Characterization of rat mammary cell types in primary culture: lectins and antisera to basement membrane and intermediate filament proteins as indicators of cellular heterogeneity. J Cell Sci 79:287-304.

Warburton MJ, Mitchell D, Ormerod EJ, Rudland PS (1982) Distribution of myoepithelial cells and basement membrane proteins in the resting, pregnant, lactating and involuting mammary gland.
J Histochem Cytochem 30:667-676

Warburton MJ, Head L, Ferns SA, Rudland PS (1982) Enhanced synthesis of basement membrane proteins during the differentiation of rat mammary tumor epithelial cells into myoepithelial-like cells in vitro. Exp Cell Res 137:373-380

Warburton MJ, Kimbell R, Rudland PS, Ferns SA, Barraclough R (1986) Control of type IV collagen production in rat mammary epithelial and myoepithelial-like cells. J Cell Physiol 128:76-84

S-100-RELATED PROTEINS IN NERVE GROWTH FACTOR-INDUCED DIFFERENTIATION OF *PC*12 CELLS

Piotr Masiakowski and Eric M. Shooter

INTRODUCTION

In the development of the nervous system, more neurons are generated than are eventually found in the adult organism. The mechanisms which determine the selective neuronal survival, the acquisition and maintenance of the differentiated phenotype, and the potential for regeneration are presently under intensive study. The pioneering work on nerve growth factor (NGF) and, more recently, also other polypeptide factors, has established the key role of target-derived neurotrophic substances in these processes (for recent review, see Snider and Johnson 1989). Much effort is now focused on understanding the molecular mechanisms triggered by the interaction of neurotrophic factors with their neuronal receptors.

A convenient and widely used model system to study these mechanisms in culture is provided by the rat adrenal pheochromocytoma-derived cell line PC12 (Greene and Tischler 1976). These cells, which show characteristics of adrenal chromaffin cells, respond to NGF by acquisition of the properties of sympathetic neurons, including neurite outgrowth. Several laboratories have attempted to characterize cellular pathways mediating the action of NGF by studying genes whose expression is affected in this system by the factor (for example, Greenberg et al. 1985; Leonard et al. 1987; Thompson and Ziff 1989). We have used the differential hybridization approach to identify mRNAs whose levels are increased in PC12 cells upon treatment with NGF (Masiakowski and Shooter 1988). The finding that the deduced protein products of these mRNAs are members of the S-100-related family leads us to consider the possible roles of these calcium-binding proteins in NGF-induced differentiation of PC12 cells and, more generally, in neuronal development.

CLONING OF 42A AND 42C cDNAS CODING FOR MEMBERS OF S-100 PROTEIN-RELATED FAMILY

To identify NGF-regulated mRNAs, we screened a cDNA library prepared from PC12 cells with radioactive cDNA probes derived from PC12 cells maintained for 7 days either in the absence or presence of NGF. Two of the cDNA clones hybridizing more strongly with the probe from NGF-induced cells, 42A and 42C, were detected in the whole 10000 clone library in two and four copies, respectively. These sequences revealed on Northern blots 700-800 base-pair long RNA species. Both mRNAs reached their maximal levels, five- to tenfold higher than control, by 24 h of treatment with NGF and remained at high levels for several days. The increase in 42C RNA was evident already after 2 h and was close to maximal at 7 h, while the level of 42A RNA showed only a slight increase at 7 h. Epidermal growth factor (EGF) also increased the levels of both RNAs at the 2 h time point, but the magnitude of this effect was much lower than with NGF, and by the time the NGF induction reached its maximum, the EGF-induced response of the two RNAs had returned to baseline levels.

Surprisingly, DNA sequencing revealed that these two RNAs are related, with 47% identity at the nucleic acid level, and 31% identity between the deduced proteins. In addition, 42A and 42C share striking homology with S-100 (Moore 1965; Isobe and Okuyama 1981) and a group of related proteins (references in Masiakowski and Shooter 1988; see also below). This group consists of seven distinct known members, including the recently discovered S-100L (Glenney et al. 1989). All these proteins are about 100 amino acids long, with identical residues in ten positions and similar features in their predicted secondary structure. In addition, two other proteins show strong preservation of the core region with two Ca^{2+}-binding domains, but differ in the C-terminal portion. Thus, intestinal calcium-binding protein is shorter than S-100 (Hofmann et al. 1979; Fullmer and Wasserman 1981), lacking the common hydrophobic segment which may be involved in interactions with membranes. The myeloid protein MRP-14, on the other hand, has a longer C-terminal sequence (Odink et al. 1987), which has been recently identified as a target for Ca^{2+}-dependent phosphorylation (Edgeworth et al. 1989).

MORPHOLOGICAL CHANGES IN *PC*12 CELLS OVEREXPRESSING 42C m*RNA*

If 42A and 42C proteins play a role in NGF-induced differentiation of PC12 cells, then overexpression of their mRNAs might mimic some effects of the factor. We have obtained preliminary evidence that this may indeed be the case by transfecting PC12 cells with 42A and 42C DNA sequences cloned into a pHßAPr-1-neo eukaryotic expression vector (Gunning et al. 1987). Following the introduction of the recombinant plasmids into PC12 cells by lipofection (Felgner et al. 1987), the cells which stably integrated the vector into their genomes were selected by their resistance to the drug G-418. Of about 100 such colonies, which were transferred to microtiter plates and cultured, only one showed significantly altered morphology, with flat cells extending processes in the absence of any added NGF. This clone, designated P42C, originated from the pool of cells transfected with the vector bearing the 42C sequence in the "sense" orientation (Masiakowski and Shooter 1990).

The original line P42C appeared to be morphologically heterogeneous. The cells were replated at low density and the resulting colonies were again picked and cultured. These lines, designated P42C.1 through P42C.17, showed some variation in morphology, but in general cells were flat, extending processes while growing at low densities (Figure 1). The processes disappeared as cells reached confluence. This phenotype was preserved during culture for up to 13 passages.

The analysis of DNA and RNA samples from these cells has linked the altered morphology to the amplification of the 42C gene, resulting in the constitutive presence of high level of its mRNA. When the DNA from PC12 cells was digested with Pst I and analyzed by Southern blotting, 42C probe detected two strong bands of 6.9 and 1.6 kb, and a weak band at 5.4 kb. These bands were also detected in P42C DNAs, but in addition a very strong band at 2.3 kb, and weak bands at 2.8, 2.7, and 1.8 kb hybridized with 42C probe (Figure 2A). The relative intensity of the new bands was remarkably similar among all P42C DNAs analyzed. The average ratio of the intensity of the 2.3 kb band (measured by densitometry) to the intensity of the 6.9 kb band was 16, and to the intensity of the 1.6 kb band was 42. Thus, it appears that the genomes of P42C lines contain about 20 copies of stably integrated 42C sequence.

Northern blot analysis revealed that this 42C gene amplification results in high levels of 42C mRNA. RNA samples were separated on denaturing agarose gels and transferred to nylon filters, which were probed with radioactively labeled 42A, 42C and 4A3 sequences (Figure 2B). 4A3 probe, corresponding to an unidentified mRNA not affected by NGF in PC12 cells, detected a band whose intensity remained constant in PC12 and P42C lines, either in absence or presence of NGF. Hybridization with 42C probe shows that 42C mRNA is induced by NGF in PC12 cells, and in P42C lines is present at high levels which are not further affected by the factor. The size of 42C RNA band in P42C lines is the same as in parental PC12 cells, about 800 base pairs, which is also the size expected for the RNA product of the transfected 42C gene. 42A probe reveals a RNA species induced by NGF in PC12 cells, but not in P42C lines. While the relative amount of 42C was always at least fourfold higher than 42A, the levels of 42A and 42C mRNAs varied considerably in RNA samples from P42C lines. The concentration of 42C RNA varied from similar to the one in NGF-stimulated PC12 cells up to about fourfold higher, and the concentration of 42A RNA varied from low amounts seen in uninduced PC12 cells up to the amounts comparable to those found in PC12 cells grown in the presence of NGF.

P42C cells resemble NGF-treated PC12 cells in the outgrowth of processes and their lack of further response to NGF. On the other hand, the short, thick processes of P42C cells are considerably different from the long, fine neurites of PC12 cells subjected to long-term treatment with NGF; in addition, P42C cells retain the ability to proliferate. This is consistent with the apparent role of 42C as a "late" gene in NGF response, possibly induced through the action of the products of some "early" genes: many elements of the cellular machinery required for the NGF response may be already present in the naive PC12 cells and thus allow certain aspects of the differentiation to occur, while others need to be induced to cooperate in the production of the fully differentiated phenotype, and thus are bypassed in experiments such as the one reported above. While morphological differentiation of PC12 has been induced by microinjection of the H-ras oncogene protein (Bar-Sagi and Feramisco 1985), a product of a gene acting early in the growth factor response, this work represents the first attempt to partially mimic this effect by introduction of a "late" gene into the PC12 cells.

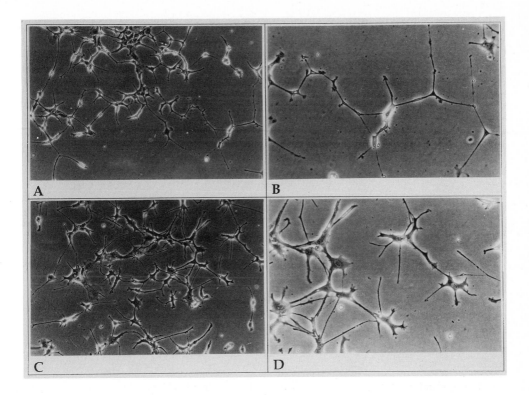

Fig. 1A-D. Altered morphology of PC12 cells transfected with 42C DNA and grown in the absence of NGF. **A** and **B** P42C.2 cells, passage 13; **C** and **D** P42C.17 cells, passage 17. **A** and **C** x100; **B** and **D** x200 (Masiakowski and Shooter 1990)

The substantiation of the above conclusions will require further work, but the results reported here suggest the possibility of dissecting the complex process of cellular differentiation into individual molecular interactions.

POTENTIAL ROLE FOR CALCIUM SIGNALLING IN *NGF*-INDUCED DIFFERENTIATION OF *PC*12 CELLS

The possibility that NGF may increase levels of calcium-binding proteins, which in turn would affect cellular processes, is interesting in view of the potential role of calcium in differentiation of PC12 cells. Initially, Schubert et al. (1978) proposed that calcium mobilization, induced by NGF via cyclic AMP and followed by efflux of the calcium from PC12 cells, produces membrane changes which enhance cell-substratum adhesion and neurite extension.

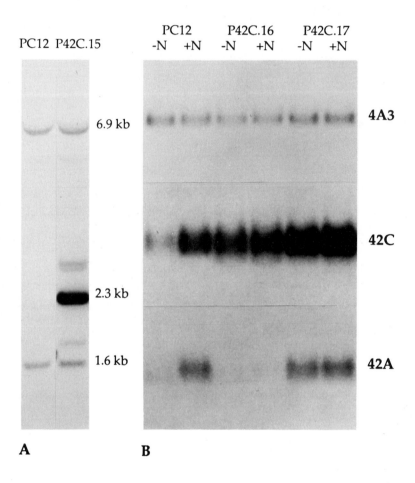

Fig. 2A,B. Amplification of 42C sequences in P42C genome is accompanied by increased concentration of cytoplasmic 42C RNA. **A** 20 μg genomic DNA samples from PC12 and P42C.15 cells were digested with Pst I, fractionated on 1% agarose gel and transferred to nylon filter. 42C sequences were visualized by hybridization with ^{32}P-labeled DNA probe, followed by autoradiography. **B** PC12, P42C.16 and P42C.17 cells were maintained for 25 h in the absence (-N) or presence (+N) of 100 ng/ml NGF. Northern blots of RNA aliquots (20 μg) from these cells were hybridized with ^{32}P-labeled 4A3, 42C, and 42A DNA probes. (Masiakowski and Shooter 1990)

However, other investigators were not able to reproduce the reported effects of NGF on cyclic AMP accumulation (Hatanaka et al. 1978) or on Ca^{2+} fluxes over a 10 to 20 min period (Landreth et al. 1980). More recently, the use of sensitive fluorescence techniques made possible the detection of a rise of cytosolic Ca^{2+} induced in PC12 cells by NGF, but not by EGF (Pandiella-Alonso et al. 1986). After a lag phase of 30 to 40 s cytosolic calcium levels increase slowly by calcium influx to 50-75% above control levels and remain elevated for at least 10 min, suggesting a possible linkage to NGF receptor occupation. Calcium has also been implicated in specific actions of NGF. Thus, NGF appears to control the phosphorylation of Nsp100 protein by altering the intracellular economy of calcium (Hashimoto et al. 1986). Influx of Ca^{2+} ions caused by K^+ depolarization, which can stimulate neurite extension in PC12 cells (Schubert et al. 1978), also mimics the action of NGF in inducing expression of proto-oncogene *c-fos* (Kruijer et al. 1985; Greenberg et al. 1985; Morgan and Curran 1986). Similarly, stimulation of nicotinic acetylcholine receptor (Greenberg et al. 1986) induces *c-fos* in an extracellular Ca^{2+}-dependent fashion. However, induction of *c-fos* by NGF is not dependent on extracellular Ca^{2+} ions (Morgan and Curran 1986; Greenberg et al. 1986).

Binding of calcium by 42A protein is suggested by the excellent conservation of the regions involved in coordinating Ca^{2+} ions in calcium-binding protein from bovine intestine and S-100. The three-dimensional structure of intestinal calcium-binding protein (Szebenyi et al. 1981) contains two calcium-binding domains, each consisting of a loop flanked by two helices. The calcium-coordinating residues in the III-IV loop are identical to those found at corresponding positions in 42A; S-100α and ß have one aspartate for asparagine substitution. The Ca^{2+}-coordinating residues in the I-II loop of intestinal calcium-binding protein show less similarity to the corresponding residues of S-100. These residues, believed to form a low affinity Ca^{2+}-binding site in S-100, are identical in S-100 and 42A.

42C peptide shows less homology to S-100 proteins and intestinal calcium-binding protein at the high affinity site, and has a three amino acid deletion at the low affinity site. In addition, 42C appears to be the rat version of p11 (light chain of calpactin I), and direct testing failed to provide evidence for Ca^{2+} binding to porcine (Gerke and Weber 1985b) and bovine (Glenney 1986) p11 protein. However, 42C is likely to be involved in intracellular calcium signalling indirectly, through its interactions with other proteins.

POSSIBLE MECHANISMS OF ACTION OF S-100 PROTEINS IN *PC*12 CELLS

The large published work concerning S-100-related proteins has provided an insight into the potentially important developmental role of these proteins, but does not allow simple generalizations. For example, the role for p11 (42C) has been discussed in the context of its possible regulation of calpactin I. Calpactin I is found in cells as either a p36 monomer, or a tetramer composed of two p36 and two p11 subunits (Gerke and Weber 1984, 1985a). The tetramer is localized under the membrane in a cytoskeletal meshwork, while the p36 monomer is also present in a soluble form (Zokas and Glenney 1987). The p11 binding site has been localized to the N-terminal tail of p36, which is also the site for tyrosine and serine phosphorylation (Glenney and Tack 1985; Glenney 1986; Glenney et al. 1986; Johnsson et al. 1986, 1988). The core region of calpactin interacts with phospholipid, actin, nonerythroid spectrin, and Ca^{2+} ions (Gerke and Weber 1984; Glenney 1985, 1986). The interactions with various ligands are interdependent. Clearly, a structural role for p11-p36 complex in the membrane skeleton could be compatible with the morphological changes induced by overexpression of 42C, described above. Of particular interest for our work is the suggestion that the amount of light chain, which stabilizes the heavy chain of calpactin I, is limiting in cells; increasing the amount of the light chain increases the steady state level of the heavy chain, which correlates with spreading behavior of cells (Zokas and Glenney 1987). Also relevant may be the recent report that calpactin I is essential for a calcium-triggered catecholamine exocytosis from bovine adrenal chromaffin cells (Ali et al. 1989).

While the involvement of calpactin I in the differentiation of PC12 remains to be established, the work on S-100 proteins provides further examples of potential regulatory interactions with cellular components. S100 protein interacts with brain-specific form of fructose-1,6-bisphosphate aldolase, stimulating in vitro the enzymatic activity (Zimmer and Van Eldik 1986), and with tau proteins (Baudier and Cole 1988). For the latter case, it was shown that binding of S-100ß to tau protein inhibits mode I phosphorylation of tau by the Ca^{2+}/calmodulin-dependent protein kinase II. This phosphorylation, apparently involved in Alzheimer's disease, interferes with promotion of microtubule formation by tau. Since microtubules are essential in both the extension and stabilization of neuronal processes, inhibition of mode I phosphorylation of tau by S-100 proteins could play a role in the regulation of NGF-induced neurite outgrowth. On the other hand, S-100 proteins have been reported to inhibit *in vitro* the assembly, and promote the disassembly, of brain microtubules in a Ca^{2+}-dependent manner (Donato 1986). Finally, the evidence is accumulating that S-100ß protein is released from the glial cells during development and may act as a neurite extension factor (Kligman and Marshak 1985; Winningham-Major et al. 1989), possibly through the interaction with a cell surface receptor.

Another controversy concerns the precise developmental role of S-100 proteins, whose induction in various systems appears to correlate either with cellular differentiation or proliferation. At least four groups have independently studied the same gene as our 42A in other systems: p9Ka which is more abundant in elongated myoepithelial-like cells than in parental cuboidal epithelial stem cells (Barraclough et al. 1987) shares with 42A the induction during the morphological transition of proliferating compact cells into elongated differentiated cells. On the other hand, 18A2 mRNA is induced after serum stimulation of quiescent mouse fibroblasts (Jackson-Grusby et al. 1987). Similarly, the two other 42A equivalents, pEL98 and mts1, are preferentially expressed under conditions associated with enhanced cellular growth: pEL98 is more abundant in established cell lines than in the corresponding parental primary cultures, and is further elevated in transformed cells (Goto et al. 1988). Expression of mts1 strongly correlates with the metastatic potential of tumor cells (Ebralidze et al. 1989).

Interestingly, another member of the S-100 family shows the same dual control. Calcyclin (clone 2A9; Calabretta et al. 1986) has been discovered as an mRNA induced when quiescent fibroblasts are stimulated to proliferate by serum, platelet-derived growth factor and EGF, and in human acute myeloid leukemia. Recently, clone 63 whose sequence is induced by NGF in PC12 cells with similar time course and magnitude as 42A and 42C (Leonard et al. 1987) was identified as the rat version of calcyclin (Thompson and Ziff 1989).

S-100 PROTEIN FAMILY AS A MODEL TO STUDY THE ROLE OF CALCIUM IN COORDINATION OF DEVELOPMENTAL PROCESSES

In view of these diverse, sometimes apparently contradictory results, it is not easy to define a consistent model for the action of S-100-like proteins in neuronal development. From a strictly deterministic point of view, generation of hierarchies of specific patterns in eukaryotic development may seem to require the existence of cascades of very specific signals. From such a point of view only some of the discussed interactions might be relevant to the developmental process, while others might be considered as irrelevant "side effects". On the other hand, one may argue that each developmental transition is an outcome of a tightly linked network of all cellular interactions, where a particular interaction may have a different end effect, depending on the cellular context. Cellular microenvironment, including Ca^{2+} fluxes, may play an important role in providing such context and linking individual interactions.

In accordance with such view, Kater et al. (1988) postulated that Ca^{2+} may act as a common integrator of environmental cues that influence neurite outgrowth and synaptogenesis, by controlling the growth cone behaviors. These authors pointed out that growth cone motility and neurite elongation require a narrow, optimum range of Ca^{2+} acting on cytoskeletal and vesicular systems; these processes are inhibited by Ca^{2+} concentrations below or above that range. Similarly, the trophic factor dependence of sympathetic neurons *in vitro* may be determined by internal Ca^{2+}. According to the "Ca^{2+} set-point hypothesis" (Koike et al. 1989), intracellular free Ca^{2+} levels are low in immature neurons which are acutely dependent on trophic factors, and raise in more mature neurons to an intermediate optimum level at which neuronal survival is independent of trophic factors; at higher levels Ca^{2+} is toxic. Within the framework of such models, S-100-related proteins, interacting with various cellular components, may act as mediators in the cellular pathways through which calcium affects the neurite outgrowth, trophic factor dependence, and other aspects of neuronal development. In addition, their differential expression might contribute to the determination of the Ca^{2+} buffering "personalities" (McBurney and Neering 1987) of different neuronal types. Judging from the progress made in the recent years, the work on S-100-related proteins will bring in the near future better understanding of the physiological roles played by the individual members of the S-100 family, and may also expand our knowledge of the molecular mechanisms coordinating developmental processes in eukaryotic cells.

REFERENCES

Ali SM, Geisow MJ, Burgoyne RD (1989) A role for calpactin in calcium-dependent exocytosis in adrenal chromaffin cells. Nature 340:313-315

Barraclough R, Savin J, Dube SK, Rudland PS (1987) Molecular cloning and sequence of the gene for p9Ka, a cultured myoepithelial cell protein with strong homology to S-100, a calcium-binding protein. J Mol Biol 198:13-20

Bar-Sagi D, Feramisco JR (1985) Microinjection of the *ras* oncogene protein into PC12 cells induces morphological differentiation. Cell 42:841-848

Baudier J, Cole RD (1988) Interactions between the microtubule-associated τ proteins and S100b regulate τ phosphorylation by the Ca^{2+}/calmodulin-dependent protein kinase II. J Biol Chem 263:5876-5883

Donato R (1986) S-100 proteins. Cell Calcium 7:123-145

Ebralidze A, Tulchinsky E, Grigorian M, Afanasyeva A, Senin V, Revazova E, Lukanidin E (1989) Isolation and characterization of a gene specifically expressed in different metastatic cells and whose deduced gene product has a high degree of homology to a Ca^{2+}-binding protein family. Genes Dev 3:1086-1093

Edgeworth J, Freemont P, Hogg N (1989) Ionomycin-regulated phosphorylation of the myeloid calcium-binding protein p14. Nature 342:189-192

Felgner PL, Gadek TR, Holm M, Roman R, Chan HW, Wenz M, Northrop JP, Ringold GM, Danielsen M (1987) Lipofection: a highly efficient, lipid-mediated DNA-transfection procedure. Proc Natl Acad Sci USA 84:7413-7417

Fullmer CS, Wasserman RH (1981) The amino acid sequence of bovine intestinal calcium-binding protein. J Biol Chem 256:5669-5674

Gerke V, Weber K (1984) Identity of p36K phosphorylated upon Rous sarcoma virus transformation with a protein purified from brush borders: calcium-dependent binding to non-erythroid spectrin and F-actin. EMBO J 3:227-233

Gerke V, Weber K (1985a) Calcium-dependent conformational changes in the 36-kDa subunit of intestinal protein I related to the cellular 36-kDa target of Rous sarcoma virus tyrosine kinase. J Biol Chem 260:1688-1695

Gerke V, Weber K (1985b) The regulatory chain in the p36-kd substrate complex of viral tyrosine-specific protein kinases is related in sequence to the S-100 protein of glial cells. EMBO J 4:2917-2920

Glenney JR (1985) Phosphorylation of p36 in vitro with pp60^{src}: regulation by Ca^{2+} and phospholipid. FEBS Lett 192:79-82

Glenney J (1986) Phospholipid-dependent Ca^{2+} binding by the 36-kDa tyrosine kinase substrate (calpactin) and its 33-kDa core. J Biol Chem 261:7247-7252

Glenney JR, Tack BF (1985) Amino-terminal sequence of p36 and associated p10: identification of the site of tyrosine phosphorylation and homology with S-100. Proc Natl Acad Sci USA 82:7884-7888

Glenney JR, Boudreau M, Galyean R, Hunter T, Tack B (1986) Association of the S-100-related calpactin I light chain with the NH_2-terminal tail of the 36-kDa heavy chain. J Biol Chem 261:10485-10488

Glenney JR, Kindy MS, Zokas L (1989) Isolation of a new member of the S100 protein family: amino acid sequence, tissue, and subcellular distribution. J Cell Biol 108:569-578

Goto K, Endo H, Fujiyoshi T (1988) Cloning of the sequences expressed abundantly in established cell lines: Identification of a cDNA clone highly homologous to S-100, a calcium binding protein. J Biochem 103:48-53

136

Greenberg ME, Greene LA, Ziff EB (1985) Nerve growth factor and epidermal growth factor induce rapid transient changes in proto-oncogene transcription in PC12 cells. J Biol Chem 260:14101-14110

Greenberg ME, Ziff EB, Greene LA (1986) Stimulation of neuronal acetylcholine receptors induces rapid gene transcription. Science 234:80-83

Greene LA, Tischler AS (1976) Establishment of a noradrenergic clonal line of rat adrenal pheochromocytoma cells which respond to nerve growth factor. Proc Natl Acad Sci USA 73:2424-2428

Gunning P, Leavitt J, Muscat G, Ng S–Y, Kedes, L (1987) A human b-actin expression vector system directs high-level accumulation of antisense transcripts. Proc Natl Acad Sci USA 84:4831-4835

Hashimoto S, Iwasaki C, Kuzuya H, Guroff G (1986) Regulation of nerve growth factor action on Nsp100 phosphorylation on PC12h cells by calcium. J Neurochem 46:1599-1604

Hatanaka H, Otten U, Thoenen H (1978) Nerve growth factor mediated selective induction of ornithine decarboxylase in rat pheochromocytoma: a cyclic AMP-independent process. FEBS Lett 92:313-316

Hofmann T, Kawakami M, Hitchman AJW, Harrison JE, Dorrington KJ (1979) The amino acid sequence of porcine intestinal calcium-binding protein. Can J Biochem 57:737-748

Isobe T, Okuyama T (1981) The amino-acid sequence of the a subunit in bovine brain S-100a protein. Eur J Biochem 116:79-86

Jackson-Grusby LL, Swiergiel J, Linzer DIH (1987) A growth-related mRNA in cultured mouse cells encodes a placental calcium binding protein. Nucleic Acids Res 15:6677-6690

Johnsson N, Van PN, Söling HD, Weber K (1986) Functionally distinct serine phosphorylation sites of p36, the cellular substrate of retroviral protein kinase: differential inhibition of reassociation with p11. EMBO J 5:3455-3460

Johnsson N, Marriott G, Weber K (1988) p36, the major cytoplasmic substrate of src tyrosine protein kinase, binds to its p11 regulatory subunit via a short amino-terminal amphiphatic helix. EMBO J 7:2435-2442

Kater SB, Mattson MP, Cohan C, Connor J (1988) Calcium regulation of the neuronal growth cone. Trends Neurosci 11:315-321

Kligman D, Marshak DR (1985) Purification and characterization of a neurite extension factor from bovine brain. Proc Natl Acad Sci USA 82:7136-7139

Koike T, Martin DP, Johnson EM (1989) Role of Ca²⁺ channels in the ability of membrane depolarization to prevent neuronal death induced by the trophic-factor deprivation: evidence that levels of internal Ca²⁺ determine nerve growth factor dependence of sympathetic ganglion cells. Proc Natl Acad Sci USA 86:6421-6425

Kruijer W, Schubert D, Verma IM (1985) Induction of the proto-oncogene fos by nerve growth factor. Proc Natl Acad Sci USA 82:7330-7334

Landreth G, Cohen P, Shooter EM (1980) Ca²⁺ transmembrane fluxes and nerve growth factor action on a clonal cell line of rat pheochromocytoma. Nature 283:202-204

Leonard DG, Ziff EB, Greene LA (1987) Identification and characterization of mRNAs regulated by nerve growth factor in PC12 cells. Mol Cell Biol 7:3156-3167

Masiakowski P, Shooter EM (1988) Nerve growth factor induces the genes for two proteins related to a family of calcium-binding proteins in PC12 cells. Proc Natl Acad Sci USA 85:1277-1281

Masiakowski P, Shooter EM (1990) Changes in PC12 cell morphology induced by transfection with 42C DNA, a member of the S-100 family. J Neurosci Res (in press)

McBurney RN, Neering IR (1987) Neuronal calcium homeostasis. Trends Neurosci 10:164-169

Moore BW (1965) A soluble protein characteristic of the nervous system. Biochem Biophys Res Commun 19:739-744

Morgan JI, Curran T (1986) Role of ion flux in the control of c-fos expression. Nature 322:552-555

Odink K, Cerletti N, Brüggen J, Clerc RG, Tarcsay L, Zwadlo G, Gerhards G, Schlegl R, Sorg C (1987) Two calcium-binding proteins in infiltrate macrophages of rheumatoid arthritis. Nature 330:80-82

Pandiella-Alonso A, Malgaroli A, Vicentini LM, Meldolesi J (1986) Early rise of cytosolic Ca²⁺ induced by NGF in PC12 and chromaffin cells. FEBS Lett 208:48-51

Schubert D, LaCorbiere M, Whitlock C, Stallcup W (1978) Alterations in the surface properties of cells responsive to nerve growth factor. Nature 273:718-723

Snider WD, Johnson EM (1989) Neurotrophic molecules. Ann Neurol 26:489-506

Szebenyi DME, Obendorf SK, Moffat K (1981) Structure of vitamin D-dependent calcium-binding protein from bovine intestine. Nature 294:327-332

Thompson MA, Ziff EB (1989) Structure of the gene encoding peripherin, an NGF-regulated neuronal-specific type III intermediate filament protein. Neuron 2:1043-1053

Winningham-Major F, Staecker JL, Barger SW, Coats S, van Eldik LJ (1989) Neurite extension and neuronal survival activities of recombinant S100b proteins that differ in the content and position of cysteine residues. J Cell Biol 109:3063-3071

Zimmer DB, van Eldik LJ (1986) Identification of a molecular target for the calcium-modulated protein S100: Fructose-1,6-bisphosphate aldolase. J Biol Chem 261:11424-11428

p11, A MEMBER OF THE S-100 PROTEIN FAMILY, IS ASSOCIATED WITH THE TYROSINE KINASE SUBSTRATE p36 (ANNEXIN II)

Volker Gerke

IDENTIFICATION AND CHARACTERIZATION OF p11

While the S-100 proteins have long been known as Ca^{2+}-binding proteins containing typical EF-hand structures (Moore 1965; Isobe et al. 1977), a third member of the still growing S-100 protein family was discovered only a few years ago. Due to its association with the Ca^{2+}-and phospholipid-binding protein p36 (annexin II), p11 was purified from mammalian tissues and shown by sequence analysis to share a high degree of homology with S-100α and ß.

The p36-p11 complex was originally isolated from chicken embryo fibroblasts and porcine intestinal epithelial cells (Erikson et al. 1984; Gerke and Weber 1984). It is easily prepared by exploiting the fact that p36 and the associated p11 are specifically extracted from membrane and/or cytoskeletal structures prepared in the presence of Ca^{2+}, once the divalent cation is chelated by the addition of EGTA. The purified complex is a heterotetramer (originally called protein I), consisting of two 36 kDa (p36) and two 11 kDa (p11) chains (for review see Gerke 1989). *In vitro*, $p36_2p11_2$ interacts in a Ca^{2+}-dependent manner with negatively charged phospholipids and elements of the cytoskeleton such as F-actin and nonerythroid-spectrin (Gerke and Weber 1985a; Glenney 1986a,b). Therefore, the name calpactin I has also been given to the $p36_2p11_2$ complex (reflecting the interaction with Ca^{2+}, lipid, and actin; Glenney 1986a).

Following denaturation of the complex in 9 M urea or 6 M guanidine-HCl, the two subunits can be separated and subsequently renatured (Gerke and Weber 1985a; Glenney 1986a). This approach has led to the identification of p36 as the subunit harboring the binding sites for Ca^{2+}, phospholipid, and cytoskeletal elements (see below). In contrast to the renatured p36 subunit, which is a monomeric molecule, renatured p11 is a dimer and is not able to bind any of the ligands mentioned above (Gerke and Weber 1985a,b; Glenney 1986a). When mixed with monomeric p36, p11 induces rapid and tight formation of the $p36_2p11_2$ complex indicating that the p11 dimer forms a cross-bridge in the native complex.

Protein sequencing of p11 (Gerke and Weber 1985b) or its N-terminal half (Glenney and Tack 1985; Hexham et al. 1986) revealed a remarkable similarity to the S-100 proteins. Figure 1 shows that the introduction of a gap of three consecutive residues in the p11 sequence (96 amino acids) allows the alignment of 43 identically placed residues when compared to S-100α and 33 identical amino acids when compared to S-100ß. Interestingly, the p11 sequence shows significant alterations in the helix-loop-helix regions (EF-hands) that are believed to represent the two Ca^{2+}-binding sites in S-100α and S-100ß. While the first EF-hand is destroyed by the three amino acid deletion, the second Ca^{2+}-binding loop (residues 62-73 of S-100α) also differs in the p11 sequence as it contains several characteristic amino acid replacements, e.g., a serine (position 70) in a location so far occupied in typical EF-hands by a glutamic acid (for review on EF-hands see Kretsinger 1987). Several results confirmed the lack of an intact Ca^{2+}-binding site in p11. While S-100 proteins show Ca^{2+}-induced differences in spectroscopical properties and an increased reactivity of cysteine residues towards 5,5'-dithiobis (2-nitrobenzoic acid) in the presence of Ca^{2+} (Baudier and Gerard 1983; Callisano et al. 1976), p11 fails to exhibit these Ca^{2+}-dependent effects (Gerke and Weber 1985b). In addition, equilibrium dialysis did not reveal any direct Ca^{2+}-binding to p11 (Glenney 1986a), supporting the now widely accepted view that p11 is a member of the S-100 family containing two defective Ca^{2+}-binding loops.

Both p11 and the S-100 proteins typically form dimers under physiological conditions. While the p11 dimer has two identical polypeptide chains, two forms of S-100 dimers are found: S-100a, comprising one copy of the S-100α and one copy of the S-100ß chain, and S-100b, a ßß dimer (Isobe et al. 1981). For both proteins, p11 and S-100, the dimers are likely to represent the biologically important forms. S-100 dimers were shown to exhibit neurite extension activity (a disulfide-bonded S-100ß dimer; Kligman and Marshak 1985; Van Eldik et al. 1988) and to inhibit the phosphorylation of protein kinase C substrates (Kligman and Patel 1985). In addition, S-100 binds to and modulates the activity of a possible cellular target, the enzyme aldolase (Zimmer and Van Eldik 1986). Similarly, the p11 dimer is the species that interacts with its cellular target, p36. The interaction with cellular protein ligands could parallel the situation observed for other proteins containing EF-hand-type Ca^{2+}-binding sites like calmodulin and troponin C. In their Ca^{2+}-bound state, these proteins bind to and by that regulate their cellular targets, which include different protein kinases in the case of calmodulin. In analogy, p11, which lacks an intact EF-hand, might be frozen in a permanently active state, which is Ca^{2+}-independent but allows the interaction with p36.

```
                1                                                       49
p11        P S QMEH AME TMMF T FHK FA ---GDKGY LT KEDLR V LMEKE F PGF LENQKDP L
S-100α     G SEL ET AMET L INV FHAHSGKEGDKY KL SKKE LKELLQTEL SGFLDAQKDAD
S-100β      SELEKAVVAL IDV FHQYSGREGDKHKLKKSELKEL INNELSHFLEE IKEQE
18A2       ARPLEEALDV IVS TFHKYSGKEGDKFKLNKTELKELLTRELPSFLFKRTDEA
S-100L     S SPLEQALAVMVATFHKYSGQEGDKFKLSKGEMKELLHKELPSFVGEKVDEE
MRP 14    TCKMSQLERN IETI INTFHQYSVKLGHPDTLNQGEFKELVRKDLQNFLKKENKNEK
2A9        ACPLDQAIGLLVAI FHKYSGREGDKHTLSKKELKELIQKEL-TIGSKLQD-A
MRP 8      LTELEKALNSI IDVYHKYSLIKGNFHAVYRDDLKKLLETECPQYI-RKK---
```

```
              50                                                    96
p11        AVDK IMKDLDQCRDGKVGFQSFFSLIAGLTIACNDYF-VVHMKQKGKK
S-100α     AVDKVMKELDEDGDGEVDFQEYVVLVAALTVACNNFFWENS
S-100β     VVDKVMETLDSDGDGECDFQEFMAFVAMITTACHEFFEHE
18A2       AFQKVMSNLDSNRDNEVDFQEYCVFLSCIAMMCNEFFEGCPDKEPRKK
S-100L     GLKKLMGDLDENSDQQVDFQEYAVFLALITIMCNDFFQGSPARS
MRP 14    VIEHIMEDLDTNADKQLSFEEFIMLMARLTWASHEKMHEGDEGPGHHHKPGLGEGTP
2A9        EIARLMEDLDRNKDQEVNFQEYVTFLGALALIYNEALKG
MRP 8      GADVWFKELDINTDGAVNFQEFLILVIKMGVAAHKKSHEESHKE
```

Fig. 1. Sequence comparison of p11 with other members of the S-100 protein family. Amino acid sequences of bovine S-100α (Isobe and Okuyama 1981), bovine S-100ß (Isobe and Okuyama 1978), murine 18A2 (Jackson-Grusby et al. 1987), bovine S-100L (Glenney et al. 1989), human 2A9 (Calabretta et al. 1986), human MRP 8 and MRP 14 (Odink et al. 1987) are compared with the sequence of porcine p11 (Gerke and Weber, 1985b). *Gaps* are introduced for better alignment. Residues identical in p11 and the different S-100 like proteins are given in *bold letters*. *Lines above* the sequences indicate the positions of the two Ca^{2+}-binding loops in S-100

THE p36₂p11₂ COMPLEX: BIOCHEMICAL AND STRUCTURAL PROPERTIES

The interaction between p11 and p36 has been studied in great detail since complex formation seems to regulate not only the physical state but also some biochemical properties of p36. p36 itself has received much attention since it represents a major cellular substrate for the tyrosine kinase encoded by the *src* oncogene and is also phosphorylated by different serine/threonine-specific protein kinases (for review see Brugge 1986).

Initial structural data on the p36 subunit were obtained by limited proteolysis (Glenney and Tack 1985; Johnsson et al. 1986a) which revealed the existence of two distinct domains: a N-terminal region of around 30 amino acids, which is susceptible to mild proteolytic treatment, and a 33 kDa fragment, which remains intact under these conditions (Figure 2). The 33 kDa core retains the ability to interact with F-actin and phospholipids in a Ca^{2+}-dependent manner but has lost the ability to bind p11. The N-terminal region (also called tail), on the other hand, contains the major serine and tyrosine phosphorylation sites: tyrosine 23 as the site for pp60^{src} phosphorylation (Glenney and Tack 1985) and serine 25 as target for protein kinase C (Gould et al. 1986; Johnsson et al. 1986b). While these sites cluster in the more C-terminal portion of the p36 tail, detailed proteolysis and binding experiments revealed that the binding site for p11 resides within the 12 amino terminal residues of p36 (Glenney et al. 1986; Johnsson et al. 1988). This sequence describes an amphiphatic α-helix and shows a high degree of conservation among species (Johnsson et al. 1988). Thus, the p11 binding site on p36 is restricted to a small portion of the molecule that is likely to protrude from the globular p36 since it is easily accessible for proteases and protein kinases. A synthetic peptide covering only the N-terminal 18 residues of p36 also shows strong binding to p11 with a dissociation constant of less than 30 nM. Besides this highly conserved stretch of amino acids, an intact p11 binding site requires the N-acetyl blocking group at the aminoterminal serine residue of p36 (Johnsson et al. 1988).

The amino acid sequence of p36, which was obtained by cDNA cloning (Saris et al. 1986; Huang et al. 1986; Kristensen et al. 1986) and partially by direct protein sequencing (Weber and Johnsson 1986), shows a repetitive structure that is characterized by four segments displaying internal homologies (Figure 2).

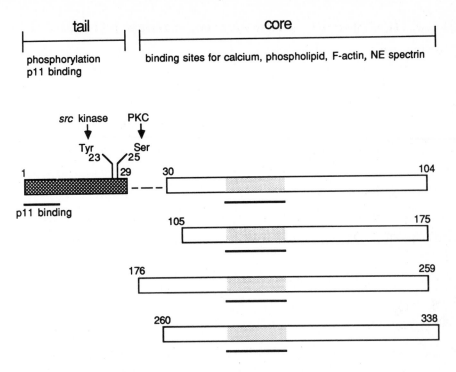

Fig. 2. Structural organization of p36 (annexin II). Highlighted in each repeat segment of the protein core is the position of the 17 amino acid consensus sequence known as endonexin fold

The repeated segments are between 70 and 80 amino acids long and together comprise the protein core, i.e., the domain responsible for Ca^{2+}-and phospholipid binding. The amino acid sequence of p36 is 50% identical with that of a protein called lipocortin I (Wallner et al. 1986), calpactin II, or p35 (De et al. 1986), which was described independently as a phospholipase A2 inhibitor (for review see Hirata 1984; Flower et al. 1984) and as a cellular substrate for the tyrosine kinase associated with the epidermal growth factor (EGF) receptor (Fava and Cohen 1984). Besides the high overall homology, lipocortin I shows the same repetitive structure and similar *in vitro* properties as p36, i.e., direct Ca^{2+} binding with a low affinity and a Ca^{2+}-dependent interaction with phospholipids.

Several other Ca^{2+}/phospholipid-binding proteins were identified in the recent past and shown to have primary structures similar to lipocortin I and p36 (for review see Klee 1988; Crompton et al. 1988). Based on the membrane binding properties common to these proteins the name annexins was introduced for this new multigene family. Like lipocortin I (annexin I) and p36 (annexin II), most other members are composed of four repeat segments, all exhibiting intra- as well as inter-molecular homologies. So far, only one annexin (annexin VI, p68) has been identified that contains eight repeat segments probably arisen by gene duplication of the fourfold structure (Crompton et al. 1988; Suedhof et al. 1988).

While the physiological function(s) of the annexins are not known precisely, it has been suggested that some members, most notably lipocortin I, participate in the anti-inflammatory response since they inhibit phospholipase A2 *in vitro*. However, the significance of these data is unclear since inhibition of phospholipase A2 by annexins seems to arise by substrate depletion, i.e., a coating of the phospholipid substrate by the annexins (Davidson et al. 1987; Haigler et al. 1987). Another annexin, annexin V (also known as endonexin II), has been shown to inhibit blood coagulation in vitro and consequently was termed PAP (placental anticoagulant protein; Funakoshi et al. 1987, Iwasaki et al. 1987; Maurer-Fogy et al. 1988; Kaplan et al. 1988; Grundmann et al. 1988). Again, it seems likely that this inhibition is based on the Ca^{2+}-dependent binding of annexin V to phospholipids present in the coagulation assay.

Although members of the annexin family can bind Ca^{2+} directly, a close examination of their primary structures does not reveal a typical EF-hand. It is discussed, however, that a highly conserved stretch of 17 amino acids, which is present in each repeat and is known as the endonexin fold (Figure 2), could form a helix-loop structure that is involved in Ca^{2+}- or Ca^{2+}/lipid-binding (Geisow et al. 1986). Recently, it has also been noted that the endonexin fold in the second repeat segment of p36 shows almost 50% homology to a putative mutant EF-hand present in a Ca^{2+}-binding protein from *Streptomyes erythraeus* (Moss and Crumpton 1990). However, future experimental data have to reveal whether this region is indeed involved in Ca^{2+}-binding.

Among the annexins only p36 has been shown to interact specifically with another cellular protein ligand, p11. Since this interaction is mediated by the N-terminal tail, it is not surprising that the tail portions of different annexins are divergent, both in sequence and length. While some are as short as 10 or 12 amino acids (annexins IV and V; Weber et al. 1987; Funakoshi et al. 1987), lipocortin I has a protease-sensitive N-terminus that is even longer than that of p36 (39 as compared to 30 residues). On paper, the N-terminal portion of this lipocortin I tail can also form an amphiphatic α-helix similar to the p11 binding domain on p36. Thus, it remains possible that lipocortin I interacts with a protein ligand (possibly of the S-100/p11 family) within the cell. If such a putative interaction, in contrast to the $p36_2p11_2$ complex formation, would depend on the presence of Ca^{2+}, it would have been overlooked so far since the isolation of lipocortin I is carried out in the presence of EGTA.

Interestingly, the N-terminal tail region of lipocortin I was found recently to be the site of an intermolecular cross-link that is generated by transglutaminase and leads to a lipocortin I homodimer (Ando et al. 1989; Pepinsky et al. 1989). The cross-linking enhances the Ca^{2+} sensitivity of phospholipid binding and thus is reminiscent of the p11-induced dimerization of p36. The $p36_2p11_2$ complex also binds phospholipids at much lower Ca^{2+} concentrations when compared to monomeric p36 (Powell and Glenney 1987). In addition, the complex promotes liposome and chromaffin granule aggregation at micromolar Ca^{2+} levels, an activity not observed for the p36 monomer (Drust and Creutz 1988). Recent experimental evidence argues against the view that this difference is based simply on a duplication of the phospholipid binding site(s) because of annexin-dimerization. Drust and Creutz (1988) could show that proteolytic derivatives of p36, which are without the N-terminal tail (and thus the p11 binding domain), still promote chromaffin granule aggregation albeit with an altered Ca^{2+} requirement. Thus, each p36 chain must contain enough phospholipid binding sites to promote the aggregation but it seems that an intact N-terminus interferes with this activity. In the $p36_2p11_2$ complex, this interfering region might be sequestered by p11-binding.

TISSUE DISTRIBUTION AND INTRACELLULAR LOCATION OF p11

As revealed by immunoblotting using monoclonal antibodies, the tissue distribution of p11 is essentially identical with that of p36 (Zokas and Glenney 1987). Highest protein levels are found in lung and intestine while moderate expression is observed in spleen, adrenal gland, kidney, and skeletal and cardiac muscle. p11, as well as p36, is undetectable or present at very low levels in red blood cells, liver, and brain. Generally, the same tissue distribution is found for the p11 mRNA, indicating that p11 synthesis is regulated at the transcriptional level (Saris et al. 1987). One exception, however, is the brain. It shows moderate levels of p11 mRNA but no detectable synthesis of the p11 protein. Whether this is due to instability of the p11 protein in brain or reflects different sensitivities of the particular Northern and Western blot analyses remains to be seen.

Although not studied in great detail, the situation in cultured cells is slightly different. Among the few cell lines tested at least two do not show a coordinate expression of p11 and p36 mRNAs. While human A 431 cells have an apparent excess of p36 transcript, mouse F9 cells contain moderate levels of p11 mRNA but no or very little p36 mRNA (Saris et al. 1987). However, these mRNA levels are not necessarily a true reflection of the actual protein levels within the cell. In A431 cells, for example, immunoblotting revealed the existence of similar amounts of p11 and p36 protein (Zokas and Glenney 1987). Again, the reason for this apparent discrepancy is not known.

The intracellular location of p11 was revealed by immunofluorescence as well as immuno-electron microscopical studies. At the light microscope level, p11 is found in the submembranous region (Figure 3) showing a distribution indistinguishable from p36 (Zokas and Glenney 1987; Osborn et al. 1988). Even when cells are treated with detergent in the presence of Ca^{2+} before fixation, p11 stays in the so-called submembranous cytoskeleton (presumably due to its interaction with p36, which is a typical constituent of this network underlying the plasma membrane). Co-localization of p11 and p36 is also found at the electron microscope level (Semich et al. 1989). Fibroblasts processed with gold labeled antibodies show a very regular staining at the cytoplasmic face of the plasma membrane by p11 antibodies (Figure 3). The staining pattern is identical with the labeling observed with antibodies against the $p36_2p11_2$ complex and is strictly dependent on the presence of Ca^{2+} (at least 10^{-7} M) during sample preparation. In all tissues and cells studied so far, p11 is found in the submembranous region, indicating again that it is present within the cell in a tight complex with p36. Even when microinjected into a living cell, p11 binds to the endogenous p36 since it assumes the submembranous location characteristic for the $p36_2p11_2$ complex (Osborn et al., 1988). However, if p11 is carboxy-methylated before injection (a modification that destroys the ability of p11 to interact with p36 *in vitro*), the molecule does not travel to the cell periphery and remains diffusely distributed in the cytoplasm.

Several lines of evidence suggest that p36, in contrast to p11, can occur in a free, i.e., monomeric, form in certain cells. Biochemical fractionation revealed that chicken as well as human fibroblasts contain approximately 50% of their p36 in the monomeric form, i.e., not associated with p11 (Erikson et al. 1984; Zokas and Glenney 1987). Interestingly, this fraction shows a higher turnover rate and different subcellular distribution than p36 complexed with p11 (Zokas and Glenney 1987).

While the latter remains in the cortical cytoskeleton in the presence of Ca^{2+}, monomeric p36 is solubilized under identical conditions, i.e., upon detergent treatment of cultured cells in the presence of Ca^{2+}.

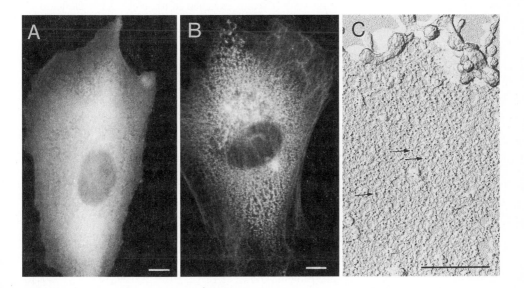

Fig. 3A-C. Distribution of p11 in rat mammary cells (**A, B**) and human fibroblasts (**C**) as revealed by immunofluorescence (**A, B**) and immuno-electron microscopy (**C**). Cells were fixed and stained with monoclonal antibodies to p11 either directly (**A**) or after extraction with Triton X-100 in the presence of Ca^{2+}. (**B**) Immunogold staining of the cytoplasmic surface of the plasma membrane (**C**) was obtained with p11 antibodies followed by 12 nm gold-conjugated second antibodies. Plasma membranes (**C**) were prepared by the lysis-squirting technique. *Arrows* shows the positions of some gold particles. Note the dense and regular staining of the plasma membrane (**C**). *Bars* 10 μm (**A, B**), 1 μm (**C**)

SIMILARITIES BETWEEN p11 AND OTHER MEMBERS OF THE S-100 PROTEIN FAMILY

In addition to p11, other proteins which belong to the S-100 family have been described in the recent past (Figure 1, see also chapt. 4-6 and 8-12 in this Vol.; for review see Kligman and Hilt 1988). These include two proteins, which are present in elevated quantities in the serum of patients suffering from cystic fibrosis (CF; Dorin et al. 1987; Odink et al. 1987). Both polypeptides, known as calgranulin A (also called MRP 8 and originally CF antigen) and calgranulin B (MRP 14), are also found in chronic inflammatory lesions and are expressed in infiltrate macrophages during chronic inflammations but not in normal tissue macrophages. Interestingly, the calgranulins seems enriched in the cortical cytoskeleton of a squamous cell carcinoma line, thus showing an intracellular distribution very similar to p11 (Wilkinson et al. 1988). A different distribution within the cell is observed for two other members of the S-100 family. S-100b (Hidaka et al. 1983) and a protein termed S-100L (Glenney et al. 1989), which was recently isolated from bovine lung, are present throughout the nucleus and the cytoplasm in adipocytes and MDBK cells, respectively.

Another set of proteins, which show a high degree of sequence similarity to S-100 and p11, is preferentially expressed when cells are stimulated to grow or differentiate. Calcyclin, a Ca^{2+}-binding protein recently isolated from Ehrlich ascites tumour cells (Kuznicki et al. 1989) and different human tissues (Gabius et al. 1989), was originally identified as a cDNA clone whose corresponding mRNA (called 2A9) is expressed in a cell cycle and growth factor-dependent manner (Calabretta et al. 1986). The synthesis of another mRNA, which encodes a S-100 like protein (known as 18A2, p9Ka, pEL 98, or 42A), is also regulated during the cell cycle (Barraclough et al. 1987; Jackson-Grusby et al. 1987; Goto et al. 1988; Masiakowski and Shooter 1988). However, while calcyclin is preferentially expressed when the cell enters G1, the expression of 18A2 increases in the S phase. Synthesis of the 18A2 mRNA is also increased in epithelial cells upon conversion of a cuboidal stem cell into myoepithelial-like cells and in rat pheochromocytoma (PC12) cells, which have been induced by nerve growth factor (NGF) to differentiate and adopt a neuronal phenotype. In PC12 cells, another message (termed 42C) was found to be induced upon NGF application (Masiakowski and Shooter 1988). Sequence analysis revealed that the 42C mRNA encodes rat p11. In the induced PC12 cells, both mRNA species (18A2 and p11) reach maximal levels by 24 h of treatment with NGF but the initial response is more rapid in the case of p11 mRNA. Here, a substantial increase is already evident after 2 h of incubation with the growth factor (Masiakowski and Shooter 1988). At present, it is not known how p11 transcription is regulated in PC12 cells. Identification of possible cis-acting elements at the transcriptional level has to await the characterization of the p11 gene. The analysis of possible transcriptional control elements also should clarify whether the expression of p11 and p36 is coordinately regulated at the transcriptional level, a tempting speculation, since both proteins form a tight complex within the cell and show a very similar if not identical tissue distribution. In this respect it is interesting that similar to p11 the expression of p36 is also increased upon NGF-induced differentiation of PC12 cells (V. Gerke, unpub.). However, additional or different regulatory mechanisms are likely to be involved in cells that contain a substantial population of monomeric p36 (see above).

POSSIBLE PHYSIOLOGICAL ROLES

Since p11 resides in a tight complex with p36 and regulates properties displayed by p36 *in vitro*, it seems likely that it carries out a similar function *in vivo*. This assumption is particularly tempting since other members of the EF-hand superfamily, e.g., calmodulin and troponin C, are known to act by modulating the activities of cellular target proteins by direct interaction. Cellular functions of p36 that might be regulated by p11 binding could include an involvement in the membrane-cytoskeletal linkage and/or the control of membrane fusion events during exocytosis (for review see Burgoyne and Geisow 1989). Both the association of p36 with the submembranous cytoskeleton and its ability to induce aggregation and fusion of chromaffin granule membranes depend on p11-induced formation of the heterotetrameric complex.

A direct participation of the $p36_2p11_2$ complex in the exocytotic process is in line with the subcellular localization of the complex in chromaffin cells after stimulation with acetylcholine. Under these conditions, $p36_2p11_2$ is closely associated with the inner face of the plasma membrane and seems particularly enriched at sites where exocytotic vesicles attach to the membrane (Nakata et al. 1990). Additional circumstantial evidence (e.g., from liposome binding experiments) points to a direct role of the $p36_2p11_2$ complex in cross-linking these vesicles to the plasma membrane in activated chromaffin cells. Interestingly, this process seems accompanied by a conformational change in the protein complex from a globular molecule to a thin strand that forms the cross-link between the adjacent membranes (Nakata et al. 1990).

The idea that $p36_2p11_2$ has a general structural role and serves as an important link between the plasma membrane and the underlying cytoskeleton stems from several *in vitro* and *in situ* observations: the Ca^{2+}-dependent binding to phospholipids and nonerythroid-spectrins (typical proteins of the submembranous network), and the co-localization with nonerythroid-spectrins at the light and electron microscope level (for review see Gerke 1989). However, different microinjection experiments argue against a tight intracellular association of $p36_2p11_2$ with nonerythroid-spectrins. Injection of monoclonal antibodies to p36 or p11 leads to an intracellular patching of the $p36_2p11_2$ complex and the formation of large $p36_2p11_2$ aggregates without affecting the distribution of nonerythroid-spectrins (Zokas and Glenney 1987). On the other hand, when cells are injected with antibodies to nonerythroid-spectrins aggregation of nonerythroid-spectrins is observed without a disturbing effect on the p36 distribution (Mangeat and Burridge 1984).

Since p36 is a substrate for various protein kinases *in vivo* and *in vitro*, it has been suggested that it is involved in signal transduction pathways which regulate cellular growth and differentiation. While the verification of this hypothesis has to await more detailed physiological studies, it is interesting that p11 binding interferes with some phosphorylation events and vice versa (Johnsson et al. 1986b). Phosphorylation of monomeric p36 with different protein kinases at (a) certain serine residue(s), for example, markedly reduces the affinity for p11. p11 binding to p36, on the other hand, interferes with the phosphorylation of p36 by several protein kinases *in vitro*.

Recent studies analyzing prolactin-induced events in certain lymphoma cells suggest that p11 and p36 could function as a signal transduction system, which is involved in the control of cell proliferation (Hughes and Prentice pers. commun.). In Nb2 cells, a T. lymphoma that proliferates in response to prolactin, the level of p11 mRNA decreases upon application of prolactin. Interestingly, agents capable of disrupting the $p36_2p11_2$

complex also promote growth of the Nb2 cells and thus mimic the prolactin effect. Both an antisense oligonucleotide complementary to p11 mRNA and a synthetic peptide comprising the N-terminal tail of p36 (and thus the p11 binding domain) induce Nb2 cell proliferation. It seems likely that both agents reduce the amount of p11 that is available for the interaction with p36 and thus lead to an altered ratio of monomeric to complexed p36. This, in turn, could be a signal directly involved in the induction of cell proliferation.

Given the high affinity and specificity of the interaction between p11 and the N-terminal tail of p36 it seems that this interaction is also involved in a cellular transformation event that is mediated by the *c-raf-1* oncogene product. Analysis of the *c-raf-1* gene from the GL-5-JCK human glioblastoma, which underwent rearrangement during transfection experiments, revealed that a fusion of the N-terminal 16 residues of p36, i.e., the entire p11 binding domain, with the *c-raf-1* kinase domain leads to the oncogenic activation (Mitsunobo et al. 1989). It is tempting to speculate that binding of the p11 dimer to the p36-raf fusion protein of 44 kDa leads to the formation of $p44_2p11_2$ and/or $p36p44p11_2$ complexes and subsequently to an unusual intracellular location of the *c-raf-1* kinase activity. This in turn could act as the transforming stimulus since the *c-raf-1* kinase will now be able to phosphorylate a certain set of substrate(s) in a well defined subcellular compartment, i.e., the cortical cytoplasm.

It will be interesting to learn in the future whether p11 generally acts on and regulates the activities of cellular targets by anchoring them in the submembranous region of the cell. Our detailed structural knowledge of the p11 molecule and the p11-p36 interaction will certainly aid the understanding of such functional aspects.

REFERENCES

Ando Y, Imamura S, Owada MK, Kakanuga T, Kannigi R (1989) Cross-linking of lipocortin I and enhancement of its Ca²⁺-sensitivity by tissue transglutaminase. Biochem Biophys Res Commun 163:944-951

Barraclough R, Savin J, Dube K, Rudland PS (1987) Molecular cloning and sequence of the gene for p9Ka, a cultured myoepithelial protein with strong homology to S-100, a calcium binding protein. J Mol Biol 198:13-20

Baudier J, Gerard D (1983) Structural changes induced by calcium and zinc on S100a and S100b proteins. Biochemistry 22:3360-3369

152

Brugge JS (1986) The p35/p36 substrates of protein-tyrosine kinases as inhibitors of phospholipase A₂.
Cell 46:149-150

Burgoyne RD, Geisow MJ (1989) The annexin family of calcium-binding proteins. Cell Calcium 10:1-10

Calabretta B, Battini R, Kaczmarek L, de Riel JK, Baserga R (1986) Molecular cloning of the cDNA for a
growth factor-inducible gene with strong homology to S-100, a calcium-binding protein.
J Biol Chem 261:12628-12632

Callisano P, Mercanti D, Levi A (1976) Ca²⁺, K⁺-regulated intramolecular crosslinking of S-100 protein via
disulfide bond formation. Eur J Biochem 71:45-52

Crompton MR, Moss SE, Crumpton MJ (1988) Diversity in the lipocortin/calpactin family. Cell 55:1-3

Crompton MR, Owens RJ, Totty NF, Moss SE, Waterfield MD, Crumpton MJ (1988) Primary structure of the
human membrane-associated Ca²⁺-binding protein p68. EMBO J 7:21-27

Davidson FF, Dennis EA, Powell M, Glenney JR Jr (1987) Inhibition of phospholipase A₂ by 'lipocortins' and
calpactins. J Biol Chem 262:1698-1705

De BK, Misono KS, Lukas TJ, Mroczkowski B, Cohen S (1986) A calcium-dependent 35-kilodalton substrate
for epidermal growth factor receptor/kinase isolated from normal tissue.
J Biol Chem 261:13784-13792

Dorin JR, Novak M, Hill RE, Brock DJH, Secher DS, van Heyningen (1987) A clue to the basic defect in
cystic fibrosis from cloning the CF antigen gene. Nature 326:614-617

Drust DS and Creutz CE (1988) Aggregation of chromaffin granules by calpactin at micromolar levels calcium.
Nature 331:88-91

Erikson E, Tomasiewicz HG, Erikson RL (1984) Biochemical characterization of a 34-kilodalton normal cellular
substrate of pp60v-src and an associated 6-kilodalton protein. Mol Cell Biol 4:77-85

Fava RA, Cohen S (1984) Isolation of a calcium-dependent 35-kilodalton substrate for the epidermal growth
factor receptor/kinase from A-431 cells. J Biol Chem 259:2636-2645

Flower RJ, Wood JN, Parente L (1984) Macrocortin and the mechanism of action of the glucocorticoids.
Adv Inflam Res 7:61-69

Funakoshi T, Hendrickson LE, McMullen BA, Fujikawa K (1987) Primary structure of human placental
anticoagulant protein. Biochemistry 26:8087-8092

Gabius H-J, Bardosi A, Gabius S, Hellmann KP, Karas M, Kratzin H (1989) Identification of a cell
cycle-dependent gene product as a sialic acid-binding protein.
Biochem Biophys Res Commun 163:506-512

Geisow MJ, Fritsche U, Hexham JM, Dash B, Johnson T (1986) A consensus amino-acid sequence repeat in
Torpedo and mammalian Ca²⁺-dependent membrane-binding protein. Nature 320:636-638

Gerke V (1989) Tyrosine protein kinase substrate p36: a member of the annexin family of Ca²⁺/phospholipid-
binding proteins. Cell Motil Cytoskeleton 14:449-454

Gerke V, Weber K (1984) Identity of p36K phosphorylated upon Rous sarcoma virus transformation with a
protein purified from brush borders; calcium-dependent binding to nonerythroid spectrin and F-actin.
EMBO J 3:227-233

Gerke V, Weber K (1985a) Calcium-dependent conformational changes in the 36-kDa subunit of intestinal
protein I related to the cellular 36-kDa target of Rous sarcoma virus tyrosine kinase.
J Biol Chem 260:1688-1695

Gerke V, Weber K (1985b) The regulatory chain in the p36-kd substrate complex of viral tyrosine-specific
protein kinases is related in sequence to the S-100 protein of glial cells. EMBO J 4:2917-2920

Glenney JR Jr (1986a) Phospholipid-dependent Ca²⁺-binding by the 36-kDa tyrosine kinase substrate (calpactin)
and its 33-kDa core. J Biol Chem 261:7247-7252

Glenney JR Jr (1986b) Two related but distinct forms of the Mr 36.000 tyrosine kinase substrate (calpactin) that
interact with phospholipid and actin in a Ca²⁺-dependent manner.
Proc Natl Acad Sci USA 83:4258-4262

Glenney JR Jr, Tack BF (1985) Amino-terminal sequence of p36 and associated p10: identification of the site
of tyrosine phosphorylation and homology with S-100. Proc Natl Acad Sci USA 82:7884-7888

Glenney JR Jr, Boudreau M, Galyean R, Hunter T, Tack B (1986) Association of the S-100-related calpactin
I light chain with the NH₂-terminal tail of the 36-kDa heavy chain. J Biol Chem 261:10485-10488

Glenney JR Jr, Kindy MS, Zokas L (1989) Isolation of a new member of the S-100 protein family: amino acid
sequence, tissue and subcellular distribution. J Cell Biol 108:569-578

Goto K, Endo H, Fujiyoshi T (1988) Cloning of the sequences expressed abundantly in established cell lines: identification of a cDNA clone highly homologous to S-100, a calcium-binding protein. J Biochem 103:48-53

Gould KL, Woodgett JR, Isacke CM, Hunter T (1986) The protein-tyrosine kinase substrate, p36, is also a substrate for protein kinase C in vivo and in vitro. Mol Cell Biol 6:2738-2744

Grundmann U, Abel K-J, Bohn H, Löbermann H, Lottspeich F, Küpper H (1988) Characterization of cDNA encoding human placental anticoagulant protein (PP4): homology with the lipocortin family. Proc Natl Acad Sci USA 85:3708-3712

Haigler HT, Schlaepfer DD, Burgess WH (1987) Characterization of lipocortin I and an immunologically unrelated 33-kDa protein as epidermal growth factor receptor/kinase substrates and phospholipase A2 inhibitors. J Biol Chem 262:6921-6930

Hexham JM, Totty NF, Waterfield MD, Crumpton MJ (1986) Homology between the subunits of S100 and a 10 kDa polypeptide associated with p36 of pig lymphocytes. Biochem Biophys Res Commun 134:248-254

Hidaka H, Endo T, Kawamoto S, Yamada E, Umekawa H, Tanabe K, Hara K (1983) Purification and characterization of adipose tissue S-100b protein. J Biol Chem 258:2705-2709

Hirata F (1984) Roles of lipomodulin: a pospholipase inhibitory protein in immunoregulation. Adv Inflam Res 7:71-78

Huang K, Wallner B, Mattaliano RJ, Tizard R, Burne C, Frey A, Hession C, McGray P, Sinclair LK, Chow EP, Browning JL, Ramachandran KL, Tang J, Smart JE, Pepinsky RB (1986) Two human 35 kd inhibitors of phospholipase A_2 are related to substrates of pp60^{v-src} and the epidermal growth factor receptor/kinase. Cell 46:191-199

Isobe T, Okuyama T (1981) The amino-acid sequence of the alpha subunit in bovine brain S-100a protein. Eur J Biochem 116:79-86

Isobe T, Nakajima T, Okuyama T (1977) Reinvestigation of the extremely acidic proteins in bovine brain. Biochim Biophys Acta 494:222-232

Isobe T, Okuyama T (1978) The amino-acid sequence of S-100 protein (PAP 1-b protein) and its relation to the calcium binding proteins. Eur J Biochem 89:379-388

Isobe T, Ishioka N, Okuyama T (1981) Structural relation of two S-100 proteins in bovine brain, subunit composition of S-100a protein. Eur J Biochem 115:469-474

Iwasaki A, Suda M, Nakao H, Nagoya T, Saina Y, Arai K, Mizoguchi T, Sato F, Yoshizaki H, Hirata M, Miyata T, Shidara Y, Murata M, Maki M (1987) Structure and expression of cDNA for an inhibitor of blood coagulation isolated from human placenta: a new lipocortin-like protein. J Biochem 102:1261-1273

Jackson-Grusby LL, Swiergiel, J, Linzer DIH (1987) A growth related mRNA in cultured mouse cells encodes a placental calcium binding protein. Nucl Acids Res 15:6677-6690

Johnsson N, Vandekerckhove J, van Damme J, Weber K (1986a) Binding sites for calcium, lipid and p11 on p36, the substrate of retroviral tyrosine-specific protein kinases. FEBS Lett 198:361-364

Johnsson N, van Nguyen P, Soeling H-D, Weber K (1986b) Functionally distinct serine phosphorylation sites of p36, the cellular substrate of retroviral protein kinase; differential inhibition of reassociation with p11. EMBO J 5:3455-3460

Johnsson N, Marriott G, Weber K (1988) p36, the major cytoplasmic substrate of src tyrosine protein kinase, binds to its p11 subunit via a short amino-terminal amphiphatic helix. EMBO J 7:2435-2442

Kaplan R, Jaye M, Burgess WH, Schlaepfer DD, Haigler HT (1988) Cloning and expression of cDNA for human endonexin II, a Ca^{2+} and phospholipid binding protein. J Biol Chem 263:8037-8043

Klee CB (1988) Ca^{2+}-dependent phospholipid- (and membrane-) binding proteins. Biochemistry 27:6645-6653

Kligman D, Marshak DR (1985) Purification and characterization of a neurite extension factor from bovine brain. Proc Natl Acad Sci USA 82:7136-7139

Kligman D, Patel J (1985) A protein modulator stimulates C kinase-dependent phosphorylation of a 90 K substrate in synaptic membranes. J Neurochem 47:298-303

Kligman D, Hilt DC (1988) The S-100 protein family. Trends Biochem Sci 13:437-443

Kretsinger RH (1987) Calcium coordination and calmodulin fold: divergent versus convergent evolution. Cold Spring Harbor Symp Quant Biol 52:499-510

Kristensen T, Saris CJM, Hunter T, Hicks LJ, Noonan DJ, Glenney, JR Jr, Tack BF (1986) Primary structure of bovine calpactin I heavy chain (p36), a major cellular substrate for retroviral protein-tyrosine kinases: homology with the human phospholipase A₂ inhibitor lipocortin. Biochemistry 25:4497-4503

Kuznicki J, Filipek A, Hunziker PE, Huber S, Heizmann CW (1989) Calcium-binding protein from Ehrlich ascites-tumour cells is homologous to human calcyclin. Biochem J 263:951-956

Mangeat PH, Burridge K (1984) Immunoprecipitation of nonerythroid spectrin within live cells following microinjection of specific antibodies: relation to cytoskeletal structures. J Cell Biol 98:1363-1377

Masiakowski P, Shooter EM (1988) Nerve growth factor induces the genes for two proteins related to a family of calcium-binding proteins in PC12 cells. Proc Natl Acad Sci USA 85:1277-1281

Maurer-Fogy I, Reutelingsperger CPM, Pieters J, Bodo G, Stratowa C, Hauptmann R (1988) Cloning and expression of cDNA for human vascular anticoagulant, a Ca²⁺-dependent phospholipid-binding protein. Eur J Biochem 174:585-592

Mitsunobo F, Fukui M, Oda T, Yamamoto T, Toyoshima K (1989) A mechanism of c-raf-1 activation: fusion of the lipocortin II amino-terminal sequence with the c-raf-1 kinase domain. Oncogene 4:437-442

Moore B (1965) A soluble protein characteristic of the nervous system. Biochem Biophys Res Commun 19:739-744

Moss SE, Crumpton MJ (1990) The lipocortins and the EF-hand proteins: Ca²⁺-binding sites and evolution. Trends Biochem Sci 15:11-12

Nakata T, Sobue K, Hirokawa N (1990) Conformational change and localization of calpactin I complex involved in exocytosis as revealed by quick-freeze, deep-etch electron microscopy and immunocytochemistry. J Cell Biol 110:13-25

Odink K, Cerletti N, Brüggen J, Clerc R, Tarcsay L, Zwadlo G, Gerhards G, Schlegel R, Sorg C (1987) Two calcium-binding proteins in infiltrate macrophages of rheumatoid arthritis. Nature 330:80-82

Osborn M, Johnsson N, Wehland J, Weber K (1988) The submembranous location of p11 and its interaction with the p36 substrate of pp60 src kinase in situ. Exp Cell Res 175:81-96

Pepinsky RB, Sinclair LK, Chow EP, O'Brine-Greco B (1989) A dimeric form of lipocortin I in human placenta. Biochem J 263:97-103

Powell MA, Glenney JR Jr (1987) Regulation of calpactin I phospholipid binding by calpactin I light-chain binding and phosphorylation by pp60^{v-src}. Biochem J 247:321-328

Saris CJM, Tack BF, Kristensen T, Glenney JR Jr, Hunter T (1986) The cDNA sequence for the protein-tyrosine kinase substrate p36 (calpactin I heavy chain) reveals a multidomain protein with internal repeats. Cell 46:201-212

Saris CJM, Kristensen T, D'Eustachio P, Hicks LJ, Noonan DJ, Hunter T, Tack BF (1987) cDNA sequence and tissue distribution of the mRNA for bovine and murine p11, the S100-related light chain of the protein-tyrosine kinase substrate p36 (calpactin I). J Biol Chem 262:10663-10671

Semich R, Gerke V, Robenek H, Weber K (1989) The p36 substrate of pp60^{src} kinase is located at the cytoplasmic surface of the plasma membrane of fibroblasts: an immunoelectron microscope analysis. Eur J Cell Biol 50:313-323

Suedhof TC, Slaughter CA, Leznicki I, Barjon P, Reynolds GA (1988) Human 67-kDa calelectrin contains a duplication of four repeats found in 35-kDa lipocortins. Proc Natl Acad Sci USA 85:664-668

van Eldik LJ, Staecker JL, Winningham-Major F (1988) Synthesis and expression of a gene coding for the calcium-modulated protein S100ß and designed for cassette-based, site-directed mutagenesis. J Biol Chem 263:7830-7837

Wallner BP, Mattaliano RJ, Hession C, Cate RL, Tizard R, Sinclair LK, Foeller C, Chow EP, Browning JL, Ramachandran KL, Pepinsky RB (1986) Cloning and expression of human lipocortin, a phospholipase A2 inhibitor with potential anti-inflammatory activity. Nature 320:77-81

Weber K, Johnsson N (1986) Repeating sequence homologies in the p36 target protein of retroviral protein kinases and lipocortin, the p37 inhibitor of phospholipase A₂. FEBS Lett 203:95-98

Weber K, Johnsson N, Plessmann U, van Nguyen P, Soeling H-D, Ampe C, Vandekerckhove J (1987) The amino acid sequence of protein II and its phosphorylation site for protein kinase C; the domain structure Ca²⁺-modulated lipid binding proteins. EMBO J 6:1599-1604

Wilkinson MM, Busuttil A, Hayward C, Brock DJH, Dorin J, van Heyningen V (1988) Expression pattern of two related cystic fibrosis-associated calcium-binding proteins in normal and abnormal tissues. J Cell Sci 91:221-230

Zimmer DB, van Eldik LJ (1986) Identification of a molecular target for the calcium-modulated protein S100. Fructose-1,6-biphosphate aldolase. J Biol Chem 261:11424-11428

Zokas L, Glenney JR Jr (1987) The calpactin light chain is tightly linked to the cytoskeletal form of calpactin I: studies using monoclonal antibodies to calpactin subunits. J Cell Biol 105:2111-2121

CALCYCLIN, FROM GENE TO PROTEIN

Jacek Kuźnicki

DISCOVERY AND EXPRESSION OF THE CALCYCLIN GENE

Serum depletion leads to growth arrest of rodent fibroblasts that remain in the G_0 or quiescent phase of the cell cycle. When serum is added, the cells enter the G_0-S transition phase (Figure 1). Five cDNA clones representing sequences that were inducible by serum were isolated by Baserga and coworkers (Hirschorn et al. 1984).

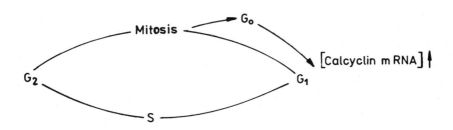

Fig. 1. Cell cycle-associated expression of calcyclin

The cDNA library from which these clones were isolated was obtained from ts13 hamster fibroblasts. Specifically, after serum stimulation, steady state levels of the cognate RNAs were very low in the G_0 cells, but increased three- to sixfold, between 6 and 16 h. The same effect was observed when the cells were stimulated to proliferate by either serum (Hirschorn et al. 1984; Rittling et al. 1986), platelet-derived growth factor or epidermal growth factor, but not by platelet-poor plasma or insulin (Calabretta et al. 1986a).

Calcyclin is the name given to a gene product that was identified as one (2A9) of the five cDNA clones that were stimulated by serum (Calabretta et al. 1986a). The increase in steady-state levels of cytoplasmic mRNA recognized by a calcyclin probe also occurs when the cells are stimulated by serum in the presence of cycloheximide at concentrations that completely suppress protein synthesis (Table 1) (Rittling et al. 1985). This shows that the calcyclin gene can be induced even in the absence of the products of other growth factors inducible genes.

Table 1. Characteristics of the calcyclin gene

Induced by	serum, PDGF, EGF, or serum plus cycloheximide
Induced in	WI-38 human diploid fibroblasts BALB/c/3T3 mouse fibroblasts ts13 (Syrian baby hamster kidney cells)
Not inducible by	platelet-poor plasma, insulin or adenovirus infection
Overexpressed in	human acute myeloid leukemia human breast cancer cell lines
Structure	3 exons (44, 160 and 232 base pairs)
Regulation	SV40-like enhancer near cap site
Localization	human chromosome 1, near ski oncogene

The level of calcyclin mRNA in different mouse tissues was estimated by Northern blotting with full length calcyclin cDNA as a probe. The highest level of calcyclin mRNA was observed in Ehrlich ascites tumor cells and a much weaker signal was detected in RNA isolated from smooth muscle. No signal was detected in RNA from liver, brain, kidney, skeletal muscle, and spleen (Kuźnicki et al. 1989b). These observations are in good agreement with the results obtained for rat tissues by Murphy et al. (1988). These authors found that the level of calcyclin mRNA is low in rat liver, brain, testes, and ovary, moderate in skeletal muscle, and high in lung, kidney, and uterus. The authors also showed that calcyclin is expressed in most, but not all, human breast cancer cell lines.

Calcyclin sequences from ts13 fibroblast cell line (BHK) hybridized to human genomic DNA giving (with three different restriction enzymes) band patterns similar to those reported for Syrian hamster genomic DNA (Calabretta et al. 1985). It was also found that in human diploid fibroblasts (WI-38 cells) the mRNA sequence cognate to calcyclin is expressed in a cell-cycle-dependent manner: the level of calcyclin mRNA was undetectable in quiescent cells, but was high in serum-stimulated cells (Table 1).

Calcyclin mRNA cannot be detected in resting or stimulated human lymphocytes, but is found in the cells obtained from myeloid leukemia patients (Calabretta et al. 1985). The overexpression of calcyclin (and other growth-regulated gene sequences) in chronic myelogenous leukemia is paralleled by increased expression of histone H3 (Calabretta et al. 1985; 1986b). This reflects, most likely, an increase of the proliferative activity of these neoplastic cells. In some cases of acute myelogenous leukemia, calcyclin and c-myc are overexpressed with respect to histone H3 suggesting a true deregulation of these two genes (Table 1).

STRUCTURE OF THE CALCYCLIN GENE; HOMOLOGY TO THE S-100 PROTEIN

The full-length human cDNA corresponding to the Syrian hamster cDNA calcyclin clone has been isolated from an Okayama-Berg library, and its complete nucleotide sequence has been determined (Calabretta et al. 1986a). There is a 90% homology between the nucleotide sequence of the insert from the original Syrian hamster calcyclin gene clone and that from the human calcyclin clone.

Using this full-length cDNA as a probe several, genomic clones containing calcyclin sequences have been isolated from human lambda phage library. One of these clones contains the entire calcyclin gene, plus extensive flanking sequences (Figure 2) (Ferrari et al. 1987). The 5' flanking sequence has been characterized, both structurally and functionally. Calcyclin promoter seems modulated by both positive and negative elements (Ghezzo et al. 1988). A 164-base pair fragment contains serum inducible sequences just upstream of the cap site. These 164 base pairs include a TATAA box (-29 upstream of the cap site), GC-rich sequences and the enhancer-like sequences that have a strong homology to the enhancer core of the SV40 promoter (Ferrari et al. 1987). This 5' flanking region of the calcyclin gene has promoter function, i.e., it can drive linked reporters in a transient expression assay after transfection of the appropriate dimeric plasmids. The sequences from -1371 to -1194 upstream of the cap site contain an element that is negatively regulated by an epidermal growth factor (Ghezzo et al. 1988). The termination codon TGA is at nucleotide 1295 and the polyadenylation signal is at nucleotide 1370.

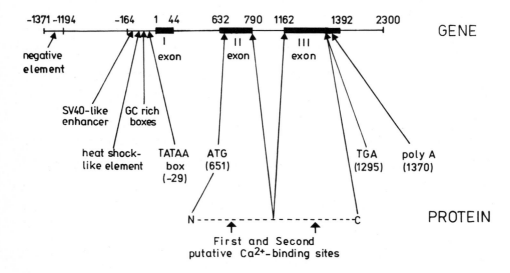

Fig. 2. Calcyclin gene and protein. The figure shows the structure of calcyclin gene schematically. The numbers refer to the nucleotides in the calcyclin clone isolated from human lambda phage library. (Ferrari et al. 1987)

The calcyclin gene is a unique copy gene and has three exons. The first ATG codon in the cDNA that is followed by an open reading frame of 270 nucleotides is at nucleotide 651 of the gene, so that the first exon of the calcyclin gene seems untranslated (Ferrari et al. 1987). The human gene codes for a putative polypeptide of 90 amino acids (including the initial methionine) that has 55% homology with the beta-subunit of a calcium-binding protein, the S-100 protein (Calabretta et al. 1986a). The homology between calcyclin and the beta-subunit of S-100 is particularly striking in the two regions that code for the calcium-binding sites in S-100 protein (between nucleotides 710-751 and 1202-1240) (Figure 3). The first putative Ca^{2+}-binding domain of calcyclin is in the second exon, while the second binding site is in the third exon.

Human calcyclin:	L M E D L D R N K D Q E V N F Q E
A peptide from EAT cells:	L M D D L D R N K D Q E V N
Beta S-100 protein:	M E T L D S D G D G E C D F Q E

Fig. 3. The partial amino acid sequence of the protein isolated from EAT cells protein is aligned with the corresponding sequence of human calcyclin and bovine brain S-100 protein (beta subunit). Potential Ca^{2+} ligands in the second putative Ca^{2+}-binding site are *underlined*

DISCOVERY AND PURIFICATION OF THE CALCYCLIN PROTEIN

A novel (10.5 kDa) calcium binding protein has been isolated from Ehrlich ascites tumor (EAT) cells (Kuźnicki and Filipek 1987). The protein was purified to homogeneity and shown to be different from S-100 protein, calbindin D9k, parvalbumin, and oncomodulin by several criteria, including electrophoretic mobility in SDS- or urea-polyacrylamide gels, amino acid composition, and the lack of cross-reactivity with antibodies specific to these Ca^{2+}-binding proteins. The partial amino acid sequence of the 10.5 kDa protein from EAT cells appeared to be identical with that of rat calcyclin and highly homologous to human calcyclin (Figure 3). We therefore called the protein from EAT cells a mouse calcyclin (Kuźnicki et al. 1989a). The complete amino acid sequence of calcyclin from EAT cells calcyclin will be published (Filipek, Gerke, Kuźnicki in prep.).

ION BINDING AND DIMER FORMATION IN CALCYCLIN FROM *EAT* CELLS

The Ca^{2+}-binding parameters of the protein purified from EAT cells has been analyzed by several methods (Kuźnicki and Filipek 1987; Kuźnicki et al. 1989a). Binding of Ca^{2+} to calcyclin was established directly when the protein was subjected to SDS-PAGE, blotted onto nitrocellulose and incubated with $^{45}Ca^{2+}$. Using gel filtration of the protein in the presence of $3 \times 10^{-5}M$ $^{45}CaCl_2$, we showed that it binds one calcium ion per molecule. The binding of calcium ions to calcyclin was also studied by several indirect methods. For example, tyrosine fluorescence intensity reversibly increases by 18% upon Ca^{2+} binding. The titration curve of fluorescence intensity plotted against the molar ratio of added Ca^{2+}, suggests the existence of two Ca^{2+}-binding sites per molecule. Since gel filtration indicates only one Ca^{2+}-binding site per molecule, it was concluded that the two binding sites differ in their relative affinities for Ca^{2+} (Kuźnicki and Filipek 1987).

The calcyclin from EAT cells changes its conformation upon Ca^{2+} binding as indicated by Ca^{2+}-dependent changes in tyrosine fluorescence, UV absorbance, mobility in urea PAGE, and hydrophobicity. The change in exposure of one or more hydrophobic domains has been demonstrated by Ca^{2+}-dependent binding to phenyl-Sepharose and to fluphenazine-Sepharose (Kuźnicki and Filipek unpubl.). Differential binding to phenyl-Sepharose is used in the purification of the protein. The binding of calcyclin to fluphenazine suggests that the protein may bind to other calmodulin inhibitors such as trifluoperazine or W-7.

Calcyclin from EAT cells binds not only Ca^{2+}, but also Zn^{2+} (Filipek et al. 1990), similarly to S-100 protein from bovine brain. Zn^{2+} binding is unique to these proteins as other calcium-binding proteins, such as calmodulin and parvalbumin, do not bind Zn^{2+} under the same conditions. Competitive binding experiments on nitrocellulose filters suggests that zinc binds to a site different from that engaged in Ca^{2+} binding. This conclusion is confirmed by the results showing that only Ca^{2+}, but not Zn^{2+}, can induce changes in intrinsic tyrosine fluorescence, and that Zn^{2+} binding does not block this effect. The location of the zinc-binding site in calcyclin and the structure of the sites is not known. There are neither clusters of acidic amino acids, nor characteristic sequences of cysteine and histidine residues that could form the well-known "zinc finger" motif. We assume that the Zn^{2+} binding sites of calcyclin are formed by amino acid residues that are located at some distance from each other in the primary sequence.

Calcyclin forms monomers and disulfide-linked dimers, which can be separated by reverse-phase HPLC in the presence of EGTA. Both forms may be physiologically active since both bind Ca^{2+} and Zn^{2+} in radioactive-ion-transblot electrophoresis.

TISSUE-SPECIFIC DISTRIBUTION OF CALCYCLIN

Expression of calcyclin in different cell lines and mouse tissues was determined with polyclonal antibodies raised against calcyclin from EAT cells (Kuźnicki et al. 1989b). The antibody recognized a calcyclin-like polypeptide in immunoblots of a variety of cells and mouse tissues. The strongest reactions are observed in mouse stomach, and in rat and chicken fibroblasts. The protein was also detected in mouse skeletal and cardiac muscle, in lung, kidney, and spleen. No positive reaction was found in mouse brain, liver, and intestine. The antibodies against calcyclin from EAT cells did not detect any protein in PtK_2 cells, Ag8 or V2 rabbit carcinoma cells. A calcyclin-like protein was purified from mouse stomach (using the procedure developed for EAT calcyclin) and is very similar to the EAT protein (Kuźnicki et al. 1989b).

A Ca^{2+}-dependent sialic acid-binding protein has been purified from several human tissues using fetuin-Sepharose (Gabius et al. 1989). The yield of protein was 1.2 µg/g from skeletal muscle, 2.4 µg/g from kidney cortex, 4.3 µg/g from heart muscle, 6.1 µg/g from kidney parenchyma, and 12.4 µg/g from Hodgkin tumor cells grown as tumors in nude mice. Partial sequence analysis of this protein revealed identity to human calcyclin. These results suggest that calcyclin binds sialic acid in a Ca^{2+}-dependent manner (Gabius et al. 1989).

When cross-sections of mouse skeletal muscle were incubated with calcyclin antibodies and inspected by light and electron microscopy, it was found that only fibroblasts were stained, but not the muscle fibers (Kuźnicki et al. 1989b).

TARGET PROTEINS OF CALCYCLIN

Two proteins with molecular mass of 36 kDa have been purified from Ehrlich ascites tumor cells by calcyclin-Sepharose affinity chromatography (Kuźnicki and Filipek 1990). Both proteins bind to calcyclin in the presence of calcium ions, and dissociate in their absence. One of them binds to phospholipids in a Ca^{2+}-dependent manner. This protein was recognized by the antibody against intestinal calpactin (Gerke and Weber 1984). The partially estimated amino acid sequence of this protein (including a fragment from the N-terminal end) is identical with that of pig calpactin. The second 36 kDa protein from EAT cells that binds to calcyclin does not bind to phospholipids, and is not recognized by the antibody against calpactin. The amino acid sequence of the five peptides derived from this protein appears to be identical with the sequence of glyceraldehyde 3-phosphate dehydrogenase (Filipek, Gerke, Weber, Kuźnicki in preparation).

The binding between calcyclin and calpactin was further studied using calcyclin-Sepharose column chromatography. It was found that not only the heavy chain of calpactin (36 kDa subunit), but also a 33 kDa proteolytic fragment (which lost the binding site for p11) bind to calcyclin in a Ca^{2+}-dependent manner. In addition, protein I, a complex between the 36 kDa calpactin subunit and 11 kDa subunit, also exhibits Ca^{2+}-dependent binding to calcyclin. This suggests that calcyclin and p11, two distinct proteins that belong to the S-100 protein family, bind simultaneously to different sites of the 36 kDa subunit of calpactin.

To verify the above results, the binding between calcyclin and glyceraldehyde 3-phosphate dehydrogenase was studied on glycerol gradient. It appears that the commercially available enzyme forms a complex with calcyclin in the presence of calcium ions, but not in their absence. Although these studies suggest that calpactin and glyceraldehyde 3-phosphate dehydrogenase might be the target proteins of calcyclin, it is possible that this interaction is of no physiological significance. However, purification of calpactin and glyceraldehyde 3-phosphate dehydrogenase on calcyclin-Sepharose column is highly reproducible and specific. They are the only two proteins in the soluble fraction of EAT cells that bind to calcyclin.

THE PROLACTIN RECEPTOR-ASSOCIATED PROTEIN IS A CALCYCLIN

Antibodies against a partially purified rabbit mammary gland prolactin receptor have been raised in rabbits. One of these antisera was shown to inhibit prolactin binding to rabbit mammary gland membranes and to inhibit selectively the biological activity of prolactin on rabbit mammary gland explants. Using this antiserum, a cDNA which encodes an approximately 10 kDa putative Ca^{2+}-binding protein has been isolated (Murphy et al. 1988). The protein identified by the antibody was originally named a prolactin receptor-associated protein, but from the nucleotide sequence of the cDNA, it appears to be identical with calcyclin.

The association of calcyclin with prolactin receptor in the rabbit mammary gland is unlikely to be merely coincidental. Calcyclin protein may be a part of the prolactin signal transduction mechanism (since Ca^{2+} seems involved in prolactin action) or alternatively, it may share sequence similarity to the prolactin receptor.

CALCYCLIN AS A MEMBER OF S-100 PROTEIN FAMILY

The amino acid sequence and the physico-chemical properties of calcyclin indicate that it belongs to the S-100 protein family (reviewed by Kligman and Hilt 1988). It is an acidic calcium binding protein with molecular mass of about 10 kDa (Table 2). It forms disulfide linked dimers and binds zinc ions, as observed for the beta subunit of S-100 protein from brain. Upon calcium binding calcyclin changes its conformation, exposing hydrophobic domain(s) that allows it to interact with proteins and hydrophobic matrices. The calcyclin sequence contains two EF-hand elements that may be responsible for Ca^{2+}-binding. There is a basic domain containing the N-terminal EF-hand, and an acidic domain (a typical calmodulin fold) containing the C-terminal EF-hand (Figure 3).

The S-100 proteins were initially purified from bovine brain, but they are present in a variety of organisms and tissues. Recently, many new members of the S-100 protein family, in addition to calcyclin, have been described: cystic fibrosis antigen (MRP-8), p9Ka, 18A2

(42A), MRP-14, p11 (42C) (a subunit of calpactin) and vitamin D-induced intestinal protein (reviewed by Kligman and Hilt 1988).

Table 2. Properties of calcyclin protein

M_r	About 10000
Ions binding	Two ions of Ca^{2+}, Zn^{2+}
Dimer	Disulfide-linked
Ca^{2+}-induced changes of conformation	UV spectrum, fluorescence intensity shift of electrophoretic mobility hydrophobicity
Ca^{2+}-dependent binding to	Phenyl-Sepharose, fluphenazine-Sepharose, fetuin-Sepharose (sialic acid), calpactin (p36 or p36-p11 complex) glyceraldehyde 3-phosphate dehydrogenase
Association with	Prolactin receptor
Distribution	Tissue and cell specific

To date, S-100 proteins have no demonstrable enzymatic functions; they probably exert their biological effects by binding to, and modulating effector proteins in a manner similar to calcium-modulated proteins such as calmodulin. The interesting feature of these proteins is that they appear to be cell-specific and that they are expressed either in a particular phase of the cell cycle (like calcyclin, S-100 beta or 18A2) or are linked to differentiation events. Also, S-100 proteins have often been described in association with human diseases: cystic fibrosis antigen, cystic fibrosis: MRP-8 and MRP-14, associated with rheumatoid arthritis: alpha subunit of S-100, in serum of patients with renal cell carcinoma: calcyclin, in acute myeloid leukemia, Hodgkin tumor cells and breast cancer cell lines.

The function of calcyclin is unknown at present, but several speculations have been made. Baserga and his coworkers suggest that it may play a role in cell cycle progression. We have found that in vitro, calcyclin interacts with calpactin and glyceraldehyde 3-phosphate dehydrogenase. It is possible that such interaction occurs in vivo. The association of calcyclin with prolactin receptor and its ability to bind sialic acid may indicate another function(s).

REFERENCES

Calabretta B, Kaczmarek L, Mars W, Ochoa D, Gibson CW, Hirschorn RR, Baserga R (1985) Cell-cycle-specific genes differentially expressed in human leukaemias. Proc Natl Acad Sci USA 82:4463-4467

Calabretta B, Battini R, Kaczmarek L, de Riel JK, Baserga R (1986a) Molecular cloning of the cDNA for a growth factor-inducible gene with strong homology to S-100, a calcium binding protein. J Biol Chem 261:12628-12632

Calabretta B, Venturelli D, Kaczmarek L, Narni F, Talpaz M, Anderson B, Beran M, Baserga R (1986b) Altered expression of G_1-specific genes in human malignant myeloid cells. Proc Natl Acad Sci USA 83:1495-1498

Ferrari S, Calabretta B, de Riel JK, Battini R, Ghezzo F, Lauret E, Griffin C, Emanuel BS, Gurrieri F, Baserga R (1987) Structural and functional analysis of a growth-regulated gene, the human calcyclin. J Biol Chem 262: 8325-8332

Filipek A, Heizmann CW, Kuźnicki J (1990) Calcyclin is a calcium and zinc binding protein. FEBS Lett 264:263-266

Gabius H-J, Bardosi A, Gabius S, Hellmann KP, Karas M, Kratzin H (1989) Identification of a cell cycle-dependent gene product as a sialic acid-binding protein. Biochem Biophys Res Commun 163:506-512

Gerke V, Weber K (1984) Identity of p36K phosphorylated upon Rous sarcoma virus transformation with a protein purified from brush borders. Calcium-dependent binding to non-erythroid spectrin and actin. EMBO J 3:227-233

Ghezzo F, Lauret E, Ferrari S, Baserga R (1988) Growth factor regulation of the promoter for calcyclin, a growth-regulated gene. J Biol Chem 263:4758-4763

Hirschorn RR, Aller P, Yuan Z-A, Gibson CW, Baserga R (1984) Cell-cycle-specific cDNAs from mammalian cells temperature sensitive for growth. Proc Natl Acad Sci USA 81:6004-6008

Kligman D, Hilt DC (1988) The S-100 protein family. TIBS 13: 437-443

Kuźnicki J, Filipek A (1987) Purification of a novel Ca^{2+} binding protein (10.5 kDa) from Ehrlich ascites tumor cells. Biochem J 247:663-667

Kuźnicki J, Filipek A (1990) Calcyclin-like protein from Ehrlich ascites tumor cells, Ca^{2+}-binding properties, distribution and target protein. In: Lawson DEM, Pochet R (eds) Calcium binding proteins in normal and transformed cells. Plenum Publishing Corporation, New York, p 145-148

Kuźnicki J, Filipek A, Hunziker PE, Huber S, Heizmann CW (1989a) Calcium-binding protein from mouse Ehrlich ascites tumor cells is homologous to human calcyclin. Biochem J 263:951-956

Kuźnicki J, Filipek A, Heimann P, Kaczmarek L, Kaminska B (1989b) Tissue specific distribution of calcyclin, 10.5 kDa Ca^{2+}-binding protein. FEBS Lett 254:141-144

Murphy LC, Murphy LJ, Tsuyuki D, Duckworth ML, Shiu RPC (1988) Cloning and characterization of a cDNA encoding a highly conserved, putative calcium binding protein, identified by an anti-prolactin receptor antiserum. J Biol Chem 263:2397-2401

Rittling SR, Gibson CW, Ferrari S, Baserga R (1985) The effect of cycloheximide on the expression of cell cycle dependent genes. Biochem Biophys Res Commun 132:327-335

Rittling SR, Brooks KM, Cristofalo VJ, Baserga R (1986) Expression of cell cycle dependent genes in young and senescent WI-38 fibroblasts. Proc Natl Acad Sci USA 83: 3316-3320

CLONING AND CHARACTERIZATION OF A *cDNA* ENCODING A HIGHLY CONSERVED Ca^{2+}-BINDING PROTEIN, IDENTIFIED BY AN ANTI-PROLACTIN RECEPTOR ANTIBODY

L.C. Murphy, Y. Gong, and R.E. Reid

INTRODUCTION

Prolactin receptors are found in many tissues (Posner et al. 1974), which suggests that prolactin may have multiple functions. Certainly prolactin has been shown to exert a wide variety of activities in different species (Nicoll and Bern 1972). The molecular mechanisms which mediate the physiological actions of prolactin are, however, unknown. Although the specific interaction of the hormone with its cell surface receptor on target cells is the first step, the subsequent mechanism(s) of signal transduction is unknown. The recent cloning of the prolactin (Boutin et al. 1988) and growth hormone (Leung et al. 1987) receptors has revealed that they is a new class of membrane receptors without sequence similarities to other known membrane receptors. Moreover, there appear to be at least two different forms of the prolactin receptor, a short and a long form. The major mRNA species for the prolactin receptor in the rat liver is 2.2 kb, whereas the major species in the rat mammary gland is 4 kb (Boutin et al. 1988). Prolactin receptor cDNAs have also been isolated from rabbit mammary gland (Edery et al. 1989) and T-47D human breast cancer cells (Boutin et al. 1989) and these cDNAs predict mRNAs which encode 592 and 598 amino acid residue mature proteins, respectively. These are in contrast to the 291 amino acid residue mature protein predicted from the rat liver cDNA. The difference between the two forms of the prolactin receptor resides primarily in the cytoplasmic domain, 57 residues for the rat liver prolactin receptor versus 358 for the rabbit mammary gland receptor (Edery et al. 1989; Boutin et al. 1989). Due to this variability in the size of the mature prolactin receptor protein, it is possible that multiple mechanisms of signal transduction may occur for prolactin receptors, in a species- and tissue-specific manner.

In the course of studying the molecular mechanism of prolactin action, antibodies had been raised in guinea pigs to a partially purified pregnant rabbit mammary gland prolactin receptor preparation (Shiu and Friesen 1976). One of these antisera, Ab 7, was shown to inhibit prolactin binding to rabbit mammary gland membranes and to inhibit selectively the biological activity of prolactin on rabbit mammary gland explants (Shiu and Friesen 1976a). This antiserum was also effective in inhibiting the binding of prolactin to membranes prepared from several other tissues from a number of species and in particular human mammary carcinomas (Shiu and Friesen 1976). Furthermore, we have shown that this antiserum inhibits the binding of ^{125}I-human prolactin to the prolactin receptor in T-47D human breast cancer cells (Murphy et al. 1988).

The T-47D human breast cancer cell line possesses a high level of prolactin receptor and responds biologically to prolactin (Shiu and Paterson 1984; Shiu and Iwasiow 1985). In the process of screening a lambda GT 11 cDNA expression library prepared from hydro-cortisone and human growth hormone-treated T-47D mRNA (Murphy et al. 1987) with antiprolactin receptor antiserum, Ab 7, we have isolated a cDNA which encodes an approximately 10 kDa, putative calcium-binding protein (Murphy et al. 1988). This gene has been recently characterized in human fibroblasts (Calabretta et al. 1986) as a cell cycle-regulated gene, calcyclin. Since calcyclin obviously co-purified with the prolactin receptor from the rabbit mammary gland to which the antibodies were raised, it is tempting to speculate that the copurification might be of functional significance.

CHARACTERIZATION AND SEQUENCING OF THE HUMAN cDNA ISOLATED WITH ANTI-PROLACTIN RECEPTOR ANTIBODIES

Antiserum Ab-7 was used to screen a lambda GT 11 cDNA expression library prepared from T-47D cells and two clones containing inserts of different sizes (approx. 500 bp and 400 bp) were isolated. The larger cDNA was used for subsequent analyses. The cDNA recognized a single species of mRNA of around 600 bases (Figure 1) in poly(A$^+$) RNA isolated from pregnant rabbit mammary gland and T-47D human breast cancer cells.

The entire length of both strands of this cDNA was sequenced. The nucleotide sequence of this human cDNA and the predicted amino acid sequence starting from the initiator methionine are shown in Figure 2. The presumptive initiator codon ATG is located at nucleotide position 103 and the termination codon TGA is located at position 373, thus providing an open reading frame for a protein of 90 amino acids and a calculated molecular mass of 10.1 kDa. No glycosylation signals are present. The polyadenylation signal AATAAA is located at nucleotide position 447 and the polyadenylation begins 18 nucleotides downstream.

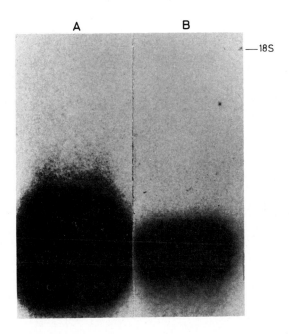

Fig. 1A,B. Northern blot analysis of RNA from T-47D human breast cancer cells (A) and pregnant rabbit mammary gland RNA (B). Five micrograms of poly(A$^+$) RNA were resolved by electrophoresis and transferred to nitrocellulose membrane. The pattern of hybridization with the PRA/calcyclin cDNA is shown, and the position of the 18S ribosomal marker is indicated by the *arrow*

172

The protein encoded by this cDNA was characterized by *in vitro* translation of the hybrid-selected mRNA. One major protein band of approximately 10 kDa was produced (Figure 3, lane C). This protein was precipitable with Ab-7 (Figure 3, lane E) but not with Ab-1 (raised to vehicle alone Figure 3, lane D). The size of the protein encoded by the cDNA suggests that it is not the prolactin receptor, since the prolactin receptor has been estimated considerably larger, 41-88 kDa (Mitani and Dufau 1986; Katoh et al. 1987). However, since the protein encoded by this cDNA co-purified with the prolactin receptor, we tentatively named this protein a prolactin receptor-associated protein (PRA).

```
           10              20              30              40              50
GGG  ACC  GCT  ATA  AGG  CCA  GTC  GGA  CTG  CGA  CAT  AGC  CCA  TCC  CCT  CGA  CCG  CTC  GCG

 60              70              80              90             100             110
TCG  CAT  TTG  GCC  GCC  TCC  CTA  CCG  CTC  CAA  GCC  CAG  CCC  TCA  GCC  ATG  GCA  TGC  CCC
                                                                      Met  Ala  Cys  Pro
          120             130             140             150             160             170
CTG  GAT  CAG  GCC  ATT  GGC  CTC  CTC  GTG  GCC  ATC  TTC  CAC  AAG  TAC  TCC  GGC  AGG  GAG
Leu  Asp  Gln  Ala  Ile  Gly  Leu  Leu  Val  Ala  Ile  Phe  His  Lys  Tyr  Ser  Gly  Arg  Glu
               180             190             200             210             220
GGT  GAC  AAG  CAC  ACC  CTG  AGC  AAG  AAG  GAG  CTG  AAG  GAG  CTG  ATC  CAG  AAG  GAG  CTC
Gly  Asp  Lys  His  Thr  Leu  Ser  Lys  Lys  Glu  Leu  Lys  Glu  Leu  Ile  Gln  Lys  Glu  Leu
230             240             250             260             270             280
ACC  ATT  GGC  TCG  AAG  CTG  CAG  GAT  GCT  GAA  ATT  GCA  AGG  CTG  ATG  GAA  GAC  TTG  GAC
Thr  Ile  Gly  Ser  Lys  Leu  Gln  Asp  Ala  Glu  Ile  Ala  Arg  Leu  Met  Glu  Asp  Leu  Asp
     290             300             310             320             330             340
CGG  AAC  AAG  GAC  CAG  GAG  GTG  AAC  TTC  CAG  GAG  TAT  GTC  ACC  TTC  CTG  GGG  GCC  TTG
Arg  Asn  Lys  Asp  Gln  Glu  Val  Asn  Phe  Gln  Glu  Tyr  Val  Thr  Phe  Leu  Gly  Ala  Leu
          350             360             370             380             390
GCT  TTG  ATC  TAC  AAT  GAA  GCC  CTC  AAG  GGC  TGA  AAA  TAA  ATA  GGG  AAG  ATG  GAG  ACA
Ala  Leu  Ile  Tyr  Asn  Glu  Ala  Leu  Lys  Gly
400             410             420             430             440             450
CCT  CTG  GGG  GTC  CTC  TCT  GAG  TCA  AAT  CCA  GTG  GTG  GGT  AAT  TGT  ACA  ATA  AAT  TTT
     460             470             480             490
TTT  TGG  TCA  AAT  TTA  AAA  AAA  AAA  AAA  AAA  AAA  AA
```

Fig. 2. Nucleotide sequence and derived amino acid sequence of human PRA/calcyclin. The *numbers above* the sequence refer to the nucleotide positions. The initiation (*ATG*) and termination (*TGA*) codons and the polyadenylation signal (*AATAAA*) are *underlined*

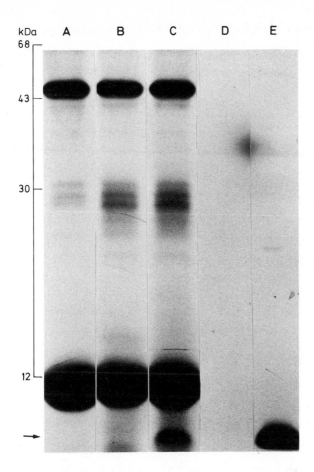

Fig. 3. Hybrid-select translation. Twenty micrograms of poly(A⁺) RNA were subjected to hybrid-selection analysis using pBR322 DNA or pPRA cDNA. Hybrid-selected RNA was translated in vitro using rabbit reticulocyte lysate. The total in vitro translation products obtained from the hybrid-selected RNA were subjected to immunoprecipitation with Ab-1 and Ab-7. The resulting products were analyzed by electrophoresis on 15% polyacrylamide/SDS gels. The molecular mass marker ^{14}C-proteins were bovine serum albumin (68 kDa), ovalbumin (43 kDa), carbonic anhydrase (30 kDa), and cytochrome c (12 kDa). *Lane A* blank (no added RNA); *lane B* in vitro translation of hybrid-selected RNA using pBR322 DNA; *lane C* in vitro translation of hybrid-selected RNA using pPRA cDNA (the *arrow* indicates the protein of interest); *lane D* immunoprecipitation using Ab-1 (raised to vehicle alone) of the in vitro translation products of RNA hybrid-selected using pPRA cDNA; *lane E* immunoprecipitation with Ab-7 (antiprolactin receptor antibodies)

SIMILARITY OF THE *PRA* GENE TO OTHER KNOWN GENES AND IDENTIFICATION OF A POSSIBLE FUNCTION OF THE *PRA* GENE PRODUCT

Comparison of the nucleotide sequence of the PRA cDNA with those stored in Genbank indicated that this gene has been recently cloned and characterized from human fibroblasts by Calabretta et al. (1986) as a cell cycle-regulated gene, calcyclin. The identity of the PRA gene with the calcyclin gene and the similarities of the predicted amino acid sequence of this gene to S-100 subunits (Kuwaro et al. 1984) and the vitamin D-dependent intestinal calcium binding protein (Fulmer and Wasserman 1981) suggest that the PRA/calcyclin gene may encode a protein which belongs to the EF-hand-like class of calcium-binding proteins which are thought to be important in calcium-mediated signal transduction in a number of biological systems (Van Eldik et al. 1982). Other proteins which appear to have significant sequence similarity to the PRA/calcyclin gene product are the 18A2, a mouse placental calcium-binding protein, which is also a serum-inducible mRNA (Jackson-Grusby et al. 1987), MRP-8 and MRP-14, migration inhibitory factor-related proteins which form a complex to become the cystic fibrosis antigen (Odink et al. 1987; Bruggen et al. 1988), and the p11 (or p10) subunit of a protein complex that serves as an *in vivo* substrate of various tyrosine-specific protein kinases (Gerke and Weber 1985). Furthermore, the putative calcium-binding domains of these proteins, of which there appear to be two, and the immediate flanking residues are highly conserved.

Fig. 4. Amino acid sequences of human prolactin receptor associated protein (calcyclin) and related calcium binding proteins. The calcium-binding features are based on the refined crystal structure of bovine calbindin. The alpha-helical regions in the presence of calcium are represented by *horizontal bars*. The unconventional calcium binding site (designated Unit II) interacts with calcium through the peptide carbonyl oxygens of the residues designated *X*, *Y*, *Z* and *-Y*, and the side chain oxygens of the residue designated *-Z*. The *-X* coordinate (not shown) is occupied by a water molecule. The conventional EF-hand calcium binding site (designated *Unit I*) interacts with the calcium cation through the side chain oxygens of residues designated *X*, *Y*, *Z*, *-X* and *-Z*. The interaction of the side chain of the residue in the *-Y* position with the cation is mediated by a water molecule. The *numbering system* is based on human prolactin receptor associated protein. Amino acids are designated using the *one letter code*. The sequence alignments are arbitrarily based on maximizing the alignment of residues in the calcium binding sites. Superscripts *1-8* refer to references as follows: *1* Murphy et al. 1988; *2* Barraclough et al 1987; *3* Odink et al. 1987; *4* Fullmer and Wasserman 1981; *5* Isobe and Okuyama 1981; *6* Saris et al. 1987; *7* Gerke and Weber 1985; *8* Masiakowski and Shooter 1988.
Asterisk: identical in amino acid sequence to 2A9 (Calabretta et al. 1986)
Two asterisks: identical in amino acid sequence to 42C (Masiakowski and Shooter 1988) and 18A2 (Jackson et al. 1987)

In the PRA/calcyclin amino acid sequence these calcium-binding sites consist of residues 20 to 33 and residues 61 to 72 (Figure 4, Murphy et al. 1988).

```
                                                        UNIT II


                       ------------------------X   Y   Z  -Y        -Z------------
            1               10            20                      30
 *h-PRA¹       M A C P L D Q A I G L L V A I F H K Y S G R E G D K H T L S K K E L K E L I Q
  r-PRA¹       M A C P L D Q A I G L L V A I F H K Y S G K E G D K H T L S K K E L K E L I Q
 **p9Ka²       M A R P L E E A L D V I V S T F H K Y S G N E G D K F K L N K T E L K E L L T
  MRP-8³       M L T E L E K A L N S I I D V Y H K Y S L I K G N G H A V Y R D D L K K L L E
  MRP-14³  M T C K M S Q L E R N I E T I I N T F H Q Y S V K L G H P D T L N Q G E F K E L V R
 b-CALBINDIN⁴            K S P E E L K G I F E K Y A A K E G D P N Q L S K E E L K L L L Q
 b-S-100a⁵        G S E L E T A M E T L I N V F H A H S G K E G D K Y K L S K E E L K E L L Q
 b-S-100b⁵          S E L E K A V V A L I D V F H Q Y S G R E G D K H K L K K S E L K E L I N
 b-p11⁶        M P S Q M E H A M E T M M F T F H K F A G D K G - - - Y L T K E D L R V L M E
 m-p11⁶        M P S Q M E H A M E T M M L T F H R F A G D K - D H - - L T K E D L R V L M E
 p-p11⁷        M P S Q M E H A M E T M M F T F H K F A G D K G - - - Y L T K E D L R V L M E
 42c⁸          M P S Q M E H A M E T M M L T F H R F A G - - - E K N Y L T K E D L R V L M E


                                                        UNIT I


            ------                ---------------------------X   Y   Z  -Y  -X  -Z---------
            40              50                        60                    70
 h-PRA         K E L T I G S K L Q D A E I A R L - - - - M E D L D R N K D Q E V N F Q E Y V T F
 r-PRA         K E L T I G A K L Q D A E I A R L - - - - M D D L D R N K D Q E V N F Q E Y V A F
 p9Ka          R E L P S F L G R R T D E A F Q K - - - L M N N L D S N R D N E V D F Q E Y C V F
 MRP-8         T E C P - - Q Y I R - K K G A D - V - W - F K E L D I N T D G A V N F Q E F L I L
 MRP-14        K D L Q N F L K K E N K E K V I E - - H I M E D L D T N A D K Q L S F E E F I M L
 b-CALBINDIN   T E F P S L L K G P S T L D - E - - - - L F E E L D K N G D G E V S F E E F Q V L
 S-100a        T E L S G F L D A Q K D - - A D A V D K V M K E L D E D G D G E V D F Q E Y V V L
 S-100b        N E L S H F L E E I K E - - Q E V V D K V M E T L D S D G D G E C D F Q E F M A F
 b-p11         T E F P G F L E N Q K D P L - - A V D K I M K D L D Q C R D G K V G F Q S F F S L
 m-p11         R E F P G F L E N Q K D P L - - A V D K I M K D L D Q C R D G K V G F Q E F L S L
 p-p11         K E F P G F L E N Q K D P L - - A V D K I M K D L D Q C R D G K V G F Q S F F S L
 42C           R E F P G F L E N Q K D P - - - A V D K I M K D L D Q C R D G K V G F Q S F L A L


            ---------
            80                90
 h-PRA         - - L G A L A - - - L I Y N E A L K G
 r-PRA         - - L G A L A - - - L I Y N E A L K G
 p9Ka          - - L S C I A - - - M M C N E F F E G C P D K E P R K K
 MRP-8         V I K M F V A - - - A A H K K S H E E S H - K E
 MRP-14        M A R L T W A S H E K M H E G D E G P G H H H K P G L G E G T P
 b-CALBINDIN   V K K I S Q
 S-100a        V A A L T V A C N N F F W E N S
 S-100b        V A M I T T A C H E F F - E H E
 b-p11         I A G L T I A C N D Y F V V H M K Q K G K K
 m-p11         V A G L T I A C N D Y F V V N M K Q K G K K
 p-p11         I A G L T I A C N D Y F V V H M K Q K G K
 42C           V A G L I I A C N D Y F V V H M K Q K - K
```

Fig. 4.

IS *PRA*/CALCYCLIN A CALCIUM BINDING PROTEIN?

We speculate using the available sequence analyses and homologies that both putative calcium-binding sites may actually bind calcium. Certainly the calcium-binding site closest to the carboxy terminus, i.e., the 12 residues from 61-72 inclusive and 9 to 10 residues on each side of this section, contains amino acids which would produce the characteristic helix-loop-helix conformation of the EF-hand type in the presence of calcium (Kretsinger and Nockold 1973).

The amino acid residues which occupy the X, Y, Z, -Y, -X, -Z coordinates (i.e., X=residue 61=Asp; Y=residue 63=Asn; Z=residue 65=Asp; -Y=residue 67=Glu; -X=residue 69=Asn; -Z=residue 72=Glu) all have the appropriate structure to allow for side chain or, in the case of residue 67, the peptide carbonyl oxygen interaction with the calcium cation.The NH_2-terminal putative calcium-binding site of 14 residues, i.e., amino acid 20 to 33, is less characteristic of the EF-hand-type site. In keeping with the crystal structure of a similar calcium-binding site in intestinal calcium-binding protein (Szebenyi and Moffat 1986), residues 20, 23, 25, 28, and 33 would be considered the chelating residues. Residues 20, 23, 25, and 28 may interact with calcium via the peptide carbonyl oxygen while the two carbonyl oxygens of the side chain of Glu 33 provide the remaining chelating oxygen atoms. Alternatively, this site may be involved in binding other cations. Interestingly, the S-100 protein has been shown to have a high affinity zinc binding site distinct from its calcium binding site (Baudier and Gerard 1983). Although the S-100 zinc-binding site is not yet defined, it lies in the NH_2-terminal region. This site probably involves histidine residues and it may be significant that histidine-17 in PRA/calcyclin is conserved in S-100-alpha and S-100-beta (Figure 4, Murphy et al. 1988). Recently, it has been demonstrated that the mouse calcyclin homolog does indeed bind Ca^{2+} (Kuznicki and Filipek 1987; Kuznicki et al. 1989) and possibly zinc ions (Filipek et al. 1989).

A sensitivity to pH is a characteristic of the S-100 protein subunits which bear sequence similarity to PRA/calcyclin and titration in vitro of the S-100 protein from pH 8.5 to 6.5 results in loss of the higher affinity calcium binding site (Szebenyi et al. 1981). Interestingly, PRA/calcyclin has a titratable histidine residue in the middle of the first putative calcium binding site at position 27, which could in principle, bestow similar pH sensitivity on PRA/calcyclin as occurs in the S-100 protein. If true, this may have potential physiological relevance, since growth factor and hormone interaction with respective target cells has been shown to be associated with significant changes in intracellular pH (Hesketh et al. 1985).

DISTRIBUTION OF *PRA* EXPRESSION IN HUMAN BREAST CANCER CELLS

Since the PRA cDNA was obtained from a human breast cancer cell library, the expression of PRA mRNA in different human breast cancer cell lines was investigated. PRA mRNA was present in all human breast cancer cell lines studied, with the exception of MCF 7 cells (Figure 5). Little, if any, PRA mRNA is detectable in RNA isolated from MCF 7 cells.

The human PRA cDNA isolated from the T-47D human breast cancer expression library contains 37 more nucleotides upstream in the 5'-untranslated region compared to the human calcyclin cDNA (Figure 6; Murphy et al. 1988). Ferrari et al. (1987) have shown by S1 nuclease mapping and genomic cloning that the cap site is indeed at the most 5' portion of their cDNA, i.e., position 38-40 of our cDNA. The extra 37 nucleotides in our cDNA are identical in nucleotide sequence (with the exception of nucleotide position 1, 2, and 33 of the PRA cDNA) to the genomic sequence of the human calcyclin gene from positions -1 to -37. Therefore, the PRA cDNA contains the nucleotide sequences of the calcyclin gene TATA box in its 5' untranslated region. This suggested that different start sites of transcription of the calcyclin gene were being used in breast cancer cells versus the fibroblasts. Using RNase protection assays we have now confirmed that different start sites of transcription are being used in T-47D as well as other human breast cancer cell lines compared to human fibroblasts (Figure 7). The pattern of start sites of transcription does not appear to be an vitro artifact since RNA isolated from a range of human breast cancer biopsy samples (Figure 7B) was found to display essentially the same pattern as seen for human breast cancer cells in culture (Figure 7A).

178

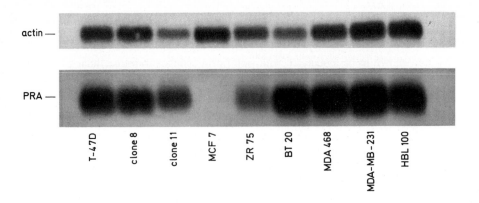

Fig. 5. Northern blot analysis of human breast cancer cell RNA. Ten micrograms of poly(A⁺) RNA were resolved by electrophoresis and transferred to nitrocellulose paper. The *upper panel* shows the pattern of hybridization obtained with the chicken ß-actin cDNA, while the *lower panel* shows the pattern of hybridization obtained with the PRA cDNA

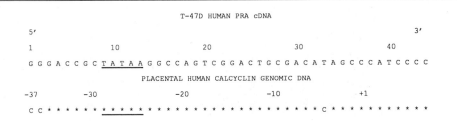

Fig. 6. Comparison of the 5'-untranslated nucleotide sequence of the T-47D human PRA mRNA with that of the transcription start site and 37 nucleotides upstream in the 5'-flanking region of the calcyclin gene (Ferrari et al. 1987). The TATA box is *underlined*

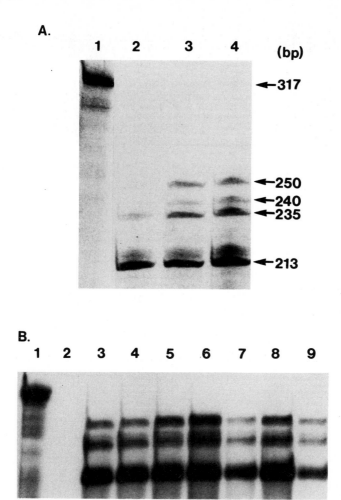

Fig. 7A,B. RNase protection assays: **A** A transcription vector consisting of the first 250 base pairs (from the 5'-end) of the human PRA/calcyclin cDNA isolated from T-47D cells, in pSP64 was constructed. P^{32}-labeled antisense riboprobe was prepared and hybridized (80% formamide, 40 mM PIPES pH 6.4, 0.4M NaCl, 1 mM EDTA, 42 °C) overnight with 5 μg of total RNA. After hybridization the samples were digested with RNase A and RNase T. The resulting protected fragments were separated and sized on a 6% PAGE/8M urea sequencing gel. The results were visualized by autoradiography. When yeast tRNA was hybridized with the riboprobe, no protected fragment was detected (data not shown). *Lane 1* probe alone; *Lane 2* 5 μg human fibroblast RNA; *Lane 3* 5 μg T-47D RNA; *Lane 4* 5 μg MDA 468 human breast cancer cell RNA. The *arrows* indicate the size of the protected fragments

B Twenty-five micrograms of total RNA from a range of human breast cancer biopsy samples were treated as described above. *Lane 1* probe alone; *Lane 2* yeast t-RNA; *Lanes 3-9* human breast cancer biopsy samples

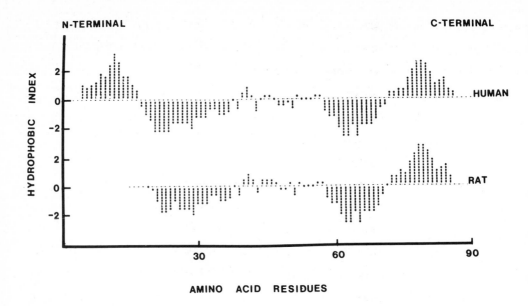

Fig. 8. Comparison of the hydropathy plots of the derived amino acid sequences for human PRA and rat PRA. Residues *above the center line* represent hydrophobic residues and those *below the line* are hydrophilic. Segments of nine residues were analyzed using Pustell Sequence Analysis Programs (International Biotechnologies, Inc.)

The significance of this observation remains to be investigated. It may reflect different mechanisms of transcriptional regulation of this gene which may exist between human neoplastic cells versus normal cells or alternatively between epithelial versus fibroblastic cells.

CLONING AND SEQUENCING OF THE RAT *PRA* cDNA

The sequence similarity of PRA/calcyclin between species was investigated by isolating the rat PRA cDNA from a rat kidney lambda GT 10 cDNA library (Murphy et al. 1987) by screening with the human PRA cDNA at high stringency. The rat PRA/calcyclin cDNA was sequenced and the predicted amino acid sequence was found to be identical to that of the human PRA/calcyclin except for four conservative amino acids which are due to single nucleotide changes (see Figure 3 in Murphy et al. 1988). Hydropathy analyses (Figure 8) confirm the high degree of similarity between the species and also indicate the apparent lack of a hydrophobic signal peptide sequence. The mouse PRA/calcyclin has recently been cloned (Guo et al. in press) as a serum-inducible mRNA (5B10) in mouse fibroblasts and the coding region of this protein has 94% and 97% amino acid identity with the human and rat calcyclin, respectively. The highly conserved nature of the PRA/calcyclin gene between species suggests a potentially important function for the gene product.

TISSUE DISTRIBUTION OF RAT *PRA*/CALCYCLIN EXPRESSION

The expression of PRA/calcyclin in rat tissue was studied. Northern blot analysis indicated that the PRA/calcyclin was expressed in a number of rat tissues (Figure 9a). Those tissues which expressed the PRA/calcyclin to the greatest degree were the lung, kidney and uterus. Interestingly, the postpartum (PP), lactating rat mammary gland expressed little PRA/calcyclin. Our studies indicate that while the pregnant mammary gland expresses high levels of the PRA a marked decrease in expression occurred with increasing stage of pregnancy and in the lactating gland (Figure 9b).

DEVELOPMENT OF SPECIFIC ANTIBODIES TO *PRA*/CALCYCLIN

To investigate the PRA/calcyclin protein, antibodies that are specific to this protein have been developed. The previous antibody used to screen the cDNA expression library (Ab 7), although it contained antibodies to the PRA, also contained antibodies to various epitopes of the prolactin receptor. The use of this antibody (Ab 7) would therefore result in difficulties in interpreting data from studies in which we were trying to measure the PRA alone. Synthetic peptides identical to regions of the PRA protein were synthesized. The region chosen to be synthesized encompassed residues 8-45 (Figure 4; Murphy et al. 1988).The peptides were synthesized using a total sequential protocol for solid phase synthesis developed for the preparation of large calcium-binding fragments of troponin C (Reid 1987a) and calmodulin (Reid 1987b).

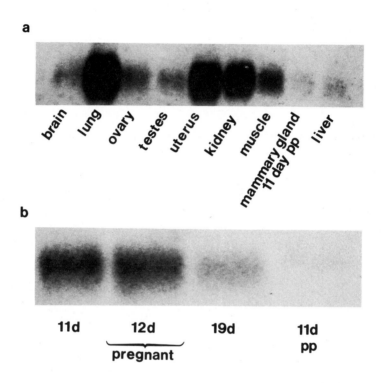

Fig. 9a,b. Northern blot analysis of RNA obtained from various rat tissues. **a** Five micrograms of poly(A⁺) RNA isolated from the tissues shown were subjected to Northern blot analysis. The pattern of hybridization obtained with the human PRA/cDNA is shown. **b** Ten micrograms of poly(A⁺) RNA, isolated from rat mammary glands at various stages of pregnancy (*d* day; *pp* postpartum) and lactation, were subjected to Northern blot analysis. The pattern of hybridization obtained with the human PRA/cDNA is shown

At preset points throughout the synthesis of the 38 residue fragment the synthesis was stopped and a large sample of peptide resin was taken prior to continuing synthesis. These samples provided peptides of smaller fragments (12, 25, and 38 residue fragments) for determination of antigenic properties. After synthesis and HF cleavage of the peptides from the resin, the peptides were purified using C8 and C18 reverse phase and cation exchange high pressure liquid chromatography both analytically and semi-preparatively. The synthetic peptides of PRA/calcyclin were conjugated to keyhole limpet hemocyanin, used to inject rabbits and high titer antibodies (specific detection of ≥ 5 ng of the 25 and 38 residue synthetic peptides at 1 in 10000 dilution of antibody) have been raised (Figure 10).

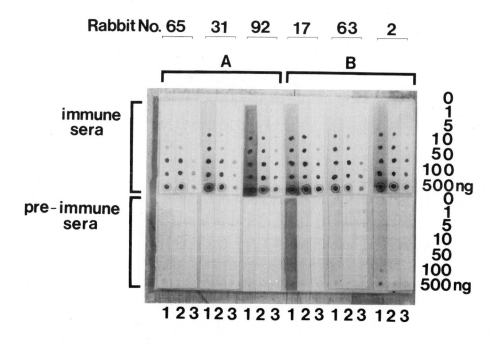

Fig. 10A,B. A The antigen used was a 38-residue peptide representing amino acids 8-45 of the human PRA (see Figure 3 in Murphy et al. 1988). **B** The antigen used was a 25-residue peptide representing amino acids 21-45 of the human PRA. 0-500 ng of antigen were immobilized on nitrocellulose and the filters were blocked with 3% bovine serum albumin. The filters were incubated for 1 h with various dilutions of antisera: **1** 1/100; **2** 1/1000; **3** 1/10000. The filters were washed and incubated for a further hour with alkaline phosphatase linked antirabbit antiserum (1/7500 dilution). After washing, the results were visualized by incubation with alkaline phosphatase substrate, following the manufacturers instructions

These antibodies have now been further characterized and do not cross-react with other EF-hand-type calcium-binding proteins such as calmodulin, troponin C and, parvalbumin. However, they do cross-react with bovine S-100 alpha and S-100 beta proteins (1 µg of S-100 α and ß proteins are detected by Western blotting, using 1/2000 dilution of the PRA/calcyclin antibody). The antibodies, at a dilution of up to 1/10000, specifically detect

184

by Western blotting a major 10 kDa protein in the soluble fraction of T-47D human breast cancer cells which have been grown under normal growth conditions (Figure 11).

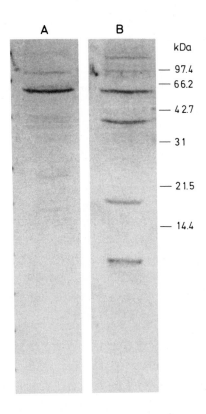

Fig. 11. Western blot analysis of 100000 g supernatant (soluble fraction) from T-47D cells. T-47D cells were homogenized and the soluble fraction was isolated. 10 μg of protein were separated on a 15% SDS/PAGE mini-gel system. The proteins were then subjected to Western blot analysis. *Lane A* filter incubated with pre-immune serum; *Lane B* filter incubated with human PRA antiserum (92-1). The results were visualized using alkaline phosphatase conjugated anti-rabbit IgG. This figure shows (for photographic purposes) results from a 1/100 dilution of hPRA (92-1), 1/1000 and 1/10000 dilution give sequentially less intense but similar results. *Arrows* indicate the standard molecular weight markers in kilodaltons (kDa)

This is the expected size for the PRA/calcyclin protein and corresponds to the protein which was detected by the Ab 7 antibody (Murphy et al. 1988). The 10 kDa protein which is detected by the PRA/calcyclin antibodies in T-47D cells is unlikely to be S-100 since anti-

S-100 antibodies (Axell rabbit polyclonal anti-bovine brain S-100, 1/200 dilution, Accurate Chemical Co.) do not detect any specific protein in T-47D cells by Western blotting.The data strongly suggest that the antibodies raised to synthetic PRA/calcyclin peptides do indeed detect the authentic PRA/calcyclin protein and can now be used to investigate the biochemical and functional characteristics of the hPRA/calcyclin protein. These antibodies also immunoprecipitate the in vitro translated protein product from 5B10 mRNA (i.e., mouse calcyclin, Guo et al. in press) synthesized by in vitro transcription. Interestingly, the antibodies also detect specifically two larger molecular weight proteins of approx. 17 and 35 kDa in the 100000 g supernatant of T-47D cells (Figure 11). These may represent oligomers of the calcyclin monomer (Kuznicki et al. 1989; Filipek et al. 1989) or may suggest that other members of this family of putative calcium-binding proteins may exist in T-47D cells.

DISCUSSION

Our observation that the PRA/calcyclin protein was recognized by antibodies prepared to the prolactin receptor suggested that the PRA/calcyclin co-purifies with the prolactin receptor under the conditions described by Shiu and Friesen (1976). The prolactin receptor preparation used to generate Ab-7 was prepared from crude membrane fractions of pregnant rabbit mammary glands by Triton-X-100 solubilization and human growth hormone-agarose affinity chromatography [note the ability of human growth hormone (hGH) to interact with the prolactin receptor is equal to that of human prolactin; Shiu 1979]. Magnesium chloride (5 M) was used to recover the prolactin receptor from the affinity matrix, with an approximately 1500-fold purification (Shiu and Friesen 1974). While it is recognized that the prolactin receptor preparation used contained other proteins, the relatively low abundance of PRA/calcyclin (as determined by mRNA in late pregnancy mammary gland see Figure 9B) suggests that it has considerable affinity for the prolactin receptor, hGH or the matrix itself (Affi-Gel 10) used in the purification procedure. Since the PRA/calcyclin is a putative calcium-binding protein, its association with the prolactin receptor in the rabbit mammary gland may not be merely coincidental. Calcium has been shown to be important in the binding of prolactin to its receptor (Shiu and Friesen 1974a) and has been implicated in the mechanism of prolactin action (Cameron and Rillema 1983), although others have found that there are no clear effects of prolactin on calcium ion regulation in target cells (Kelly et al. 1984; Houdebine et al. 1985). It is possible to speculate that the PRA/calcyclin protein may be part of the prolactin signal transduction mechanism especially since multiple mechanisms of signal transduction may occur for the apparently heterogenous prolactin receptor (Edery et al. 1989; Boutin et al. 1989). Alternatively, PRA/calcyclin may share sequence similarity to the prolactin receptor. However, comparison of sequences between the human prolactin receptor and calcyclin reveal no major sequence similarities.

The distribution of PRA/calcyclin gene expression, at least in the rat (see Figure 9A), does not necessarily follow that of the tissue distribution of prolactin receptors, e.g., the lactating rat mammary gland has low levels of PRA mRNA compared to the pregnant mammary gland, while prolactin receptor levels are higher in the lactating mammary gland compared to the tissue from pregnant rats (Shiu and Friesen 1981); the human breast cancer cell line MCF 7 has little if any PRA mRNA but contains prolactin receptors (Shiu 1979). Interestingly, MCF 7 cells, in contrast to T-47D cells, do not respond to prolactin morphologically or with increased lipid synthesis (Shiu and Paterson 1984). Furthermore, the prolactin-inducible protein (PIP) is not expressed in MCF 7 cells (Murphy et al. 1987). Since prolactin may have diverse and multiple functions in various tissues (Horrobin 1978), receptor heterogeneity may account for the failure to observe an identical correlation between PRA/calcyclin expression and the tissue distribution of prolactin receptors in the rat. Moreover, prolactin receptor heterogeneity at the mRNA as well as the protein level has now been observed (Boutin et al. 1988; Mitani and Dufau 1986; Katoh et al. 1987). The heterogeneity appears to reside in the cytoplasmic domain of the receptor (Edery et al. 1989), which suggests that different signal transduction mechanisms may exist in different tissues and species to trigger different biological responses of prolactin. Alternatively, this lack of correlation would be consistent with a more ubiquitous role of PRA/calcyclin in peptide hormone signal transduction. This would also be consistent with the observation of Calabretta et al. (1986) that some but not all mitogens increase the expression of calcyclin mRNA in fibroblasts.

The calcyclin gene has been shown by Calabretta et al. (1986) to be preferentially expressed when quiescent fibroblasts are stimulated to proliferate by serum, epidermal growth factor and platelet-derived growth factor, however, not all mitogens (e.g., insulin) increase the expression of this gene. Our tissue distribution studies (Murphy et al. 1988) are not consistent with the expression of the PRA/calcyclin gene being solely cell cycle regulated, since the highest levels of expression in the adult rat are seen in the lung, kidney, and uterus. The lung and kidney are not tissues normally regarded as possessing a high percentage of proliferating cells. It is of interest that the tissue distribution of p11 gene expression (this gene shares sequence similarity to both PRA and S-100) is very similar to that of the PRA/calcyclin gene (Saris et al. 1987). The product of the p11 gene is calpactin 1 light chain. This protein is found in a complex with p36/calpactin 1, which is a major *in vivo* substrate for retroviral and growth factor receptor protein-tyrosine kinases (Gerke and Weber 1985). However, recently it has been shown that the expression of p11 and p36 mRNAs is not always coordinated (Saris et al. 1987), suggesting that p11 may subserve a broader function of which its interaction with p36 is only a subset. This might be true in the broader sense for this family of S-100-related genes.

The similarity of the PRA/calcyclin protein to that of S-100 may provide some clues as to the possible function of PRA/calcyclin. Despite the large body of knowledge about S-100 proteins, their function remains elusive. Although originally thought to be brain-specific, S-100 has subsequently been detected in a wide range of cell types and tissues, suggesting a more generalized function. Recently, it was shown that a neurite extension factor isolated from brain is identical to S-100-beta (Kligman and Marshak 1985). Interestingly, it has also been shown that S-100 can act on microtubule protein in vitro primarily by binding to tubulin, this event being Ca^{2+}-regulated at a given pH and pH-regulated at a given free Ca^{2+} concentration (Donato 1988). This action affects microtubule assembly and dissembly *in vitro*. Since colchicine has been shown to block the action of prolactin in rabbit mammary gland explants (Djiane et al. 1981) and cell shape is a critical factor associated with prolactin regulation of milk protein gene expression in rat mammary epithelial cells (Lee et al. 1985; Haeuptle et al. 1983) it is tempting to speculate that the PRA/calcyclin protein may have a role in this function. Furthermore, T-47D human breast cancer cells undergo a marked cell shape change when treated with prolactin in the presence of hydrocortisone (Shiu and Paterson 1984). Interestingly, a gene (clone 63) which is similar if not identical to rat calcyclin (Metz et al. 1988; Leonard et al. 1987; E. Ziff pers. commun.) is one of the later genes induced in PC12 cells when they are treated with nerve growth factor (NGF). Under these conditions NGF induces the differentiation of PC 12 cells from a chromaffin-like phenotype to a sympathetic neuron-like phenotype and this is accompanied by gross morphological changes, including the outgrowth of neurite extensions (Metz et al. 1988).

ACKNOWLEDGMENTS

This work was supported by the Medical Research Council of Canada, Manitoba Health Research Council and an equipment grant from H. E. Sellers Fund. Dr. Leigh C. Murphy is a National Cancer Institute of Canada Scientist.

REFERENCES

Barraclough R, Savin J, Dube SK, Rudland PS (1987) Molecular cloning and sequence of the gene for p9Kd a cultured myoepithelial cell protein with strong homology to S-100, a calcium binding protein. J Mol Biol 198:13-20

Baudier J, Gerard D (1983) Ions binding to S-100 proteins: structural changes induced by calcium and zinc on S-100a and S-100b proteins. Biochemistry 22:3360-3369

Boutin J-M, Jolicoeur C, Okamura H, Gagnon J, Edery M, Shirota M, Banville D, Dusanter I, Djiane J, Kelly PA (1988) Cloning and expression of the rat prolactin receptor, a member of the growth hormone/prolactin receptor gene family. Cell 53:69-77

Boutin J-M, Edery M, Shirota M, Jolicoeur C, Lesueur L, Gould D, Djiane J, Kelly PA (1989) Identification of a cDNA encoding a long form of prolactin receptor in human hepatoma and breast cancer cells. Mol Endocrinology 3:1455-1461

Brüggen J, Tarcsay L, Cerletti N, Odink K, Rutishauser M, Hollander G, Sorg C (1988) The molecular nature of the cystic fibrosis antigen. Nature 331:570 (Letter)

Calabretta B, Battini R, Kaczmark L, de Riel JK, Baserga, R (1986) Molecular cloning of the cDNA for a growth factor-inducible gene with strong homology to S-100, a calcium-binding protein. J Biol Chem 261:12628-12632

Cameron CM, Rillema JA (1983) Extracellular calcium ion concentration required for prolactin to express its action on casein, ribonucleic acid and lipid biosynthesis in mouse mammary gland explants. Endocrinology 113:1596-1600

Donato R (1988) Calcium-independent pH-regulated effects of S-100 proteins on assembly-dissembly of brain microtubule protein in vitro. J Biol Chem 263:106-110

Djiane J, Houdebine LM, Kelly PA (1981) Prolactin-like activity of anti-prolactin receptor antibodies on casein and DNA synthesis in the mammary gland. Proc Natl Acad Sci USA 78:7445-7448

Edery M, Jolicoeur C, Levi-Meyrueis C, Dusanter-Fourt I, Petridou B, Boutin J-M, Lesueur L, Kelly PA, Djiane J (1980) Identification and sequence analysis of a second form of prolactin receptor by molecular cloning of complementary DNA from rabbit mammary gland.
Proc Natl Acad Sci USA 86:2112-2116

Ferrari S, Calabretta B, de Riel JK, Battini R, Ghezzo F, Lauret E, Griffen C, Emanuel BS, Gurrieri F, Baserga R (1987) Structural and functional analysis of a growth-regulated gene, the human calcyclin.
J Biol Chem 262:8325-8332

Filipek A, Heizmann CW, Kuznicki J (1989) Zinc binding and dimer formation by calcyclin-like calcium binding protein from Ehrlich ascites tumor cells. Abstract in First European Symposium on Calcium Binding Proteins in Normal and Transformed Cells. Bruxelles, April 20-22

Fulmer CS, Wasserman RH (1981) The amino acid sequence of bovine intestinal calcium-binding protein.
J Biol Chem 256:5669-5674

Gerke V, Weber K (1985) The regulatory chain in the p36-kd substrate complex of viral tyrosine-specific protein kinases is related in sequence to the S-100 protein of glial cells. EMBO J 4:2917-2920

Guo X, Chambers AF, Parfett CLJ, Waterhouse P, Murphy LC, Reid RE, Craig AM, Edwards DR, Denhardt DT (1990) Identification of a serum inducible mRNA (5B10) as the mouse homolog of calcyclin: tissue distribution and expression in metastatic, *ras*-transformed NIH 3T3 cells. Cell Growth and Differentiation (in press)

Haeuptle M-T, Yolande LMS, Bogenmann E, Reggio H, Racine L, Kraehenbuhl J-P (1983) Effect of cell shape change on the function and differentiation of rabbit mammary cells in culture. J Cell Biol 96:1425-1434

Hesketh TR, Moore JP, Morris JDH, Taylor MV, Rogers J, Smith GA, Metcalfe JC (1985) A common sequence of calcium and pH signals in the mitogenic stimulation of eukaryotic cells.
Nature 313:481-484

Horrobin DF (1978) Prolactin, Vol. 6, Eden Press, Montreal

Houdebine LM, Djiane J, Dusanter-Fourt I, Martel P, Kelly, PA, Devinoy E, Servely JL (1985) Hormonal action controlling mammary activity. J Dairy Sci 68:489-500

Isobe T, Okuyama T (1981) The amino acid sequence of the alpha subunit in bovine brain S-100a protein.
Eur J Biochem 116:79-86

Jackson-Grusby LL, Swiergel J, Linzer DH (1987) A growth-related mRNA in cultured mouse cells encodes a placental calcium-binding protein. Nucleic Acid Res 15:6677-6690

Katoh M, Raguet S, Zachwieja J, Djiane J, Kelly PA (1987) Hepatic prolactin receptors in the rat: characterization using monoclonal antireceptor antibody. Endocrinology 120:739-749

Kelly PA, Djiane J, Katoh M, Ferland LH, Houdebine L, Teyssot, B, Dusanter-Fourt I (1984) The interaction of prolactin with its receptors in target tissues and its mechanism of action.
Recent Prog Horm Res 40:379-436

Kligman D, Marshak DR (1985) Purification and characterization of a neurite extension factor from bovine brain. Proc Natl Acad Sci USA 82:7136-7139

Kretsinger RH, Nockold CE (1973) Carp muscle calcium-binding protein. II. Structure determination and general description. J Biol Chem 248:3313-3326

Kuwar R, Usui H, Maeda T, Fukui T, Yamanari N, Ohtsuka E, Ikehara M, Takahashi Y (1984) Molecular cloning and the complete nucleotide sequence of cDNA to mRNA for S-100 protein of rat brain.
Nucleic Acid Res 12:7455-7465

Kuznicki J, Filipek A (1987) Purification and properties of a novel Ca^{2+}-binding protein (10.5 kDa) from Ehrlich-ascites-tumor cells. Biochem J 247:663-667

Kuznicki J, Filipek A, Hunziker PE, Huber S, Heizmann CW (1989) Calcium-binding protein from mouse Ehrlich ascites-tumor cells is homologous to human calcyclin. Biochem J 263:951-956

Leonard DGB, Ziff EB, Greene LA (1987) Identification and characterization of mRNA's regulated by Nerve Growth Factor in PC 12 cells. Mol Cell Biol 7:3156-3167

Leung DW, Spencer SA, Cachianes G, Hammonds RG, Collins C, Henzel WJ, Barnard R, Waters MJ, Wood WI (1987) Growth hormone receptor and serum binding protein: purification, cloning and expression. Nature 330:537-543

Lee EY-H, Lee W-H, Kaetzel CS, Parry G, Bissell MJ (1985) Interaction of mouse mammary epithelial cells with collagen substrate: regulation of casein gene expression and secretion.
Proc Natl Acad Sci 82:1419-1423

Masiakowski P, Shooter M (1988) Nerve growth factor induces the genes for two proteins related to a family of calcium binding proteins in PC 12 cells. Proc Natl Acad Sci USA 85:1277-1281

Metz R, Gorham J, Siegfried Z, Leonard D, Gizang-Ginsberg E, Thompson MA, Lawe D, Kouzarides T, Vosatka R, MacGregor D, Jamal S, Greenberg ME, Ziff EB (1988) Gene regulation by growth factors. Cold Spring Harbour Symposia on Quantitative Biology 53:727-737

Mitani M, Dufau ML, (1986) Purification and characterization of prolactin receptors from rat ovary. J Biol Chem 261:1309-1315

Murphy LC, Tsuyuki D, Myal Y, Shiu RPC (1987) Isolation and sequencing of a cDNA for a prolactin-inducible protein (PIP): regulation of PIP gene expression in the human breast cancer cell line, T-47D. J Biol Chem 262:15236-15241

Murphy LC, Murphy LJ, Tsuyuki D, Duckworth ML, Shiu RPC (1988) Cloning and characterization of a highly conserved putative calcium-binding protein, identified by anti-prolactin receptor antiserum. J Biol Chem 263:2397-2401

Murphy LJ, Bell CI, Duckworth ML, Friesen HG (1987) Identification, characterization and regulation of a rat complementary deoxyribonucleic acid which encodes insulin-like growth factor-1. Endocrinology 121:684-691

Nicoll CS, Bern HA (1972) In: Lactogenic hormones, Ciba Foundation Symposium, (Wolstenholme GE and Knight J, eds), Churchill Livingstone, London

Odink K, Cerletti N, Brüggen J, Clerc RG, Tarcsay L, Zwadlo G, Gerhards A, Schlegel R, Sorg C (1987) Two calcium-binding proteins in infiltrate macrophages of rheumatoid arthritis. Nature 330:80-82

Posner BI, Kelly PA, Shiu RPC, Friesen HG (1974) Studies of insulin, growth hormone and prolactin binding: tissue distribution, species variation and characterization. Endocrinology 96:521-531

Reid RE (1987a) Total sequential solid phase synthesis of rabbit skeletal tropinin C calcium-binding site III. Internat J Pept Prot Res 30:613-621

Reid RE (1987b) A synthetic 33-residue analogue of bovine brain calmodulin calcium-binding site III: synthesis, purification and calcium-binding. Biochemistry 26:6070-6073

Saris CJM, Kristensen T, D'Eustachio P, Hicks LJ, Noonan DJ, Hunter T, Tack BF (1987) cDNA sequence and tissue distribution of the mRNA for bovine and murine p11, the S-100-related light chain of the protein-tyrosine kinase substrate p36 (Calpactin 1). J Biol Chem 262:10663-10671

Shiu RPC (1979) Prolactin receptors in human breast cancer cells in long-term tissue culture. Cancer Res 39:4381-4386

Shiu RPC, Friesen HG (1974) Solubilization and purification of a prolactin receptor from the rabbit mammary gland. J Biol Chem 249:7902-7911

Shiu RPC, Friesen HG (1974a) Properties of a prolactin receptor from the rabbit mammary gland. Biochem J 140:301-311

Shiu RPC, Friesen HG (1976) Interaction of cell-membrane prolactin receptor with its antibody. Biochem J 157:619-626

Shiu RPC, Friesen HG (1976a) Blockade of prolactin action by antiserum to its receptor. Science 192:259-261

Shiu RPC, Friesen HG (1981) In: Receptor regulation (Lefkowitz RJ, ed.) Vol. 13, Chapman and Hall, London, pp69-81

Shiu RPC, Iwasiow B (1985) Prolactin-inducible proteins in human breast cancer. J Biol Chem 260:11307-11313

Shiu RPC, Paterson JA (1984) Alteration of cell shape, adhesion and lipid accumulation in human breast cancer cells (T-47D) by human prolactin and growth hormone. Cancer Res 44:1178-1186

Szebenyi DME, Moffat K (1986) The refined structure of vitamin D-dependent calcium-binding protein from bovine intestine. Molecular details, ion binding and implications for the structure of other calcium-binding proteins. J Biol Chem 261:8761-8777

Szebenyi DME, Obendorf SK, Moffat K (1981) Structure of vitamin D-dependent calcium-binding protein from bovine intestine. Nature 294:327-332

van Eldik LJ, Zendegui JG, Marshak DR, Watterson DM (1982) Calcium-binding proteins and the molecular basis of calcium action. Int Rev Cytol 77:1-61

THE CALGRANULINS, MEMBERS OF THE S-100 PROTEIN FAMILY: STRUCTURAL FEATURES, EXPRESSION, AND POSSIBLE FUNCTION

David Longbottom and Veronica van Heyningen

STRUCTURAL FEATURES OF THE CALGRANULINS, AND OTHER MEMBERS OF THE S-100 PROTEIN FAMILY
MAJOR PATHWAYS OF SIGNAL TRANSDUCTION

Cells communicate with each other by a variety of signal molecules, such as hormones and neurotransmitters, which stimulate specific receptors on the plasma membrane of the responding cell. The effect of these agonists is to initiate a cascade of biochemical events producing an intracellular signal which ultimately causes a change in the behavior of the cell. The response of the cell may be to secrete, contract, divide, differentiate, or signal other cells to respond in a certain way. The pathways by which cells convert signals into responses are of extreme complexity and controlled at various levels. There are two main pathways involved in signal transduction: the cyclic AMP cascade (for a review see Neer and Clapham 1988) and the phosphatidyl inositol cascade (for a review see Berridge 1987). Cross-regulation allows precise coordination between the pathways. This cross-regulation may involve, for example, modulation of the receptor-G-protein-adenylate-cyclase or phospholipase C complexes. Some ligands, for example the hormone glucagon, have been shown to elicit responses in both pathways but through different receptors which respond to changes in physiological conditions (Wakelam et al. 1986).

The adenylate cyclase/cyclic AMP system has been studied in a lot of detail over the last 20-30 years and a great deal is known about it; however, it is only recently that there has been rapid progress made in understanding the receptors that generate intracellular signals from inositol lipids as well as in the understanding of the signals themselves (Berridge and Irvine 1984). Hormonal stimulation of phospholipase C causes the hydrolysis of phosphatidylinositol 4,5-bisphosphate located in the inner side of the plasma membrane into two intracellular signals, inositol 1,4,5-trisphosphate (IP3) and diacylglycerol. Diacylglycerol activates protein kinase C, whereas IP3 is released into the cytoplasm where it causes the release of calcium ions from a compartment of the endoplasmic reticulum (Berridge and Irvine 1984; Berridge 1987). Accompanying this mobilization of internal calcium is an influx of external calcium, although much less is known about the way this influx is controlled (Putney 1986; Irvine and Moor 1986, 1987; Meldolesi and Pozzan 1987; Berridge and Taylor 1988). Currently there is a lot of interest in the identification and function of proteins regulated by this rise in intracellular calcium concentration.

CALCIUM-BINDING PROTEINS WITH EF-HANDS

Many responses to calcium are mediated by the acidic, low molecular weight protein calmodulin (reviewed by Manalan and Klee 1984). Calmodulin appears to bind four calcium ions, and upon binding calcium its conformation changes exposing hydrophobic regions (La Porte et al. 1980) and increasing its helical content (Liu and Cheung 1976). It is this conformational change which allows binding of the calmodulin-calcium complex to specific sites on protein targets. However, calmodulin has been found to be ubiquitous among eukaryotes, being found in many species from both the animal and plant kingdoms, suggesting that calmodulin generates a universal type of signal. Therefore, how is specific control achieved? There are a whole series of more tissue-specific calcium-binding proteins with calcium-binding sites related to those of calmodulin. These include troponin C (van Eldick et al. 1982), parvalbumin (reviewed by Heizmann 1984) and the functionally much less well-defined members of a family of homologous proteins which include S-100 protein (reviewed by Donato 1986), vitamin D3-dependent calcium binding proteins (reviewed by Glenney et al. 1982) and cystic fibrosis antigen (van Heyningen et al. 1988; van Heyningen and Dorin submitted). Some individual members of the S-100 family have been discovered independently by different groups and so have been assigned different names. Table 1 lists the S-100-related proteins as well as any proteins which have been shown to be synonyms or

Table 1. Members of the S-100 protein family

S-100 protein family	Synonyms and species homologs	Reference	Ca^{2+}-binding
S-100α		Isobe and Okuyama (1981)	Yes; Baudier et al. (1986)[a]
S-100ß		Isobe and Okuyama (1978, 1981)	Yes; Baudier et al. (1986)[a]
S-100L		Glenney et al. (1989)	Yes; Glenney et al. (1986)[b,e]
Ca[1]	p10 p11 42C	Glenney and Tack (1985) Gerke and Weber (1985a) Masiakowski and Shooter (1988)	No; Gerke and Weber (1985b)[e], Glenney (1986)[b] and Pigault et al. (1989)[e]
CAPL	p9Ka 18A2 pEL98 42A Metastasin	Barraclough et al. (1987) Jackson-Grusby et al. (1987) Goto et al. (1988) Masiakowski and Shooter (1988) Ebralidze et al. (1989)	Yes; Barraclough and Gibbs (1989)[c]
CACY	2A9 PRA CaBP	Calabretta et al. (1986b) Murphy et al. (1988) Kuznicki and Filipek (1987)	Yes; Kuznicki and Filipek (1987)[d,e,f,g]
CaBP9K	IaBP Calbindin-D_{9k}	Desplan et al. (1983a,b) Lee et al. (1987) Perret et al. (1988)	Yes; Szebenyi and Moffat (1986)[h]
CAGA	CFAg MIF MRP8 p8 L1 Ag light chain MAC387 Ag α chain 60B8Ag	Dorin et al. (1987) Burmeister et al. (1986) Odink et al. (1987) Hogg et al. (1989) Andersson et al. (1988) Flavell et al. (1987) Nozawa et al. (1988)	Not proven
CAGB	CFAg MIF MRP14 p14 L1 Ag light chain MAC387 Ag ß chain 60B8Ag CFP	Wilkinson et al. (1988) Burmeister et al. (1986) Odink et al. (1987) Hogg et al. (1989) Andersson et al. (1988) Flavell et al. (1987) Nozawa et al. (1988) Barthe et al. (1989)	Not proven

194

Ca^{2+}-binding was deduced by [a] flow dialysis, [b] equilibrium dialysis, [c] ^{45}Ca^{2+} blotting, [d] ^{45}Ca^{2+} gel filtration, [e] fluorescence spectroscopy, [f] UV difference spectroscopy, [g] fluorescence titration or [h] by crystallographic studies. CAPL is the placental calcium-binding protein, Ca[1] the calpactin-1 light chain, CACY is calcyclin, CaBP9K is the 9 kDa vitamin D3-induced intestinal calcium-binding protein, and CAGA and CAGB are calgranulins A and B respectively.

or species homologs of any of the family members. Each of these proteins will be briefly discussed later.

A hypothesis formulated by Kretsinger states that these calcium-binding proteins, which are the probable targets and mediators of calcium acting as a second messenger, are proteins containing EF-hand structures (reviewed by Kretsinger 1980a). Parvalbumin, a calcium-binding protein found in high concentrations in muscle, has been studied in the greatest detail (Moews and Kretsinger 1975; Kretsinger 1980a) and consists of six α-helical regions denoted by letters A-F. The EF-hand is defined by the structure formed by the E and F helices, connected by a calcium-binding loop which contains the oxygen atoms that coordinate the calcium ion (Kretsinger 1980a,b): the two helices are related by an approximate two-fold axis. This EF-hand consisting of a 29-amino-acid domain is now more properly called the calmodulin fold (Kretsinger 1987). Strong homology has been found between proteins belonging to a family of calcium-binding proteins whose first described member is known as S-100 protein (Moore 1965; Moore et al. 1968). The homology is most marked for the EF region of all these molecules (Kretsinger 1987; Kligman and Hilt 1988), but it is the sequence conservation of the amino acids outside the EF-hand region that allows these proteins to be placed in the S-100 subfamily of the calmodulin superfamily of EF-hand calcium- binding proteins. Figure 1 shows the amino acid sequence similarity for the nine members of the S-100 family currently known. The amino- and carboxy-terminal EF-hands are located in regions rich in basic and acidic amino acids respectively, whereas conserved hydrophobic regions flank these domains (Kligman and Hilt 1988). Although both of these calcium-binding domains are derived from the EF-hand type, the N-terminal domain is a variant calcium-binding site differing significantly from the canonical EF-hand with the loop around the calcium ion being extended due to the insertion of two amino acids (Szebenyi and Moffat 1986; Kretsinger 1987). The variant structure of this EF-hand was established by crystallographic studies on the 9 kDa bovine vitamin D-dependent intestinal calcium binding protein (CaBP9K) (Szebenyi et al. 1981; Szebenyi and Moffat 1986). In the calmodulin fold the calcium-binding loop coordinates calcium with the side-chain oxygens which can be offered by Ser, Thr, Asx, or Glx. These can be assigned to the vertices of an octahedron: X, position 10; Y, 12; Z, 14; -Y, 16; -X, 18; -Z, 21 (see Figure 1, and Kretsinger in this volume for a diagrammatic representation) (Kretsinger 1980a,b, 1987; Szebenyi and Moffat 1986). The carboxylate ligand at -Z probably coordinates calcium with two oxygen atoms, resulting in the calcium being seven-coordinate with six amino acids involved. The oxygen at -Y comes from the carbonyl group of the main chain; the side chain is exposed to the solvent and is variable. Szebenyi and Moffat (1986) suggest that if Asp or Asn occur at the -Y+2 position, they may be hydrogen bonded to a water molecule (the actual -X ligand); if Glu (and perhaps Gln) occurs then it would be the -X ligand; if Ser or Thr occurs then water would be the ligand, with the Ser and Thr playing a part in stabilizing the C-terminal helix of the hand. In the variant EF-hand the insertion of two amino acids (one between X and Y and one between Z and -Y) alters the conformation so that the ligands at positions X, Y and Z are peptide carbonyls.

MEMBERS OF THE S-100 PROTEIN FAMILY

S-100α, S-100ß and S-100L. The founding member S-100 proteins (called S-100 because of their solubility in a 100% saturated solution of ammonium sulfate) after which the family was named are a group of low molecular weight (10 kDa) acidic proteins that were found to be highly enriched in nervous tissue (Moore 1965; Moore et al. 1968), representing approximately 0.5% of soluble proteins in the brain. The proteins are a mixture of hetero- and homodimers containing two homologous 10 kDa subunits, α and ß, in three possible combinations, αα, αß (S-100a) and ßß (S-100b) (Isobe and Okuyama 1978, 1981; Isobe et al. 1983). Since the discovery of the proteins, the α and ß subunits have also been found to exist in other tissues (Molin et al. 1984, 1985; Kato and Kimura 1985; Zimmer and van Eldick 1987). S-100α is found in muscle, kidney, and a variety of other cells, and has been reported to be present in neurons (Kuwano et al. 1987).

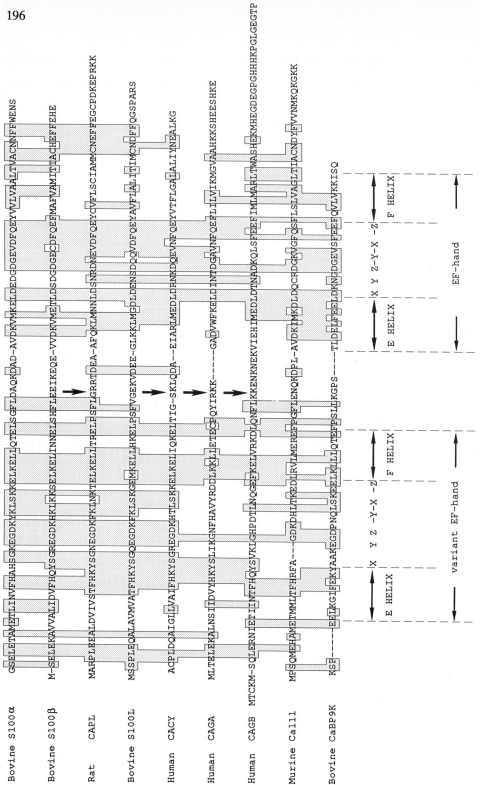

Fig. 1. The S-100 protein family. The amino acid sequences have short *gaps* introduced to maximize alignment. Amino acids which are identical for at least four of the S-100-related proteins are *boxed*. The *vertical arrows* indicate the positions where intron 2 interrupts the coding sequences of the genes where the genomic structure has been determined (Ferrari et al. 1987; Barraclough et al. 1987; Lagasse and Clerc 1988). The amino acid residues are represented by the single letter code: *A* Ala; *R* Arg; *N* Asn; *D* Asp; *C* Cys; *Q* Gln; *E* Glu; *G* Gly; *H* His; *I* Ile; *L* Leu; *K* Lys; *M* Met; *F* Phe; *P* Pro; *S* Ser; *T* Thr; *W* Trp; *Y* Tyr; *V* Val. *X, Y, Z, -Y, -X* and *-Z* represent the calcium-binding ligands (see text). Sequences were taken from Isobe and Okuyama (1978, 1981) for S-100α and S-100ß, Barraclough et al. (1987) for CAPL, Glenney et al. (1989)for S-100L, Calabretta et al. (1986b) for CACY, Dorin et al.(1987) and Odink et al. (1987) for CAGA and CAGB, Saris et al.(1987) for Ca[1], and Fullmer and Wasserman (1981) for CaBP9K. CAPL is the placental calcium-binding protein, CACY is calcyclin, Ca[1] is calpactin-1 light chain and CaBP9K is the vitamin D3-induced intestinal calcium-binding protein

S-100ß, found in glial and Schwann cells, and also in melanocytes, chondrocytes, and adipocytes (Cocchia et al. 1981; Stefansson et al. 1982; Kato et al. 1983; Kahn et al. 1983), is expressed at higher levels during the G1 phase of the cell cycle in rat C6 glioma cells (Fan 1982). Little is known about the function of the S-100 proteins, although recent studies have shown them to be highly conserved among mammals suggesting that they have potentially important roles in cells. It is only recently that definite biological activities have been demonstrated for the subunits: neurite extension activity (Kligman and Marshak 1985; Kligman and Hsieh 1987; van Eldik et al. 1988); calcium-dependent microtubule dissociation and inhibition of microtubule assembly (Baudier et al. 1982; Dunn et al. 1987; Donato 1988); calcium-dependent inhibition of tau protein phosphorylation by both protein kinase C and calmodulin kinase (Baudier and Cole 1987, 1988); inhibition of the phosphorylation of an 87 kDa protein (also known as pp80) that is the major protein kinase C substrate (Kligman and Patel 1986); calcium-dependent activation of fructose 1,6-bisphosphate aldolase (Zimmer and van Eldik 1987); stimulation of prolactin secretion from pituicytes (Ishikawa et al. 1983). A low molecular weight protein, S-100L, has recently been isolated from bovine lung by Glenney et al. (1989) which shares many of the properties of brain S-100, such as self-association and calcium binding. Immunological and sequence analysis, however, show that S-100L is clearly distinct from other members of the S-100 family. The protein is also expressed at high levels in bovine kidney: the Madin-Darby bovine kidney (MDBK) cell line was found to contain both S-100L and the calpactin light chain (p11) (Glenney et al. 1989).

Ca[1]. Another member of the family, p11 (p10), was first described as a protein that binds to and inhibits the phosphorylation of p36, also known as calpactin, a major substrate for the tyrosine-specific phosphorylation of pp60src kinase (Gerke and Weber 1985a; Glenney and Tack 1985). Three forms of calpactin have been isolated by Glenney et al. (1987) from bovine lung and human placenta of which the predominant form calpactin-1 is a hetero-tetramer consisting of two p11 (calpactin-1 light chain; Ca[1]) subunits and two p36 (calpactin-1 heavy chain) subunits. The calcium-binding sites of calpactin-1 have been assigned to the heavy chain (Gerke and Weber 1985b; Glenney 1986) and the binding of calcium is greatly enhanced by the binding of phospholipids (Glenney 1985, 1986). Calpactin-1 heavy chain, also known as lipocortin II, is one of five known members of the calpactin/lipocortin family which are all thought to be calcium- and phospholipid-binding (Pepinsky et al. 1988). There is currently no evidence to suggest any associations between other members of the S-100 family and the calpactins. Ca[1] has apparently lost the ability to bind calcium (Gerke and Weber 1985a; Glenney 1986), although it has retained its amino acid sequence similarity to the other members of the S-100 family. This can be recognized by the three amino acid deletion in the variant EF-hand calcium-binding loop (Figure 1), as well as the loss of essential calcium-binding coordinates Y (Cys-61), -X (Gly-67), and -Z (Ser-70) in the canonical EF-hand. Calpactin-1 may be involved in calcium-dependent exocytosis, cytoskeletal organization, actin bundling and coupling the cytoskeleton to the plasma membrane (Gerke and Weber 1984; Glenney and Glenney 1985; Ali et al. 1989). Masiakowski and Shooter (1988) have demonstrated that the treatment of a rat pheochromocytoma-derived cell line (PC12 cells) with nerve growth factor (NGF) induces the synthesis of two mRNA species, 42C and 42A, which encode proteins that are almost certainly the rat homologs of p11 and 18A2 (discussed below) respectively.

CAPL. 18A2 is the protein encoded by mRNA species which can be induced by the serum stimulation of quiescent mouse fibroblasts (Jackson-Grusby et al. 1987). This stimulation of growth arrested cells to produce increasing amounts of mRNA is expressed in a cell-cycle-specific manner during the S phase of the cell cycle and is dependent on protein synthesis (Jackson-Grusby et al. 1987). The 18A2 mRNA is expressed in a variety of mouse tissues, with the highest levels being detected in the placenta and the nonpregnant uterus (Jackson-Grusby et al. 1987). Subsequently, isolation of the same gene/protein has been reported in mouse and related species by other groups. Barraclough et al. (1987) have described a rat protein, named p9Ka, with only five amino acid differences from the 18A2 mouse sequence. It is induced 15-fold during the differentiation of rat mammary cuboidal epithelial stem cells to elongated myoepithelial-like cells. Goto et al. (1988) isolated a cDNA clone named pEL98 by differential colony hybridization, which allows the identification and characterization of mRNAs abundant in established cell lines but less so in the corresponding parental counterpart, using the established mouse cell line Balb/c 3T3 and single-stranded cDNAs from Balb/c 3T3 cells and mouse embryo fibroblasts. Another reported isolation of the same gene has been described by Ebralidze et al. (1989) by subtractive hybridization of mRNA from metastatic and nonmetastatic sublines of a spontaneous mouse mammary carcinoma line. This gene, named mts1, has been suggested to be involved in regulating the metastatic behavior of tumor cells. The product of the gene, metastasin, is identical to the other previously isolated mouse calcium-binding proteins, pEL98 and 18A2, as well as the two rat proteins, p9Ka and 42A. In order to simplify matters, the nomenclature committee of the Tenth International Workshop on Human Gene Mapping (1989) agreed to name the protein, represented by 18A2, pEL98, p9Ka and 42A, placental (murine homolog) calcium-binding protein (CAPL).

CACY. Another cell-cycle-associated protein of the S-100 family is calcyclin, the product of the human cell-cycle G1-specific gene 2A9 (Calabretta et al. 1986a,b; Ferrari et al. 1987). The 2A9 mRNA was found to be over-expressed but de-regulated in human acute myeloid leukemias (Calabretta et al. 1985, 1986a) and by the serum stimulation of quiescent WI-38 human diploid fibroblasts, as well as by epidermal and platelet-derived growth factors in quiescent 3T3 cells, but not by platelet-poor plasma or insulin (Calabretta et al. 1986b). These observations have led to the proposal that calcyclin is an important protein of cell cycle progression (Hirschhorn et al. 1984; Calabretta et al. 1986b). Calcyclin has also been purified from T-47D human breast cancer cells, where it was found to be co-purified with the prolactin receptor and hence was named the prolactin receptor-associated protein (PRA) (Murphy et al. 1988); and from Ehrlich ascites tumor cells (Kuznicki and Filipek 1987).

CaBP9K. The intestinal calcium-binding protein (CaBP9K), also known as calbindin-D9K, was initially described as a protein whose synthesis was stimulated by vitamin D3 (Wasserman and Taylor 1966) illustrating yet another member of the S-100 family that is inducibly expressed. CaBP9K is mainly located in the duodenum, possesses two calcium-binding domains (Wasserman et al. 1978; Baimbridge et al. 1982; Thomasset et al. 1982) and is dependent on 1,25-dihydroxy-vitamin D3 (the hormonal form of vitamin D3) for its synthesis (Thomasset et al. 1979, 1980, 1982), whereas a higher molecular weight (28 kDa) vitamin D-dependent calcium-binding protein with four calcium-binding sites (CaBP28K) was found in high concentration in chicken intestine, kidney, and cerebellum (Wasserman et al. 1978) and is not a member of the S-100 family. Recently Mathieu et al. (1989) investigated the gestational changes in CaBP9K which occur in rat uterus, yolk sac, and placenta and found good evidence for CaBP9K-induced maternal-fetal calcium transport and uterine muscle contraction. This particular S-100-related calcium-binding protein differs from the rest of the family in having less pronounced N- and C-terminal hydrophobic domains (see Figure 1).

CAGA and CAGB. The last member of the S-100 family to be discussed is the cystic fibrosis antigen. This protein was first detected on polyacrylamide isoelectric focusing gels as a pI 8.4 doublet band (Wilson et al. 1973, 1975) which was present in serum samples from cystic fibrosis (CF) homozygotes and heterozygotes but not from normals. This finding was confirmed by other groups later (Scholey et al. 1978; Nevin et al. 1981; Brock et al. 1982). In 1980 Manson and Brock raised a guinea pig antiserum against the excised pI 8.4 doublet band and demonstrated that there was a quantitative variation in this protein which they named cystic fibrosis protein (CFP) between CF homozygotes, heterozygotes and normals. CFP levels, determined by rocket immunoelectrophoresis (RIE), were shown to fall into three genotype-dependent classes (Bullock et al. 1982): homozygotes had high levels, heterozygotes intermediate and normals very low or absent. It was this pattern, particularly the intermediate levels of CFP present in the serum of clinically unaffected heterozygotes, which suggested that the increase of this protein compared to normals was related to the basic defect in the autosomal recessive disease cystic fibrosis. At the time it was thought that CFP could be the product of the CF gene itself. In 1985 van Heyningen et al. showed that peripheral leukocytes from chronic myeloid leukaemia (CML) patients and normal mature granulocytes are very good sources of CFP, which they named CFAg. They also showed by constructing mouse-human somatic cell hybrids, using parental cells of myeloid origin that express this differentiated cell product, that CFAg is encoded by a gene on chromosome 1. However, that same year, the disease locus was assigned to chromosome 7 (Knowlton et al. 1985; White et al. 1985; Wainwright et al. 1985), showing that CFAg is not the product of the CF gene. Identification of the tissue source of CFAg as granulocytes (van Heyningen et al., 1985) facilitated the production of a series of monoclonal antibodies directed to at least two different epitopes on the protein (Hayward et al. 1986). These antibodies made possible the development of a two-site sandwich enzyme-linked immunosorbent assay (ELISA) (Hayward et al. 1986, 1987) for accurately quantifying the levels of CFAg in serum samples. However, as with the guinea pig antiserum (Manson and Brock 1980; Bullock et al. 1982), there were considerable overlaps in the distribution of CFAg between CF homozygotes, heterozygotes, and normals, even though the mean values were significantly different. Therefore, this method, even when taking into account comparisons between CFAg levels and the levels of marker proteins like lactoferrin (for granulocyte turnover) and C-reactive protein (as a measure of inflammation) (Hayward et al. 1987), is not suitable for use in wide-scale heterozygote detection. The serum component was purified using the monoclonal antibodies by immunoaffinity chromatography, and sodium dodecyl sulfate polyacrylamide gel electrophoresis (SDS-PAGE) analysis revealed two components of molecular masses 8×10^3 and 14×10^3 (Novak, Walker and Secher unpubl.). Partial N-terminal amino acid sequence analysis of the purified protein, however, revealed only one polypeptide (Novak, Walker and Secher unpubl.) which allowed the synthesis of an oligonucleotide probe to isolate a cDNA clone from a library constructed from chronic myeloid leukaemia cells (Dorin et al. 1987). Southern blot analysis of somatic cell hybrids confirmed and sub-localized the gene encoding the CFAg as being on chromosome 1 (region q12-q22), showing that the cloned gene was the one whose product was recognized by the original guinea pig antiserum. Later, in 1987 Odink et al. reported the isolation of two proteins from human peripheral blood mononuclear cell cultures as part of a complex, using a monoclonal antibody directed against human macrophage migration inhibitory factor (MIF). MIF itself was the first lymphokine ever to be described (Bloom and Bennett 1966; David 1966) and the monoclonal antibody IC5 raised against MIF was found to bind three molecular weight species of 8, 14, and 28 kDa from Concanavalin A-stimulated mononuclear cells, each protein having MIF activity and with the 28 kDa species not seen on SDS-polyacrylamide gels after reduction with dithiothreitol (Burmeister et al. 1986). The 8 and 14 kDa proteins were isolated, sequenced, and stated to bear no homology to any other known protein sequence. However, on later description (Odink et al. 1987), these proteins (8 and 14 kDa) were shown not to possess MIF activity and to be homologous not only to each other but also to the S-100 family of calcium-binding proteins, contradicting the results of Burmeister et al. (1986): the two proteins have been reported to form a complex in vitro with a third component containing the 1C5 epitope and MIF activity (Zwadlo et al. 1988). Therefore, Odink et al. (1987) refer to the proteins as MIF-related proteins, MRP-8 and MRP-14. The isolation of cDNA clones for the two proteins enabled the prediction of molecular masses of 10835 and 13242 for MRP-8 and MRP-14 respectively (Odink et al. 1987): the 11 kDa protein migrates on SDS-polyacrylamide gels as an 8 kDa protein. Sequence comparisons between MRP-8 and CFAg revealed them to be the same proteins (a single nucleotide sequencing error altered the C-terminal portion of the deduced amino acid sequence of CFAg). Since Odink et al. (1987) isolated both proteins as a complex, it appeared very likely that Dorin et al. (1987) did as well and that they only obtained the sequence for the 11 kDa protein because the 14 kDa protein is N-terminally blocked (Odink et al. 1987). Subsequently this was shown to be the case (Novak, Walker and Secher unpubl.) when, using the published sequence information of Odink et al. (1987), clones encoding part of the 14 kDa polypeptide were isolated (Wilkinson et al. 1988) from the CML cDNA library used by Dorin et al. (1987). The two proteins which comprise CFAg have also been isolated by other groups. Andersson et al. (1988) have shown that they are the components of the leukocyte L1 complex. Since the initial reports of this major human leukocyte derived protein (Fagerhol et al. 1980a,b; Dale et al. 1983a,b), L1 has been characterized and its expression localized. The complex consists of three polypeptides, I, II, and III having molecular masses of 12.5, 13.3, and 8.3 respectively, as judged by two-dimensional isoelectric focusing (IEF)/SDS-PAGE. Polypeptides I and II are indistinguishable immunologically and have similar amino acid sequences, and probably arise from post-translational modifications, which are not thought to be due to glycosylation (Odink et al. 1987) since there are no potential N-glycosylation sites. Partial amino acid sequences of L1 light chain (polypeptide III) and L1 heavy chain (polypeptides I/II) are identical to the sequences of the 11 and 14 kDa CFAg proteins respectively. The two proteins p8 and p14 isolated and sequenced by Edgeworth et al. (1989a) using monoclonal antibody 5.5 (Hogg et al. 1989) raised against cells from acute monocytic leukemia patients have also been shown to be identical to the CFAg proteins. As previously observed by us as well as by other groups (Andersson et al. 1988; Edgeworth et al. 1989b), two principal forms of p14 can be seen on SDS-polyacrylamide gels. However, Edgeworth et al. (1989b) have also reported a further two less abundant forms comprising 5-10% of the p14 molecule which can be seen by two-dimensional IEF/SDS-PAGE. The mouse monoclonal antibody MAC387 raised against a purified protein fraction obtained from human monocytes recognized a protein named MAC387 antigen from monocytes and granulocytes (Flavell et al. 1987) which was later shown to be identical to L1 antigen (Brandtzaeg et al. 1988) and hence to CFAg as well. MAC387 Ag consists of an α and ß chain, found both as monomers and putatively as disulfide-linked heterodimers. However, in monocytes only heterodimers and associations of two and four heterodimers were found. Both proteins have been found in peripheral blood neutrophils and monocytes by Nozawa et al. (1988) and were named 60B8 antigen. Barthe et al. (1989) have also isolated a protein, named CFP, using the original isoelectric focusing approach. They say CFP is the same as MRP-14.

On the basis of calcium binding and the major cellular localization to granulocytes, as well as to reduce confusion due to the many different names for these proteins, Wilkinson et al. (1988) suggested the names calgranulin A and B (CAGA and CAGB) for the 11 and 14 kDa proteins respectively. These names were accepted by the nomenclature committee of the Tenth International Workshop on Human Gene Mapping (1989).

GENOMIC ORGANIZATION AND CHROMOSOMAL MAPPING OF THE S-100 PROTEIN GENE FAMILY

The genomic structure of the genes for CAGA (CAGA) and CAGB (CAGB) (Lagasse and Clerc 1988) and for CACY (CACY) (Ferrari et al. 1987) in man, as well as for CAPL (CAPL) in rat (Barraclough et al. 1987) have been determined. They have a similar genomic structure, each consisting of three exons, of which exon 1 is untranslated. Exons 2 and 3 each code for one of the calcium-binding domains, and intron 2 is located in the middle linker region (see Figure 1). The similar location of this intron strongly suggests that these genes share a common ancestry. In fact, it is very probable that the two-site S-100-related calcium-binding proteins arose from the four-site calcium-binding calmodulin ancestor by the loss of two sites and the creation of a divergent lineage (Goodman et al. 1979; Perret et al. 1988b). If this is the case and the proteins are a true subfamily of the calmodulin superfamily, then mapping of the genes might be expected to reveal their chromosomal clustering. Studies to determine the genomic structures of the remainder of the family are obviously awaited to see if they are similar and also possess an intron in the middle linker region, which would be further evidence in support of their common ancestry. Perhaps the presence, and the position, or absence of this intron between the two calcium-binding domains could be determined with the aid of PCR (polymerase chain reaction) primers.

CAGA has been assigned to human chromosome 1 q12-q22 (Dorin et al. 1987), and using the same somatic cell hybrid panel Dorin et al. (submitted) have shown that a cDNA probe for CAGB co-segregates with it. They also confirmed the in situ hybridization assignment of CACY to human chromosome 1 q21-25 (Ferrari et al. 1987), and showed that CACY and CAPL co-segregate with CAGA and CAGB. Similarly they mapped CAPL to a region of chromosome 3 in the mouse where Ca[1] has been localized (Saris et al. 1987). CACY has also recently been mapped to mouse chromosome 3 using interspecific back-crosses (Moseley and Seldin 1989). CACY and CAPL are only approximately 8 kb apart in the mouse genome (Dorin et al. submitted). The linkage of CAGA, CAGB, CACY, and CAPL in man and of CACY, CAPL, and Ca[1] in mouse suggest that these calcium-binding proteins are members of a distinct subfamily which arose by ancestral duplications and by subsequent divergence in sequence and function. It will be interesting to see if S-100ß, the gene for which has been localized to human chromosome 21q22 (Allore et al. 1988), is an isolated member when the remainder of the family (S-100α, S-100L, and CaBP9K) are mapped.

PHOSPHORYLATION OF THE CALGRANULINS AND THEIR RELATIONSHIP WITH THE NEUTROPHIL IMMOBILIZING FACTORS NIF-1 AND NIF-2

As previously mentioned, CAGB can be separated by two-dimensional electrophoresis into four distinct isoforms (Edgeworth et al. 1989b): CAGA runs electrophoretically as a single entity. The CAGB proteins, differing slightly in apparent molecular weight, are present as two abundant proteins with pIs of 5.4 and 5.6 and two less abundant forms with pIs of 5.4 and 5.3. Edgeworth et al. (1989b) have reported the phosphorylation of the two less abundant forms of CAGB in both monocytes and neutrophils: they could not demonstrate any phosphorylation of CAGA. They suggest that their results imply that the two abundant forms of CAGB, which differ by some unknown post-translational modification, are phosphorylated to produce the two less abundant forms. The level of phosphorylation was increased by elevating the intracellular calcium concentration using the ionophore ionomycin, but was not affected by the activation of protein kinase C using phorbol 12,13-dibutyrate. The phosphorylated residue is Thr 113, which is the penultimate amino acid contained in the longer 'tail' sequence of CAGB. This longer C-terminal sequence after the second calcium-binding domain of CAGB is what makes this protein unique in comparison to the other members of the family (all of which have a lower molecular weight of approximately 10 kDa) and which may be conferring on it a unique function. Indeed, residues 89-108 contained within this tail sequence are identical to the N-terminal 20 amino acids of neutrophil immobilizing factors NIF-1 and NIF-2, with the exception of an extra Ala residue at the N-terminal end of NIF-2 (Watt et al. 1983; Freemont et al. 1989). This could mean that the phosphorylation of Thr 113 positioned at the end of the NIF-like sequence of CAGB, in the CAGA-CAGB complex, may play a role in inducing the immobilization of myeloid cells at the endothelial surface membrane both during an inflammatory response and also during the normal exudation of myeloid cells into tissues, as indicated by Hogg et al. (1989). Freemont et al. (1989) also suggest that the common sequence of CAGB and NIF may serve some other function and that the actual NIF activity could reside in the C-terminal half of the NIFs. It is also possible that the NIFs are formed by the proteolytic cleavage of CAGB after Trp 88 but this would yield a C-terminal peptide of 26 amino acids (Freemont et al. 1989), whereas NIF-1 and NIF-2 are predicted from their amino acid compositions to contain 41 and 38 amino acids respectively (Watt et al. 1983). Recently Murao et al. (1989) isolated the CAGA-CAGB complex from human spleen and showed that the protein inhibits the activity of casein kinase I and II but not of cyclic-AMP-dependent protein kinase or protein kinase C. It remains to be seen if the phosphorylation of CAGB can regulate the activity of the casein kinases, which phosphorylate a variety of enzymes associated with the control of gene expression including topoisomerase II (Ackerman et al. 1985) and RNA polymerases

I and II (Stetler and Rose 1982; Rose et al. 1983). This regulation of casein kinase activity and the probable mediation of signal-transducing pathways by protein kinases, as well as the high levels of the calgranulins found in monocytes and granulocytes, suggest that the CAGA-CAGB complex may have a role in the maturation and function of myeloid cells.

EXPRESSION AND POSSIBLE RELATIONSHIP OF *CAGA* AND *CAGB* TO CYSTIC FIBROSIS

When CFAg was initially described as being a protein that was found to be elevated in the serum of both cystic fibrosis (CF) homozygotic patients and heterozygotic unaffected carriers, it immediately became a candidate for being the product of the CF gene (it was later realized that CFAg consisted of a complex of CAGA and CAGB). However, upon assignment of the genes for CAGA and CAGB to chromosome 1 and of the CF gene to chromosome 7 it became clear that this was not the case. There now remains the problem of identifying the functions of both the calgranulins and of the product of the CF gene, and establishing the relationship, if any, that exists between them.

CYSTIC FIBROSIS PATHOLOGY

The major abnormalities in the autosomal recessive disorder cystic fibrosis are those of exocrine secretion. In 10-15% of neonates with CF there is a complete obstruction of the intestine characterized by the presence of a viscous meconium-meconium ileus. The first major symptom is usually failure to thrive due to exocrine pancreatic insufficiency, but this can be corrected by treatment with pancreatic enzymes. This pancreatic dysfunction results in the maldigestion and malabsorption of fat and protein resulting in steatorrhea and malnutrition. The best-defined quantifiably demonstrable abnormality is the elevated sweat chloride resulting from defective chloride ion transport across epithelia (Case 1986). The increase in survival age to the 20's is due to improvements in antibiotic therapy to combat the progressive obstructive lung disease which is the predominant cause of mortality. This lung involvement usually occurs in later childhood and is characterized by recurrent bacterial infections and inflammations associated with macrophage infiltration. This coincides with decreased fluid secretion due to the altered chloride ion transport and accumulation of thick dehydrated mucus, resulting in the clogging of the lungs and the respiratory tract (McPherson and Dormer 1988). The high level of the calgranulins found in the serum of CF patients is most probably of granulocyte origin. It may result from aberrant granulocyte function in CF which could contribute directly to the CF-associated lung disease. Alternatively, the high CFAg levels may be derived from the high turnover of fragile granulocytes and macrophages infiltrating the lung as part of the inflammatory response to repeated infection. The latter possibility would account for the observations of elevated CFAg levels in other diseases with a strong inflammatory component (Hayward et al. 1987). However, this would not account for elevated CFAg levels in symptom-free heterozygotes, in whom there is no evidence of heightened inflammatory processes. Thus, the suggestion remains of a direct functional relationship between calgranulins and the basic genetic defect in CF.

Airway epithelia (tracheal and nasal) are relatively easily available "affected tissues" in CF and so have been the most widely studied. In 1986 the physiological parameters of isolated chloride channels on membrane fragments derived from these affected tissues were shown to be indistinguishable from those in "normal tissues"; instead it was the control of chloride channel activity that appeared to be altered (Welsh and Liedtke 1986; Frizzell et al. 1986). This altered regulation of apical membrane chloride channels in CF homozygote-derived tissues has been demonstrated since, in cultured cells from airway epithelia (Schoumacher et al. 1987; Li et al. 1988), sweat glands (Bijman and Frömter 1986) and from small intestine and colon (Berschneider et al. 1988). In similar studies, the ß-adrenergic stimulation of chloride channel activity that is observed in normal cells was absent in CF cells, despite the fact that the signal resulted in normal increases in the levels of cyclic AMP (Welsh and Liedtke 1986; Sato and Sato 1984). These results imply that the deviation from normal cellular function must occur later on in the cyclic AMP pathway and/or in other signal transduction systems, a prime candidate being the phosphatidyl inositol pathway. Indeed, channel activity has been shown to be increased by both cyclic-AMP-dependent protein kinase and protein kinase C in normal subjects but not in CF patients (Widdicombe et al. 1985; Schoumacher et al. 1987; Hwang et al. 1989; Li et al. 1989).

Therefore, the abnormality must occur somewhere in the signal transduction pathway between the eliciting of the signal and the response controlling chloride channel opening.

EXPRESSION PATTERN OF *CAGA* AND *CAGB*

Myeloid Cells. The calgranulins are expressed in myeloid cells at a very specific point in their differentiation pathway. The proteins appear to be expressed by circulating granulocytes and monocytes, but not in the leukaemias of immature myeloid cells (Linch et al. 1984). They have been shown to be inducible in the promyelocytic HL60 cell line. Dimethyl-sulfoxide- and retinoic-acid-induced differentiation along the granulocytic lineage resulted in the expression of both CAGA and CAGB mRNAs (Wilkinson et al. 1988), whereas no expression was detected by phorbol-myristate-acetate-induced monocytic differentiation (Lagasse and Clerc 1988). A study of normal bone marrow spreads (Wilkinson and van Heyningen, unpubl.) revealed that the expression of the calgranulins is turned on relatively late in the myeloid cell pathway, at around the formation of band cells. This is confirmed by the pattern of expression in chronic (CML) versus acute (AML) myeloid leukemias. The most mature cell type, as seen in CML, is usually positive, whilst some acute myeloid cells do not express the calgranulins.

The proteins have been shown to be absent from resident tissue macrophages, suggesting that as monocytes mature into macrophages with passage through the endothelium into tissues, the calgranulins are released from the cells and their expression is turned off (Hogg et al. 1989). This suggests that the proteins may play a role in the interaction of the myeloid cells with the endothelium. However, in chronic inflammatory conditions where there is a substantial infiltrate of mononuclear cells, such as in rheumatoid arthritis, macrophages in affected tissues have been shown to express both CAGA and CAGB (Odink et al. 1987; Zwadlo et al. 1988). In contrast to this, in acute inflammatory reactions, such as in gingivitis, CAGA is not seen (Zwadlo et al. 1988). Zwadlo et al. (1988) suggest from these observations that CAGA and CAGB are expressed sequentially at defined stages of monocyte/macrophage differentiation and that the dysregulation of this process, i.e., the failure in the regulation of monocyte to macrophage differentiation in chronic inflammation is paralleled by the presence of CAGA-positive macrophages in the tissue.

Epithelial cells. To try to assess any relationship between the elevated serum levels of CAGA and CAGB to the basic defect in CF, an extensive immunohistochemical tissue search was carried out on normal and fetal tissues and, where possible, on analogous CF tissue (Wilkinson et al. 1988). Monoclonal antibodies CF145 and CF557 (Hayward et al. 1986), which recognize CAGA and CAGB respectively (both native and denatured forms), were most widely used: CF145 also recognizes bacterially expressed CAGA. The two antibodies revealed completely parallel expression profiles in all the tissues studied (Wilkinson et al. 1988), as well as in extensive double-labeling analyses (Jones, pers. commun.). The calgranulins were localized to the cytoplasm of granulocytes (Figure 2A) (Wilkinson et al. 1988). In addition to their presence in mature granulocytes and in some monocytes and macrophages, the only normal adult tissues found to be calgranulin-positive were those in the nonkeratinized, stratified squamous epithelial cells of tongue, esophagus and buccal tissue. No calgranulin expression was observed in epithelial cells implicated as "affected tissue" in physiological ion transport analyses (Welsh and Liedtke 1986; Frizzell et al. 1986); sections from lung, pancreas, and sweat glands from both normal and CF individuals (adult and fetal) were shown to be negative, except for occasional positive hair follicle cells in skin. Epithelial cells of simple mucous nasal polyp tissue were shown occasionally to express both CAGA and CAGB, whereas cultured nasal polyp cells were positive at a higher frequency. Hoogeveen et al. (pers. commun.) have found calgranulin expression, using CF145 and CF557, in cultured nasal polyp and sweat gland epithelial cells with physiologically demonstrable chloride channels of the type shown to function abnormally in CF. Patchy to strong expression was found in the abnormal squamous epithelia of hyperproliferative conditions of the skin (psoriasis and eczema) and also in malignant squamous cell carcinomas of skin, lung, and buccal origin.

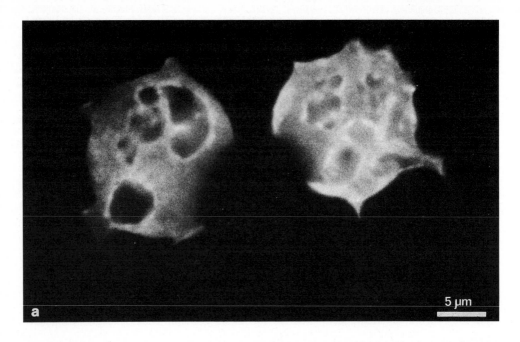

Fig. 2a,b. Immunofluorescent localization of calgranulins A and B. Immunofluorescent detection of CAGB with CF557 in peripheral leukocytes. The same pattern was detected for CAGA with CF145. **b** Immunofluorescent detection of CAGA with CF145 on TR146 cultured cells. The same pattern was detected for CAGB with CF557

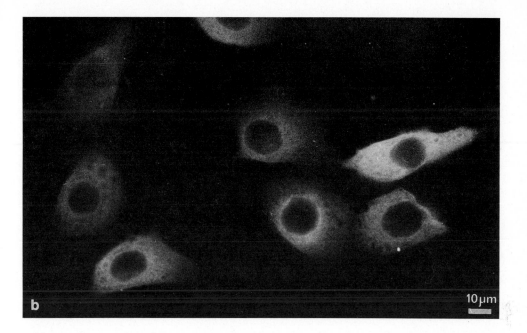

The calgranulin-positive TR146 cell line, derived from a squamous cell carcinoma of buccal origin, may prove to be a suitable cell line for investigating the function of the calgranulins in relation to chloride channel control, since a proportion of normal buccal cells also express the proteins and because an altered calmodulin activity in CF buccal epithelial cells has been described suggesting that they express the CF gene (McPherson et al. 1987). Figure 2B shows that the sensitive indirect immunofluorescence of TR146 cells that have been in prolonged culture reveals that the calgranulins have a reticular network-like disposition, illustrating a possible cytoskeletal association for the proteins.

Where RNA samples were available, RNA blot analysis using the cDNA probes for CAGA and CAGB confirmed the patterns of expression (Wilkinson et al. 1988). A very similar tissue distribution was described for the myelomonocytic L1 antigen (Brandtzaeg et al. 1988), confirming its identity to the calgranulins.

HOW MIGHT THE *CAGA-CAGB* COMPLEX BE IMPLICATED FUNCTIONALLY IN THE *CF* ABNORMALITY

The most likely subcellular function of the calgranulins is in signal transduction. It is quite possible that they participate in the communication system leading to the control of the opening of the chloride channel, which appears to be the defective part of the pathway in CF. Since neither of the proteins is the product of the CF gene, their effect must be secondary to the basic defect. The abnormality may lie in their molecular interaction with the CF gene product or with another disturbed component of the pathway, leading to an altered localization and half-life for the calgranulins. Such associations could be investigated by trying to "catch" interacting members of the pathway by controlled immunoprecipitation or by gentle cross linking or, if large amounts of the purified calgranulins are available, by affinity capture. The state of "activation" of the cells used may be critical for catching these transient associations, and various stimuli could be applied to the test cell immediately before disruption, for example ß-adrenergic stimulation or protein kinase C activation. Another approach would be to study subcellular associations, as well as location, by immunofluorescence histochemistry and confocal laser microscopy.

It would be expected from the tissue distribution results that the source of the raised serum level of the calgranulins is myeloid cells (granulocytes and monocytes), but there is no clear evidence to show that the CF defect or appropriate chloride channels are expressed in these cells. However, no physiological measurements have been tried, possibly because of the short half-lives of the cells.

The CF gene has now been cloned and its product has been named the cystic fibrosis transmembrane conductance regulator (CFTR) (Rommens et al. 1989; Riordan et al. 1989; Kerem et al. 1989). The CFTR protein has been shown to have organizational and sequence similarities to a family of ATP-dependent transport systems in bacterial and, more recently, in eukaryotic cells (Higgins 1989). The name ABC (ATP-Binding Cassette)-transporters has been suggested for the members of this family of proteins (Hyde et al. submitted). The bacterial transporters have been shown to be specific for different substrates, mostly small molecules such as an amino acid, sugar, peptide, or inorganic ion (Ames 1986), but in at least one case the substrate is the 107 kDa protein hemolysin. The first example of a eukaryotic transporter was the P-glycoprotein or MDR (multidrug resistance) protein, which is responsible for multiple drug resistance in tumors: the protein functions by pumping drugs out of cells. CFTR is structurally similar to these proteins and has two hydrophobic domains, each of which comprises six potential membrane-spanning helices, and two ATP-binding domains. The protein also has an additional large cytoplasmic domain between the two ATP-binding loops, named the R domain, since the presence of many potential protein kinase A and C phosphorylation sites on it suggests a probable regulatory role (Riordan et al. 1989). The overall structural arrangement of CFTR is similar to several cation channel proteins (Noda et al. 1986; Tanabe et al. 1987; Baumann et al. 1987) and some cation-translocating ATPases (Chen et al. 1986), as well as to the recently described adenylate cyclase of bovine brain (Krupinski et al. 1989). CFTR is itself not thought to be the chloride channel (Hyde et al. submitted), but is more probably involved in the regulation of the ion channel activity. Evidence in support of this is that the ABC-transporters driven by ATP hydrolysis can accumulate substrates against concentration gradients, whereas chloride channels are energy-independent and do not accumulate against gradients (Hyde et al. submitted). ABC-transporters function unidirectionally whereas chloride channels can function in both directions (Hyde et al. submitted). The chloride channel has a very rapid turnover number, characteristic of channels, which is far greater than has been observed for any active transporter (Hyde et al. submitted). Moreover, the predicted molecular mass of CFTR (Riordan et al. 1989) is much greater than that of the recently purified polypeptides (from trachea and kidney) that are capable of reconstituting chloride channels in lipid membranes (Landry et al. 1989).

About 70% of CF patients seem to have the same genetic defect, a deletion of three nucleotides resulting in the removal of Phe-508 from the first ATP-binding domain of the CFTR protein (Riordan et al. 1989). Unlike Riordan et al. (1989), Hyde et al. (submitted) suggest that this deletion is in a loop which can play no significant role in ATP binding or hydrolysis, but alters the interaction between the ATP-binding cassette and the membrane associated domains of the transport system, which is perhaps regulated by the high density of phosphorylation sites for protein kinases A and C.

It is necessary to determine the connection existing between CFTR and the chloride channel, as well as to attempt to look for any direct associations of CFTR with the calgranulins. The possibility of both CFTR and the calgranulins being co-expressed in the same cells needs to be determined. Appropriate cells for investigation are those implicated in the pathology of CF-epithelial cells of the airway (tracheal and nasal), colon and pancreas, and myeloid cells. So far no CFTR expression has been seen in CML cells by RNA PCR (McIntosh unpubl.). Established colon cell lines appear not to express the calgranulins when CFTR is present. Therefore, it still remains to analyze the cell-cycle variability of CFTR, although there is some suggestion that maximal expression occurs in G1-arrested cells. The cell-cycle variability of the calgranulins has been hinted at in TR146 cells (Wilkinson and van Heyningen unpubl.). If mature granulocytes were demonstrated to be an affected tissue in CF, it would suggest that the elevated serum levels of the calgranulins could be attributed to the absence in homozygotes, or the reduction in heterozygotes, of the normal product of the CF gene in these cells. Another possibility for interaction between the two systems is that the calgranulins could be the substrate for the CFTR channel, and mutation at this locus leads to elevated serum levels of calgranulins. This model does not require both genes to be co-expressed in the same cell.

We are currently trying to raise both polyclonal and monoclonal antibodies to various peptides which we chose on the basis of the predicted CFTR protein structure (Riordan et al. 1989). If the antibodies turn out to recognize CFTR protein, in Western blotting or by immunohistochemistry, they will be of great value in elucidating its molecular associations with proteins such as the calgranulins. They could also be used, in conjunction with the anti-calgranulin antibodies, to define the subcellular localization of CFTR and associated proteins in response to physiologically relevant external signals.

TRANSDUCTION OF THE CALCIUM SIGNAL

The calgranulins and the S-100 proteins in general probably exert their biological effects by modulating the activity of specific effector proteins. This could be achieved by the binding of calcium, leading to conformational change and the exposure of hydrophobic domains which could interact with hydrophobic domains of the effector protein. The effector protein would then exert its biological effect, after which the intracellular calcium levels would decrease, resulting in the dissociation of the proteins. This model assumes that the intracellular concentration of the calcium-binding proteins is lower than that of the calcium, so that the calcium sites on the proteins are not saturated. The N-terminal variant EF-hands in S-100ß and S-100α have been shown to be poor calcium-binding sites, unlike that of the C-terminal region (K_d=20-50 μM for the canonical EF-hand in contrast to 200-500 μM for the variant EF-hand, by in vitro studies) (Mani et al. 1983; Baudier et al. 1986). If this model also holds for the calgranulins it could mean that at different calcium concentrations the C- and N-terminal binding sites may interact differently with sites on the target molecule(s). Further variations in control could be mediated by Zn^{2+} and by K^+ (which increase and decrease calcium affinity respectively), providing finely tuned regulation of processes such as calcium-induced exocytosis (Baudier et al. 1986). Another possible reason for the evolution of different affinity calcium-binding domains may arise from proposed extracellular functions. The N-terminal variant EF-hand may only bind calcium upon exposure of the protein to the high extracellular calcium environment, giving rise to distinct extracellular and intracellular functions. Either of these mechanisms is possible for the calgranulins, dependent on whether the proteins are present in the serum due to the rapid turnover of the large number of granulocytes which are found at the site of infection in the lungs of CF patients

216

or because of their specific secretion or mislocalization from these cells. If the proteins are secreted, they may be targeted to the secretory pathway by allowing them to interact with membranes through an exposed hydrophobic domain. Such a mechanism may operate with S-100ß, whose release is stimulated by a variety of agonists and peptides, including isoproterenol and adrenocorticotropic hormone, suggesting that the extracellular S-100ß levels are under physiological control (Suzuki et al. 1987).

The difference in calcium-binding affinity of the two sites on each calgranulin subunit may be related to the calcium-concentration-dependent regulation of chloride channel activity by protein kinase C proposed by Li et al. (1989). They demonstrated that protein kinase C inactivates and activates chloride channels at high and low calcium concentrations respectively, and that in CF cells protein-kinase-C-dependent channel inactivation was normal, whereas it was the activation that was defective. They conclude that protein kinase C phosphorylates and regulates two different sites on the channel or on an associated membrane protein (perhaps CFTR?), one of which is defective in CF. Thus the defect in CF may be either in the ability of the channel to become phosphorylated, or in the mechanism by which the phosphorylation results in channel activation.

Clearly, we are still some way from determining the exact role of calgranulins in the cell and their relationship, if any, with the product of the cystic fibrosis gene. The abnormality in CF is most likely in signalling and the most probable function for the calcium-binding proteins is in signal transduction. The system is very complex, but the answers to the many questions are worth pursuing.

ACKNOWLEDGMENTS

We thank the Cystic Fibrosis Research Trust and the Medical Research Council for their continued and generous support. Our thanks also go to the photography department, particularly Mr. Norman Davidson, for their work on the illustrations.

REFERENCES

Ackerman P, Glover CVC, Osheroff N (1985) Phosphorylation of DNA-topoisomerase II by casein kinase II: modulation of eukaryotic topoisomerase II activity in vitro. Proc Natl Acad Sci USA 82:3164-3168

Ali SM, Geisow MJ, Burgoyne RD (1989) A role for calpactin in calcium-dependent exocytosis in adrenal chromaffin cells. Nature (London) 340:313-315

Allore R, O'Hanlon D, Price R, Neilson K, Willard HF, Cox DR, Marks A, Dunn RJ (1988) Gene encoding the ß subunit of S-100 protein is on chromosome 21: implications for Down Syndrome. Science 239:1311-1313

Ames GF-L (1986) Bacterial periplasmic transport systems: structure, mechanism, and evolution. Ann Rev Biochem 55:397-425

Andersson KB, Sletten K, Berntzen HB, Fagerhol MK, Dale I, Brandtzaeg P, Jellum E (1988) Leukocyte L1 protein and the cystic fibrosis antigen. Nature (London) 332:688

Baimbridge KG, Miller JJ, Parkes CO (1982) Calcium-binding protein distribution in the rat brain. Brain Res 239:519-525

Barraclough R, Gibbs F (1989) p9Ka, a calcium-binding protein related to S-100 protein. In: First European Symposium on Calcium Binding Proteins in Normal and Transformed Cells. 20-22 April 1989. University Press, Bruxelles, B7

Barraclough R, Savin J, Dube SK, Rudland PS (1987) Molecular cloning and sequence of the gene for p9Ka a cultured myoepithelial cell protein with strong homology to S-100, a calcium-binding protein. J Mol Biol 198:13-20

Barthe C, Carrere J, Figarella C, Guy-Crotte O (1989) Isolation of the "cystic fibrosis protein" from serum. Clin Chem 35:1901-1905

Baudier J, Cole R (1987) Phosphorylation of tau proteins to a state like that in Alzheimer's brain is catalyzed by a calcium/calmodulin-dependent kinase and modulated by phospholipids. J Biol Chem 262:17577-17583

Baudier J, Briving C, Deinum J, Haglid K, Sorskog L, Wallin M (1982) Effect of S-100 proteins and calmodulin on Ca^{2+}-induced disassembly of brain microtubule proteins in vitro. FEBS Lett 147:165-167

Baudier J, Cole RD (1988) Interactions between the microtubule-associated τ proteins and S-100b regulate τ phosphorylation by the Ca^{2+}/calmodulin-dependent protein kinase II. J Biol Chem 263:5876-5883

Baudier J, Glasser N, Gerard D (1986) Ions binding to S-100 proteins. I. Calcium- and zinc-binding properties of bovine brain S-100αα, S-100a(αß), and S-100b(ßß) protein: Zn^{2+} regulates Ca^{2+} binding on S-100b protein. J Biol Chem 261:8192-8203

Baumann A, Krah-Jentgens I, Müller R, Müller-Holtkamp F, Seidel A, Kecskemethy N, Casal J, Ferrus A, Pongs O (1987) Molecular organization of the maternal effect region of the *Shaker* complex of *Drosophila*: characterization of an I_A channel transcript with homology to vertebrate Na^+ channel. EMBO J 6:3419-3429

Berridge MJ (1987) Inositol trisphosphate and diacylglycerol: two interacting second messengers. Ann Rev Biochem 56:159-193

Berridge MJ, Irvine RF (1984) Inositol trisphosphate, a novel second messenger in cellular signal transduction. Nature (London) 312:315-320

Berridge MJ, Taylor CW (1988) Inositol trisphosphate and calcium signaling. In: Cold Spring Harbor Symposia on Quantitative Biology, Vol LIII. Cold Spring Harbor Laboratory, New York, pp927-933

Berschneider HM, Knowles MR, Azizkhan RG, Boucher RC, Tobey NA, Orlando RC, Powell DW (1988) Altered intestinal chloride transport in cystic fibrosis. FASEB J 2:2625-2629

Bijman J, Frömter E (1986) Direct demonstration of high transepithelial chloride-conductance in normal human sweat duct which is absent in cystic fibrosis. Pflügers Arch Eur J Physiol 407:S123-S127

Bloom BR, Bennett B (1966) Mechanism of a reaction in vitro associated with delayed-type hypersensitivity. Science 153:80-82

Brandtzaeg P, Jones DB, Flavell DJ, Fagerhol MK (1988) Mac 387 antibody and detection of formalin resistant myelomonocytic L1 antigen. J Clin Pathol 41:963-970

Brock DJH, Hayward C, Super M (1982) Controlled trial of serum isoelectric focusing in the detection of the cystic fibrosis gene. Hum Genet 60:30-31

Bullock S, Hayward C, Manson J, Brock DJH, Raeburn JA (1982) Quantitative immunoassays for diagnosis and carrier detection in cystic fibrosis. Clin Genet 21:336-341

Burmeister G, Tarcsay L, Sorg C (1986) Generation and characterization of a monoclonal antibody (1C5) to human migration inhibitory factor (MIF). Immunobiol 171:461-474

Calabretta B, Kaczmarek L, Mars W, Ochoa D, Gibson CW, Hirschhorn RR, Baserga R (1985) Cell-cycle-specific genes differentially expressed in human leukemias. Proc Natl Acad Sci USA 82:4463-4467

Calabretta B, Battini R, Kaczmarek L, de Riel JK, Baserga R (1986b) Molecular cloning of the cDNA for a growth factor-inducible gene with strong homology to S-100, a calcium-binding protein. J Biol Chem 261:12628-12632

Calabretta B, Venturelli D, Kaczmarek L, Narni F, Talpaz M, Anderson B, Beran M, Baserga R (1986a) Altered expression of G1-specific genes in human malignant myeloid cells. Proc Natl Acad Sci USA 83:1495-1498

Case M (1986) Chloride ions and cystic fibrosis. Nature (London) 322:407

Chen C-M, Misra TK, Silver S, Rosen BP (1986) Nucleotide sequence of the structural genes for an anion pump. The plasmid-encoded arsenical resistance operon. J Biol Chem 261:15030-15038

Cocchia D, Michetti F, Donato R (1981) Immunochemical and immunocytochemical localization of S-100 antigen in normal human skin. Nature (London) 294:85-87

Dale I, Fagerhol MK, Naesgaard I (1983a) Purification and partial characterization of a highly immunogenic human leukocyte protein, the L1 antigen. Eur J Biochem 134:1-6

Dale I, Fagerhol MK, Frigård M (1983b) Quantitation of a highly immunogenic leukocyte antigen (L1) by radioimmunoassay: methodological evaluation. J Immunol Meth 65:245-255

David JR (1966) Delayed hypersensitivity in vitro: its mediation by cell-free substances formed by lymphoid cell-antigen interaction. Proc Natl Acad Sci USA 56:72-77

Desplan C, Heidmann O, Lillie JW, Auffray C, Thomasset M (1983b) Sequence of rat intestinal vitamin D-dependent calcium-binding protein derived from a cDNA clone. Evolutionary implications. J Biol Chem 258:13502-13505

Desplan C, Thomasset M, Moukhtar MS (1983a) Synthesis, molecular cloning, and restriction analysis of DNA complementary to vitamin D-dependent calcium-binding protein mRNA from rat duodenum. J Biol Chem 258:2762-2765

Donato R (1986) S-100 proteins. Cell calcium 7:123-145

Donato R (1988) Calcium-independent, pH-regulated effects of S-100 proteins on assembly-disassembly of brain microtubule protein in vitro. J Biol Chem 263:106-110

Dorin JR, Emslie E, van Heyningen V (1990) Related calcium-binding proteins map to the same sub-region of chromosome 1q and to an extended region of synteny on mouse chromosome 3. (submitted)

Dorin JR, Novak M, Hill RE, Brock DJH, Secher DS, van Heyningen V (1987) A clue to the basic defect in cystic fibrosis from cloning the CF antigen gene. Nature (London) 326:614-617

Dunn R, Landry C, O'Hanlon D, Dunn J, Allore R, Brown I, Marks A (1987) Reduction in S-100 protein ß subunit mRNA in C6 rat gliomacells following treatment with anti-microtubular drugs. J Biol Chem 262:3562-3566

Ebralidze A, Tulchinsky E, Grigorian M, Afanasyeva A, Senin V, Revazova E, Lukanidin E (1989) Isolation and characterization of a gene specifically expressed in different metastatic cells and whose deduced gene product has a high degree of homology to a Ca^{2+}-binding protein family. Genes and Development 3:1086-1093

Edgeworth JD, Perks K, Brown R, Hogg N (1989a) Phosphorylation of the myeloid calcium binding protein p14. In: First European Symposium on Calcium Binding Proteins in Normal and Transformed Cells. 20-22 April 1989. University Press, Bruxelles, E17

Edgeworth J, Freemont P, Hogg N (1989b) Ionomycin-regulated phosphorylation of the myeloid calcium-binding protein p14. Nature (London) 342:189-192

Fagerhol MK, Dale I, Andersson T (1980a) Release and quantitation of a leucocyte derived protein (L1). Scand J Haematol 24:393-398

Fagerhol MK, Dale I, Andersson T (1980b) A radioimmunoassay for a granulocyte protein as a marker in studies on the turnover of such cells. Bull Europ Physiopath Resp 16 (suppl.):273-281

Fan K (1982) S-100 protein synthesis in cultured glioma cell is G1-phase of cell cycle dependent. Brain Res 237:498-503

Ferrari S, Calabretta B, de Riel JK, Battini R, Ghezzo F, Lauret E, Griffin C, Emanuel BS, Gurrieri F, Baserga R (1987) Structural and functional analysis of a growth-regulated gene, the human calcyclin. J Biol Chem 262:8325-8332

Flavell DJ, Jones DB, Wright DH (1987) Identification of tissue histiocytes on paraffin sections by a new monoclonal antibody. J Histochem Cytochem 35:1217-1226

Freemont P, Hogg N, Edgeworth J (1989) Sequence identity. Nature (London) 339:516

Frizzell RA, Rechkemmer G, Shoemaker RL (1986) Altered regulation of airway epithelial cell chloride channels in cystic fibrosis. Science 233:558-560

Fullmer CS, Wasserman RH (1981) The amino acid sequence of bovine intestinal calcium-binding protein. J Biol Chem 256:5669-5674

Gerke V, Weber K (1984) Identity of p36K phosphorylated upon *Rous sarcoma* virus transformation with a protein purified from brush borders; calcium-dependent binding to non-erythroid spectrin and F-actin. EMBO J 3:227-233

Gerke V, Weber K (1985a) The regulatory chain in the p36-kd substrate complex of viral tyrosine-specific protein kinases is related in sequence to the S-100 protein of glial cells. EMBO J 4:2917-2920

Gerke V, Weber K (1985b) Calcium-dependent conformational changes in the 36-kDa subunit of intestinal protein 1 related to the cellular 36-kDa target of *Rous sarcoma* virus tyrosine kinase. J Biol Chem 260:1688-1695

Glenney J (1986) Phospholipid-dependent Ca^{2+} binding by the 36-kDa tyrosine kinase substrate (calpactin) and its 33-kDa core. J Biol Chem 261:7247-7252

Glenney JR, Boudreau M, Galyean R, Hunter T, Tack B (1986) Association of the S-100-related calpactin 1 light chain with the NH_2-terminal tail of the 36-kDa heavy chain. J Biol Chem 261:10485-10488

Glenney JRJr (1985) Phosphorylation of p36 in vitro with pp60^{src}. Regulation by Ca^{2+} and phospholipid. FEBS Lett 192:79-82

Glenney JRJr, Glenney P (1985) Comparison of Ca^{++}-regulated events in the intestinal brush border. J Cell Biol 100:754-763

Glenney JRJr, Tack BF (1985) Amino-terminal sequence of p36 and associated p10: identification of the site of tyrosine phosphorylation and homology with S-100. Proc Natl Acad Sci USA 82:7884-7888

Glenney JRJr, Matsudaira PT, Weber K (1982) Calcium control of the intestinal microvillus cytoskeleton. In: Cheung WY (ed) Calcium and Cell Function. Academic Press, New York, pp 357-380

Glenney JRJr, Tack B, Powell MA (1987) Calpactins: two distinct Ca^{++}-regulated phospholipid- and actin-binding proteins isolated from lung and placenta. J Cell Biol 104:503-511

Glenney JRJr, Kindy MS, Zokas L (1989) Isolation of a new member of the S-100 protein family: amino acid sequence, tissue, and subcellular distribution. J Cell Biol 108:569-578

Goodman M, Pechere JF, Haiech J, Demaille JG (1979) Evolutionary diversification of structure and function in the family of intracellular calcium-binding proteins. J Mol Evol 13:331-352

Goto K, Endo H, Fujiyoshi T (1988) Cloning of the sequences expressed abundantly in established cell lines: identification of a cDNA clone highly homologous to S-100, a calcium binding protein. J Biochem 103:48-53

Hayward C, Chung S, Brock DJH, van Heyningen V (1986) Monoclonal antibodies to cystic fibrosis antigen. J Immunol Meth 91:117-122

Hayward C, Glass S, van Heyningen V, Brock DJH (1987) Serum concentrations of a granulocyte-derived calcium-binding protein in cystic fibrosis patients and heterozygotes. Clin Chim Acta 170:45-56

Heizmann CW (1984) Parvalbumin, an intracellular calcium-binding protein; distribution, properties and possible roles in mammalian cells. Experientia 40:910-921

Higgins C (1989) Export-import family expands. Nature (London) 340:342

Hirschhorn RR, Aller P, Yuan Z-A, Gibson CW, Baserga R (1984) Cell-cycle-specific cDNAs from mammalian cells temperature sensitive for growth. Proc Natl Acad Sci USA 81:6004-6008

Hogg N, Allen C, Edgeworth J (1989) Monoclonal antibody 5.5 reacts with p8,14, a myeloid molecule associated with some vascular endothelium. Eur J Immunol 19:1053-1061

Hwang T-C, Lu L, Zeitlin PL, Gruenert DC, Huganir R, Guggino WB (1989) Cl⁻channels in CF: lack of activation by protein kinase C and cAMP-dependent protein kinase. Science 244:1351-1353

Hyde SC, Emsley P, Hartshorn M, Mimmack MM, Gileadi U, Pearce SR, Gallagher MP, Hubbard R, Higgins CF (1990) Structural and functional relationships of ATP-binding proteins associated with cystic fibrosis, multidrug resistance and bacterial transport. (submitted)

Irvine RF, Moor RM (1986) Micro-injection of inositol 1,3,4,5-tetrakisphosphate activates sea urchin eggs by a mechanism dependent on external Ca^{2+}. Biochem J 240:917-920

Irvine RF, Moor RM (1987) Inositol (1,3,4,5)tetrakisphosphate-induced activation of sea urchin eggs requires the presence of inositol trisphosphate. Biochem Biophys Res Commun 146:284-290

Ishikawa H, Nogami H, Shirasawa N (1983) Novel clonal strains from adult rat anterior pituitary producing S-100 protein. Nature (London) 303:711-713

Isobe T, Okuyama T (1978) The amino-acid sequence of S-100 protein (PAP I-b protein) and its relation to the calcium-binding proteins. Eur J Biochem 89:379-388

Isobe T, Okuyama T (1981) The amino-acid sequence of the α subunit in bovine brain S-100a protein. Eur J Biochem 116:79-86

Isobe T, Ishioka N, Masuda T, Takahashi Y, Ganno S, Okuyama T (1983) A rapid separation of S-100-subunits by high-performance liquid-chromatography. The subunit compositions of S-100-proteins. Biochem Int 6:419-426

Jackson-Grusby LL, Swiergiel J, Linzer DIH (1987) A growth-related mRNA in cultured mouse cells encodes a placental calcium binding protein. Nucleic Acid Res 15:6677-6690

Kahn HJ, Marks A, Thom H, Baumal R (1983) Role of antibody to S-100 protein in diagnostic pathology. Am J Clin Pathol 79:341-347

Kato K, Kimura S (1985) S-100ao (αα) protein is mainly located in the heart and striated muscles. Biochim Biophys Acta 842:146-150

Kato K, Suzuki F, Nakajima T (1983) S-100 protein in adipose tissue. Int J Biochem 15:609-613

Kerem B-S, Rommens JM, Buchanan JA, Markiewicz D, Cox TK, Chakravarti A, Buchwald M, Tsui L-C (1989) Identification of the cystic fibrosis gene: genetic analysis. Science 245:1073-1080

Kligman D, Hilt DC (1988) The S-100 protein family. Trends Biochem Sci 13:437-443

Kligman D, Hsieh L-S (1987) Neurite extension factor induces rapid morphological differentiation of mouse neuroblastoma cells in defined medium. Dev Brain Res 33:296-300

Kligman D, Marshak D (1985) Purification and characterization of a neurite extension factor from bovine brain. Proc Natl Acad Sci USA 82:7136-7139

Kligman D, Patel J (1986) A protein modulator stimulates C kinase-dependent phosphorylation of a 90K substrate in synaptic membranes. J Neurochem 47:298-303

Knowlton RG, Cohen-Haguenauer O, van Cong N, Frézal J, Brown VA, Barker D, Braman JC, Schumm JW, Tsui L-C, Buchwald M, Donis-Keller H (1985) A polymorphic DNA marker linked to cystic fibrosis is located on chromosome 7. Nature (London) 318:380-382

Kretsinger RH (1980a) Structure and evolution of calcium-modulated proteins. Crit Rev Biochem 8:119-174

Kretsinger RH (1980b) Crystallographic studies of calmodulin and homologs. Annals N Y Acad Sci 356:14-19

Kretsinger RH (1987) Calcium coordination and the calmodulin fold: divergent versus convergent evolution. In: Cold Spring Harbor Symposia on Quantitative Biology, Vol. LII, Cold Spring Harbor Laboratory, New York, pp 499-510

Krupinski J, Coussen F, Bakalyar HA, Tang W-J, Feinstein PG, Orth K, Slaughter C, Reed RR, Gilman AG (1989) Adenylyl cyclase amino acid sequence: possible channel- or transporter-like structure. Science 244:1558-1564

Kuwano R, Usui H, Maeda T, Araki K, Yamakuni T, Kurihara T, Takahashi Y (1987) Tissue distribution of rat S-100α and ß subunit mRNAs. Mol Brain Res 2:79-82

Kuznicki J, Filipek A (1987) Purification and properties of a novel Ca^{2+}-binding protein (10.5 kDa) from Ehrlich-ascites-tumor cells. Biochem J 247:663-667

Lagasse E, Clerc RG (1988) Cloning and expression of two human genes encoding calcium-binding proteins that are regulated during myeloid differentiation. Mol Cell Biol 8:2402-2410

Landry DW, Akabas MH, Redhead C, Edelman A, Cragoe EJJr, Al-Awqati Q (1989) Purification and reconstitution of chloride channels from kidney and trachea. Science 244:1469-1472

La Porte DC, Wierman BM, Storm DR (1980) Calcium-induced exposure of a hydrophobic surface on calmodulin. Biochemistry 19:3814-3819

Lee YS, Taylor AN, Reimers TJ, Edelstein S, Fullmer CS, Wasserman RH (1987) Calbindin-D in peripheral nerve cells is vitamin D and calcium dependent. Proc Natl Acad Sci USA 84:7344-7348

Li M, McCann JD, Liedtke CM, Nairn AC, Greengard P, Welsh MJ (1988) Cyclic AMP-dependent protein kinase opens chloride channels in normal but not cystic fibrosis airway epithelium. Nature (London) 331:358-360

Li M, McCann JD, Anderson MP, Clancy JP, Liedtke CM, Nairn AC, Greengard P, Welsh MJ (1989) Regulation of chloride channels by protein kinase C in normal and cystic fibrosis airway epithelia. Science 244:1353-1356

Linch DC, Allen C, Beverley PCL, Bynoe AG, Scott CS, Hogg N (1984) Monoclonal antibodies differentiating between monocytic and nonmonocytic variants of AML. Blood 63:566-573

Liu YP, Cheung WY (1976) Cyclic 3/:5/-nucleotidephosphodiesterase. J Biol Chem 251:4193-4198

Manalan AS, Klee CB (1984) Calmodulin. Adv Cyclic Nucl Protein Phosph Res 18:227-278

Mani RS, Shelling JG, Sykes BD, Kay CM (1983) Spectral studies on the calcium binding properties of bovine brain S-100b protein. Biochemistry 22:1734-1740

Manson JC, Brock DJH (1980) Development of a quantitative immunoassay for the cystic fibrosis gene. Lancet i:330-331

Masiakowski P, Shooter EM (1988) Nerve growth factor induces the genes for two proteins related to a family of calcium-binding proteins in PC12 cells. Proc Natl Acad Sci USA 85:1277-1281

Mathieu CL, Burnett SH, Mills SE, Overpeck JG, Bruns DE, Bruns ME (1989) Gestational changes in calbindin-D9k in rat uterus, yolk sac, and placenta: implications for maternal-fetal calcium transport and uterine muscle function. Proc Natl Acad Sci USA 86:3433-3437

McPherson MA, Dormer RL (1988) Cystic fibrosis: a defect in stimulus-response coupling. Trends Biochem Sci 13:10-13

McPherson MA, Tiligada E, Bradbury NA, Goodchild MC (1987) Altered calmodulin activity in buccal epithelial cells from cystic fibrosis patients. Clin Chim Acta 170:135-142

Meldolesi J, Pozzan T (1987) Pathways of Ca^{2+} influx at the plasma membrane: voltage-, receptor-, and second messenger-operated channels. Exp Cell Res 171:271-283

Moews PC, Kretsinger RH (1975) Refinement of the structure of carp muscle calcium-binding parvalbumin by model building and difference Fourier analysis. J Mol Biol 91:201-228

Molin S-O, Rosengren L, Haglid K, Baudier J, Hamberger A (1984) Differential localization of "brain-specific" S-100 and its subunits in rat salivary glands. J Histochem Cytochem 32:805-814

Molin SO, Rosengren L, Baudier J, Hamberger A, Haglid K (1985) S-100 alpha-like immunoreactivity in tubules of rat kidney. A clue to the function of a "brain-specific" protein. J Histochem Cytochem 33:367-374

Moore BW (1965) A soluble protein characteristic of the nervous system. Biochem Biophys Res Commun 19:739-744

Moore BW, Perez VJ, Gehring M (1968) Assay and regional distribution of a soluble protein characteristic of the nervous system. J Neurochem 15:265-272

Moseley WS, Seldin MF (1989) Definition of mouse chromosome 1 and 3 gene linkage groups that are conserved on human chromosome 1: evidence that a conserved linkage group spans the centromere of human chromosome 1. Genomics 5:899-905

Murao S, Collart FR, Huberman E (1989) A protein containing the cystic fibrosis antigen is an inhibitor of protein kinases. J Biol Chem 264:8356-8360

Murphy LC, Murphy LJ, Tsuyuki D, Duckworth ML, Shiu RPC (1988) Cloning and characterization of a cDNA encoding a highly conserved, putative calcium binding protein, identified by an anti-prolactin receptor antiserum. J Biol Chem 263:2397-2401

Neer EJ, Clapham DE (1988) Roles of G protein subunits in transmembrane signaling. Nature (London) 333:129-134

Nevin GB, Nevin NC, Redmond AO, Young IR, Tully GW (1981) Detection of cystic fibrosis homozygotes and heterozygotes by serum isoelectrofocusing. Hum Genet 56:387-389

Noda M, Ikeda T, Kayano T, Suzuki H, Takeshima H, Kurasaki M, Takahashi H, Numa S (1986) Existence of distinct sodium channel messenger RNAs in rat brain. Nature (London) 320:188-192

Nozawa R, Kato H, Ito T, Yokota T (1988) Identification and characterization of a differentiation antigen in human neutrophils and monocytes. Blood 71:1288-1294

Odink K, Cerletti N, Brüggen J, Clerc RG, Tarcsay L, Zwadlo G, Gerhards G, Schlegel R, Sorg C (1987) Two calcium-binding proteins in infiltrate macrophages of rheumatoid arthritis. Nature (London) 330:80-82

Pepinsky RB, Tizard R, Mattaliano RJ, Sinclair LK, Miller GT, Browning JL, Chow EP, Burne C, Huang K-S, Pratt D, Wachter L, Hession C, Frey AZ, Wallner BP (1988) Five distinct calcium and phospholipid binding proteins share homology with lipocortin 1. J Biol Chem 263:10799-10811

Perret C, Lomri N, Gouhier N, Auffray C, Thomasset M (1988) The rat vitamin-D-dependent calcium-binding protein (9-kDaCaBP) gene. Complete nucleotide sequence and structural organization. Eur J Biochem 172:43-51

Perret C, Lomri N, Thomasset M (1988b) Evolution of the EF-hand calcium-binding protein family: evidence for exon shuffling and intron insertion. J Mol Evol 27:351-364

Pigault C, Follenius-Wund A, Lux B, Gérard D (1989) Fluorescence spectroscopy study of the calpactin-1 subunits. In: First European Symposium on Calcium Binding Proteins in Normal and Transformed Cells. 20-22 April 1989. University Press, Bruxelles, E12

Putney JWJr (1986) A model for receptor-regulated calcium entry. Cell Calcium 7:1-12

Riordan JR, Rommens JM, Kerem B-S, Alon N, Rozmahel R, Grzelczak Z, Zielenski J, Lok S, Plavsic N, Chou JL, Drumm ML, Iannuzzi MC, Collins FS, Tsui L-C (1989) Identification of the cystic fibrosis gene: cloning and characterization of complementary DNA. Science 245:1066-1073

Rommens JM, Iannuzzi MC, Kerem B-S, Drumm ML, Melmer G, Dean M, Rozmahel R, Cole JL, Kennedy D, Hidaka N, Zsiga M, Buchwald M, Riordan JR, Tsui L-C, Collins FS (1989) Identification of the cystic fibrosis gene: chromosome walking and jumping. Science 245:1059-1065

Rose KM, Duceman BW, Stetler D, Jacob ST (1983) RNA polymerase I in hepatoma 3924A: mechanism of enhanced activity relative to liver. Adv Enzyme Regul 21:307-319

Saris CJM, Kristensen T, D'Eustachio P, Hicks LJ, Noonan DJ, Hunter T, Tack BF (1987) cDNA sequence and tissue distribution of the mRNA for bovine and murine p11, the S-100-related light chain of the protein-tyrosine kinase substrate p36 (calpactin1). J Biol Chem 262:10663-10671

Sato K, Sato F (1984) Defective beta adrenergic response of cystic fibrosis sweat glands in vivo and in vitro. J Clin Invest 73:1763-1771

Scholey J, Applegarth DA, Davidson AGF, Wong LTK (1978) Detection of cystic fibrosis protein by electrofocusing. Pediatr Res 12:800

Schoumacher RA, Shoemaker RL, Halm DR, Tallant EA, Wallace RW, Frizzell RA (1987) Phosphorylation fails to activate chloride channels from cystic fibrosis airway cells. Nature (London) 330:752-754

Stefansson K, Wollmann RL, Moore BW, Arnason BGW (1982) S-100 protein in human chondrocytes. Nature (London) 295:63-64

Stetler DA, Rose KM (1982) Phosphorylation of deoxyribonucleic acid dependent RNA polymerase II by nuclear protein kinase NII: mechanism of enhanced ribonucleic acid synthesis. Biochemistry 21:3721-3728

Suzuki F, Kato K, Kato T, Ogasawara N (1987) S-100 protein in clonal astroglioma cells is released by adrenocorticotropic hormone and corticotropin-like intermediate-lobe peptide. J Neurochem 49:1557-1563

Szebenyi DME, Moffat K (1986) The refined structure of vitamin D-dependent calcium-binding protein from bovine intestine. Molecular details, ion binding, and implications for the structure of other calcium-binding proteins. J Biol Chem 261:8761-8777

Szebenyi DME, Obendorf SK, Moffat K (1981) Structure of vitamin D-dependent calcium-binding protein from bovine intestine. Nature (London) 294:327-332

Tanabe T, Takeshima H, Mikami A, Flockerzi V, Takahashi H, Kangawa K, Kojima M, Matsuo H, Hirose T, Numa S (1987) Primary structure of the receptor for calcium channel blockers from skeletal muscle. Nature (London) 328:313-318

Thomasset M, Cuisinier-Gleizes P, Mathieu H (1979) 1,25-dihydroxy cholecalciferol: dynamics of the stimulation of duodenal calcium-binding protein, calcium transport and bone calcium mobilization in vitamin D and calcium-deficient rats. FEBS Lett 107:91-94

Thomasset M, Cuisinier-Gleizes P, Mathieu H, DeLuca HF (1980) Intestinal calcium-binding protein (CaBP) and bone calcium mobilization in response to 1,24(R),25-(OH)₃D3. Comparative effects of 1,25-(OH)₂D3 and 24(R),25-(OH)₂D3 in rats. Mol Pharmacol 17:362-366

Thomasset M, Parkes CO, Cuisinier-Gleizes P (1982) Rat calcium-binding proteins: distribution, development, and vitamin D dependence. Am J Physiol 243:E483-E488

van Eldik LJ, Zendegui JG, Marshak DR, Watterson DM (1982) Calcium-binding proteins and the molecular basis of calcium action. Int Rev Cytol 77:1-61

van Eldik LJ, Staecker JL, Winningham-Major F (1988) Synthesis and expression of a gene coding for the calcium-modulated protein S-100ß and designed for cassette-based, site-directed mutagenesis. J Biol Chem 263:7830-7837

van Heyningen V, Dorin J (1990) Possible role for two calcium-binding proteins of the S-100 family, co-expressed in granulocytes and certain epithelia. (submitted)

van Heyningen V, Hayward C, Fletcher J, McAuley C (1985) Tissue localization and chromosomal assignment of a serum protein that tracks the cystic fibrosis gene. Nature (London) 315:513-515

van Heyningen V, Brock DJH, Dorin JR, Hayward C, Novak M, Wilkinson M (1988) Calcium binding protein homology of cystic fibrosis associated serum protein: a clue to the basic defect. In: Mastella G, Quinton PM (eds) Cellular and molecular basis of cystic fibrosis. San Francisco Press, San Francisco, pp 90-94

Wainwright BJ, Scambler PJ, Schmidtke J, Watson EA, Law H-Y, Farrall M, Cooke HJ, Eiberg H, Williamson R (1985) Localization of cystic fibrosis locus to human chromosome 7cen-q22. Nature (London) 318:384-385

Wakelam MJO, Murphy GJ, Hruby VJ, Houslay MD (1986) Activation of two signal-transduction systems in hepatocytes by glucagon. Nature (London) 323:68-70

Wasserman RH, Taylor AN (1966) Vitamin D3-induced calcium-binding protein in chick intestinal mucosa. Science 152:791-793

Wasserman RH, Fullmer CS, Taylor AN (1978) Vitamin-D-dependent calcium-binding proteins. In: Lawson DEM (ed) Vitamin D. Academic Press, New York, pp 133-136

Watt KWK, Brightman IL, Goetzl EJ (1983) Isolation of two polypeptides comprising the neutrophil-immobilizing factor of human leucocytes. Immunol 48:79-86

Welsh MJ, Liedtke CM (1986) Chloride and potassium channels in cystic fibrosis airway epithelia. Nature (London) 322:467-470

White R, Woodward S, Leppert M, O'Connell P, Hoff M, Herbst J, Lalouel J-M, Dean M, Vande Woude G (1985) A closely linked genetic marker for cystic fibrosis. Nature (London) 318:382-384

Widdicombe JH, Welsh MJ, Finkbeiner WE (1985) Cystic fibrosis decreases the apical membrane chloride permeability of monolayers cultured from cells of tracheal epithelium. Proc Natl Acad Sci USA 82:6167-6171

Wilkinson MM, Busuttil A, Hayward C, Brock DJH, Dorin JR, van Heyningen V (1988) Expression pattern of two related cystic fibrosis-associated calcium-binding proteins in normal and abnormal tissues. J Cell Sci 91:221-230

Wilson GB, Jahn TL, Fonseca JR (1973) Demonstration of serum protein differences in cystic fibrosis by isoelectric focusing in thin-layer polyacrylamide gels. Clin Chim Acta 49:79-91

Wilson GB, Fudenberg HH, Jahn TL (1975) Studies on cystic fibrosis using isoelectric focusing. I. An assay for detection of cystic fibrosis homozygotes and heterozygote carriers from serum. Pediatr Res 9:635-640

Zimmer DB, van Eldik LJ (1987) Tissue distribution of rat S-100α and S-100ß and S-100-binding proteins. Am J Physiol 252:c285-c289

Zwadlo G, Brüggen J, Gerhards G, Schlegel R, Sorg C (1988) Two calcium-binding proteins associated with specific stages of myeloid cell differentiation are expressed by subsets of macrophages in inflammatory tissues. Clin Exp Immunol 72:510-515

*MRP*8 AND *MRP*14, TWO CALCIUM-BINDING PROTEINS EXPRESSED DURING MYELOPOIESIS

Eric Lagasse

INTRODUCTION

Circulating blood cells must be regenerated continuously throughout life. Once mature, the vast majority of these cells are destined to remain functionally active only for a limited time before being removed from the circulation. This process, known as hematopoiesis, gives rise to many progenitor cells committed to distinct pathways of differentiation. A key advantage of the hematopoietic system is that blood is a liquid tissue easily separated into its unicellular components; it provides a useful experimental system to explore the function of proteins in cell differentiation and proliferation.

Myelomonocytic cells are derived from multipotential hematopoietic stem cells via a series of intermediate myeloid progenitor cells with increasingly restricted developmental potential. They give rise to two types of blood cells namely: neutrophils and monocytes, which share several differentiation markers. Recently, two of these components, the calcium-binding proteins MRP8 and MRP14, were isolated from human blood myelomonocytic cells. MRP8 and MRP14 are present in myeloid progenitor cells. Furthermore they constitute the major proteins found in the cytoplasm of neutrophils and are also abundant in monocytes.

This study reports the cloning, characterization, and expression during myelopoiesis of both human and mouse MRP8 and MRP14. The main focus will be recent data, giving special emphasis to the structural homology of MRPs with small calcium-binding proteins and their possible function during myelopoiesis.

MOLECULAR CLONING AND CHARACTERIZATION OF THE *MRPs* ISOLATION OF HUMAN *MRP8* AND *MRP14*

MRP8 and MRP14 proteins were isolated as part of a complex from human peripheral blood mononuclear cell cultures using a monoclonal antibody directed against human macrophage migration inhibitory factor (MIF) (Burmeister et al. 1986). The relative molecular masses of the proteins on SDS-PAGE are 8000 and 14000, respectively. Therefore we refer to these proteins as MIF or now more accurately, myeloid related proteins, MRP8 and MRP14. Using oligonucleotide probes based on the partial amino acid sequence, cDNA clones were isolated for both MRPs as described previously (Odink et al. 1987). The sequence of MRP8 cDNA has an open reading frame of 279 nucleotides, predicting a protein of 93 amino acids. MRP8 is identical with the cystic fibrosis antigen (CFAg), with one exception, a G at position 292 (Dorin et al. 1987) possibly reflecting a sequencing error (Andersson et al. 1988; Brandtzaeg et al. 1988; Brueggen et al. 1988). MRP14 cDNA contains an open reading frame of 352 nucleotides predicting a protein of 114 amino acids. MRP8 and MRP14 both contain a single cysteine residue, which allows them to form homo- and heterodimers. They have no signal or membrane anchor sequences and lack consensus sequence for N-linked glycosylation. MRP8 and MRP14 were later shown to be identical to the previously described L1 complex (Andersson et al. 1988) composed of three subunits, one molecule of MRP8 associated with two molecules of MRP14. Finally p8 and p14 (Palmer et al. 1987), two proteins expressed in a subset of activated macrophages were also shown to be MRP8 and MRP14 (Hogg et al. 1989).

ISOLATION OF MOUSE *MRP8* AND *MRP14* AND COMPARISON WITH THEIR HUMAN COUNTERPART

Preliminary Western blot analysis using antibodies raised against human recombinant MRP8 and MRP14 revealed that both cross reacted with analogous proteins in mouse spleen cell preparations. Mouse MRP8 and MRP14 cDNAs were isolated from a spleen expression library with anti-human MRP8 antibody and human MRP14 cDNA probe (Lagasse et al.

submitted). The sequence of the mouse MRP8 cDNA has an open reading frame of 267 nucleotides predicting a protein of 89 amino acids (M_r 10295) whereas the mouse MRP14 cDNA contains an open reading frame of 339 nucleotides predicting protein of 113 amino acids (M_r 13049).

A comparison of complete mouse and human MRP8 and MRP14 cDNAs revealed overall identities of 64% and 59% respectively (Wisconsin software). The strongest homology is found in the coding regions but additional areas of homologies are found in the untranslated regions (Figure 1).

Mouse MRP8 contains a 12 nucleotide deletion after position 267 compared to its human counterpart. Mouse and human MRP14 3' region have no more homology after position 324 but surprisingly, part of the mouse 3' coding region matches with the trailer region of human MRP14 containing repeats and palindromic structure (Lagasse and Clerc 1988), indicating that this region may be rearranged by some recombination event. This unusual 3' region structure of mouse MRP14 was confirmed by several independent clones.

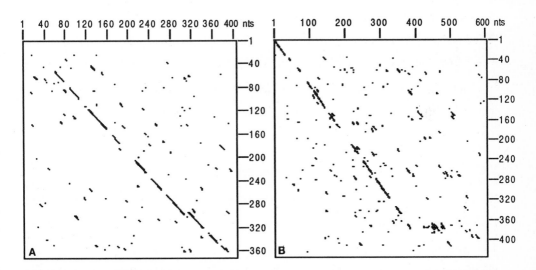

Fig. 1A,B. Dot plot analysis of human and mouse MRP cDNAs. A MRP8. B MRP14. *Horizontal* and *vertical* sequences correspond to human and mouse cDNAs, respectively. The search element length is 11 nucleotides and the maximum of allowed mismatches is three nucleotides

228

At the protein level, both mouse MRP8 and MRP14 share a 59% identity with their human counterpart (Wisconsin software). 53 residues between mouse and human MRP8 and 66 residues between mouse and human MRP14 are identically placed with respect to their position in the sequence (Figure 2).

The areas of greatest homology between mouse and human MRPs are clearly located in the calcium-binding sites of the proteins, whereas the flanking ends, particularly the C-terminal region of both MRPs, are less conserved. By alignment of mouse and human MRP protein sequences (Figure 2), a deletion of four amino acids is located at the C-terminal part of the mouse MRP8 protein, after amino acid 84.

```
                       α-helix-<----turn---->α-helix-
           1           L--LL--L+--OGO-+-LO--OL--LL--O        47
Mouse  MPSELEKALSNLIDVYHNYSNIQGNHHALYKNDFKKMVTTECPQFVQ
MRP8   | .||||||..:|||||.|| |.|| ||:|::|:||::.|||||::.
Human  MLTELEKALNSIIDVYHKYSLIKGNFHAVYRDDLKKLLETECPQYIR
           1                                            47
              α-helix-<---turn--->α-helix-
         48   L--LL--LO-O-OG-LO--OL--LL--L               89
Mouse  NINIENLFRELDINSDNAINFEEFLAMVIKVGVASHK****DSHKE
MRP8   . . : :|:|||||.|.|:||:||| :|||.|||.|| :|||| 
Human  KKGADVWFKELDINTDGAVNFQEFLILVIKMGVAAHKKSHEESHKE
         48                                          93
                       α-helix-<----turn---->α-helix-
           1           L--LL--L+--OGO-+-LO--OL--LL--O        56
Mouse  MANKAPSQMERSITTIIDTFHQYSRKEGHPDTLSKKEFRQMVEAQLATFMKKEKRN
MRP14  |. |   ||:||.|.|||:|||| | ||||||.. ||:::| :|..|:|||.:|
Human  MTCKM*SQLERNIETIINTFHQYSVKLGHPDTLNQGEFKELVRKDLQNFLKKENKN
           1                                                  55
              α-helix-<---turn--->α-helix-
         57   L--LL--LO-O-OG-LO--OL--LL--L               113
Mouse  EALINDIMEDLDTNQDNQLSFEECMMLMAKLIFACHEKLHE*NNPRGHGHSHGKGCGK
MRP14  | :|:.|||||||||.|.||||||| :||||:|.:|:|||:|| ::..|| |..| | |.
Human  EKVIEHIMEDLDTNADKQLSFEEFIMLMARLTWASHEKMHEGDEGPGHHHKPGLGEGTP
         56                                                  114
```

Fig. 2. Comparison of mouse and human MRPs using the bestfit program (University of Wisconsin). The α helices and turns of the calcium-binding sites or EF-hands are presented as predicted (Kretsinger 1980). *O* stands for residues EDQNST (oxygen-containing residues), *L* stands for residues LVIFM (hydrophobic amino acids), and + stands for inserted amino acids. (Szebenyi et al. 1981; Lagasse and Clerc 1988)

The lower molecular weight of mouse MRP8 is confirmed by the more rapid migration of the mouse protein compared to its human counterpart on a SDS gel. For mouse MRP14 an insertion and deletion of one amino acid at positions 6 and 97 respectively are detected after alignment with the human sequence. The 16 last amino acids of mouse MRP14 have less homology with its human counterpart. One striking feature of the mouse MRP14 is the three cysteines in the C terminal part of the protein (positions 80, 91, and 111) in contrast to the single cysteine at position 3 for the human protein. Both mouse and human MRP8 proteins contain one cysteine residue at amino acid 42.

ORGANIZATION OF THE *MRP* GENES

Following cloning of the human cDNAs, the corresponding genomic regions of both MRP8 and MRP14 were isolated (Lagasse and Clerc 1988). They have a strikingly similar organization, (Figure 3) both genes consisting of three exons. Exon 1 encodes part of the 5' untranslated region of the mRNA, whereas exon 2 contains the translation initiation site and exon 3 the termination codon and polyadenylation signal. The second intron is of class 0, since it interrupts the MRP8 and MRP14 coding region between codons. The similar genomic organization of the MRP genes has interesting implications with regard to the evolution of the calcium-binding proteins family.

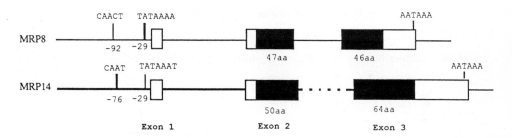

Fig. 3. Genomic organization of the human MRP8 and 14. The *fitted areas* indicate coding regions, and the open regions show the untranslated part of the genes

*MRP*8 AND *MRP*14 ARE CALCIUM MODULATED PROTEINS

Structural Homologies Between *MRP*8 and *MRP*14 Proteins Are Shared with the S-100 Protein Family

A computer search showed that MRP8 and MRP14 proteins shared extensive homology to a group of small calcium-binding proteins, the S-100-like proteins, including S-100a, S-100b, Calcyclin or 2A9, 18A2, 42A, 42C, p9Ka or ICaBP, and P11 (Lagasse and Clerc 1988). MRP8, MRP14, and S-100 (α and β subunits) are more closely related to each other than they are to other calcium-binding proteins, as they share approximatively 30 to 35% identity when compared using the bestfit program of the Wisconsin University. The homology is mostly concentrated to the calcium-binding domains of these proteins. An interesting feature of the evolution of calcium-binding proteins is the structure of their calcium-binding site. They contain the same characteristic configuration: a 12 amino acid loop containing six calcium-coordinating ligands and two alpha helical regions of 8 amino acids each on either side of the loop (see Figure 2). This helix-turn-helix conformation of 28 amino acids is termed the EF-hand (Kretsinger 1980). So far more than 100 different calcium-binding proteins have been isolated which bear two to six EF-hand sites (Persechini et al. 1989), and it is believed that these proteins have evolved from a common ancestor. MRP8 and MRP14 contain two calcium-binding sites (Figure 2); one, a 28 amino acid region in the C-terminal part of the protein, is very similar to the EF-hands found in members of the calcium-binding protein family (calmodulin-like protein family). In contrast, the N-terminal EF hand differs from the canonical calcium-binding site in that it contains 30 amino acids. The two additional residues inserted in the loop are serine/histidine for MRP8 or serine/aspartic acid for MRP14. Interestingly, S-100ß and calcyclin contain also inserted serine/histidine residues.

Comparison of MRP8 and MRP14 proteins by hydropathy profile shows that the overall patterns are highly conserved and similar to the other the S-100-like proteins (Kligman and Hilt 1988). Conserved hydrophobic amino acid domains flank the ends of both MRP proteins, suggesting that these regions are functionally significant. It is known that calcium-modulated proteins, such as calmodulin, change in hydrophobicity after calcium binding, and interact with their target proteins through their hydrophobic domains (Johnson et al. 1986; Kilhoffer et al. 1983). This suggests that MRP8 and MRP14, like the S-100-like proteins, exert specific effector proteins in a manner similar to other calcium-binding proteins.

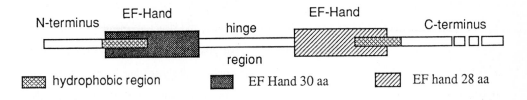

Fig. 4. Domain model for MRP8 and MRP14 proteins. The hydrophobic regions interact largely with the calcium-binding sites

However, MRP8, MRP14, and the S-100-like proteins lack substantial sequence similarity in the hinge region as well as in the distance that separates the two EF-hands; i.e., 9 amino acids in MRP8, 14 amino acids in MRP14, and 13 amino acids in S-100. Since the hinge regions are seemingly free to evolve with less constraint compared to the calcium-binding domains, this suggests that divergence in the hinge region between different members of the S-100 protein family results from a unique function of these proteins related to their specific interactions with effector proteins in different tissues. Finally, the C-terminal segment of MRP14 is unusually long, being approximatively 20 amino acids longer than the other members of the S-100 protein family, and consequently may have some special function related to this unique sequence. Recently, a polypeptide which constitutes the neutrophil-immobilizing factor (NIF) was isolated from human neutrophils (Watt et al. 1983). Partial amino acid sequence determination of the 20 C-terminal amino acids of NIF was identical to amino acid 89 to 108 of human MRP14 suggesting that in vitro the C-terminus of MRP14 has a NIF activity (Freemont et al. 1989). In addition, the threonine at position 113 is phosphorylated, both in human monocytes and neutrophils, indicating that the phosphorylation event could have a role in the generation of NIF activity in MRP14 protein (Edgeworth et al. 1989). However, the mouse MRP14 has a low homology with the human sequence in the C-terminal part of the protein and has no threonine. Therefore the physiological role of mouse MRP14 protein in neutrophil function remains to be determined.

*MRP*8 AND *MRP*14 BELONG TO A DISTINCT FAMILY OF CALCIUM-BINDING PROTEINS

As mentioned above, MRP8 and MRP14 share many structural homologies in their primary and secondary amino acid sequences with the S-100-like proteins. In addition, MRP8 and MRP14 have a similar genomic organization and because the structure of the calcyclin gene was shown to be identical (Ferrari et al. 1987), we postulated that other genes of the S-100-like proteins would have a similar genomic structure (Lagasse and Clerc 1988). This hypothesis was confirmed when more genes of this family were described (Allore et al. 1988; Krisinger et al. 1988). They all share with MRP8 and MRP14 the following genomic organization: the gene is composed of three exons, exon 1 is untranslated and exon 2 and 3 each delimit an entire calcium-binding domain. In addition the second intron which separate the two EF-hands is of class 0. This pattern of structural organization is unique to the S-100-like proteins and is completely different from other calcium-binding proteins such as the calmodulin-like proteins indicating that all these closely related S-100-like proteins belong to a new distinct group of calcium-binding proteins.

EXPRESSION OF THE *MRP* GENES DURING MYELOPOIESIS

The cloning and characterization of human MRP cDNAs, as well as polyvalent monospecific antisera raised against MRP8 and MRP14, allowed studies on the biological properties of these proteins to be done. These proteins are predominantly expressed in cells of myeloid origin; in peripheral blood they appear in monocytes and neutrophils; in tissue, the expression of these proteins is in part inflammation dependent. MRP8 is expressed by infiltrating macrophages only in chronically inflamed tissues, whereas MRP14 expressing macrophages are found in most inflammations (Odink et al. 1987; Palmer et al. 1987). However, in the human biopsies examined so far, normal tissue macrophages never express these proteins at a level detectable with immunohistochemical methods (Zwadlo et al. 1988). Finally, in many patients with infectious (e.g., viral, bacterial, or parasitic) or malignant conditions (e.g., pulmonary or gastrointestinal cancers) increased concentration of MRP proteins (L1 complex or CFAg) are found in the plasma (Dale et al. 1983; Sander et al. 1984; Dorin et al. 1987).

cis CONTROL OF HUMAN *MRP*8 AND *MRP*14 DURING IN VITRO MYELOID DIFFERENTIATION

To study the regulation of the MRP genes during myeloid differentiation we used the leukemia cell line HL60. Proliferating HL60 cells can be stimulated to terminally differentiate into nonproliferating monocytic or granulocytic lineages by treatment with phorbol ester or dimethylsulfoxide, respectively. Within 24 h of induction of HL60 to the monocytic lineage, the low levels of MRP8 and MRP14 mRNA essentially disappear. In contrast, after 24 h of induction of HL60 to the granulocytic lineage, both MRP8 and MRP14 are highly expressed. After mapping the transcription initiation site of MRP8 and MRP14 genes by primer extension, transient expression experiments were devised to investigate the regulatory elements responsible for tissue-specific regulation of both MRP genes. The chloramphenicol acetyltransferase (CAT) gene was inserted in the coding region of MRP8 and MRP14, and tested in transient expression assay with HL60 cell lines. We found that our two chimeric constructs both express CAT activity in a way similar to the endogenous MRPs (Lagasse and Clerc 1988). In addition this expression was tissue-specific, as no expression was observed in the human epithelial cell line L132 under the same conditions. The result of these investigations demonstrated a strong cis control regulation of both MRP genes during myeloid differentiation.

ANALYSIS OF THE MOUSE *MRP*8 AND *MRP*14 EXPRESSION

Analysis of mouse tissues by Northern and Western blot demonstrated that the major site of MRP8 and MRP14 expression is the bone marrow. Low levels of both MRPs are also found in spleen and lung (Lagasse et al. submitted). Both bone marrow and spleen are the major sites of myelopoiesis, and as the lung is a primary site for invasion by microorganisms, it is a tissue rich in myelomonocytic cells.

To identify cells expressing MRP8 and MRP14, murine bone marrow cells and spleen cells were fixed and immunostained with rabbit anti-human MRP8 and MRP14 antibodies. A significant proportion (70%) of the bone marrow cells and to a lesser extent spleen cells stained positive with both MRP antibodies (Figure 5).

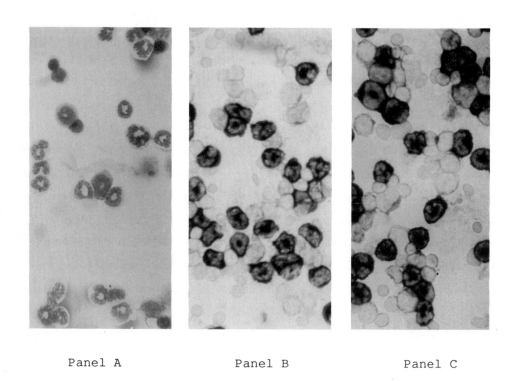

Panel A   Panel B   Panel C

Fig. 5A-C. Photomicrographies of bone marrow smears. **A** Giemsa and May-Gruenwald staining showing numerous myeloid and neutrophil cells with characteristic ring and horseshoe-shaped nucleus. **B** and **C** Immunostaining with anti-MRP8 and anti-MRP14 antibodies. Most of the myeloid cells expressed MRP8 and MRP14 proteins

These cells are of myelomonocytic morphology and most probably represent mixed populations of myeloid precursor cells, neutrophils and monocytes. Finally, during embryogenesis MRP8 and MRP14 expression appears early in development associated with generation of the fetal myelopoietic system (Lagasse unpubl.).

HAVE *MRP*8 AND *MRP*14 A ROLE IN MYELOPOIESIS?

The results reported above clearly indicate an association of MRP8 and MRP14 expression with myelopoiesis. It is known that expression of some of the S-100-like proteins is also associated with cell differentiation. In the RT4 rat neuroblastoma cell system, a precursor stem cell gives rise to neuronal or glial daughter cells. The stem cell and the glial daughter cell lines express S-100b, whereas the neuronal cell lines do not (Freeman and Sueoka 1987). Similarly, mammary cuboidal stem cells are p9Ka negative but become p9Ka-positive after differentiation into myoepithelial-like cells (Barraclough et al. 1987). Nerve growth factor (NGF) treatment of PC12 cell, a rat pheochromocytoma cell line, induces synthesis of 42A and 42C after the appearance of a neural phenotype (Masiakowski and Shooter 1988). All these systems involve cell differentiation in association with the expression of S-100-like proteins. However, little is known about the function of all these S-100-like proteins, although by analogy with other calcium-binding proteins they could be molecules involved in intracellular signalling that are activated by an increase in the intracellular calcium concentration. Recently, a protein complex (PC), identified as MRP8 and MRP14, was shown to inhibit a kinase-mediated stimulation of RNA polymerase activity (Murao et al. 1989). This preliminary result suggests that the MRP8 and MRP14 complex could modulate RNA polymerase function and consequently may be associated with myeloid cell functions.

REFERENCES

Allore R, O'Hanlon D, Price R, Neilson K, Willard HF, Cox DR, Marks A, Dunn RJ (1988) Gene encoding the beta subunit of S-100 protein is on chromosome 21: implication for Down syndrome. Science 239:1311-1313

Andersson KB, Sletten K, Berntzen HB, Dale I, Brandtzaeg P, Jellum E, Fagerhol MK (1988) The leukocyte L1 protein: Identity with the cystic fibrosis antigen and the calcium-binding MRP8 and MRP14 macrophage components. Scan J Immunol 28:241-245

Barraclough R, Savin J, Dube SK, Rudland PS (1987) Molecular cloning and sequence of the gene for p9Ka. A culture myoepithelial cell protein with strong homology to S-100, a calcium-binding protein. J Mol Biol 198:13-20

Brandtzaeg P, Jones DB, Flavell DJ, Fagerhol MK (1988) Mac 387 antibody and detection of formalin resistant myelomonocytic L1 antigen. J Clin Pathol 41:963-970

Brüggen J, Tarcsay L, Cerletti N, Odink K, Rutishauser M, Hollander G, Sorg C (1988) The molecular nature of the cystic fibrosis antigen. Nature 331:570

236

Burmeister G, Tarcsay L, Sorg C (1986) Generation and characterization of a monoclonal antibody (1C5) to human migration inhibition factor (MIF). Immunol 171:461-474

Dale I, Fagerhol MK, Naesgaard I (1983) Purification and partial characterization of a highly immunogenic human leukocyte protein, the L1 antigen. Eur J Biochem 134:1-6

Dorin JR, Novak M, Hill RE, Brock DJH, Secher DS, van Heyningen V (1987) A clue to the basic defect in cystic fibrosis from cloning the CF antigen. Nature 326:614-617

Freemont P, Hogg N, Edgeworth J (1989) Sequence identity. Nature 339:516

Edgeworth J, Freemont P, Hogg N (1989) Ionomycin-regulated phosphorylation of the myeloid calcium-binding protein p14. Nature 342:189-192

Ferrari S, Calabretta B, de Riel JK, Battini R, Ghezzo F, Lauret E, Griffin C, Emanuel BS, Gurrieri F, Baserga R (1987). Structural and functional analysis of a growth-regulated gene, the human Calcyclin. J Biol Chem 262:8325-8325

Freeman MR, Sueoka N (1987) Induction and segregation of glial intermediate filament expression in the RT4 family of peripheral nervous system cell lines. Proc Natl Acad Sci USA 84:5808-5812

Hogg N, Allen C, Edgeworth J (1989) Monoclonal antibody 5.5 reacts with p8, 14, a myeloid molecule associated with some vascular endothelium. Eur J Immunol 19:1053-1061

Johnson JD, Mills JS (1986) Calmodulin. Med Res Rev 6:341-363

Kilhoffer MC, Haiech J, Demaille JG (1983) Ion binding to calmodulin. Mol Cell Biochem 51:33-54

Kligman D, Hilt DC (1988) The S-100 protein family. TIBS 13:437-443

Kretsinger RH (1980) Structure and evolution of calcium-modulated proteins. CRC Crit Rev Biochem 8:119-174

Krisinger J, Darwish H, Maeda N, Deluca H (1988) Structure and nucleotide sequence of the rat intestinal vitamin D-dependent calcium binding protein gene. Proc Natl Acad Sci USA 85:8988-8992

Lagasse E and Clerc RG (1988) Cloning and expression of two human genes encoding calcium-binding proteins that are regulated during myeloid differentiation. Mol Cell Biol 8:2402-2410

Masiakowski P, Shooter EM (1988) Nerve growth factor induces the genes for two proteins related to a family of calcium-binding proteins in PC12 cells. Proc Natl Acad Sci USA 85:1277-1281

Murao S, Collart FR and Huberman E (1989) A protein containing the cystic fibrosis antigen is an inhibitor of protein kinases. J Biol Chem 264:8356-8360

Odink K, Cerletti N, Brüggen J, Clerc RG, Tarcsay L, Zwadlo G, Gerhards G, Schlegel R, Sorg C (1987) Two calcium-binding proteins in infiltrate macrophages of rheumatoid arthritis. Nature 330:80-82

Palmer DG, Hogg N, Allen CA, Highton J, Heesian PA (1987) A mononuclear phagocyte subset associated with cell necrosis in rheumatoid nodules: identification with monoclonal antibody 5.5. Clin Immunol Immunopath 45:17-28

Persechini A, Moncrief N D, Kretsinger RH (1989) The EF-hand family of calcium-modulated proteins. TINS 12:462-467

Sander J, Fagerhol MK, Bakken JS, Dale I (1984) Plasma levels of the leukocyte L1 protein in febrile conditions: relation to aetiology, number of leukocytes in blood, blood sedimentation reaction and C-reactive protein. Scand J Clin Lab Invest 44:357-362

Szebenyi DME, Obendorf SK, Moffat K (1981) Structure of vitamin D-dependent calcium-binding protein from bovine intestine. Nature 294:327-332

Watt KWK, Brightman I, Goetzl EJ (1983) Isolation of two polypeptides comprising the neutrophil-immobilizing factor of human leukocytes. Immunol 48:79-86

Zwadlo G, Brüggen J, Gerhards G, Schlegel R, Sorg C (1988) Two proteins associated with specific stages of myeloid cell differentiation are expressed by subsets of macrophages in inflammatory tissues. Clin Exp Immunol 72:510-515

DISTRIBUTION OF THE CALCIUM-BINDING PROTEINS *MRP*-8 AND *MRP*-14 IN NORMAL AND PATHOLOGICAL CONDITIONS: RELATION TO THE CYSTIC FIBROSIS ANTIGEN

Josef Brüggen and Nico Cerletti

INTRODUCTION

Eukaryotic cells require calcium ions for optimal growth and functioning. The intracellular actions of calcium as a biological second messenger appear to be the result of its interaction with a set of calcium-binding proteins referred to as calcium modulated proteins. (Persechini et al. 1989).

Recently we described the isolation, molecular cloning and recombinant expression of two calcium-binding proteins MRP-8 and MRP-14 (Odink et al. 1987). The proteins were isolated from human peripheral blood mononuclear cell cultures as part of a complex, using a monoclonal antibody directed against the human macrophage inhibitory factor (MIF). The relative molecular masses (M_r) of the proteins on SDS-PAGE are 8 kDa and 14 kDa, respectively. Therefore we relate to these proteins as MIF related proteins, MRP-8 and MRP-14. MRP-8 and MRP-14 are predominantly expressed intracellularly by cells of the myeloid lineage, i.e., granulocytes, monocytes and macrophages (Zwadlo et al. 1988) and in certain epithelial tissues (Andersson et al. 1988; Wilkinson et al. 1988). The sequence of MRP-8 is identical to the sequence of the cystic fibrosis (CF) antigen (Dorin et al. 1987). After publishing the sequences of MRP-8 and MRP-14 it was reported to be identical to the two polypeptides chains of the leucocyte L1 protein (Anderson et al. 1988). These proteins are members of a family of calcium binding proteins, which include S-100α and ß, calcyclin, p11 and others (Lagasse and Clerc 1988).

In this article we will emphasize studies addressing the distribution of the MRP proteins in tissues and body fluids under normal conditions and in various diseases.

Knowledge of the in vivo distribution of these proteins could provide insights into potential functions, which are still not clear.

STRUCTURAL AND BIOCHEMICAL PROPERTIES OF *MRP*-8 AND *MRP*-14

The sequence of MRP-8 cDNA has an open reading frame of 279 nucleotides, predicting a protein of 93 amino acids and M_r 10835.

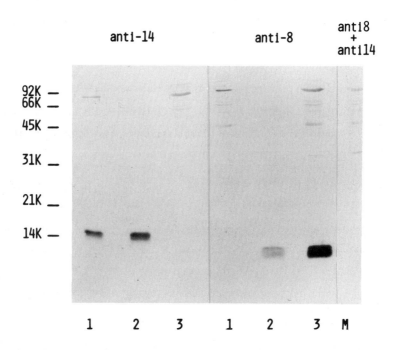

Fig. 1. Western blot analysis. Reactivity of affinity-purified anti MRP-8 and anti MRP-14 antibodies with natural and recombinant MRP. Crude cell lysates of stable transformed L-132-clones (human embryonic lung cells) producing MRP-8 (*lane 3* 10μg protein/lane) or MRP-14 (*lane 1* 10μg protein/lane) and of human blood monocytes (*lane 2* 50μg protein/lane) were separated by SDS-PAGE and electroblotted to an Immobilon membrane. The blot was developed with anti MRP-8/14 and visualized with immuno alkaline phosphatase. Antisera were raised in rabbits using recombinant MRP-8 and MRP-14 and subsequently affinity purified

The sequence of MRP-8 is identical to the sequence of the cystic fibrosis (CF) antigen (Dorin et al. 1987), after correction of a sequencing error for the latter. MRP-14 cDNA contains an open reading frame of 352 nucleotides predicting a protein of 114 amino acids and M_r 13242.

MRP-8 and MRP-14 both contain a single cysteine residue, have no signal or membrane anchor sequence. Both MRP-8 and MRP-14 contain two Ca^{2+}-binding domains, which are quite distinct. The two proteins can form homo- and heterodimers, as known for the S-100 proteins. Recombinant proteins produced in *E. coli* or in mammalia cells comigrate with the natural proteins (Odink et al. 1987).

Affinity purified rabbit antibodies raised against the recombinant proteins were proven to be monospecific by Western blot analysis of the recombinant and natural proteins (Figure 1). The antibodies did not cross react and recognize both natural and recombinant MRP protein.

EXPRESSION OF *MRP*-8 AND *MRP*-14 IN PERIPHERAL BLOOD CELLS AND IN TISSUES UNDER NORMAL AND ABNORMAL CONDITIONS

One way to understand the biological role of MRP-8 and MRP-14 in vivo is to determine their precise expression in cells and tissues under normal and abnormal conditions.

The monospecific polyvalent rabbit antibodies raised against the recombinant proteins were used to detect the proteins in fixed cells of the peripheral blood and cryostat sections of human biopsies. MRP-8 and MRP-14 were visualized immunohistochemically using the indirect alkaline phosphatase or peroxidase technique (Zwadlo et al. 1989).

Within *peripheral blood cells* both proteins were expressed in monocytes and granulocytes, but were absent from platelets and lymphocytes. While the number of positive granulocytes was constantly high (85-100%) the percentage of positive monocytes varied greatly, depending on the status of the donor. Freshly prepared monocytes of normal donors in the presence of cycloheximide were positive to varying degrees (4%-20%); upon cultivation the percentage of MRP-8/14 positive cells rose to 50%-80% (Zwadlo et al. 1988). The blood monocytes of patients suffering from various diseases (Cystic fibrosis, rheumatoid arthritis) were up to 100% positive for both proteins (unpubl. observ.).

In sections of *normal human tissues* neither protein was detectable in resident macrophages. While the skin was largely negative, the placenta and the lung contained MRP-14 positive cells (monocytes, granulocytes) only. The liver contained cells positive for both proteins. The MRP-14 positive monocytes are found only in the lumen of capillary vessels (Zwadlo et al. 1988).

In *inflammatory tissues* a differentiated picture is seen. In conditions of inflammation containing a substantial infiltrate of mononuclear cells a large proportion of the cells expressed MRP-8 and MRP-14. However, MRP-8 and MRP-14 were differentially expressed depending on the type of inflammation. In acutely inflamed tissues like gingivitis, psoriasis, neurodermitis MRP-14 was present in macrophages of the perivascular infiltrate, whereas MRP-8 was absent. In contrast, in chronically inflamed tissues like primary chronic polyarthritis macrophages were found staining positive for MRP-8 and MRP-14 (Odink et al. 1987; Zwadlo et al. 1988). In a recent study investigating the differential expression of MRP-8 and MRP-14 in granulomatous conditions it was described, that non-phagocytic mononuclear cells in delayed hypersensitivity type granulomas express MRP-14 only, whereas phagocytic mononuclear phagocytes in non-hypersensitivity and non-immunological granulomas express MRP-8 and MRP-14 (Delabie et al. 1990).

The absence of the MRP-8 and MRP-14 proteins in resident macrophages of normal tissues and their presence in the mononuclear infiltrate of activated or inflamed tissues like rheumatoid arthritis emphasizes, that these two proteins are intimately involved in the activation and differentiation of macrophages. It is clear that MRP-8 and MRP-14 can be selectively expressed, alone or in combination, depending on the type of inflammatory condition (Zwadlo et al. 1988; Delabie et al. 1990). This is in contrast to the findings of others that describe a coexpression of these proteins under any conditions (Anderson et al. 1988; Wilkinson et al. 1988).

Calcium-binding proteins are involved in cell regulatory processes (Wright et al. 1985). However, as it was found recently (Freemont et al. 1989), MRP-14 seems to display additional functions: the C-terminal part of MRP-14 was shown to be identical to a polypeptide sequence originally described as the neutrophil-immobilizing factor (Watt et al. 1983). Another interesting observation was, that the MRP8-MRP14 complex specifically blocks casein kinase II activity (Murao et al. 1989).

ELEVATED *MRP*-14 PLASMA LEVELS IN INFLAMMATORY DISORDERS: RELATION TO THE CYSTIC FIBROSIS ANTIGEN

MRP-8 and MRP-14 are considered to be soluble cytoplasmic proteins and lack a typical signal sequence, either N-terminal or internal (Odink et al. 1987). Nevertheless, these proteins are released into biological fluids under certain pathological conditions. The L1 complex, which has been shown to consist of MRP-8 and MRP-14, can be found at increased concentrations in patients with infectious and malignant disorders (Andersson et al. 1988). Recently the molecular cloning of the *cystic fibrosis (CF) antigen* was described, which is present at elevated levels in the plasma of CF patients and carriers (Dorin et al. 1987). The sequence for the CF antigen, after correction of a sequencing error, is identical to *MRP-8* (Odink et al. 1987). However, as we suggested from a preliminary study the plasma levels of *MRP-14* and *not of MRP-8* are elevated in patients and carriers of the cystic fibrosis (CF) trait (Brüggen et al. 1988). Elevated *MRP-14* levels are not unique to cystic fibrosis, but found in other pathological conditions as well.

Cystic fibrosis is the most common autosomal recessive disease among Caucasians. About 1 in 25 people, though healthy, is a carrier, and 1 in 2500 newborns is affected by the disease. In order to examine the possibility of detecting carriers in a random population we extended our preliminary study (Brüggen et al. 1988) to a statistical relevant number of CF patients, carriers, normal controls and other disorders (Brüggen et al. 1989; Brüggen et al. 1990).

The proteins were detected in heparinized plasma by use of a two-site enzyme linked immunosorbent assay (ELISA) using the affinity purified monospecific rabbit antibodies to MRP-8/MRP14 and highly purified recombinant MRP-8 or MRP-14 as a standard (Brüggen et al. 1990). The characteristics of the donor groups are summarized in Table 1.

MRP-8 plasma levels are low in all test groups

MRP-8 was detectable only in a range of zero to 10 - 30 ng/ml plasma. Recombinant MRP-8 mixed with blood one hour prior to sample preparation was fully detectable, excluding a selective loss of MRP-8 during the isolation. However, the levels of MRP-8 were elevated in blood samples containing lysed cells.

Table 1. Test groups of the cystic fibrosis study

Test groups	number	age (range)
CF study I: Switzerland		
Normal controls		
- adults	39	20 - 49
- children	55	1 - 15
Cystic fibrosis patients	27	1 - 21
Cystic fibrosis carriers	49	5 - 62
(parents of patients and genetically determined siblings)		
Other diseases	31	
CF study II: Denmark		
Normal controls		
- adults*(non B haplotype)	55	19 - 60
- siblings of CF patients	13	11 - 26
Cystic fibrosis patients	20	7 - 26
Cystic fibrosis carriers	48	11 - 49
(parents of patients and genetically determined siblings)		

*Individuals of the non-B-haplotype carry a low risk for the CF trait
(M. Schwartz, pers. commun.)

Our findings are in contrast to recent reports on the molecular nature of the CF antigen that described MRP-8 as the main molecule released into the plasma of CF patients and CF carriers (Dorin et al. 1987). These authors probably isolated the CF antigen from granulocytes as a complex of MRP-8 and MRP-14 as well. If so, they obtained only the MRP-8 sequence because, as shown by our data (Odink et al. 1987), MRP-14 is N-terminally blocked.

MRP-14 plasma levels are elevated in Cystic fibrosis

Table 2 summarizes the plasma levels of MRP-14 in CF patients, CF carriers (parents and genetically analyzed siblings) and healthy random controls collected in Switzerland and in Denmark. Although in the Swiss study the median plasma levels of normal controls and CF carriers are quite different, the overlap between the groups is too large to allow random carrier detection on the basis of MRP-14 levels. The CF patients contained elevated plasma levels up to 5.7 µg/ml.

Table 2. MRP-14 plasma levels of cystic fibrosis patients, cystic fibrosis carriers and normal controls from Switzerland and Denmark

Diagnosis	number (n)	MRP-14(µg/ml)[a] median	range
CF study I: Switzerland[b]			
Normal controls	94	0.048	0.003-0.230
CF carriers	49	0.180	0.060-0.390
CF patients	27	0.480	0.190-5.780
CF study II: Denmark[c]			
Normal controls	63	0.132	0.054-0.241
CF carriers	48	0.139	0.054-0.329
CF patients	20	0.476	0.168-2.590

[a] Two-site ELISA, two independent determinations were done on each sample; the inter-assay coefficient of variation is 10%.
[b] CF carriers include parents of CF patients (n=40) and genetically analyzed siblings (n=8).
[c] The normal controls are of the non-B-haplotype. CF carriers include parents of CF patients (n=34) and genetically analyzed siblings (n=13).

Some high values in the control group might come from unidentified carriers (5% among caucasians). In a study with Danish donors, carriers could virtually be excluded on the basis of genetic analysis (i.e., non-B-haplotype). The data are shown in table 2. In this study the MRP-14 values of CF carriers and normal controls were within the same range with a median plasma level of 0.132 and 0.139 µg/ml respectively.

MRP-14 plasma levels are elevated in a variety of pathological conditions

Table 3 shows that MRP-14 levels are elevated in donors suffering from a variety of diseases. Especially high values can be found in rheumatoid arthritis (up to 1.48 µg/ml). However, there are also disorders having MRP-14 values within the normal range like psoriasis and sarcoidosis.

Table 3. MRP-14 plasma levels in patients suffering from other diseases than cystic fibrosis

Diagnosis	number	MRP-14 (µg/ml)	
		median	range
Rheumatoid arthritis	13	0.618	0.093-1.480
Asthma	2	0.357	
		0.476	
T-cell lymphoma	2	0.240	
		2.040	
Neurodermitis	1	1.840	
Psoriasis	6	0.027	0.014-0.036
Miscellaneous*	7	<0.03	

* Mycosis fungoides, sarcoidosis, leprosy, contact dermatitis

It is clear from our results that the MRP-14 and not the MRP-8 plasma levels correlate with the cystic fibrosis trait. Although the cystic fibrosis gene has been cloned recently (Kerem et al. 1989) an additional test for carrier detection would still be desirable because the deletion identified is responsible for approximately 70% of the mutations found in CF patients (Wilfond et al. 1990). In our two independent studies the MRP-14 plasma levels showed a large overlap (Switzerland) to no difference (Denmark) between CF carriers and normal donors. Therefore, in agreement with others (Hayward et al. 1987), MRP-14 cannot be used to detect CF carriers in a random population.

Elevated MRP-14 levels are not unique to cystic fibrosis. Like described for L1, which is a complex of MRP-14 and MRP-8 (Andersson et al. 1988), we find that high plasma levels correlate with inflammatory conditions. In disagreement with the data on L1,

MRP-8 is present only at lower concentrations or virtually absent. As the granulocyte and the monocyte/macrophage are the major cells producing MRP-14 and MRP-8 (Dorin et al. 1987; Andersson et al. 1988; Zwadlo et al. 1988), increased turnover of leukocytes is probably the cause of the higher MRP-14 values in inflammatory conditions. How MRP-8 from the serum is cleared is not known.

CONCLUSIONS

MRP-8 and MRP-14 are calcium binding proteins that are predominantly expressed by granulocytes, monocytes, macrophages, histiocytes (Dorin et al. 1987; Zwadlo et al. 1988; Andersson et al. 1988) and in certain epithelial tissues (Andersson et al. 1988; Wilkinson et al. 1989; Hogg et al. 1989). The isolation and the molecular cloning of this pair of proteins and the cloning of the corresponding genes has been described recently (Odink et al. 1987; Lagasse et al. 1988). Here we discussed the distribution of these proteins in tissues and plasmas under normal and pathological conditions. Normal tissue macrophages are virtually negative for MRP-8 and MRP-14. Whereas most inflammatory macrophages and histiocytes express MRP-14, MRP-8 is expressed mainly in chronic inflammation like rheumatoid arthritis (Odink et al. 1987). A differential expression of MRP-8 and MRP-14 was also described for granulomas of the delayed hypersensitivity type and non-hypersensitivity type respectively (Delabie et al. 1990).

Although localized intracellularly and lacking a typical signal sequence these proteins are released into biological fluids under certain pathological conditions often involving inflammatory processes like cystic fibrosis and rheumatoid arthritis (Hayward et al. 1987; Andersson et al. 1988; Brüggen et al. 1988; Brüggen et al. 1990). We could detect only elevated plasma levels of *MRP-14*, and not of *MRP-8*, which was described as cystic fibrosis antigen (Dorin et al. 1987), on screening a large number of plasma samples. The plasma levels of these proteins are likely to reflect the turnover of leukocytes and , therefore, could be a valuable parameter in inflammatory diseases (Berntzen et al. 1989). Interestingly, the release of another pair of calcium-binding proteins, S-100 α+ß, has been described recently into urine of patients with kidney diseases (Kato et al. 1987) and into patient sera during acute myocardial infarction (Usui et al. 1990).

MRP-8 and MRP-14 belong to the family of S-100-like calcium-binding proteins. Little is known about their function, although they are likely to be intimately involved in cell regulatory processes and intracellular signaling (Wright et al. 1985). MRP-8 and MRP-14 are closely linked to the activation and differentiation of myeloid cells like granulocytes and monocytes/macrophages (Zwadlo et al. 1988; Hogg et al. 1988). The complex of MRP8-MRP14 was reported to block specifically casein kinase II activity (Murao et al. 1989). MRP-14 may even display additional functions: the C-terminal part is identical to a polypeptide sequence originally described as the neutrophil-immobilizing factor (Freemont et al. 1989).

The expression and the activities of these two proteins and their interaction with each other and with other proteins seem to be important in the regulation of macrophage differentiation and thus to our understanding of the mechanisms underlying inflammation.

MRP-8 and MRP-14 have been referred to as cystic fibrosis antigen, calgranulin A and B (Wilkinson et al. 1989), L1 heavy and light chain (Anderson et al. 1988) or p8 and p14 (Hogg et al. 1988). We propose to use the names MRP-8 and MRP-14, which are assigned to the sequence in the EMBL database, until a clear function for these proteins has been found.

REFERENCES

Andersson KB, Sletten K, Berntzen HB, Dale I, Brandtzaeg P, Jellum E, Fagerhol MK (1988) The leukocyte L1 protein: identity with the cystic fibrosis antigen and the calcium-binding MRP-8 and MRP-14 macrophage components. Scand J Immunol 28:241-245

Berntzen HB, Munthe E, and MK Fagerhol (1988) The major leukocyte protein as an indicator of inflammatory joint disease. Scand J Immunol suppl 76:251-256

Brüggen J, Tarcsay L, Cerletti N, Odink K, Rutishauser M, Holländer G, Sorg C (1988) The molecular nature of the cystic fibrosis antigen. Nature 331:570

Brüggen J, Tarcsay L, Cerletti N, Wiesendanger W, Odink K, Rutishauser M, Holländer G, Schwartz M, Sorg C (1989) The plasma level of the calcium binding protein MRP-14 is elevated in patients suffering inflammatory disorders: Relation to the cystic fibrosis antigen. In: 1-st European Symposium on Calcium Binding Proteins in Normal and Transformed Cells. Presses Universitaires A.S.B.L., Bruxelles E 16 (Abstract)

Brüggen J, Holländer G, Tarcsay L, Cerletti N, Odink K, Rutishauser M, Schwartz M, Sorg C (1990) The plasma level of the calcium-binding protein MRP-14 is elevated in inflammatory disorders. (submitted)

Delabie J, De Wolf-Peeters C, Van den Oord JJ & VJ Desmet (1990) Differential expression of the calcium-binding proteins MRP8 and MRP14 in granulomatous conditions: an immunohistological study. Clin exp Immunol (in press)

Dorin JR, Novak M, Hill RE, Brock DJH, Secher DS, van Heyningen V (1987) A clue to the basic defect in cystic fibrosis from cloning the CF antigen. Nature 326:614-617

Freemont P, Hogg N, Edgeworth J (1989) Sequence identity. Nature 339:516

Hayward C, Glass S, Van Heyningen V, Brock DJH (1987) Serum concentrations of a granulocyte-derived calcium-binding protein in cystic fibrosis patients and heterozygotes. Clinica Chimica Acta 170:45-56

Hogg N, Allen C, Edgeworth J (1989) Monoclonal antibody 5.5 reacts with p8, 14, a molecule associated with some vascular endothelium. Eur J Immunol 19:1053-1061

Kato K, Haimoto H, Shimizu A, Tanaka J (1987) Enzyme immunoassay for measurement of the α subunit of S-100 protein in human biological fluids. Biomedical Research 8:119-125

Kerem BS, Rommens JM, Buchanan JA, Markiewicz D,Cox TK, Chakravarti A, Buchwald M, Tsui LC (1989) Identification of the cystic fibrosis gene: genetic analysis. Science 245:1073-1080

Lagasse E, Clerc R (1988) Cloning and expression of two human genes encoding calcium-binding proteins that are regulated during myeloid differentiation. Molec Cell Biology 8:2402-24110

Murao S, Collart FR and Hubermann E (1989) A protein containing the cystic fibrosis antigen is an inhibitor of protein kinases. J Biol Chem 164:8356-8360

Odink K, Cerletti N, Brüggen J, Clerc R, Tarcsay L, Zwadlo G, Gerhards G, Schlegel R, Sorg C. Two calcium-binding proteins in infiltrate macrophages of rheumatoid arthritis. Nature 330:80-82

Persechini A, Moncrief ND, Kretsinger RH (1989) The EF-hand family of calcium-modulated proteins. TINS 12:462-467

Usui A, Kato K, Sasa H, Minaguchi K, Abe T, Murase M, Tanaka M, Takeuchi E (1990) S-100α_0 protein in serum during myocardial infarction. Clin Chem 36/4:639-641

Watt KWK, Brightman I, Goetzl EJ (1983) Isolation of two polypeptides comprising the neutrophil-immobilizing factor of human leukocytes. Immunol 48:79-86

Wilfond BS, Fost N (1990) The cystic fibrosis gene: Medical and social implications for heterozygote detection. JAMA 263:2777-2782

Wilkinson MA, Busuttil A, Hayward C, Brock DJH, Dorin JR, van Heyningen V (1988) Expression pattern of two related cystic fibrosis-associated calcium-binding proteins in normal and abnormal tissues. J Cell Science 91:221-230

Wright B, Zeidman I, Greig R, Poste G (1985) Inhibition of macrophage activation by calcium channel blockers and calmodulin antagonists. Cell Immunol 95:46-52

Zwadlo G, Brüggen J, Gerhards G, Schlegel R, Sorg C (1988) Two calcium-binding proteins associated with specific stages of myeloid cell differentiation are expressed by subsets of macrophages in inflammatory tissues. Clin Exp Immunol 72:510-515

Section III

Novel Members of the EF-Hand Calcium-Binding Protein Family in Normal and Transformed Cells

CALRETININ

John H. Rogers

INTRODUCTION

Calretinin is a calcium-binding protein of 29-31 kilodaltons, expressed mainly if not exclusively in nerve cells. It is most homologous to the 28-kilodalton avian intestinal CaBP, calbindin-D28 (hereinafter calbindin). Calretinin and calbindin were both probably neuronal proteins in origin, and they are expressed in largely separate sets of neurons both in birds and in mammals.

DISCOVERY AND RELATIONSHIP TO CALBINDIN-D28K

In retrospect, calretinin was first noticed as a 29-kDa band cross-reacting with antisera against chick calbindin on immunoprecipitations and on Western blots of brain tissue (Pochet al. 1985; Parmentier et al. 1987; Rogers 1989a). The cross-reaction was variable and sometimes weak on Western blots, but strong on immunoprecipitation (Pochet et al. 1985; Rogers 1987), presumably because immunoprecipitations are less sensitive to differences in affinity. This cross-reacting protein was detected in brains of mammals, birds, reptiles, and amphibia, although not in fish (Parmentier et al. 1987). While the affinity of the antibody for the 27-kDa calbindin diminished from birds to amphibia, the affinity for the 29-kDa band was almost unchanged, consistent with its representing an independently evolving protein (Parmentier 1989). As described below, the 29-kDa band is calretinin, and some of the published immunohistochemical surveys performed with antisera against calbindin also detected cross-reaction with calretinin.

Chick calretinin mRNA was cloned serendipitously as one of several cDNA clones with abundant expression in chick retina (Rogers 1987). It was named calretinin because it encoded a CaBP (see below) which was particularly abundant in retina,although it is also abundant in various other parts of the central nervous system. Calretinin and calbindin are encoded by separate genes, and on Southern blots the chick cDNA probe detected a putative calretinin gene in rat as well as chick.

The calretinin gene product was identified as a 29-kDa protein (as sized by SDS-PAGE) by several lines of evidence:

1. The cDNA clone hybrid-selected a mRNA from chick retina which was translated in vitro to give a 29-kDa product (Rogers 1987).

2. Antisera were raised against ß-galactosidase-calretinin fusion proteins (see below), and were used to immunoprecipitate an in vitro translation of mRNA from chick retina; a 29-kDa protein was specifically precipitated (unpublished results).

3. The same antisera were used on Western blots of extracts of chick retina and brain, and labelled the same 29-kDa protein (Rogers 1989a).

The following evidence showed that calretinin was in fact the same as the 29-kDa band detected by anti-calbindin antisera:

1. In the three experiments described above, calretinin comigrated with this band.

2. The sequence is 59% identical to calbindin (see below).

3. A survey of chick brain by in situ hybridization revealed that several nuclei previously reported as immunoreactive for calbindin are in fact negative for calbindin mRNA but positive for calretinin mRNA (Rogers 1987).

4. A survey of rat brain by immunohistochemistry showed that some calbindin immunoreactivity could be eliminated by pre-absorption of the antiserum with rat calretinin (purified by immunoaffinity with calretinin antiserum) (Résibois et al. 1990 and in preparation).

5. Calbindin antiserum directly detects purified calretinin on Western blots (Pasteels et al. 1990).

The human calretinin gene was independently cloned, in a more deliberate manner, by Parmentier (1989, 1990), who used antisera against calbindin to screen a λgt11 cDNA library for clones expressing the cross-reacting protein. The gene encodes a protein with 86%

homology to chick calretinin. This degree of homology strongly suggests that the human and chick genes are orthologous.

Calretinin protein has been independently identified and purified from guinea-pig brain by Winsky et al. (1989a,b) (see also Jacobowitz this Vol.). It was identified as a CaBP on two-dimensional protein gel blots from brain, especially the cochlear nuclei. These researchers named it protein 10 but its molecular weight (29 kDa) and isoelectric point (5.3) were very close to those of mammalian calretinin (Table 1), and partial sequences of five peptides from it were identical to human calretinin at 82 out of 86 positions. (A sixth reported peptide sequence, N-E-P-A-I-L, cannot obviously be aligned with other calretinin sequences.)

In confirmation that the human gene and the guinea-pig protein are orthologous to chick calretinin, a survey of the rat brain using immunohistochemistry with antisera against chick calretinin (see below) is congruent with preliminary results from surveys using antisera against the guinea-pig protein (Winsky et al. 1989b) and using in situ hybridization with the human cDNA probe (cited in Pochet et al. 1989).

EVOLUTION AND NEURONAL SPECIFICITY OF CALRETININ AND CALBINDIN

It is clear that calretinin and calbindin have been separately conserved throughout tetrapod evolution, as indicated by the existence of separate, orthologous genes for them in birds and mammals, and by the detection of a calretinin-like band on Western blots as far away as amphibia (see above), and by immunohistochemical results in amphibia (see below).

Calretinin in birds and mammals appears to be expressed mainly, and perhaps exclusively, in neurons. A survey of chick tissues by RNA gel blotting showed the calretinin mRNA only in brain, retina, and spinal cord; there was none in non-neural tissues including intestine, kidney, and muscle (Rogers 1987). Other surveys have also detected calretinin in brain but not in non-neural tissues such as intestine or kidney, viz. Northern blots with calretinin cDNA probes on dog tissues (Parmentier and Lefort 1990) and guinea-pig tissues (Winsky et al. 1989), and Western blots with calbindin antisera on rat, human, and amphibian tissues (Parmentier et al. 1987).

However, similar Western blots on two reptilian species did detect the 29-kDa band (putative calretinin) in intestine and kidney, in addition to 27-kDa calbindin (Parmentier et al. 1987).

Calbindin may have been originally neuron-specific, as Western blots of fish tissues detect it only in brain; in all higher vertebrates it is also present in kidney and often in intestine (Parmentier et al. 1987). These authors inferred that when amphibia moved onto land, a CaBP became necessary in renal and intestinal epithelia, and calbindin was co-opted for this purpose in most vertebrate classes. In mammalian intestine, however, a distantly related protein (calbindin-D9 or ICaBP-9) is used instead, and in reptilian tissues it is possible that calretinin may also be used.

SEQUENCE AND STRUCTURE
STRUCTURES OF mRNAs, cDNAs AND PROMOTERS

The sequence of chick calretinin has now been completed from a combination of cDNA and genomic clones (Rogers et al. 1990, and Figure 1), while the sequence of human calretinin has been deduced from cDNA clones (Parmentier 1989, 1990).

All the chick cDNA clones obtained (Rogers et al. 1990) had been cut at an internal EcoRI site, but some of them, bizarrely, were circularly permuted; the cDNA appeared to have been circularized during synthesis then cleaved at the internal EcoRI site. However, the validity of the sequence was established by the identity of the coding sequences in two cDNA clones, and by overlap with a genomic exon sequence across the EcoRI site. The partial sequence first published (Rogers 1987) contained an error in that nucleotides 24 and 25 were transposed. The correct sequence in this region is shown in Figure 1 as GAGCTC (SacI site, encoding Glu-Leu) instead of GACGTC (AatII site, encoding Asp-Val). The misreading arose in a region of compression where the 'minus' strand was ambiguous and the 'plus' strand appeared to give the published reading; however, using Sequenase II and dITP (U.S. Biochemicals) to eliminate secondary structure, the correct reading became clear.

Additional secondary structure at the 5' end of the chick mRNA probably explains why the 5' end has not yet been found in cDNA clones. It has been sequenced from a chick genomic clone (Figure 2) and there is a strong inverted repeat where all the cDNA clones terminate. The initiation codon is so identified because any translation from further upstream would make the protein much larger than 29 kDa, and there appear to be no RNA splice sites in the vicinity. Upstream of the initiation codon the sequence is extremely (G,C)-rich with a TATA box embedded in it; these are typical features of promoters, but the actual start site has not yet been located.

RNA gel blots of chick brain and retina reveal two major calretinin mRNA species, of 1.2 and 1.4 kb. The cDNA sequenced amounts to 1.1 kb plus poly(A) (Rogers et al. 1990). It is not yet known whether the two species differ only in the 3' untranslated region (like calbindin mRNAs; Hunziker 1986) or elsewhere.

The chick calretinin cDNA sequence encodes 268 amino acids (not including the initiator methionine) and the deduced molecular weight is 31077, compared to the estimate of 29000 from SDS-PAGE. A similar difference was noted for the calbindin sequence (Table 1).

Table 1. Properties of calretinin compared with calbindin

	Mammal calretinin	Chick calretinin	Chick calbindin	Mammal calbindin
Molec. weight (SDS-PAGE)	29000	29000	27–28000	27–28000
Iso-electric point[a]	*(mouse)* 5.1	–	–	*(mouse)* 4.6
Molec. weight (sequence)[b]	*(human)* 31389	30946	30036	*(rat)* 29863
Number of amino acids[b]	270	268	261	261
Homology	86%		59%	78%

[a] Iso-electric points from Parmentier and Lefort (1990).
[b] Sequence information from references in Figure 1; initiator methionine excluded.

The human calretinin sequence has been deduced from cDNA clones which were essentially full-length (Parmentier 1989, 1990; Parmentier and Lefort 1990).

```
Standard EF-hand:              αα..φD.D.NGK I...EL ..αα
                                        D GY
                               x.y.z.y.x..z
1. MAGPQQQPPYLHLAELTASQFLEIWKH FDADGNGYIEGKEL ENFFQELEKARKGSGMM
   ***   *  * ****** ***** * *  *************** ******** **** *
2. MAG--QRAPHLHLAELSASQFLDVWRH FDADGNGYIEGKEL ENFFQELESARKGTGVD
      **   * ** **   * *   * ***** ****   ** *** *** * *
3.    MTAETHLQGVEISAAQFFEIWHH YDSDGNGYMDGKEL QNFIQELQQARKKAGLD
      ** ***   * * ****** *  * ** ** **  **** ** **** ********
4.    MAESHLQSSLITASQFFEIWLH FDADGSGYLEGKEL QNLIQELLQARKKAGLE

   SKSDNFGEKMKEFMQK YDKNSDGKIEMAEL AQILPTEENFLLCFR-
   ** *   * ****** * ****  ********* **************
   SKRDSLGDKMKEFMHK YDKNADGKIEMAEL AQILPTEENFLLCFR-
      *   ** *  * *   **** ** ** ********* **
   -----LTPEMKAFVDQ YGKATDGKIGIVEL AQVLPTEENFLLFFRC
        * **** **** **  ******** * ********** ***
   -----LSPEMKTFVDQ YGQRDDGKIGIVEL AHVLPTEENFLLLFRC

   QHVGSSAEFMEAWRK YDTDRSGYIEANEL KGFLSDLLKKANRPY
   ****** ******* ************** ***************
   QHVGSSSEFMEAWRR YDTDRSGYIEANEL KGFLSDLLKKANRPY
   * ** ** ** * ** *  ** * ** *** ***
   QQLKSSEDFMQTWRK YDSDHSGFIDSEEL KSFLKDLLQKANKQI
   *****   ** **** ** ****** *** * ****** ****
   QQLKSCQEFMKTWRK YDTDHSGFIETEEL KNFLKDLLEKANKTV

   DEPKLQEYTQTILRM FDLNGDGKLGLSEM SRLLPVQENFLLKFQ
   ** ************ ** ********** ***************
   DEAKLQEYTQTILRM FDMNGDGKLGLSEM SRLLPVQENFLLKFQ
   ** *** ***  ** **   *  **** * ********** ***
   EDSKLTEYTEIMLRM FDANNDGKLELTEL ARLLPVQENFLIKFQ
   * ** *** ** ** ****** *** *********** ***
   DDTKLAEYTDLMLKL FDSNNDGKLQLTEM ARLLPVQENFLLKFQ

   GMKLTSEEFNAIFTF YDKDRSGYIDEHEL DALLKDLYEKNKK
   **** ******** * **** ** ****** *************
   GMKLSSEEFNAIFAF YDKDGSGFIDEHEL DALLKDLYEKNKK
   * *    *** * ** ** * *** ** ******* *****
   GVKMCAKEFNKAFEM YDQDGNGYIDENEL DALLKDLCEKNKK
   * *** ******** ************** ************
   GIKMCGKEFNKAFEL YDQDGNGYIDENEL DALLKDLCEKNKQ

   EINIQQLTNYRKSVM SLAEA-GKLYRKDL EIVLCSEPPM.
   *  ******** * * *     ****** * * *******
   EMSIQQLTNYRRSIM NLSDG-GKLYRKEL EVVLCSEPPL.
   *  *  *  *  *** ****  ***** **  ** *
   ELDINNLATYKKSIM ALSDG-GKLYRAEL ALILCAEEN.
   ****** **** ** ***** *****  * **** *  *
   ELDINNISTYKKNIM ALSDG-GKLYRTDL ALILSAGDN.
                               x.  .y.z.  .y.x..z
S-100 EF-hand:             αα..φHβφS. .EGDK ..L.K.DL β.Lα
                                   A           E
```

(α, hydrophobic; φ, Phe/Tyr; β, basic)

Fig. 1. Sequences of calretinin and calbindin from birds and mammals. The six domains are aligned are aligned vertically, with consensus sequences for standard EF-hands (*top*) and for the variant found in the S-100/ICaBP-9 family (*bottom*). *Hyphens* are gaps introduced to maximize homology. Sequences are: *1* human calretinin (Parmentier 1990); *2* chick calretinin (this Chap.); *3* chick calbindin (Wilson et al. 1985; Hunziker 1986); *4* rat calbindin (Yamakuni et al. 1986).

Again, the first few codons were not represented directly in cDNA clones, but were found in a cDNA clone as part of a 5' inversion, whose sequence was confirmed from a genomic clone. These authors also sequenced the promoter region, and as in chick it is very (G,C)-rich with an embedded TATA box. On RNA gel blots from dog brain, the human cDNA probe detects a mRNA 1.85 kb long (Parmentier and Lefort 1990).

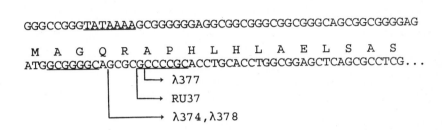

Fig. 2. Sequence of the 5' part of the chick calretinin gene. The exact 5' end of the mRNA has not yet been located. The first 15 nucleotides shown have only been determined from one strand but are included to show the location of an apparent TATA box (*underlined*). *Also underlined* is an inverted repeat, and the 5' ends of four cDNA clones are marked.

SEQUENCE HOMOLOGIES

Chick and human calretinin have 86% amino acid homology (Figure 1). In comparison, chick and rat calbindin have 78% homology, while chick calretinin and chick calbindin have 59% homology. Calbindin is the only other CaBP with spacers between the EF-hands that are as long as those in calretinin, and is the only one with which calretinin can be aligned almost continuously. Almost all the differences between the four sequences in Figure 1 are chemically conservative.

The degree of conservation is remarkable. The evolutionary rates, in amino acid substitutions per 10^9 years, are only 0.27 for calretinin and 0.30 for calbindin (Parmentier 1990), compared to 0.04 for calmodulin, 0.14 for troponin C, 0.30 for cytochrome c, and 1.20 for hemoglobin α. The rates for EF-hand proteins, which are thought to be merely calcium-buffering proteins, are 0.62-0.68 for parvalbumins and 1.30 for ICaBP-9. The strong sequence conservation may imply that calretinin and calbindin do more than merely binding calcium.

FEATURES OF CALRETININ PROTEIN SEQUENCE

Calretinin, like calbindin, consists essentially of six tandem EF-hand domains which represent possible calcium-binding sites. The structure clearly evolved by triplication of a two-domain structure. In calbindin, sites II and VI are clearly aberrant (Wilson et al. 1985; Hunziker 1986; Takagi et al. 1986), and presumably do not bind calcium as only four ions are bound per calbindin molecule (Bredderman and Wasserman 1974). In calretinin, site VI has the same features as in calbindin, but site II appears normal; all the characteristic amino acids are present. A computer secondary structure prediction using the algorithm of Garnier et al. (1978) (data not shown) is consistent with the helix-loop-helix structure for all the EF-hands in calretinin and calbindin except calretinin sites II and VI. In calretinin domain II, the program predicts continuous helix structure, but in view of the close adherence of this domain to the EF-hand consensus, it may be that the prediction is in error.

The largest single insertion in calretinin compared to calbindin is between domains I and II, and this region contains calretinin-specific sequences which might be recognized by cyclic AMP-dependent protein kinase: RKGS in human calretinin, and RKGT and KRDS in chick calretinin. It remains to be seen whether these sites really are phosphorylatable.

The most conserved segments between calretinin and calbindin are also those which are most conserved between chick and rat calbindin, and they are not the calcium-binding sites. They are the segments following sites II, IV, and V, and the altered site VI is also quite highly conserved. Perhaps the most interesting sequences are those following sites II and IV, where homologous stretches of 14-15 amino acids are almost invariant. Each of these stretches begins in the second helix of the EF-hand and is predicted to be essentially α-helical, even though there is an invariant proline near the beginning of each stretch. This will probably merely kink the α-helix; kinked helices are observed in equivalent positions in parvalbumin and ICaBP-9 (Strynadka and James 1989). Although these conserved stretches may form α-helices that are longer than the typical second helix, they are not predicted to run continuously into the first helices of the following domains in the manner of the calmodulin structure, except possibly between calbindin domains IV and V. The conserved stretches following site II in calretinin and calbindin, and following site IV in calbindin only, all have a cysteine near the end, but these cysteines are not in corresponding positions, and therefore were probably introduced independently into each of these domains in the course of evolution. The short sequence following site VI in chick calbindin also resembles an incomplete copy of the conserved sequence, and includes a cysteine. The function of these conserved stretches is unknown, but they should be accessible at the protein surface, where they might interact with some other protein.

EF-hands generally come in pairs, and the pairs form structural units in which the calcium-binding loops are linked back-to-back by ß-sheet bonding (Strynadka and James 1989). A feature of this arrangement is that odd-numbered EF-hands tend to have a tyrosine or phenylalanine in the -y position where even-numbered ones tend to have a lysine. This pattern is strictly adhered to in calretinin and calbindin, implying that domains I and II, III and IV, and V and VI, form three successive structural units.

Site VI is aberrant but quite strongly conserved, the differences in Figure 1 mostly being chemically conservative. Its C-terminal half is typical of an EF-hand but its N-terminal half is completely different. In this it is reminiscent of the variant calcium-binding site I of the two-domain family of CaBPs (ICaBP-9, S-100, etc.: Szebenyi and Moffat 1986; Kligman and Hilt 1988). As indicated in the last line of Figure 1, domain VI of calretinin and calbindin may resemble this variant type of site; it has a similar pattern of hydrophobic and basic amino acids in the putative first α-helix, but the hydrophilic amino acids that include the x and y calcium ligands of ICaBP-9 are replaced by bulky hydrophobic residues. One can thus speculate that domain VI resembles the ICaBP-9 domain with the calcium-binding site plugged by hydrophobic side chains.

In EF-hand proteins whose structures have been determined (Strynadka and James 1989), the four α-helices of each pair of EF-hands form a large hydrophobic pocket. In the calcium-bound forms of calmodulin and troponin C, this pocket is open to bind other proteins, but in the calcium-buffering proteins parvalbumin and ICaBP-9, it is blocked by extra N-terminal sequences. The lengthy interdomain linkers in calretinin and calbindin sequences might be long enough to block the equivalent pockets if they could fold appropriately, but present information does not allow a prediction as to whether the pockets will be open or closed.

CHROMOSOMAL GENE STRUCTURE: THE GENE HAS INTRONS IN DISCORDANT POSITIONS

The 5' two-thirds of the chick calretinin gene was cloned from a chick genomic library and partially sequenced, thus identifying the positions of eight introns (Wilson et al. 1988). In the regions sequenced, all the intron positions are in exactly the same places as in the calbindin gene. The first four exons of the human calretinin gene have also been cloned from a genomic library (Parmentier 1990). Again, the intron positions are the same as in the calbindin gene.

It was interesting to learn whether the introns were consistently placed with respect to the coding regions for the tandemly repeated EF-hand domains. Although some introns in these genes fall in similar positions between domains, others fall within the domains, at different positions in each domain, sometimes interrupting conserved coding sequences. A comparison with other available gene sequences from the CaBP superfamily shows few coincident intron positions, and implies that the introns were inserted separately in different branches of the superfamily (Wilson et al. 1988; Perret et al. 1988).

The human calretinin gene is on chromosome 16, whereas the human calbindin gene is on chromosome 8 (Parmentier et al. 1989).

EXPRESSION, PURIFICATION, PROPERTIES AND IMMUNIZATIONS

In order to obtain chick calretinin sequences in the form of fusion proteins, fragments of a cDNA clone were inserted into the pUR289 vector and expressed in bacteria, thus producing fusion proteins which contained ß-galactosidase coupled to either two or four domains from chick calretinin. On a protein gel blot, both fusion proteins bound ^{45}Ca, proving that calretinin is a CaBP (Rogers 1987). The entire human calretinin sequence has also been expressed in bacteria (Parmentier and Lefort 1990), and similarly shown to bind ^{45}Ca.

PURIFICATION OF NATURAL CALRETININ

Calretinin was purified from guinea-pig brain by Winsky et al. (1989b) (see also Jacobowitz this Vol.). They used ammonium sulfate precipitations (which may partially separate calretinin from calbindin), followed by DEAE cellulose columns loaded in the presence or absence of calcium, and finally a gel filtration column. The purified protein migrated as a single spot on two-dimensional gels. These methods were similar to those used by Maruyama et al.(1985) for calbindin from cerebellum, and a 30-kDa contaminant copurified by the latter authors may have been calretinin.

RAISING ANTISERA

The chick calretinin fusion proteins were used to immunize rats and a rabbit in order to raise antisera (Rogers 1989a). The resulting antisera were tested on protein gel blots of extracts from chick retina or brain, where they label only calretinin (apart from two bands which are labelled by all sera including nonimmune sera). On immunohistochemistry to chick retina or brain, all the antisera give the same distinctive staining patterns, which are generally consistent with the pattern of calretinin mRNA distribution as deduced from in situ hybridization. The antisera do not detect calbindin (Rogers 1989a; Pasteels et al. 1990).

The antisera against the chick proteins react well with the homologous proteins in mammals, and the specificities appear to be the same as in chick: anti-calretinin does not detect calbindin, whereas anti-calbindin does detect calretinin, as described before.

This pattern of cross-reaction might be due to the strongly conserved sequences of domains V and VI, which were present in the natural calbindin used to raise the calbindin antisera, but not in the fusion proteins used to raise calretinin antisera. However, antisera has also been raised against natural guinea-pig calretinin by Winsky et al. (1989b), and these sera are also entirely specific for calretinin.

DISTRIBUTION OF CALRETININ AND CALBINDIN IN CHICK BRAIN

The chick brain was first surveyed by in situ hybridization with cDNA probes for calretinin and calbindin (Rogers 1987) The two mRNAs were found to be present in many types of neurons, with little overlap in their distributions.

Immunohistochemical surveys of the chick brain with antisera against calbindin have been published (Jande et al. 1981; Roth et al. 1981), but in some regions these antisera appear to have detected calretinin also, as described before. They stain some of the brain nuclei in which in situ hybridization detects mRNA for calretinin but not calbindin. Fortunately, the calbindin antisera stain calretinin sufficiently weakly in our immunohistochemistry conditions that the two can usually be distinguished by two-color immunofluorescence. It was therefore possible to re-examine the chick brain by this method with antisera against both proteins, to distinguish the neurons that contained one or the other or both (Rogers 1989a,b; Rogers et al. 1990).

In the chick central nervous system, the calretinin antisera give striking staining of many types of neurons (Figures 3,4,5). Glia are not stained. The distribution generally agrees with the results from in situ hybridization, with the surprising exception of the molecular layer of the cerebellum (see below). Two-color immunofluorescence revealed that few neurons are positive for both calretinin and calbindin, except in a few regions.
Most of the positive neurons occur in the following areas.

SENSORY NUCLEI

All the sensory modalities are associated with populations of neurons that are conspicuously positive for one or other protein (Rogers 1989a), either secondary sensory neurons (e.g., calretinin in the auditory pathways) or interneurons (e.g., calbindin in the optic tectum). Only in the nucleus solitarius are there many double-positive neurons.

The intense expression of calretinin in the auditory nuclei angularis, magnocellularis, and laminaris is especially striking (Rogers 1987, 1989a; Figure 4a). This is probably the explanation for the calbindin-immunoreactivity of the same nuclei in the barn owl (Carr et al. 1985; Takahashi et al. 1987). These neurons are remarkable for the temporal precision of their signals: those of nucleus magnocellularis have action potentials locked to the incoming sound waves and transmit bilaterally to nucleus laminaris, where timing differences as small as 10 µs are detected to build up an auditory map of space (Knudsen 1981; Sullivan and Konishi 1986). Calretinin is also abundant, along with calbindin, at earlier stages in the pathway, viz. the cochlear hair cells and the large ganglion cells which innervate the outer hair cells (though apparently not in many of the auditory nerve fibers) (Rogers 1989a). It has been suggested that the CaBP might responsible for the temporal precision of these responses (Takahashi et al. 1987; Rogers 1987).

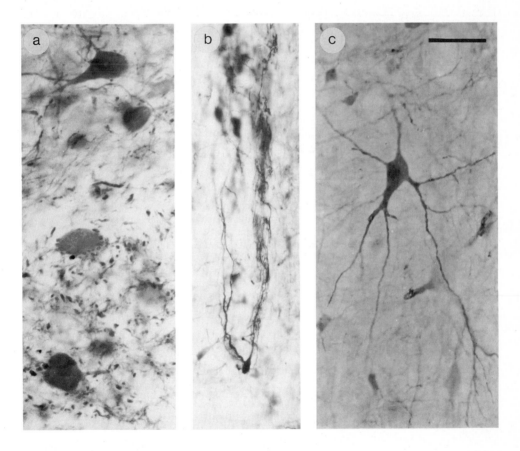

Fig. 3a-c. A variety of neurons in chick medulla stained with calretinin antiserum (peroxidase technique). **a** Calretinin-positive vestibular nerve terminals on positive and negative neurons in the vestibular nuclei (border of superior and lateral vestibular nuclei). **b** Comet-shaped neuron with fibers running longitudinally in medulla. *Bar* = 50 µm

Many peripheral sensory neurons also contain calretinin and/or calbindin. In contrast to central neurons, peripheral neurons often contain both proteins, for example in many of the ganglion cells of dorsal root ganglia, inner ear, and retina (Rogers 1989a). Many other cell types in the retina are positive for calretinin (Figure 6) or for calbindin, but not for both.

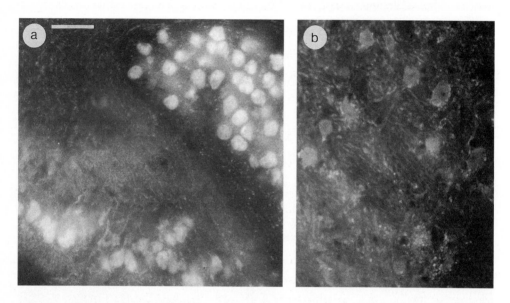

Fig. 4a,b. Ventral cochlear nucleus of chick and rat, stained with calretinin antiserum by immunofluorescence. **a** Chick: nucleus magnocellularis (*top right*, equivalent of ventral cochlear nucleus) and nucleus laminaris (*bottom left*, equivalent of medial superior olivary nucleus). All these cells are positive for calretinin and negative for calbindin. **b** Rat: ventral cochlear nucleus. Unlike chick, many of these cells and fibers are positive for calbindin (data not shown) as well as calretinin. *Bar* = 50 μm

CEREBELLUM

It was surprising to see abundant calretinin immunoreactivity not only in ascending mossy and climbing fibers, but also in stellate cells, although there is no detectable calretinin mRNA in the cerebellar cortex (Rogers 1989b). The origin of this immunoreactivity is still unresolved. Purkinje cells contain no calretinin although they do have large amounts of calbindin and parvalbumin (Jande et al. 1981b; Roth et al. 1981; Braun et al. 1986).

BASAL FOREBRAIN AND BRAINSTEM

Up to 20% of neurons are positive for calretinin in some regions of the reticular formation. In the septal nuclei, separate regions are strikingly positive for calretinin or for calbindin (data not shown). The hypothalamus, unlike other regions of the brain, contains many cells positive for calretinin and calbindin together, as well as for each separately (Rogers et al. 1990).

CEREBRAL HEMISPHERES

Immunoreactivity appears to be confined to scattered cells, probably interneurons. Most of the cerebral hemispheres is crisscrossed by a network of calbindin-positive multipolar cells, except for the paleostriatum where there are multipolar cells positive for calretinin and parvalbumin instead (Rogers et al. 1990).

DISTRIBUTION OF CALRETININ AND CALBINDIN IN RAT BRAIN

Surveys of the rat brain for calbindin immunoreactivity have been published (Jande et al. 1981a; Baimbridge and Miller 1982; Feldman and Christakos 1983; Garcia-Segura et al. 1984). In order to distinguish the distributions of calretinin and calbindin, a complete survey of the rat brain has now been done with antisera to both proteins (Résibois et al. 1990; Rogers et al. 1990; and in prep.). The data agree with preliminary independent results on rat and guinea-pig brains from Winsky et al. (1989b). Different fixation and detection methods give essentially the same results.

   In comparison with the chick, many general features of the CaBP distribution are conserved in the rat, but there are also many differences. Cells expressing both proteins seem to be more common in the rat brain.

266

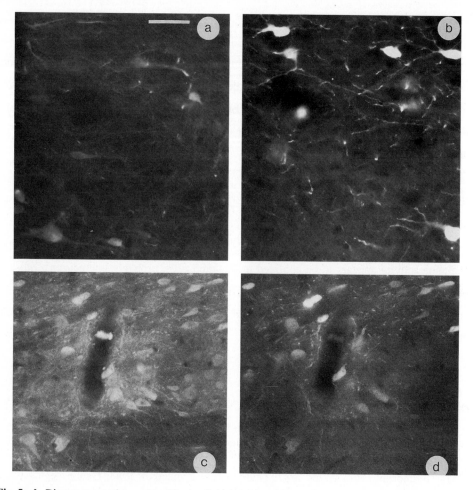

Fig. 5a-d. Diverse neuronal types in midbrain of chick (**a,b**) and rat (**c,d**), seen in sections doubly stained for calretinin (**a,c**) and calbindin (**b,d**). In chick (**a,b**), unidentified region, various cells are positive for one or other protein but not both. In rat (**c,d**), substantia nigra pars compacta, many cells are positive for both proteins and many for calretinin only, but none for calbindin only. *Bar* = 50 μm

Some regions of rat brain have also been examined with antisera against CBP-18 (gift of Dr. Claude Klee: Manalan and Klee 1984), GABA (mAb3A12, gift of Dr. Peter Streit: Matute and Streit 1986), and various neuropeptides (commercial antisera), using two-color immunofluorescence to look for colocalisation with calretinin.

SENSORY NUCLEI

These show much immunoreactivity in the rat, as in the chick, but more neurons are double-positive, for example in the spinal cord (Rogers et al. 1990) and in the auditory nuclei (Pochet et al. 1989; Winsky et al. 1989b; Résibois et al. 1990 and in preparation; Figure 4b). In the auditory system the dorsal and ventral cochlear nuclei contain conspicuous calretinin-positive cells and fibers like their equivalents in the chick, but in contrast to the chick, some of the neurons of the ventral cochlear nucleus are also calbindin-positive, while those of the medial superior olivary nucleus are negative for both proteins. The patterns in the vestibular nuclei are essentially the same as in the chick, and strong calretinin immunoreactivity extends up to the vestibulo cerebellum, which contains strongly calretinin-positive mossy fibers and granule cells absent from other parts of cerebellum (Rogers 1989b).

The thalamus is devoid of calretinin, except for a shell of positive cells and fibers in the reticular nucleus and around the periphery of the whole thalamus, which does not respect standard anatomical divisions.

In the olfactory bulb, antisera for calretinin and calbindin strongly stain many periglomerular cells (local inhibitory neurons), and so does antiserum for a third CaBP, CBP-18 (Manalan and Klee 1984; data not shown). The calretinin is in different cells from the calbindin and from the CBP-18, and calretinin immunoreactivity is also seen in granule cells.

CEREBELLUM

Whereas the distribution of calbindin and parvalbumin is conserved (Jande et al. 1981a,b; Baimbridge and Miller 1982; Celio and Heizmann 1981), that of calretinin is totally different in rat from in chick. In the rat, calretinin is seen in the Lugaro cells and in the granule cells, the latter staining being weak except in the vestibulo cerebellum (Rogers 1989b).

BASAL FOREBRAIN AND MIDBRAIN

The triangular septal nucleus is strikingly positive for calretinin and calbindin. Parts of the hypothalamus, ventral tegmental area, and substantia nigra (Figure 5), also contain many cells positive for both proteins together, as well as for one or other separately.

CEREBRAL HEMISPHERES

The cerebral cortex and hippocampus show scattered bipolar interneurons positive for calretinin or calbindin but never both. The calretinin-positive neurons have been compared with those positive for vasoactive intestinal peptide, somatostatin, or GABA, by two-color immunofluorescence (unpubl. data). In each case the distributions overlap slightly (and extensively with GABA in the ventral hippocampus), but not completely.

DISTRIBUTION OF CALRETININ AND CALBINDIN IN RETINAE OF VERTEBRATE SPECIES

Calretinin is abundant in the retinae of all species examined (e.g., Figure 6). An extensive phylogenetic survey has been done on retinae, covering not only chick and rat, but also cat and salamander (as there is extensive physiological data on these species), and sheep, pig, and monkey (Pasteels et al. 1990). Calretinin and calbindin are detectable as separate immunoreactivities in all these species, and some features of their distributions are conserved, although there are also many variations between species.

Calbindin is present in at least some cones in all these species except rat, and photoreceptors are negative for calretinin except for the cones in monkey. In each species, there are many amacrine cells and ganglion cells positive for one or other protein or for both. Most strikingly, in each species there is conspicuous labeling of the horizontal cells and their network of fibers, but there is no consistency as to which CaBP is present: it is calretinin in chick and salamander, calbindin in rat, and both proteins plus parvalbumin in all the horizontal cells in cat.

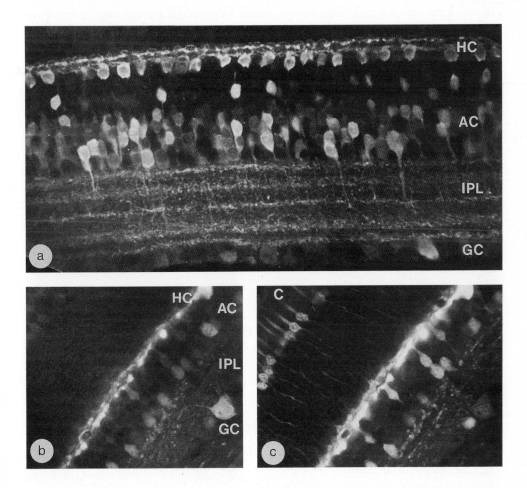

Fig. 6a-c. Retinae of chick and cat, viewed with the confocal laser scanning microscope (White et al. 1987). **a** Chick, stained for calretinin only. **b** Cat, stained for calretinin with rhodamine. **c** Cat, same section, stained for calbindin with fluorescein. Because of the filters used for confocal microscopy, some of the purely calbindin-positive cells in **c** are also faintly visible in **b**; however, the horizontal cells and most of the amacrine cells are genuinely double-positive. Layers are labeled as follows: *C* cones; *HC* horizontal cells; *AC* amacrine cells; *IPL* inner plexiform layer; *GC* ganglion cells.

The significance of these striking distribution patterns remains unknown. Dr. P. McNaughton has suggested that calbindin in cones may serve to delay the calcium changes that occur during photoreception, so that they have a slow timecourse suitable for mediating adaptation (Pasteels et al., 1990). However, the fact that the CaBP in cones is almost always calbindin suggests that calbindin may be performing some more particular function there. In contrast, it may not matter which CaBP is present in horizontal cells as long as they have a lot of it. Conversely the rods, and the rod bipolar cells which are identifiable morphologically in the cat, are always negative for the known CaBPs.

Retinae of chick embryos were examined to see when calretinin and calbindin appeared during development (Rogers et al. 1990; Ellis et al. in prep.). In most cell types, the appropriate immunoreactivity appears several days after the cells differentiate (horizontal cells being the last to turn on calretinin), and the CaBP pattern appears virtually mature a few days before hatching. There also seems to be an early phase of calbindin immunoreactivity in amacrine cells, since there are many more calbindin-immunoreactive amacrine cells at early stages than later, and they include cells of a conspicuous calretinin-positive type that later contains no calbindin. There may therefore be a special role for calbindin during the development of some neurons.

DISCUSSION: FUNCTIONS OF CaBPS IN NEURONS

It seems likely that calretinin and calbindin were originally neuronal proteins, as described before, and calretinin may still be an exclusively neuronal protein in birds and mammals.

Immunohistochemistry indicates that both proteins are generally present throughout the cytosol of the cells that contain them. Some possible exceptions have been observed, of which the clearest is the inferior olivary nucleus; its cells are calbindin-positive, in chick and in rat, but their axons (the climbing fibers in the cerebellum) are calbindin-negative (Rogers 1989b and references therein). However, such discrepancies are unusual. Counterexamples where the cell body, dendrites, long axons, and axonal terminals are all stained include Purkinje cells (calbindin), auditory nuclei (calretinin), and retinal ganglion cells (one or both proteins).

The concentrations of these proteins can be very high. Calbindin comprises 1.5% of soluble protein in the rat cerebellum (Baimbridge et al. 1982), and calretinin comprises

≈0.5% of total protein in the mouse olfactory bulb (Parmentier and Lefort 1990); the concentrations must be several times higher in the individual positive cells and fibers.

INFERENCES FROM NEUROANATOMICAL DISTRIBUTION

Unfortunately, the surveys of CaBP-positive neuronal types have not revealed obvious common features among cells positive for one or other protein. These proteins are present in many neurons of the sensory systems, in particular, but the positive cells include a wide range of electrophysiological types: fast-responding and slow-responding, excitatory and inhibitory, long-range and short-range.

For example, calretinin is present in many projection neurons of the auditory and vestibular pathways, both in chick and in rat. In the avian auditory pathway calretinin is conspicuous in the sensory hair cells and in the secondary and tertiary projection neurons, all of which have rapidly and temporally precise responses (see above). Similarly, in the avian vestibular system calretinin is present in hair cells, vestibular ganglion cells and their axons, and some (though not all) neurons of the vestibular nuclei, while in the rat it is also present up to the vestibulo cerebellum. On the other hand, calretinin and/or other CaBPs are conspicuous in various local inhibitory cell types which are involved mainly in dendrodendritic communication to control activity in adjacent sensory projections, often with slow timecourses: the periglomerular and granule cells of the olfactory system, and the horizontal and amacrine cells of the retina. Other possible examples of conspicuously labeled local inhibitory neurons include the calretinin-positive cells of rat cerebral cortex and of chick cerebellar molecular layer, and a layer of calbindin-positive cells in the optic tectum of chick.

Jande et al. (1981) noted that several conspicuously calbindin-positive cell types can produce calcium-dependent action potentials in their dendritic trees: the cerebellar Purkinje cells, the inferior olivary neurons, and the hippocampal pyramidal cells. It is not known whether other CaBP-containing neurons also have this property. These may form a subset of CaBP-containing neurons in which calbindin has a particular function related to their high local calcium fluxes. Purkinje cells are also notable for containing enormous quantities of the inositol triphosphate receptor protein (Gill 1989), which may imply a correspondingly active system for calcium signaling by intracellular calcium release.

The inter-species comparisons described above show that expression of a single CaBP is conserved in some cell types, such as cones and Purkinje cells, but that other neuronal types contain different CaBPs in different species.

INFERENCES FROM BIOCHEMISTRY

The functions of calretinin and calbindin in neurons are still unknown, although many possibilities have been discussed, in view of the widespread importance of calcium in neuronal signaling. Although calbindin's basic biochemical properties are known, no quantitative studies of calretinin's calcium-binding or structural properties have yet been done. The two proteins will therefore be discussed together, with the acknowledgement that there may be important differences which are so far unknown.

The apparently simpler case of parvalbumin may be instructive for comparison (reviewed by Heizmann and Berchtold 1987). Parvalbumin is mostly located in fast-firing, usually GABA-ergic inhibitory neurons, as well as in fast-twitch muscle fibers. In muscle, it is believed to terminate contraction by soaking up the free calcium; it does so with a slow (≈ 1 s) timecourse because the calcium has to displace bound magnesium. Parvalbumin may have a similar role in neurons, both terminating the immediate actions of calcium and redistributing it in the cell, and possibly protecting the neuron against toxic effects of raised intracellular calcium.

Similar functions have been proposed for calbindin and might apply to calretinin as well, but the calcium-binding kinetics may be significantly different. In contrast to parvalbumin, calbindin does not bind magnesium significantly in the cell (Bredderman and Wasserman 1974), and therefore should be able to bind calcium rapidly under physiological conditions. Possible consequences can be grouped into the following three categories. The biggest unknown is whether these proteins merely bind calcium, or whether they then undergo structural changes that lead to interactions with other macromolecules.

1. Short-Term Calcium Buffering. For calbindin, the in vitro calcium affinity (≈0.5 μM for each site: Bredderman and Wasserman 1974) and the in vivo abundance (up to ≈0.1 mM in some neurons: Baimbridge et al. 1982) suggest that the protein should bind ≈99% of the free calcium entering the cytosol; thus it should be a very effective calcium buffer. This would presumably be a short-term effect, while the cell's higher-affinity, lower-capacity calcium pumps begin to restore resting levels of calcium (McBurney and Neering 1987). This could have many effects on the activities of neurons. The most straightforward would be protection against the toxic effects of excess calcium influx (Sloviter 1989; Baimbridge and Kao 1988). Damping out endogenous calcium oscillations (Berridge et al. 1988) might also be a useful consequence which could serve to protect the temporal fidelity of sensory signals. With regard to the effect on calcium transients, one cannot predict exactly what the effects might be without knowing the kinetics of the calcium channels and the types of calcium-sensitive molecules present in the individual neuron, as well as the concentration and affinity of calbindin or calretinin in the relevant cytosolic compartment of that neuron. At low concentrations of protein and of calcium, the main effect might be to slow the changes in free calcium, as proposed for retinal cones (see above). At high concentrations of protein and of calcium, the effect might be to "wall in" local regions of saturating calcium concentrations around individual calcium channels. Thus there might be a wide range of effects on neuronal activity, such as damping out or delaying agonist-induced calcium fluxes, reducing or prolonging calcium-stimulated transmitter release, reducing the cell's liability to seizure activity (Jande et al. 1981; Baimbridge and Miller 1982), and changing the length and shape of action potentials (Miller and Baimbridge 1983; Nicoll 1988). Careful experiments will be needed to find out what difference these proteins make in particular neuronal types.

2. Facilitating Calcium Diffusion. Given that the free cytosolic calcium must not rise too high (to avoid cytotoxic effects), and is kept low by homeostatic calcium pumps, a CaBP can increase the diffusional flux of calcium through the cytosol because the flux of protein-bound calcium is added to the flux of free calcium (Feher 1984). This may be the role of calbindin or ICaBP-9 in intestinal calcium absorption (Bronner et al. 1986). In neurons, such an effect could be seen as an extension of the protein's calcium-buffering capability, and might serve to redistribute calcium away from sites of influx or release. A calcium buffer facilitates diffusion most effectively where its dissociation constant is equal to the average calcium concentration along a gradient (Speksnijder et al. 1989); this might provide a rationale for the existence of multiple CaBPs.

3. Regulating Other Proteins. It is conceivable that calbindin and calretinin might act as calcium-dependent regulators of other proteins, in the manner of calmodulin. Because the amount of CaBP in at least some neurons greatly exceeds likely increments in cytosolic calcium, it seems unlikely that the average calcium occupancy of the CaBP would change significantly under physiological conditions, but one cannot rule out local saturation of the protein with calcium around individual calcium channels. So far, attempts to find specific calbindin-binding or calbindin-regulated proteins have not met with undisputed success.

In view of the conserved diversity of expression in neurons, and the conserved amino-acid sequences in regions which are not primarily calcium-binding sites, it increasingly seems probable that calretinin and calbindin perform unknown functions beyond the simple buffering of calcium. Their roles may well be different in different cell types. There is little evidence for any functional difference between calbindin and calretinin, but the conserved expression of calbindin in some cell types suggests that some functions may be specific to only one of these proteins.

ACKNOWLEDGMENTS

I thank D.E.M. Lawson and P. Wilson for the gift of calbindin clones and antisera; A. Résibois, B. Pasteels, and R. Pochet for collaboration on immunohistochemical surveys; and numerous colleagues for discussions about the possible functions of CaBPs. Work in my laboratory is supported by the Wellcome Trust.

REFERENCES

Baimbridge KG, Kao J (1988) Calbindin-D28K protects against glutamate-induced neurotoxicity in rat CA1 pyramidal neuron cultures. Soc Neurosci Abstr 14:507.1
Baimbridge KG, Miller J (1982) Immunohistochemical localization of CaBP in the cerebellum, hippocampal formation and olfactory bulb of the rat. Brain Res 245:223-229
Baimbridge KG, Miller JJ, Parkes CO (1982) CaBP distribution in the rat brain. Brain Res 239:519-525
Braun K, Schachner M, Scheich H, Heizmann CW (1986) Cellular localization of the CaBP parvalbumin in the developing avian cerebellum. Cell Tissue Res 243:69-78

Bredderman PJ, Wasserman RH (1974) Chemical composition, affinity for calcium, and some related properties of the vitamin D-dependent CaBP. Biochem 13:1687-1694

Bronner F, Pansu D, Stein WD (1986) An analysis of intestinal calcium transport across the rat intestine. Am J Physiol 250(G):561-569

Carr C, Brecha N, Konishi M (1985) Organization of nucleus laminaris in the barn owl. Soc Neurosci 11:735

Celio MR, Heizmann CW (1981) CaBP parvalbumin as a neuronal marker. Nature 293:300-302

Feldman SC, Christakos S (1983) Vitamin D-dependent CaBP in rat brain: biochemical and immunocyto-chemical characterization. Endocrinol 112:290-302

Feher JJ (1984) Measurement of facilitated calcium diffusion by a soluble calcium-binding protein. Biochim et Biophys Acta 773:91-98

Garcia-Segura LM, Baetens D, Roth J, Norman AW, Orci L (1984) Immunohistochemical mapping of CaBP immunoreactivity in the rat CNS. Brain Res 296:75-86

Garnier J, Osguthorpe DJ, Robson B (1978) Analysis of the accuracy and implications of simple methods for predicting the secondary structure of globular proteins. J Mol Biol 120:97-120

Gill DL (1989) Calcium signaling: receptor kinships revealed. Nature 342:16-18

Heizmann CW, Berchtold MW (1987) Expression of parvalbumin and other Ca^{2+}-binding proteins in normal and tumor cells: a topical review. Cell Calcium 8:1-41

Heizmann CW, Celio MR (1987) Immunolocalization of parvalbumin. Methods in Enzymology 139:552-570

Hunziker W (1986) The 28-kDa vitamin D-dependent calcium-binding protein has a six-domain structure. Proc Natl Acad Sci USA 83:7578-7582

Jande SS, Maler L, Lawson DEM (1981) Immunohistochemical mapping of vitamin D-dependent CaBP in brain. Nature 294:765-767

Kligman D, Hilt DC (1988) The S-100 protein family. Trends in Biochem Sci 13:437-443

Knudsen EI (1981) The hearing of the barn owl. Sci Am 245:82-91

Manalan AS, Klee CB, (1984) Purification and characterization of a novel CaBP (CBP-18) from bovine brain. J Biol Chem 259:2047-2050

Maruyama K, Ebisawa K, Nonamura Y (1985) Purification of vitamin D-dependent 28000-M_r CaBP from bovine cerebellum and kidney by calcium-dependent elution from DEAE-cellulose DE-52 column chromatography. Anal Biochem 151:1-6

Matute C, Streit P, (1986) Monoclonal antibodies demonstrating GABA-like immunoreactivity. Histochem 86:147-157

McBurney RN, Neering IR (1987) Neuronal calcium homeostasis. Trends in Neurosciences 10:164-169

Miller JJ, Baimbridge KG (1983) Biochemical and immunohistochemical correlates of kindling-induced epilepsy: role of CaBP. Brain Res 278:322-326

Nicoll RA (1988) The coupling of neurotransmitter receptors to ion channels in the brain. Science 241:545-551

Parmentier M (1989) The human calbindins: cDNA and gene cloning. In: Hidaka H, Carafoli E, Means AR, Tanaka T (eds) Calcium protein signaling (Proceedings of the Sixth International Symposium on CaBPs, 1988), Plenum, pp 233-240

Parmentier M (1990) Structure of the human cDNAs and genes coding for calbindin D28k and calretinin. In: Pochet R, Lawson DEM, Heizmann CW (eds) CaBPs in normal and transformed cells. Plenum, New York, pp 27-34

Parmentier M, Lefort A (1990) Structure of the human brain-specific CaBP calretinin, and expression in bacteria. (submitted)

Parmentier M, Ghysens M, Rypens F, Lawson DEM, Pasteels JL Pochet R (1987) Calbindin in vertebrate classes: immunohistochemical localization and Western blot analysis. Gen Comp Endocrinol 65:399-407

Parmentier M, Szpirer J, Levan G, Vassart G (1989) The human genes for calbindin 27 and 29 kDa proteins are located on chromosomes 8 and 16, respectively. Cytogenet Cell Genet 52:85-87

Pasteels B, Rogers J, Blachier F, Pochet R (1990) Calbindin and calretinin localization in retina from different species. Visual Neuroscience (in press)

Perret C, Lomri N, Thomasset M (1988) Evolution of the EF-hand CaBP family: evidence for exon shuffling and intron insertion. J Mol Evol 27:351-364

Pochet R, Parmentier M, Lawson DEM, Pasteels JL (1985) Rat brain synthesizes two 'vitamin D-dependent' CaBPs. Brain Res 345:251-256

Pochet R, Blachier F, Malaisse W, Parmentier M, Pasteels B, Pohl V, Résibois A, Rogers J, Roman A (1989) Calbindin-D28 in mammalian brain, retina, and endocrine pancreas: immunohistochemical comparison with calretinin. In: Hidaka H, Carafoli E, Means AR, Tanaka T (eds) Calcium protein signaling (Proceedings of the Sixth International Symposium on CaBPs, 1988), Plenum, pp 435-443

Résibois A, Blachier F, Rogers JH, Lawson DEM, Pochet R (1990) Comparison between rat brain calbindin and calretinin immunoreactivities. In: Pochet R, Lawson DEM, Heizmann CW (eds) CaBPs in normal and transformed cells. Plenum, New York, pp 211-214

Rogers JH (1987) Calretinin: a gene for a novel CaBP expressed principally in neurons. J Cell Biol 105:1343-1353.

Rogers JH (1989a) Two CaBPs mark many chick sensory neurons. Neurosci 31:697-709

Rogers JH (1989b) Immunoreactivity for calretinin and other CaBPs in cerebellum. Neurosci 31:711-721

Rogers JH, Khan M, Ellis J (1990) Calretinin and other CaBPs in the nervous system. In: Pochet R, Lawson DEM, Heizmann CW (eds) CaBPs in normal and transformed cells. Plenum, New York, pp 195-203

Roth J, Baetens D, Norman AW, Garcia-Segura L-M (1981) Specific neurons in chick central nervous system stain with an antibody against chick intestinal vitamin D-dependent CaBP. Brain Res 222:452-457

Sloviter RS (1989) Calcium-binding protein (calbindin-D28) and parvalbumin immunocytochemistry: localization in the rat hippocampus with specific reference to the selective vulnerability of hippocampal neurons to seizure activity. J Comp Neurol 280:183-196

Speksnijder JE, Miller AL, Weisenseel MH, Chen T-H, Jaffe LF (1989) Calcium buffer injections block fucoid egg development by facilitating calcium diffusion. Proc Natl Acad Sci 86:6607-6611

Strynadka NCJ, James MNG (1989) Crystal structures of the helix-loop-helix calcium-binding proteins. Annu Rev Biochem 58:951-98

Sullivan WE, Konishi M (1986) Neural map of interaural phase difference in the owl's brainstem. Proc Natl Acad Sci USA 83:8400-8404

Szebenyi DME, Moffat K (1986) The refined structure of vitamin D-dependent calcium-binding protein from bovine intestine. J Biol Chem 261:8761-8777

Takagi T, Nojiri M, Konishi K, Maruyama K, Nonomura Y (1986) Amino acid sequence of vitamin D-dependent calcium-binding protein from bovine cerebellum. FEBS Lett 201:41-45

Takahashi T, Carr CE, Brecha N, Konishi M (1987) CaBP immunoreactivity labels the terminal field of nucleus laminaris of the barn owl. J Neurosci 7:1843-1856

White JG, Amos WB, Fordham M (1987) An evaluation of confocal versus conventional imaging of biological structures by fluorescence light microscopy. J Cell Biol 105:41-48

Wilson PW, Harding M Lawson DEM (1985) Putative amino acid sequence of chick CaBP deduced from a complementary DNA sequence. Nucl Acids Res 13:8867-8881

Wilson PW, Rogers J, Harding M, Pohl V, Pattyn G, Lawson DEM (1988) Structure of chick chromosomal genes for calbindin and calretinin. J Mol Biol 200:615-625

Winsky L, Nakata H, Jacobowitz DM (1989a) Identification of a CaBP highly localized to the cochlear nucleus. Neurochem Int 15:381-389

Winsky L, Nakata H, Martin BM, Jacobowitz DM (1989b) Isolation, partial amino acid sequence, and immunohistochemical localization of a brain-specific calcium-binding protein. Proc Natl Acad Sci USA 86:10139-10143

Yamakuni T, Kuwano R, Odani S, Kiki N, Yamaguchi Y, Takahashi Y (1986) Nucleotide sequence of cDNA to mRNA for a cerebellar CaBP, spot 35 protein. Nucl Acids Res 14:6768

PURIFICATION, IDENTIFICATION AND REGIONAL LOCALIZATION OF A BRAIN-SPECIFIC CALRETININ-LIKE CALCIUM BINDING PROTEIN (PROTEIN 10)

Lois Winsky and David M. Jacobowitz

INTRODUCTION

We have employed two-dimensional gel electrophoresis to examine and compare the relative amounts of both known and unidentified proteins across auditory brain regions in rabbits (Winsky et al. 1989a). While several proteins were found to vary considerably in amounts between regions, one prominent protein (protein 10), having an apparent molecular mass (M_r) of 29 kDa and isoelectric point (pI) of 5.3, was found to be particularly abundant within the cochlear nucleus (Figure 1) and some other brainstem areas. In more rostral brain regions (e.g., medial geniculate and cerebral cortex) this protein was less prominent.

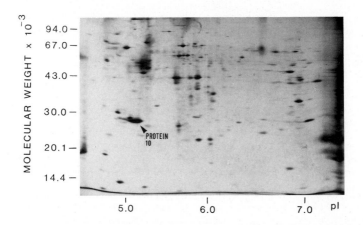

Fig. 1. Silver stain of proteins on a two dimensional polyacrylamide gel showing the location of protein 10 (*arrowhead*). Sample was from 24 µg protein in a micropunch sample of the ventral cochlear nucleus of a guinea pig as previously described. (Winsky et al. 1989a)

Subsequent studies (Winsky et al. 1989b) described a similar distribution of protein 10 across the central auditory regions of rat and guinea pig brains.

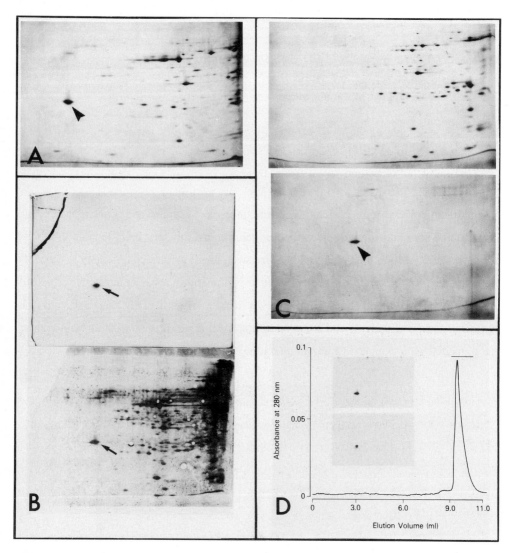

Fig. 2A-D. A Silver stain of a two dimensional gel including proteins in the pass-through fraction of a DEAE column in the absence of EDTA. **B** Autoradiogram (*top*) demonstrating $^{45}Ca^{2+}$ binding on a nitrocellulose blot including proteins of the first DEAE cellulose column pass-through fraction. Colloidal gold stain of same blot on bottom. **C** Silver stain of a two-dimensional gel including proteins present in the pass-through fraction from the second DEAE column in the presence of 1 mM EDTA (*top*). Protein 10 was eluted from the second column by increasing NaCl from 50 to 100 mM (*bottom*). Position of protein 10 in **A-C** is indicated by *arrows*. **D** Elution profile of protein 10 on a gel filtration column. Silver stain of a gel (*top insert*) and autoradiogram showing $^{45}Ca^{2+}$ binding on a blot (*bottom insert*) from samples of the peak fraction (indicated by the *line*)

The abundance of this protein in the cochlear nucleus suggested that it may be associated with molecular events relatively unique to sensory processing within the auditory system and led us to undertake the purification and identification of protein 10.

PROTEIN 10 PURIFICATION

Protein 10 was purified from guinea-pig brain by precipitation with 60-80% ammonium sulfate (Winsky et al. 1989b). This precipitate was resuspended in buffer, desalted, and applied onto a DEAE cellulose column in 20 mM TRIS buffer (pH=8.0) containing 20 mM NaCl. Two-dimensional gel electrophoresis of protein fractions eluted from this column revealed an interesting pattern in which protein 10 appeared as one of the few acidic proteins found in the pass-through (nonbinding) fraction (Figure 2A).

Based on this finding, we hypothesized that the charged portion of protein 10 was somehow blocked from binding to the DEAE ion exchange column. Our laboratory previously identified a calcium-binding protein in rat (#47) having a pI and Mr similar to protein 10 which was also precipitated by 60-80% ammonium sulfate (Santer et al. 1988). To test whether protein 10 might also bind calcium, proteins from a two-dimensional gel of the DEAE pass-through fraction were electrophoretically transferred onto nitrocellulose and incubated in a buffer containing radiolabeled calcium. The resulting autoradiogram revealed protein 10 as a calcium-binding protein (Figure 2B, top). With this information we were able to purify protein 10 to homogeneity by incubating the pass-through fraction of the first DEAE cellulose column in 1 mM EDTA and then reapplying this fraction onto a second DEAE cellulose column in the presence of EDTA (Winsky et al. 1989c). While the addition of EDTA did not affect the elution of most proteins in the pass-through fraction (Figure 2C, top), the protein 10 remained bound to the column until the NaCl concentration was increased from 50-100 mM NaCl (Figure 2C, bottom). Application of the protein 10 fraction onto a TSK gel filtration column produced one protein peak with a calculated Mr of 35 kDa which retained calcium binding activity on a nitrocellulose blot of a two-dimensional gel (Figure 2D). We have employed similar procedures for the purification of protein 10 from rat brain.

Attempts at sequential Edman degradation suggested that protein 10 was blocked at the N-terminus. A partial amino acid sequence was determined by amino acid sequence analysis of proteolytic fragments of protein 10 following digestion by trypsin, Asp-N, and V8 proteases (Winsky et al. 1989c). A comparison of the amino acid sequences of these protein 10 fragments with other known proteins revealed an 85% sequence homology between protein 10 from guinea pig and calretinin, a calcium-binding protein encoded by a cDNA clone from chicken retina (Rogers 1987, 1990).

Table 1. Comparison of amino acid sequence of chicken calretinin and proteolytic digestion fragments of protein 10[a]

```
CR:  . . A P H L H L A D V S A S Q F L D V W R H F D A D G N G Y
10:                                      E I W K H F D A D E N G Y
CB:  M A E S H L Q S S L I T A S Q F F E I W L H F D A D G S G Y

CR:  I E G K E L E N F F Q E L E S A R K G T G V D S K R D S L G
10:  I E G K
CB:  L E G K E L Q N L I Q E L L Q A R K K A G L E - - - - - L S

CR:  D K M K E F M H K Y D K N A D G K I E M A E L A Q I L P T E
10:            F M Q K Y D K N S D     N E P A I L A Q
CB:  P E M K T F V D Q Y G Q R D D G K I G I V E L A H V L P T E

CR:  E N F L L C F R - Q H V G S S S E F M E A W R R Y D T D R S
10:                  - Q H V G S S A E F M E A W R
CB:  E N F L L L F R C Q Q L K S C Q E F M K T W R K Y D T D H S

CR:  G Y I E A N E L K G F L S D L L K K A N R P Y D E A K L Q E
10:
CB:  G F I E T E E L K N F L K D L L E K A N K T V D D T K L A E

CR:  Y T Q T I L R M F D M N G D G K L G L S E M S R L L P V Q E
10:                                      D M S R L L P V Q E
CB:  Y T D L M L K L F D S N N D G K L Q L T E M A R L L P V Q E

CR:  N F L L K F Q G M K L S S E E F N A I F A F Y D K D G S G F
10:  N F L L K F Q G M K L T S E E F N A I F T F Y D K
CB:  N F L L K F Q G I K M C G K E F N K A F E L Y D Q D G N G Y

CR:  I D E H E L D A L L K D L Y E K N K K E M S I Q Q L T N Y R
10:                                      E M N I Q Q L T N Y
CB:  I D E N E L D A L L K D L C E K N K Q E L D I N N I S T Y K

CR:  R S I M N L S D G G K L Y R K E L E V V L C S E P P L
10:
CB:  K N I M A L S D G G K L Y R T D L A L I L S A G D N
```

[a] *CR* chicken calretinin sequence from Rogers (1990); *10* sequence of proteolytic digestion fragments of protein 10 (Winsky et al. 1989); *CB* sequence of rat calbindin-D28k (Yamakuni et al. 1986). *Boxes* highlight amino acid sequence homologous with protein 10 fragments

In comparison, the amino acid sequence homology between the protein 10 fragments and rat or bovine calbindin D-28K (Yamakuni et al. 1986; Takagi et al. 1986) was 54% (Table 1). These data strongly suggest that protein 10 is a member of the superfamily of calcium-binding proteins, distinct from calbindin D-28K, but probably represents the mammalian counterpart of chicken calretinin.

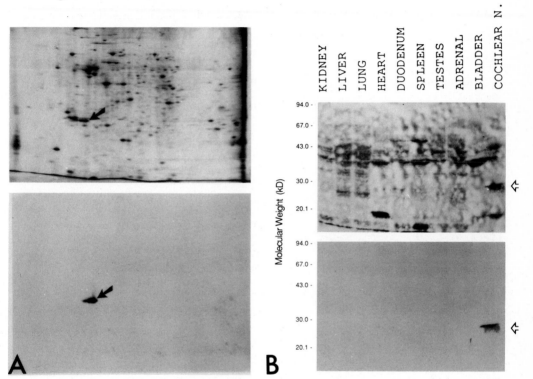

Fig. 3A,B. **A** Silver stain of a two-dimensional gel (*top*) and specific antibody reaction of protein 10 antisera (1:1000) on an immunoblot developed from a two-dimensional gel (*bottom*) from guinea-pig cochlear nucleus (24 μg protein each). **B** Autoradiogram of $^{45}Ca^{2+}$ binding (*top*) and reaction of protein 10 antisera (*bottom*) on a single blot developed from a one-dimensional gel containing 48 μg of protein from peripheral tissues and 24 μg of cochlear nucleus from a guinea pig. Position of protein 10 is indicated by *arrows*

IMMUNOHISTOCHEMICAL LOCALIZATION OF PROTEIN 10

A polyclonal antibody was raised in a rabbit against the purified protein 10. The selectivity and sensitivity of this antibody for protein 10 was confirmed on immunoblots of two-dimensional gels including proteins from cochlear nucleus (Figure 3A) and cerebellum (Winsky et al. 1989c) where protein 10 was the only protein recognized at concentrations of 1:500 to 1:100000. In addition, this antibody showed no cross-reactivity with several other calcium-binding proteins from peripheral tissues on immunoblots developed from a one dimensional gel (Figure 3B). Specific protein 10 antibody label was observed in formalin-fixed brain samples from several species including rat, guinea pig, cow, monkey, and human at dilutions of 1:500-1:8000. Similar label was not apparent in either pre-immune serum or preabsorption control sections using 0.1 µM of purified protein 10. Immunohistofluorescent examination of protein 10 antibody label (1:2500) confirmed our previous two-dimensional gel analyses of protein 10 within the brainstem auditory system.

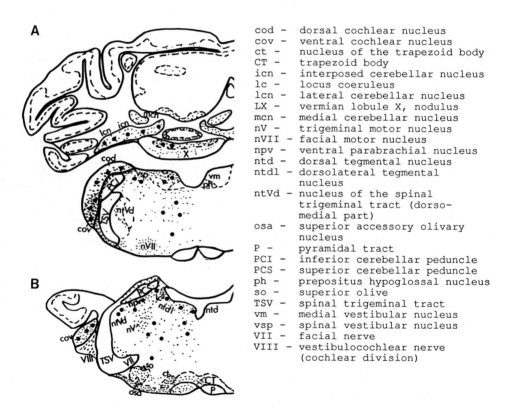

cod –	dorsal cochlear nucleus
cov –	ventral cochlear nucleus
ct –	nucleus of the trapezoid body
CT –	trapezoid body
icn –	interposed cerebellar nucleus
lc –	locus coeruleus
lcn –	lateral cerebellar nucleus
LX –	vermian lobule X, nodulus
mcn –	medial cerebellar nucleus
nV –	trigeminal motor nucleus
nVII –	facial motor nucleus
npv –	ventral parabrachial nucleus
ntd –	dorsal tegmental nucleus
ntdl –	dorsolateral tegmental nucleus
ntVd –	nucleus of the spinal trigeminal tract (dorso-medial part)
osa –	superior accessory olivary nucleus
P –	pyramidal tract
PCI –	inferior cerebellar peduncle
PCS –	superior cerebellar peduncle
ph –	prepositus hypoglossal nucleus
so –	superior olive
TSV –	spinal trigeminal tract
vm –	medial vestibular nucleus
vsp –	spinal vestibular nucleus
VII –	facial nerve
VIII –	vestibulocochlear nerve (cochlear division)

Fig. 4A,B. Schematic diagram of protein 10 levels **A** P4.5 and **B** P2.8 of rat (Palkovits and Jacobowitz 1974). Cell body fluorescence is indicated by *large black circles*, fibers by *smaller dots*. Abbreviations for this and subsequent figures is as indicated on the *right*

A schematic diagram of protein 10 antibody label of cell bodies and fibers at the brainstem levels described below is presented in Figure 4. A low fluorescence intensity was observed in what appears to be the large globular bushy cells (Moore 1986) of the ventral cochlear nucleus (Figures 4, 5A). Many of these labeled cells and some large unlabeled cells were enveloped by intensely fluorescent fibers. These thick fibers were similar in appearance to the VIIIth nerve axons shown entering the cochlear nucleus in Figure 5A. The protein 10 antibody label of the dorsal cochlear nucleus was most prominent in the deep nucleus region and consisted primarily of immunofluorescence in small round or oval cells and in both heavy and fine fiber bundles and fiber complexes similar in appearance to the glomeruli of the cerebellum (see below). The immunoreactivity observed in the cochlear nuclei and other regions examined was generally similar in rat and guinea pig. The superior olive was clearly delineated by a dense plexus of fine fibers and numerous immunofluorescent cells were visualized in the accessory olivary nucleus (Figures 4B, 5B). The protein 10 antibody also revealed intensely fluorescent fibers in the lateral portion of the inferior cerebellar peduncle, pyramidal tract and trapezoid body with large varicosities enveloping the unstained cell bodies of the nucleus of the trapezoid body (Figures 4, 5B). The varicosities surrounding the trapezoid cells were similar in appearance to the labeled fibers outlining the large cells of the ventral cochlear nucleus in that the fluorescent label appeared to be absent from the center of these large axons.

Examination of the vestibular nuclei revealed numerous fluorescent cell bodies and fibers in the medial vestibular nucleus (Figures 4A, 6A) and in some large cell bodies and fibers of the adjacent spinal vestibular nucleus. Numerous fine and thick processes were also observed in the nucleus of the spinal trigeminal tract. The nucleus of the trigeminal motor nerve (V) was revealed by fluorescence of a dense plexus of fine varicose fibers, some of which enveloped the large motor cells (Figures 4B, 6B). Abundant labeled fibers enveloping the locus coeruleus cell bodies (Figure 4B) were also noted. A small cluster of moderately intense fluorescent cells and numerous fibers were observed in the dorsal tegmental nucleus (Figure 4B). In the reticular formation, large cells and processes were moderately stained (Figure 6C) while smaller cells were more intensely fluorescent (Figure 4). Numerous, thick fibers were also revealed throughout the reticular formation.

The predominant localization of protein 10 antibody label in the rat cerebellum was observed in the glomeruli (Figure 7A-C). These glomeruli (rosettes) were most pronounced in the granule cell layers of lobes I (lingula), X (nodulus), IXc (uvula), flocculus and paraflocculus. Moderately intense fluorescent fibers emanated from the white matter and appeared to terminate in the glomeruli (Figure 7A-C). This pattern of glomerular label is most likely due to mossy fibers which are components of the vestibulocerebellar pathway (Mehler and Rubertone 1985).

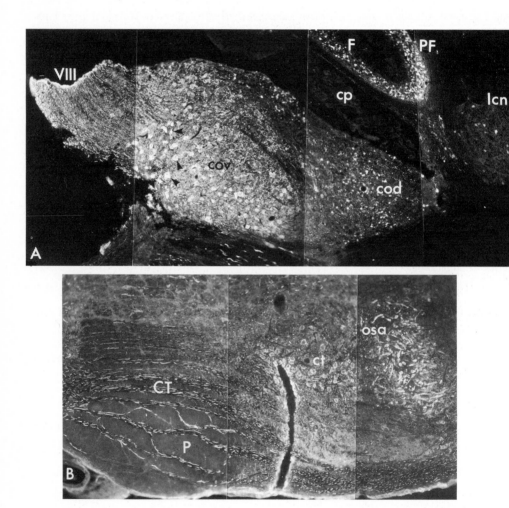

Fig. 5A,B. Protein 10 antibody label (1:2500) of brainstem regions including: **A** The ventral and dorsal cochlear nucleus of a rat (dorsal is to the *right*, x155); and **B** The trapezoid body and accessory olivary nucleus of guinea pig (x130); *cp* choroid plexus; *F* flocculus; *PF* paraflocculus; other abbreviations as in Figure 4

We have previously observed (Duchesne et al. 1990) intense protein 10 antibody label in a portion of the vestibular ganglion cells and in vestibular nerve fibers which are components of the peripheral origin of mossy fibers (Mehler and Rubertone 1985).

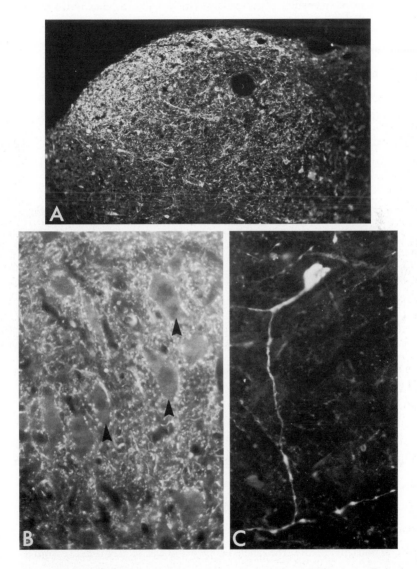

Fig. 6A-C. Immunofluorescent label by protein 10 antisera in rat brain. **A** Fibers and cells in the medial vestibular nucleus. Midline is on the *left* (x155). **B** Fibers surrounding large cells (*arrowheads*) of the trigeminal motor nucleus (x800). **C** A single large cell in the reticular formation (x850)

It thus appears that protein 10 serves as a marker for the innervation pattern of a subpopulation of mossy fibers which emanate from some vestibular ganglion cells. This localization pattern supports those studies which conclude that the ventral paraflocculus and lingula receive primary vestibular nerve fibers (Brodal and Hoivik 1964; Carpenter 1960).

In addition to the mossy fibers, numerous intensely fluorescent round to ovoid granule cells with unstained nuclei had extended processes (dendrites) which also joined the glomeruli. These intensely fluorescent cells were most noticeable in those regions with intensely fluorescent glomeruli (lobes I, X, IX, flocculus and paraflocculus) and may represent a subpopulation of granule cells (Figure 7). Small numbers of positive granule cells, glomeruli and mossy fibers were also observed at various levels of lobes II to VII and in the lateral lobules. The granule cell layer also contained an occasional cell, larger than the granule cell, which was adjacent to the Purkinje cells with processes that ramified in a horizontal direction along the Purkinje cells and into the molecular layer (Figures 7C,E). Whether these are the Lugaro cells as reported by Rogers (1989b) cannot be ascertained. In the molecular layer, particularly the ventral lobes (I, II, III, IX, X, flocculus and paraflocculus), many long, moderately intense fluorescent fibers were observed which had the appearance of parallel fibers with processes traversing horizontally and vertically (Figure 7D). These labeled fibers which occupy that half of the molecular layer adjacent to the Purkinje cells are presumed to emanate from the intensely stained granule cells. Finally, a dense innervation of fine fibers and a small number of immunoreactive cells were noted in the cerebellar nuclei (Figures 4A, 5A).

Immunocytochemical studies of both peripheral organs and the CNS revealed that protein 10 was specifically localized to brain, spinal cord and associated sensory ganglia. A significant observation was the presence of protein 10 within all first order sensory neurons, i.e., optic, olfactory, vestibulocochlear, lingual, and somesthetic nerves from the trigeminal and dorsal root ganglia (Figure 8). In the retina, amacrine cells were immunofluorescent in the inner nuclear layer and in three distinct fiber layers of the inner plexiform region (Figure 8B). The labeled optic nerve was seen emanating from the ganglion cells of the retina. Within the forebrain, the optic tract could be followed to labeled terminal fields in the lateral geniculate body and superior colliculus (Figure 9B).

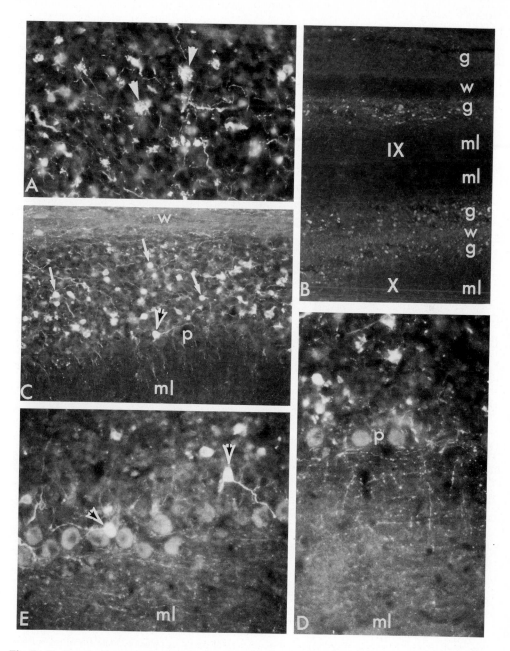

Fig. 7A-E. Protein 10 immunofluorescence in the rat cerebellum. **A** High power magnification of glomeruli (*arrows*) from lobe X of the granule cell layer (x705). **B** Pattern of positive structures in lobules X and IX of the cerebellum (x180). **C** Positive fibers seen entering the granule cell layer from the white matter (*W*). *White arrows* indicate granule cells. *Black arrows* indicate Lugaro-like cells (x435). **D** Parallel fibers entering the molecular layer (*ml*) presumably from granule cells (x465; *p* Purkinje cell layer). **E** Lugaro-like cells (*arrows*) adjacent to negative Purkinje cells (x465)

288

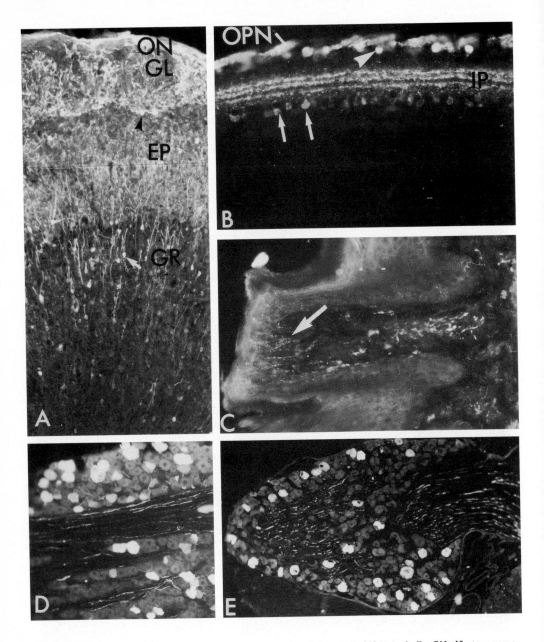

Fig. 8A-E. Protein 10 immunofluorescence in sensory cells of the rat. **A** Olfactory bulb; *ON* olfactory nerve layer; *GL* glomerular layer; *EP* external plexiform layer. *White arrow* points to a granule cell, *black arrowhead* to a periglomerular cell (x300). **B** Retina; *IP* inner plexiform layer; *OPN* optic nerve. *White arrowhead* points to a ganglion cell, *white arrows* to amacrine cells in inner nuclear layer (x645). **C** Thin lingual nerve terminals (*arrow*) in a taste bud of the rat tongue (x715). **D** Trigeminal ganglion and **E** Dorsal root ganglion intensely fluorescent cell bodies and axons (x210)

Within the olfactory bulb, the olfactory nerve was observed as it projected to the glomeruli which also contained many positive periglomerular cells (Figure 8A). The most prominent feature of the olfactory bulb was a dense population of immunoreactive granule cells which projected numerous fibers into the external plexiform layer. The lateral olfactory tract was labeled within the olfactory bulb and along the entire extent of its projection path coinciding with the basal aspect of the pyriform cortex (Figure 10A). These axons of the lateral olfactory tract most likely emanate from the mitral cells which contained only a low intensity of fluorescence. Fine delicate lingual nerve endings were immunofluo-rescent in the taste buds of the tongue (Figure 8C). The trigeminal and dorsal root ganglia contained label in a subpopulation of large diameter cell bodies and in a few smaller cells (Figures 8D,E). Immunofluorescent fibers emanating from the dorsal root ganglion were observed as they entered the substantia gelatinosa (laminae II and III) of the spinal cord.

In addition to these sensory systems, a variety of other cells and neuronal pathways were positively stained for protein 10. An extensive array of fluorescent interneurons was noted within the cerebral cortex (Figure 9A). While these positive cell bodies and processes were visible in all layers of the cortex, most somata were seen in layers II and III. Scattered immunoreactive interneurons were visible in all layers of the hippocampus (Figure 9C). Numerous positive cells in the septofimbrial nucleus were enmeshed within the axonal bundles of the ventral hippocampal commissure (Figure 10B). The significance of this system is unknown.

The periventricular nucleus of the thalamus contained an extensive system of immunoreactive cell bodies and fibers (Figure 11A). Immuno-fluorescent label in the habenula could be traced within the entire neuronal circuit from the perikarya and along the fasciculus retroflexus into the interpeduncular nucleus (Figures 11A, 12). The mamillothalamic tract was part of another neuronal circuit exposed by the protein 10 antibody. Numerous positive cells were observed in the lateral nucleus of the mammillary body (Figure 11B). These cells projected axons into the developing mamillothalamic tract where the immunofluorescent fibers appeared to innervate the medial thalamic nucleus. A third circuit revealed by protein 10 antisera was that of the nigrostriatal/mesolimbic projections which emanate from the A9/A10 dopaminergic cells (see Jacobowitz and Palkovits 1974; Palkovits and Jacobowitz 1974). Protein 10 appeared to be present within the

290

substantia nigra compacta and ventral tegmental area cells known to contain dopamine (Figure 13). Dendrites of the substantia nigra compacta cells were labeled as they entered into the reticulata. These protein 10/dopamine-containing axonal tracts could be seen projecting rostrally through the globus pallidus to innervate the caudate-putamen. Scattered immunofluorescent cells were also observed in the hypothalamus and amygdala.

PROTEIN 10 RADIOIMMUNOASSAY

Purified guinea-pig protein 10 (5 µg) was iodinated by incubation (15 min) in 50 mM phosphate buffered saline (pH = 7.4) containing 1 mCi ^{125}I and 2 iodobeads (Pierce). The solution was applied onto a G-50 (Pharmacia) column previously equilibrated with PBS containing 1% BSA. Radioactive fractions with maximal binding to the protein 10 antibody were used for RIA. Tissue samples from discrete regions of rat brain were obtained by micropunch (Palkovits 1973), sonicated in PBS and duplicate aliquots were taken for protein determination (Lowry et al. 1951). The remaining homogenates were centrifuged for 30 min at 18000 rpm. Duplicate aliquots of tissue supernatants were taken for RIA. Standards, blanks, and unknowns were incubated 48-72 h in a total 0.40 ml of RIA buffer (50 mM PBS, pH = 7.6, 0.2% BSA, 0.1% gelatin, 0.02% methiolate, 0.01% Triton-X100) containing 1% normal rabbit serum, 1:1200 protein 10 antisera and labeled protein 10 (20000-30000 cpm/tube) at 4 °C. Immunoprecipitation was initiated by the addition of goat anti-rabbit IgG. Following 1 h incubation (4 °C), precipitates were collected by centrifugation at 5000 g for 1 h. Supernatants were aspirated and radioactivity in pellets was determined. Specific binding was calculated by subtracting radioactivity precipitated in the absence of protein 10 antisera from total binding. Counts were converted to ng protein 10 using a LOGIT/LOG analysis. Total binding (without added antigen) ranged from 30-50%. As shown in Figure 14, the protein 10 standards displaced the radioactive protein 10 in a dose-dependent manner with assay sensitivity ranging from 2-60 ng protein 10. Dilutions of rat superior colliculus showed a parallel displacement of radioactive label with the purified protein 10 from rat brain (Figure 14). Table 2 presents the distribution of protein 10 immunoreactivity in discrete samples of rat brain.

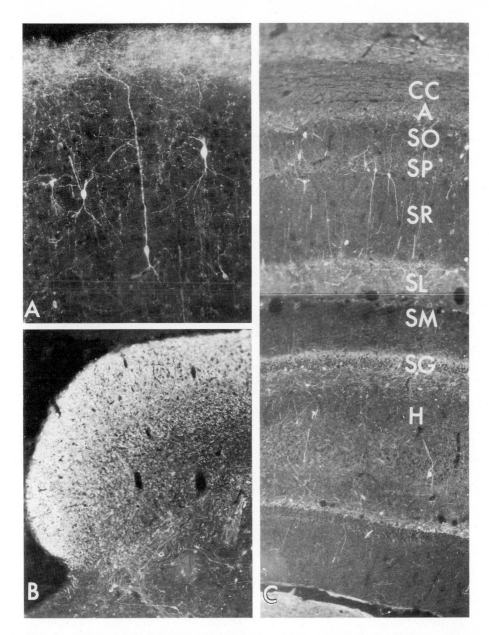

Fig. 9A-C. Protein 10 immunolabel of rat forebrain regions (x210). **A** Label of a multipolar cell in layer III of cerebral cortex as it projects a long varicose fiber to layer I. Note high density of immunofluorescent fibers in layer I. **B** Dense stain in the superior colliculus (stratum greseum superficiale). **C** Immunofluorescent cells and fibers in the hippocampus. *A* alveolus; *CC* corpus callosum; *H* hilum of the dentate gyrus; *SG* stratum granulosum; *SL* stratum lacunosum; *SM* stratum moleculare; *SO* stratum oriens; *SP* stratum pyramidalis; *SR* stratum radiatum

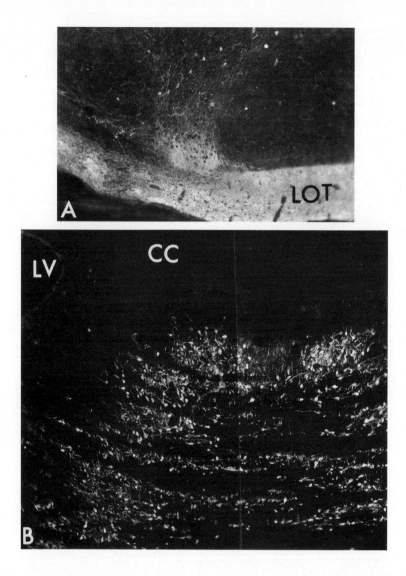

Fig. 10A,B. **A** Protein 10 antibody label of the rat lateral olfactory tract on the base of the pyriform cortex.
B Abundant labeled cells of the rat septofimbrial nucleus; *CC* corpus callosum; *LV* lateral ventricle (x210)

In agreement with the two-dimensional gel and immunohistochemical analyses, the ventral cochlear nucleus contained the greatest concentration of protein 10 (6.38 ± 0.19 ng/μg protein).

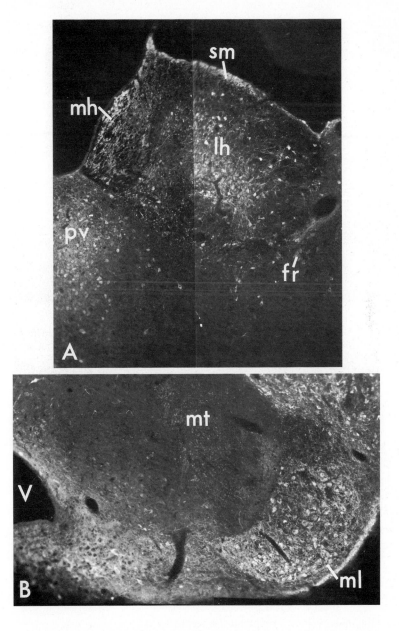

Fig. 11A,B. **A** Immunohistofluorescence of rat habenula (*fr* fasciculus retroflexus; *lh* lateral habenula; *mh* medial habenula; *pv* periventricular nucleus of the thalamus; *sm* stria medularis). **B** Protein 10 antibody label of rat lateral mammillary nucleus (*ml*) and mamillothalamic tract (*mt*); *V* third ventricle (x210)

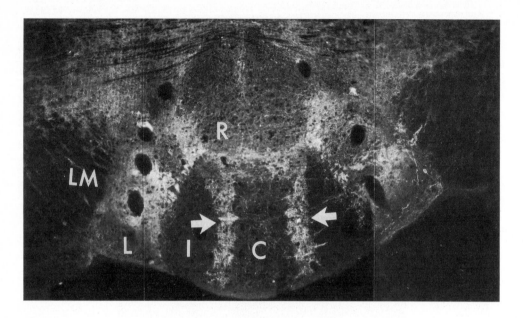

Fig. 12. Protein 10 antibody stain of rat interpeduncular nucleus; subnuclei: *C* central; *L* lateral; *I* intermediate; *R* rostral. Note the two vertically oriented bands (*arrows*) of dense fibers in the lateral aspect of the central subnucleus; *LM* medial lemniscus (x285)

High concentrations (>2 ng/μg protein) of protein 10 were also found in the periventricular nucleus of the thalamus, medial habenula, optic chiasm, superior colliculus, vestibular nuclei, dorsal cochlear nucleus and in the lateral lobules of the cerebellum (Table 2). With the exception of the central and medial amygdala, the concentration of protein 10 was low in telencephalic regions including the cerebral cortex and hippocampus. The somewhat higher concentrations of protein 10 in the lateral cerebellar lobules versus the vermis was in contrast with the more intense protein 10 immunoreactivity of glomeruli and granule cells in the vermis as previously discussed. This apparent discrepancy is probably not due to cross-reactivity of the protein 10 antisera with calbindin D-28K since such an effect would result in inflated protein 10 values throughout the cerebellar cortex. However, we have noticed a diffuse fluorescent label that is more prominent in the molecular layer of the lateral lobules which is not completely preabsorbed by a concentration of protein 10 (0.1 μM) which effectively preabsorbed other specific label in the cerebellum.

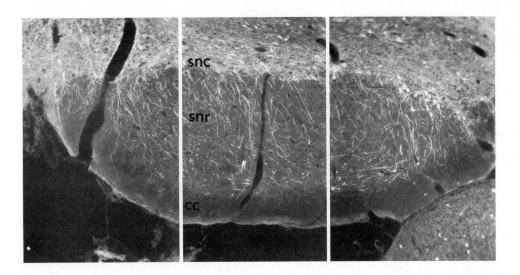

Fig. 13. Protein 10 antibody reaction in rat substantia nigra; (*snc* substantia nigra compacta; *snr* substantia nigra reticulata). Note cells in snc and dendrites in snr; *cc* crus cerebri (x130)

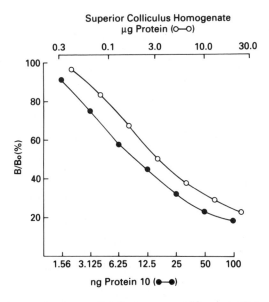

Fig. 14. RIA standard curve showing parallel displacement of ^{125}I-protein 10 from the antiserum (1:1200) by increasing concentrations of unlabeled rat protein 10 (*full circles*) and dilutions of rat superior colliculus (*open circles*)

Table 2. Regional distribution of protein 10 in rat brain as determined by radioimmunoassay (n=3-6 per region)

Region of brain	Protein (10 ng/µg protein ± S.E.M.)
Telencephalon:	
Olfactory tubercle	0.39 ± .13
Prefrontal cortex	0.19 ± .10
Cingulate cortex	0.06 ± .01
Caudate-putamen	0.23 ± .01
Central amygdaloid nucleus	1.54 ± .08
Medial amygdaloid nucleus	1.42 ± .13
Cortical amygdaloid nucleus	0.12 ± .02
Septal fimbrial nucleus	0.84 ± .16
Hippocampus	0.02 ± .01
Dentate gyrus	0.03 ± .02
Diencephalon:	
Medial preoptic nucleus	1.30 ± .06
Anterior hypothalamic nucleus	1.61 ± .19
Supraoptic nucleus	1.61 ± .17
Paraventricular nucleus	1.62 ± .26
Dorsomedial nucleus	1.89 ± .19
Ventromedial nucleus	1.47 ± .14
Arcuate nucleus	0.60 ± .10
Lateral mammillary nucleus	1.25 ± .20
Medial habenula	3.42 ± .08
Periventricular nucleus of the thalamus	4.40 ± .51
Optic chiasm	3.64 ± .29
Lateral geniculate nucleus	0.97 ± .21
Mesencephalon:	
Substantia nigra compacta	0.95 ± .21
Substantia nigra reticulata	0.03 ± .02
Superior colliculus	2.45 ± .20
Inferior colliculus	0.88 ± .09
Medial geniculate body	0.44 ± .11
Rhombencephalon:	
Ventral cochlear nucleus	6.38 ± .19
Dorsal cochlear nucleus	2.05 ± .46
Medial vestibular nucleus	2.55 ± .28
Lateral vestibular nucleus	1.36 ± .12
Superior olive	0.38 ± .12
Cerebellar lobules	2.07 ± .22
Cerebellar nuclei	0.28 ± .01
Cerebellar vermis	1.04 ± .22
Substantia gelatinosa	0.17 ± .03
Nucleus tractus spinalis N. trigemini	0.17 ± .03
Nucleus tractus solitarius	1.36 ± .08
Hypophysis:	
Neurohypophysis	0.22 ± .02
Adenohypophysis	0.19 ± .08

Possibly the antisera to protein 10 recognizes some unidentified antigen which is localized to this region but which is not revealed on two dimensional immunoblots of cerebellum (Winsky et al. 1989c).

It should be noted that the protein 10 values in Table 2 are presented as ng/μg of total protein. When these values are expressed in terms of soluble protein (60% of total using PBS tissue buffer), the concentration of protein 10 in cerebellum lateral lobules changes to 3.45 ng and vermis to 1.73 ng protein 10/μg soluble protein. These protein 10 values in cerebellum are lower than the calbindin D-28k concentrations (15-37 ng/μg soluble protein) reported for rat cerebellum (Baimbridge et al. 1982; Sonnenberg et al. 1984). However, the concentration of protein 10 in the ventral cochlear nucleus (10 ng/μg soluble protein) is within the range of those highest calbindin D-28k values reported for cerebellum.

COMPARISON WITH CALRETININ DISTRIBUTION

The results reported here represent a summary of a detailed study of protein 10 localization within the rat and guinea pig brain. Rogers (1989a,b) has reported the localization of calretinin in the PNS and brain of chicks and in the cerebellum of both rats and chicks. A comparison of the localization of protein 10 immunoreactivity with that of calretinin supports other data suggesting a common identity for these proteins. Both proteins appear to be specifically localized to neural tissue with a extensive distribution within components of the major sensory systems. In agreement with Rogers' (1989a) description of calretinin in the chick, we have found protein 10 antibody label in the spiral, vestibular, retinal, and dorsal root ganglion cells and in some hair cells of the cochlea and vestibule but not in the photoreceptors of the retina (Winsky et al. 1989; Duchesne et al. 1990). In the rat cerebellum, both calretinin and protein 10 antisera stain mossy fibers in the white matter and reveal some intensely staining granule cells, particularly in lobe X. However, while Rogers (1989b) has reported that calretinin antibody label is most distinct in the rat Lugaro cells adjacent to Purkinje cells, we have noted the Lugaro-like cells to be less pronounced using the protein 10 antisera. Instead we find the glomeruli to be the most prominent structures stained by the protein 10 antisera in the rat and guinea pig cerebellum. Both antibodies appear to label a small population of cells in the deep cerebellar nuclei. Rogers (1989b) has suggested that the calretinin-positive cell bodies in chick cerebellum may represent cross-reactivity with some unknown antigen since these cells are not revealed by *in situ* hybridization. Additional studies are required to determine whether this may also be true for the calretinin and protein 10 immunoreactive cells in rat cerebellum.

SUMMARY AND CONCLUSIONS

The high degree of amino acid sequence homology and similar patterns of distribution between immunoreactive protein 10 and calretinin in some brain regions strongly suggest that they represent the same protein. Future studies will be directed toward unraveling the function of protein 10. Hints as to which avenues to pursue in this regard may be found by comparing the anatomical and subcellular localization of protein 10 and calretinin with putative neurotransmitters/neuromodulators and intracellular enzymes. For example, there is evidence to suggest that an excitatory amino acid such as glutamate may represent a putative neurotransmitter of the eighth nerve inputs from the organ of Corti to the cochlear nuclei (Wenthold and Martin 1984; Altschuler et al. 1989). The globular bushy cells of the ventral cochlear nucleus which are revealed by the protein 10 antibody (Figure 5A) receive large calyceal auditory nerve endings on their somata (Moore 1986). These thick processes may be represented in Figure 5A by the intensely fluorescent thick fibers surrounding the large cells. It is interesting to note that the bushy cells of the ventral cochlear nucleus have been shown to respond to the application of excitatory amino acids in an atypical manner which is characterized by either no stimulation or a hyperpolarization (Martin and Adams 1979). Possibly, calretinin/protein 10 is involved in the pre- and/or postsynaptic actions mediating this effect. For example, this calcium binding protein could serve to modulate the large influx of calcium that might accompany neural activation by excitatory amino acids.

The near-complete amino acid sequence of chick calretinin (Rogers 1990) may also suggest enzyme systems with which this protein might interact. The similarity between calbindin D-28k and both protein 10 and calretinin suggests similar functions and thus, studies of the action of calbindin D-28k in the periphery might also suggest potential roles for these calcium binding proteins in brain.

Regardless of the approach taken, the function of calcium binding proteins such as protein 10/calretinin will most likely be related to their ability to bind calcium. While calcium binding and/or transport may be primary functions of these proteins in brain as suggested by Baimbridge et al. (1982), it is doubtful that such sequestering of calcium can occur in the absence of other intracellular metabolic change.

REFERENCES

Altschuler RA, Sheridan CE, Horn JW, Wenthold RJ (1989) Immunocytochemical localization of glutamate immunoreactivity in the guinea pig cochlea. Hearing Res 42:167-174

Baimbridge KG, Miller JJ, Parkes CO (1982) Calcium-binding protein distribution in the rat. Brain Res 239:519-525

Brodal A, Hoivik B (1964) Site and mode of termination of primary vestibulo-cerebellar fibers in the cat. An experimental study with silver impregnation methods. Arch Ital Biol 102:1-21

Carpenter MB (1960) Experimental anatomical-physiological studies of the vestibular nerve and cerebellar connections. In: Rasmussen GL, Widle W (eds) Neural mechanisms of the auditory and vestibular systems. CC Thomas, Springfield, pp 297-323

Duchesne CJ, Winsky L, Kim HM, Goping G, Vu TD, Wenthold RJ, Jacobowitz DM (1990) Identification and localization of the calcium binding protein, protein 10 in the guinea pig inner ear. Abstract to 13th meeting of the Association for Research in Otolaryngology.

Jacobowitz DM, Palkovits M (1974) Topographical atlas of catecholamine and acetylcholinesterase-containing neurons in the brain. I. Forebrain (telencephalon, diencephalon). J Comp Neurol 157:13-28

Lowry OH, Rosebrough NJ, Farr AL, Randall, RJ (1951) Protein measurement with the folin phenol reagent. J Biol Chem 193:265-275

Martin MR, Adams JC (1979) Effects of DL-alpha-amino adipate on synaptically and chemically evoked excitation of anteroventral cochlear nucleus neurons of the cat. Neurosci 4:1097-1105

Moore JK (1986) Cochlear nuclei: relationship to the auditory nerve. In: Altschuler RA, Hoffman DW, Bobbin RP (eds) Neurobiology of hearing: The cochlea. Raven Press, New York, pp 283-301

Palkovits M (1973) Isolated removal of hypothalamic or other brain nuclei of the rat. Brain Res 59:449-450

Palkovits M, Jacobowitz DM (1974) Topographic atlas of catecholamine and acetylcholinesterase-containing neurons in the rat brain. II. Hindbrain (mesencephalon, rhombencephalon). J Comp Neurol 157:29-42

Rogers JH (1987) Calretinin: a gene for a novel calcium-binding protein expressed principally in neurons. J Cell Biol 105:1343-1353

Rogers JH (1989a) Two calcium-binding proteins mark many chick sensory neurons. Neuroscience 31:697-709

Rogers JH (1989b) Immunoreactivity for calretinin and other calcium binding proteins in cerebellum. Neuroscience 31:711-721

Rogers JH, Khan M, Ellis J (1990) Calretinin and other CaBPs in the nervous system. In: Proceedings of the first European symposium on calcium binding proteins. Plenum, New York, London (in press)

Santer DM, Heydorn WE, Creed GJ, Fukuda T, Jacobowitz DM (1988) Localization of Ca^{2+}-binding proteins of rat cortex on two-dimensional gels-II. Analysis of Ca^{2+} binding proteins in ammonium sulfate ractions of rat brain. Neurochem Int 12:225-236

Sonnenberg J, Pansini AR, Christakos S (1984) Vitamin D-dependent rat renal calcium-binding protein: development of a radioimmunoassay, tissue distribution, and immunologic identification. Endocrinology 115:640-648

Takagi T, Nojiri M, Konishi K, Maruyama K, Nonomura Y (1986) Amino acid sequence of vitamin D-dependent calcium-binding protein from bovine cerebellum. FEBS Lett 201:41-45

Wenthold RJ, Martin MR (1984) Neurotransmitters of the auditory nerve and central auditory system. In: Berlin C (ed) Hearing science. College Hill Press, San Diego, pp 341-369

Winsky L, Harvey JA, McMaster SE, Jacobowitz DM (1989a) A study of proteins in the auditory system of rabbits using two-dimensional gels: identification of glial fibrillary acidic protein and vitamin D-dependent calcium binding protein. Brain Res 496:136-146

Winsky L, Nakata H, Jacobowitz DM (1989b) Identification of a calcium binding protein highly localized to the cochlear nucleus. Neurochem Int 15:381-389

Winsky L, Nakata H, Martin BM, Jacobowitz DM (1989c) Isolation, partial amino acid sequence and immunohistochemical localization of a brain-specific calcium binding protein. Proc Natl Acad Sci USA 86:10139-10143

Yamakuni T, Kuwano R, Odani S, Kiki N, Yamaguchi Y, Takahashi Y (1986) Nucleotide sequence of cDNA to mRNA for a cerebellar CaBP, spot 35 protein. Nucl Acids Res 14:6768

*LSP*1 IS A NEW LYMPHOCYTE-SPECIFIC Ca^{2+}-BINDING PROTEIN WHICH CO-CAPS WITH SURFACE IMMUNOGLOBULIN

J. Jongstra, J. Jongstra-Bilen, S. Galea, and D.P. Klein

INTRODUCTION

The formation of mature functional lymphocytes from a single type of pluripotent stem cell involves a precisely regulated process of differential gene expression as cells mature along the developmental pathway. The understanding of lymphocyte growth and development is critically dependent on our knowledge of the genes and gene products which are involved in the regulation of these processes. Starting with the premise that such important regulatory genes are, at least in part, expressed in a lymphocyte-specific manner, we undertook a systematic search for cDNA clones which represent lymphocyte-specific genes. To do this we made use of a subtractive hybridization technique which had been used previously to isolate lymphocyte-specific genes such as the T-cell antigen receptor beta-chain or a gene closely linked to the *xid* defect in mice (Cohen et al. 1985; Hedrick et al. 1984).

To identify a gene as a possible regulatory gene, several of its characteristics should be considered in addition to its lineage or cell-type specific expression. First, the nucleotide sequence of the isolated cDNA clone should be used to determine its coding potential and to predict certain characteristics of the encoded protein. For instance, it should be possible to predict whether a particular cDNA codes for a secreted or membrane-bound protein (von Heyne 1983) or for certain classes of proteases (Gershenfeld 1986), Ca^{2+}-binding proteins (Kretsinger and Nockolds 1973), or certain types of DNA-binding proteins (Berg 1986). Secreted or membrane-bound proteins are good candidates for being lymphokine or lymphokine receptor molecules involved in cell-to-cell interactions, while nuclear proteins might be transcriptional regulators. Protein kinases and Ca^{2+}-binding proteins could be involved in the regulation of specific steps in a signal transduction pathway.

Normal resting B-lymphocytes which express immunoglobulin molecules on the cell surface as antigen receptors (sIg$^+$ B-cells) undergo a transition of the G_0 to G_1 stage after treatment with antibodies directed against sIg (anti-Ig stimulation). High concentrations of anti-Ig (50-100 µg/ml) will induce entry into S-phase while cells treated with low concentrations of anti-Ig need the additional presence of the lymphokine interleukin 4 (IL-4) to progress to DNA synthesis (DeFranco et al. 1982; Howard et al. 1982). Some of the initial events in the cell membrane following treatment of resting B-cells and B-lymphoma lines with anti-Ig have been well documented and include an increase in membrane fluidity (Mizuguchi et al. 1988), a rapid increase in phosphatidylinositol turnover with subsequent production of inositol 1,4,5-triphosphate (IP3) and diacylglycerol (DG) (Bijsterbosch et al. 1985; Myers et al. 1987; Coggeshall and Cambier 1984). Protein kinase C (PKC) is then translocated from the cytoplasm to the cytoplasmic face of the plasma membrane where it phosphorylates an as yet little characterized set of membrane and cytoskeletal proteins (Hornbeck and Paul 1986). The increased production of IP3 leads to an elevation of intracellular free Ca^{2+} ($[Ca^{2+}]_i$) mainly from intracellular stores (Ransom et al. 1986). Increases in $[Ca^{2+}]_i$ may regulate B-cell physiology through Ca^{2+}-binding proteins such as calmodulin but little is known about other proteins which can respond to the transient Ca^{2+} fluxes which occur after activation of B-cells through the sIg antigen receptor. Other biological consequences of anti-Ig treatment of normal B-cells or B-lymphoma cells include the rapid association of a fraction of sIg molecules with the cytoskeleton and patching and capping of sIg molecules (Albrecht and Noelle 1988; Braun and Unanue 1980; Braun et al. 1982), RNA synthesis as measured by increased uridine incorporation (Bijsterbosch et al. 1985) or with specific probes for IgH and IgL chains (Nakanishi et al. 1984) or for c-fos (Monroe 1988) and finally in normal B-cells DNA synthesis.

Thus, the biochemical consequences of anti-Ig treatment of B-cells include rapid increase of PI turnover, activation of PKC and increased $[Ca^{2+}]_i$ and protein phosphorylation. Anti-Ig treatment also leads to structural rearrangements of cytoskeletal components and to patching and capping of sIg molecules.

We have reported the isolation of several cDNA clones which represent genes expressed in all major lymphocyte populations or in certain populations only (Jongstra et al. 1987; Jongstra and Davis 1988; Jongstra et al. 1988; Jongstra et al. 1988a; Schall et al.

1988). Here we describe the LSP1 cDNA and its protein product, which is a new lymphocyte-specific Ca^{2+}-binding protein. Based on its biochemical characteristics, its subcellular localization and its interaction with sIg, we propose that the LSP1 protein is part of the sIg signal transduction pathway.

EXPRESSION OF THE MOUSE AND HUMAN *LSP*1 GENES

Northern analysis reveals that the mouse LSP1 gene is expressed as a 1.8 Kb RNA species in normal and transformed B-cells, in normal T-cells, such as functional T-helper and T-cytotoxic cells, but not in the ten transformed T-lymphoma cell lines tested. Other cell types such as the related peripheral blood granulocytes or non-hematopoietic cells such as kidney, heart or lung are also negative for LSP1 expression (Table 1, Jongstra and Davis 1988; Jongstra et al. 1988). We noted that functional human T-cell lines expressed a 1.8 Kb RNA species which cross-hybridized with our mouse LSP1 cDNA. Furthermore, no cross-hybridizing RNA could be detected in transformed human T-lymphoma cell lines, suggesting the existence of a human LSP1 gene with a DNA sequence and expression pattern similar to those of the mouse LSP1 gene. Indeed, subsequent analysis of human cDNA clones from a functional human T-helper cell line revealed that the mouse and human LSP1 genes code for a very similar protein (see below).

The complete coding sequences of the mouse and human LSP1 cDNA clones were transferred to the prokaryotic expression vector pET-3c (Rosenberg et al. 1987) and the LSP1 proteins were expressed in *E. coli* as fusion proteins containing the N-terminal 11 amino acid residues of the T7 phage major capsid protein followed by the complete mouse or human LSP1 protein. These recombinant LSP1 proteins (rLSP1) were then used to generate LSP1-specific polyclonal rabbit antisera with which we studied the expression of the LSP1 proteins by Western analysis and immunoprecipitation techniques.

Western blot analysis of whole cell protein extracts prepared from a series of transformed mouse B-lymphoma cell lines representing different stages of the B-lymphocyte developmental pathway showed the presence of a prominent doublet of proteins with apparent molecular masses of 52 and 50.5 kDa (Figure 1a, Klein et al. 1989). This doublet

304

was not present when similar Western blots were developed using pre-immune rabbit sera (not shown).

Table 1. Summary of LSP1 mRNA expression

Sample	Lymphoid Cells	
Bone marrow	+	
Spleen	+	
Thymus	+	
Blood lymphocytes	+	
	Myeloid Cells	
Blood granulocytes	-	
MEL cells	-	
Mastocytoma P815	-	
Monomyelocytic cells WEHI-3	±	
	Nonhemopoietic Cells	
L cells	-	
Liver	-	
Kidney	-	
Heart	-	
	B Lineage Cells	
LPS blasts	+	
Cultured bone marrow	+	
Transformed B cell lines	+	(10/10)
	T lineage Cells	
Thy-1$^+$ thymocytes	+	
IL-2-dependent T-cells	+	(7/7)
Transformed T cell lines	-	(8/9)
	±	(1/9)

Table 1. Northern blots of RNA from the indicated cells were probed with radiolabeled LSP1 cDNA as described (Jongstra et al. 1988). Cells were considered negative for LSP1 RNA expression when after prolonged exposure (up to 10 days) of the Northern blot no LSP1 RNA could be detected. The cell lines EL4 and WEHI-3 contain trace amounts of LSP1 RNA comparable to approximately 1% of the amount of LSP1 RNA in BAL17 cells (indicated with ±). Where appropriate, the number of positive or negative cell lines over the total number of cell lines tested has been indicated in *parentheses*

We also performed experiments to compare extracts from the mouse pre-B cell line RAW112, the mature sIg$^+$ B-cell line BAL17 and the plasma cell line J558L, in which each lane contained equal amounts of protein (as judged by Coomassie Blue staining). These experiments showed that BAL17 cell extracts contain five- to tenfold more LSP1 protein per unit total protein than RAW112 or J558L cells. This is in close agreement with our Northern analysis which showed that mature B-cell lines contain more LSP1 RNA than pre-B cell lines or plasma cell lines. Figure 1b,c shows that normal T-cells from thymus express LSP1 protein but transformed T-cell lines do not, again in close agreement with our previous Northern analysis of RNA extracted from a similar panel of cells.

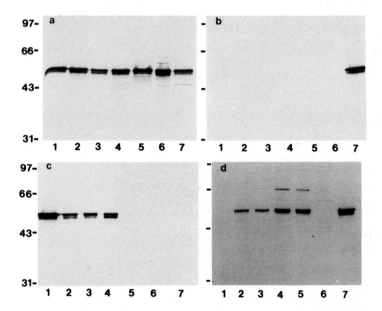

Fig. 1a-d. LSP1 protein expression analyzed by Western blotting. Whole cell extracts from 5 x 10^5 cells prepared by solubilizing cells in Laemmli sample buffer were separated using 10% SDS-PAGE. Western blots were performed using a BioRad Immun-Blot Assay Kit according to the manufacturer's specifications. The anti-mouse LSP1 serum was used at a dilution of 1:2000. **a** Immunoblots of B-lineage cells. *Lanes 1* and *3* pre-B cell lines RAW112 and L12; *lane 2* pre-pre-B cell line R8; *lanes 4-6* sIg$^+$ B-cell lines WEHI231, WEHI279 and BAL17; *lane 7* plasma cell line J558L. **b** Immunoblots of transformed T-lineage cells. *Lanes 1-6* T-lymphoma lines YAC-1, KKT-2, BAL13, VL3, C6VL1 and BW5147; *lane 7* B-lymphoma BAL17. **c** Immunoblots of T-lineage cells. *Lane 1* BAL17; *lanes 2-4* normal thymocytes from C57Bl6, BALB/c and AKR/J mice; *lanes 5-7* T-lymphoma lines BAL9, RBL-5 and EL4. **d** Immunoblots of BW5147/LSP1 transfectants. *Lanes 1-5* transfectants 23, 1, 9, 10, 2; *lane 6*, BW5147; *lane 7* BAL17

However, to unequivocally establish the relation between the presence of the 52/50.5 kDa protein doublet and expression of the LSP1 gene, we transfected the T-lymphoma cell line BW5147 which does not express the LSP1 gene with the mouse LSP1 cDNA clone inserted in the eukaryotic expression vector pECE-B (Ellis et al. 1986), and isolated several transfectants which expressed varying levels of LSP1 RNA (Klein et al. 1989). Western analysis of total protein extracts showed a close correlation between the amounts of the 52 kDa protein present (Figure 1d, lanes 2-5) and the level of LSP1 RNA in these transfectants (not shown). No 52 kDa protein was detected in the parental BW5147 cells (lane 6) or in a transfectant which did not express LSP1 RNA (lane 1). No 50.5 kDa protein was found in any of the transfectants. From these experiments we conclude that the LSP1 gene is expressed as a 52/50.5 kDa protein doublet in lymphoid cells.

In the experiment shown in figure 1, extracts from 5×10^5 cells were layered in each lane. The exact nature of the doublet in these cells and the absence of the 50.5 kDa species in transfected BW5147 cells is not known at present. Experiments to establish whether this is due to post-transcriptional modifications, differential splicing of the primary LSP1 transcript, or the presence of a second LSP1 related gene are now in progress. Western analysis of extracts from a series of human transformed B-cells and functional T-cell lines with the anti-mouse LSP1 serum or with a recently developed anti-human LSP1 serum shows that the human LSP1 gene is expressed as a 52 kDa protein which co-migrates with the 52 kDa mouse LSP1 species (Jongstra-Bilen et al. 1990).

STRUCTURE OF THE MOUSE AND HUMAN *LSP*1 PROTEINS

We first isolated a partial LSP1 cDNA clone from the sIg⁺ mouse B-lymphoma cell line BAL17 during a systematic search for cDNA clones which represent lymphocyte-specific genes. A full-length clone was isolated subsequently from the mouse pre-B cell line 220.2 (Jongstra et al. 1988). We also isolated a full-length cDNA clone representing the human LSP1 gene from a cDNA library made from RNA from a functional human T-helper cell line using the mouse cDNA as a probe (Jongstra et al. 1987; Jongstra-Bilen et al. 1990). Both the mouse and the human LSP1 cDNA clones contain a single long open reading frame. The mouse LSP1 protein is predicted to contain 330 amino acid residues, while its human counterpart has 339 amino acid residues. Both proteins have a predicted molecular mass of approximately 37 kDa but, probably due to the high percentage of charged residues, migrate with different apparent molecular weight during SDS-PAGE.

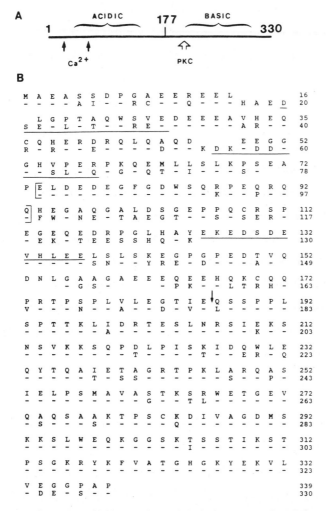

Fig. 2A,B. Structure and sequence of LSP1 proteins. **A** Schematic representation of the mouse LSP1 protein showing the proposed acidic and basic domain structure. The *two small vertical arrows* indicate the positions of two putative EF-hand-like Ca^{2+}-binding sites in the acidic domain and the *large arrow* denotes a putative PKC phosphorylation site in the basic domain. **B** Comparison of the predicted human (*top line*) and mouse (*bottom line*) LSP1 protein sequences. *Gaps* are introduced for maximal alignment. The acidic and basic domains of the human LSP1 protein are separated by a *vertical arrow*. A highly homologous stretch of amino acids within the acidic domain is marked between *brackets*. The putative Ca^{2+}-binding sites of the mouse and human LSP1 proteins are underlined. - indicates identical residues

Comparison of the amino acid sequences of the mouse and human LSP1 proteins has led to several interesting predictions about the structure of these proteins. Both the mouse and the human proteins contain an N-terminal domain of approximately 175 residues with a high content of acidic glutamic acid or aspartic acid residues, while the C-terminal half of both proteins contain an excess of basic residues. This uneven distribution of acidic and basic residues is so strikingly conserved that we predict that the LSP1 proteins contain two domains of opposite charge (Figure 2a).

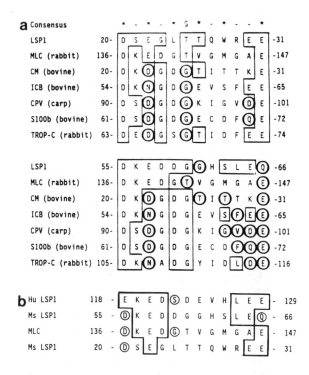

Fig. 3a,b. The mouse and human LSP1 protein sequences identify possible Ca^{2+}-binding sites. **a** The sequences of two possible Ca^{2+}-binding sites in the mouse LSP1 protein are compared with the sequences of Ca^{2+}-binding sites of myosin light chain (MLC, Frank and Weeds 1974), calmodulin (CM, Babu et al. 1985), intestinal calcium-binding protein (ICB, Fullmer and Wasserman 1981), parvalbumin (CPV, Kretsinger and Nockolds 1973), the beta chain of S-100 (taken from Baudier and Gerard 1983) and troponin-c (trop-c, Collins et al. 1977; Wilkinson et al. 1980). **b** The sequence of a possible Ca^{2+}-binding site in the human LSP1 protein is compared with the possible binding sites in the mouse LSP1 protein and in MLC. The *numbers directly on the left and right* of the sequences shown indicate the positions of the amino and carboxy terminal residues of the Ca^{2+}-binding sites. Amino acids which are identical to those in the LSP1 protein are boxed and chemically conservative substitutions are circled. The *top line* indicates the consensus sequence for a Ca^{2+}-binding loop taken from Baudier and Gerard 1983. *Asterisk* oxygen containing residues (D, E, N, Q, S or T) involved in Ca^{2+}-binding.

Alignment of the predicted amino acid sequence of the mouse and human LSP1 proteins shows that 67% of the residues in both proteins are identical (Figure 2b). However, this homology is not evenly distributed along the LSP1 protein. The basic C-terminal domains show a striking degree of identity. Out of the 153 C-terminal residues in the human LSP1 protein 130 residues (85%) are identical to those in the corresponding part of the mouse protein. The remaining 186 N-terminal amino acid residues are 53% identical. Interestingly, within this acidic domain there is a stretch of 20 amino acids, 18 of which are identical in the mouse and human LSP1 proteins (Jongstra-Bilen et al., 1990).

Inspection of the primary amino acid sequence of the mouse protein revealed the presence of two possible EF-hand-like Ca^{2+}-binding motifs (Kretsinger 1977; Kretsinger and Nockolds 1973) located in the acidic domain at positions 20 and 55 (Figure 3a). These domains do not appear to be present in the acidic domain of the human LSP1 protein, but a stretch of amino acid residues present in the acidic domain of the human protein but not in the mouse LSP1 protein appears to contain one potential EF-hand structure (Figure 3b). This points to a functional conservation of the acidic domain rather than a strict sequence conservation, suggesting an important role of the Ca^{2+}-binding property for the function of LSP1. The sequence of the basic domains of the mouse and human LSP1 proteins predict the presence of several conserved phosphorylation sites, most notably a putative PKC site around position 200 near the N-terminus of the basic domain.

*LSP*1 IS Ca^{2+}-BINDING

The presence of two EF-hand-like Ca^{2+}-binding motifs in the amino acid sequence of the mouse LSP1 protein and one such motif in the human LSP1 protein predicts that LSP1 is Ca^{2+}-binding. To test this prediction, graded amounts of mouse rLSP1 protein were immobilized on a nitrocellulose filter using a slot-blot apparatus (Maruyama et al. 1984; Klein et al. 1989). The known Ca^{2+}-binding protein troponin-C (as part of the intact troponin complex) was used as a positive control, while lysozyme was used as an example of a protein which does not bind Ca^{2+}. The protein containing filter was then incubated with 1 μM $^{45}CaCl_2$ and then washed briefly in 50% methanol. Specific binding of $^{45}Ca^{2+}$ was determined

310

by autoradiography of the dried filter. After autoradiography the filter was stained with amido black to control for the amounts of protein retained on the filter.

Figure 4 shows that when equal amounts of protein are compared, rLSP1 binds five-to tenfold less $^{45}Ca^{2+}$ than intact troponin. No binding of $^{45}Ca^{2+}$ to lysozyme was detected. The use of a filter binding assay and of rLSP1 protein purified under denaturing conditions does not allow a rigorous quantitative analysis of the binding characteristics of Ca^{2+} to LSP1 protein. This must await the availability of LSP1 protein purified under nondenaturing conditions from lymphoid cells. However, the significant point of the above experiments is that we can detect Ca^{2+}-binding at a concentration of 1 μM Ca^{2+}. Since the free intracellular concentration of Ca^{2+} in lymphocytes varies between approximately 0.1 μM in resting cells and 1 μM in activated cells, this suggests that the LSP1 protein might respond to the transient increase in $[Ca^{2+}]_i$ which occurs after activation of lymphocytes through their respective antigen receptors (Pozzan et al. 1982; Ransom et al. 1986; Weiss et al 1984; Weiss et al. 1984).

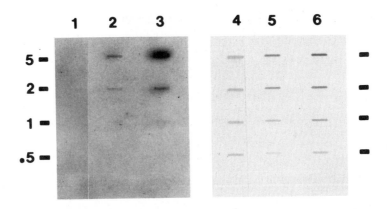

Fig. 4. Mouse rLSP1 binds Ca^{2+}. The amounts of lysozyme (*lanes 1, 4*), mouse rLSP1 (*lanes 2, 5*), or troponin (*lanes 3, 6*) loaded in each slot of a slot blot apparatus and transferred to a nitrocellulose filter are indicated at the *left* in micrograms. *Lanes 1-3* are autoradiograms of the filters after incubation with $^{45}CaCl_2$ and washing. *Lanes 4-6* are the same filters stained with amido black after autoradiography

Ca²⁺-MODULATED INTERACTION OF *LSP*1 WITH A 43 kDa PROTEIN

To investigate whether the LSP1 protein interacts with other cellular proteins and whether such interactions might be modulated by Ca^{2+}, we precipitated LSP1 protein from ^{35}S-methionine-labeled cell lysates using the anti-mouse LSP1 serum, to determine whether in addition to LSP1 protein other proteins were precipitated from these lysates as a result of a specific association with LSP1. The possible effect of Ca^{2+} was measured by lysing cells in buffers containing NP-40 and varying concentrations of free Ca^{2+}.

Figure 5, lane 1, shows that when BAL17 cells are lysed in the presence of 2 mM EGTA a strong band with an apparent M_r of 43000 (p43) co-precipitates with the 52 kDa LSP1 species.

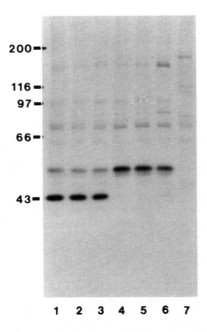

Fig. 5. LSP1 interacts with a 43000 dalton protein. BAL17 cells were labeled with ^{35}S-methionine and lysed in buffer containing 0.5% NP-40 and 2 mM EGTA or varying $[Ca^{2+}]$. Lysates were then incubated with anti-mouse LSP1 serum (*lanes 1-6*) or pre-immune rabbit serum (*lane 7*) and the immunoprecipitates were collected using protein A-Sepharose beads, washed and analyzed by PAGE and autoradiography. *Lanes 1* and 7 2mM EGTA; *lanes 2-6* 0.01 μM, 0.1 μM, 1.0 μM, 10 μM, and 100 μM free Ca^{2+} respectively

On longer exposure of this autoradiogram, the 50.5 kDa LSP1 species is also present (not shown). Similar results are obtained when the [Ca^{2+}] in the lysis buffer is raised to 0.01 μM (lane 2) or 0.1 μM (lane 3). However, in the presence of 1.0 μM, 10 μM or 100 μM Ca^{2+} the 52/50.5 kDa doublet is efficiently precipitated but no or little p43 is present in the immunoprecipitates (lanes 4-6). Pre-immune rabbit serum does not precipitate either the 52/50.5 kDa LSP1 doublet or p43 from a lysate prepared in the presence of 2 mM EGTA (lane 7), showing that the presence of p43 in the precipitates analyzed in lanes 1-3 is not due to the fact that it is insoluble in buffers containing 0.1 μM Ca^{2+} or less.

To exclude the possibility that p43 was precipitated because of a fortuitous cross-reaction with antibodies in our anti-mouse LSP1 serum, we prepared a ^{35}S-methionine labeled NP-40 extract from the LSP1[-] BW5147 cells and its LSP1[+] transfectant T2-2 (see Figure 1d, lanes 5,6). No LSP1 protein or p43 was precipitated from the BW5147 extract after incubation with the anti-mouse LSP1 serum in the presence of 0.1 μM or 1 μM Ca^{2+}. Both proteins were precipitated from the T2-2 transfectant in 0.1 μM Ca^{2+} but only LSP1 was precipitated in 1.0 μM Ca^{2+} (not shown). From these experiments we conclude that LSP1 protein can form a specific protein complex with an as yet unidentified 43000 Da protein and that this complex is dissociated at [Ca^{2+}] of 1 μM or higher. At present we assume that this LSP1/p43 complex exists in intact B-cells and is not formed after lysis of these cells with NP-40. However, rigorous proof for the existence of the LSP1/p43 complex in vivo involves the use of cell-permeable cross-linking agents. These experiments are presently underway.

*LSP*1 PROTEIN CO-CAPS WITH *SIG*

Several lines of evidence argue for a strictly intracellular localization of the LSP1 protein. First, using an indirect immunofluorescence assay, anti-LSP1 antibodies stain only permeabilized but not intact cells. Second, no ^{125}I-labeled LSP1 protein can be precipitated from surface iodinated cells, and thirdly, in vitro translation assays show that the in vitro generated LSP1 protein cannot translocate through microsomal membranes (Klein et al. 1989 and results not shown). These data are in agreement with the highly hydrophilic character of the predicted amino acid sequence of the LSP1 protein and the absence of any predicted hydrophobic regions which could serve as a membrane spanning domain or signal sequence.

When cytospin preparations of the sIg⁺ B-lymphoma cell line BAL17 are fixed and permeabilized with methanol or acetone and then treated with anti-LSP1 serum followed by incubation with FITC-conjugated goat anti-rabbit Ig, prominent staining of the periphery of the cells and weak intracellular staining is observed (Figure 6a, see Klein et al. 1989 for other examples and specificity controls). Peripheral staining is also observed in the mouse pre-B cell line RAW112, the B-cell lines WEHI231 and WEHI279 and in normal thymocytes (not shown). These fluorescence results show that part of the total intracellular pool of LSP1 protein is associated at or near the cytoplasmic side of the plasma membrane.

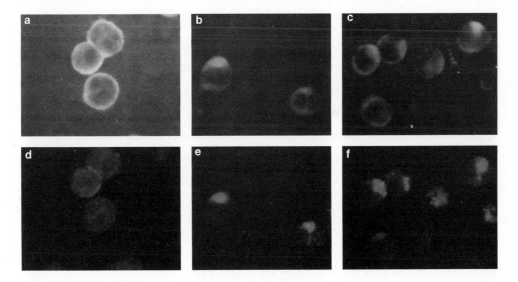

Fig. 6a-f. LSP1 co-caps with sIg. BAL17 cells were first stained for sIg using TRITC-conjugated anti-mouse IgM and then for intracellular LSP1 protein using anti-mouse LSP1 and FITC- conjugated anti-rabbit Ig as described in the text. **a** and **d** are BAL17 cells stained with anti-Ig under noncapping conditions; **b, c, e, f** are cells stained under capping conditions. *Top panels* show the LSP1-specific fluorescence while the *bottom panels* show the sIg-specific fluorescence of the same field as the panel directly above

The peripheral localization of part of the cellular LSP1 protein prompted the question whether LSP1 interacts with the sIg molecules present on mature B-lymphocytes. We took advantage of the observation that sIg molecules assemble in large caps after treatment of

B-cells with divalent anti-Ig antibodies (Braun and Unanue 1980), and asked the question whether the intracellular LSP1 protein would co-cap with sIg. BAL17 cells were incubated with TRITC-conjugated F(ab')$_2$ fragments of a goat anti-mouse IgM at 37 °C to allow capping of sIg molecules or under conditions where capping cannot occur (4 °C, 0.1% sodium azide).

The cells were then spun onto a glass slide, fixed in cold methanol and stained for LSP1 as described (Klein et al. 1989), but with the omission of the Evans' Blue counterstaining. Figure 6b shows the staining pattern for sIg on BAL17 cells incubated under noncapping conditions. The staining is mostly diffuse with a faint peripheral staining. This peripheral staining is much more prominent when FITC-conjugated anti-Ig reagents are used (not shown). The peripheral intracellular LSP1 staining is also clearly visible (panel a). Staining BAL17 cells with TRITC-conjugated anti-Ig under capping conditions resulted in the formation of large sIg caps on greater than 95% of the cells (panels e,f). Approximately 25% of the anti-Ig-treated cells showed the presence of a clearly distinguishable intracellular LSP1 cap (panels b,c). In all cases where we have examined the sIg and the LSP1 cap in the same cell the LSP1 cap was located directly underneath of the sIg cap. Few BAL17 cells treated with anti-Ig under noncapping conditions had clear LSP1 caps, while no Ig caps were present in those cells. These experiments show that there exists an interaction between the transmembrane Ig molecule which serves as the antigen receptor on B-cells and the intracellular LSP1 protein.

SUBCELLULAR LOCALIZATION OF *LSP*1 PROTEIN

The peripheral staining of lymphoid cells with anti-LSP1 serum can be explained by a submembranous localization of the LSP1 protein or might indicate that the LSP1 protein is associated with the membrane cytoskeleton, or both.

To investigate these possibilities we performed the following experiments. To quantitate the fraction of the total intracellular LSP1 pool which is associated with the plasma membrane and the cytoplasm, cells were ruptured using nitrogen cavitation (Quigley 1976; Klein et al. 1989). A soluble cytoplasmic fraction and a particulate fraction containing plasma membrane fragments were then prepared from 10^8 BAL17 cells of which a fraction had been surface labeled with ^{125}I using the lactoperoxidase-glucose oxidase method. The addition of ^{125}I-labeled cells allowed us to estimate the yield of plasma membrane fragments in the particulate fraction to be approximately 40% by determining the recovery of immunoprecipitable ^{125}I-labeled sIg molecules.

Similarly, using the recovery of the cytoplasmic enzyme lactate dehydrogenase we estimated the yield of soluble cytoplasmic proteins to be approximately 90%. Analysis of graded amounts of protein from the cytoplasmic and particulate fractions compared to the protein present in a total cell extract showed that, correcting for the yield of plasma membrane fragments and soluble cytoplasmic proteins, approximately 30% of the total cellular pool of LSP1 protein is associated with the cytoplasmic side of the plasma membrane (Figure 7, lanes 2, 8) and approximately 55% is present in the cytoplasm (lanes 1, 3).

To determine whether part of the intracellular pool of LSP1 protein is associated with the cytoskeleton, BAL17 cells were lysed in a buffer containing 0.5% of the nonionic detergent NP-40 and the NP-40 insoluble residue was collected by low-speed centrifugation. The resulting pellet has been shown to contain the cytoskeleton and nuclear material (Braun et al. 1982; Osborn and Weber 1977). This pellet was then solubilized in Laemmli buffer and the amount of LSP1 protein was determined by Western blotting. By analyzing graded amounts of the NP-40 insoluble material and of a total protein extract of the same BAL17 cells, we determined that approximately 10% of the total cellular pool of LSP1 protein is associated with the insoluble pellet (Figure 8, lanes 1-6). To rule out that the LSP1 was present in the nuclear material present in the NP-40 insoluble pellet we prepared nuclei essentially free of cytoskeletal material by rupturing the cells using nitrogen cavitation. The amount of nuclear material in this fraction was estimated relative to that present in the NP-40 insoluble pellet by comparing the intensity of the Coomassie Blue stained histone bands after PAGE analysis of appropriate aliquots. As can be seen in Figure 8, lanes 7-9, nuclear pellets prepared from as much as 1.5×10^7 cells prepared by nitrogen cavitation contain only a barely detectable amount of LSP1 protein. This suggests that even after correcting for the fourfold higher content of nuclear material in the NP-40 insoluble fraction, this nuclear material does not contribute significantly to the LSP1 signal seen in lanes 2-6. We thus conclude that approximately 10% of the total cellular pool of LSP1 protein is associated with

316

the cytoskeleton. This cytoskeletal pool of LSP1 protein appears to be different from the membrane-associated LSP1 protein, since further experiments have shown that the membrane associated LSP1 protein is soluble in NP-40 and thus does not appear to be linked to the cytoskeleton (not shown).

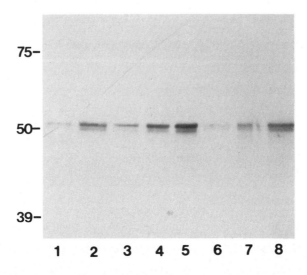

Fig. 7. Subcellular localization of LSP1. BAL17 cells were ruptured by nitrogen cavitation and total cell, cytoplasmic and plasma membrane containing fractions were prepared essentially as described (Klein et al. 1989). The amount of LSP1 was then determined in varying amounts of each fraction by Western blot analysis. *Lanes 1, 2* total cell extracts from 1 x 10^5 and 2 x 10^5 cells; *lanes 3 to 5* cytoplasmic fractions from 2 x 10^5, 4 x 10^5 and 6 x 10^5 cells; *lanes 6 to 8* plasma membrane containing fractions from 5 x 10^5, 1 x 10^6 and 2 x 10^6 cells. The *numbers on the left* are molecular sizes of standard proteins in kilodaltons

Thus we conclude that there exist three intracellular pools of LSP1 protein: a cytoplasmic pool, which accounts for approximately 55% of the total LSP1 protein, a membrane-associated pool constituting approximately 30% of the total LSP1 protein, and a cytoskeleton-associated pool which accounts for approximately 10% of the total cellular LSP1 protein.

The activation of resting B-cells through cross-linking sIg molecules leads to rapid biochemical changes in the plasma membrane, which include enhanced phosphoinositide

317

turnover, cleavage of PIP_2 by phospholipase c which results in the production of IP_3 and diacylglycerol. IP_3 mediates the release of Ca^{2+} from intracellular stores into the cytoplasm which results in a transient increase of $[Ca^{2+}]_i$ from approximately 0.1 µM in resting cells to approximately 1 µM in activated B-cells. The membrane-soluble DG activates the enzyme PKC at the inner face of the plasma membrane. Neither the substrates for activated PKC nor the ultimate fate of the free cytoplasmic Ca^{2+} is well known. Ca^{2+} is an essential co-factor for PKC and Ca^{2+}/calmodulin-dependent protein kinases; however, protein substrates for these kinases which might be involved in the regulation or mediation of the sIg signal have not been described.

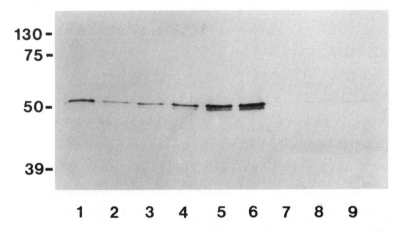

Fig. 8. LSP1 is associated with the cytoskeleton. The LSP1 content of varying amounts of different BAL17 fractions, prepared as described in the text was determined by Western blot analysis. *Lane 1* total cell extract from 1×10^5 cells; *lanes 2 to 6* NP-40 insoluble pellets from 5×10^5, 7.5×10^5, 1×10^6, 1.5×10^6 and 2×10^6 cells; *lanes 7 to 9* nuclear material from 5×10^6, 1×10^7 and 1.5×10^7 cells

Activation of B-cells by cross-linking the sIg molecules also results in a number of structural changes. An early response appears to be the attachment of the sIg molecule to the underlying cytoskeleton. The precise nature of this attachment is unknown, but does not appear to involve direct interaction between sIg and actin. When cross-linked cells are incubated at 37 °C the sIg will readily form a cap on one end of the cell in an active energy-dependent process. This capping phenomenon involves a rearrangement of the underlying cytoskeleton and several experiments argue in favor of the view that this rearrangement is a necessary intermediate in the sIg signal transduction pathway. First, both the capping of sIg and the activation of enhanced phosphoinositide turnover with its resultant increase in $[Ca^{2+}]_i$ and activation of PKC are dependent on the use of bivalent anti-Ig reagents. Second, it was shown that only those monoclonal anti-IgD antibodies which were able to cap sIgD were mitogenic, while those antibodies which could not cap sIgD were not.

Although at this point we can only speculate about the precise biological function of the LSP1 protein, several of its characteristics point towards a role in the sIg signal transduction pathway. First, its lymphocyte-specific expression pattern suggests that it is involved in a lymphocyte-specific process, which argues against a structural role of the LSP1 protein. Second, the subcellular localization of LSP1 on the inside of the plasma membrane and on the underlying cytoskeleton is significant, since several proteins known to be involved in signal transduction are located at this site. For instance, activated PKC and adenylate cyclase and the family of G proteins are located at the cytoplasmic side of the plasma membrane. Thirdly, the co-capping of LSP1 protein with sIg and its Ca^{2+}-modulated interaction with p43 suggests that LSP1 is a protein that can respond to the transient rise in intracellular Ca^{2+} shortly after activation of B-cells by cross-linking sIg. In this model, the association of LSP1 and p43 will be transiently disrupted after anti-Ig (and possibly antigen) stimulation. This in turn suggests a movement, or flow of information, with the LSP1 molecule being the mediating molecule. Thus we speculate that we have identified three components of a signal transduction pathway: the extracellular receptor, sIg; the second messenger, Ca^{2+}; and an intracellular mediator or effector molecule, LSP1.

Many questions remain to be answered before we have a clear understanding of the complete pathway. For instance, what is the identity of p43 and what are the consequences of the disruption of the LSP1/p43 complex? What is the ultimate fate of the LSP1 protein released from the complex, does it bind to other cellular proteins, and most importantly is the LSP1 protein a mediator or a regulator of sIg signaling? To answer some of these questions we are currently in the process of expressing mutant forms of the LSP1 protein in B-lymphoma lines and to determine the effect of these mutations on sIg capping, anti-Ig induced phosphoinositide turnover, induction of transient increases in $[Ca^{2+}]_i$ or activation of

PKC. These mutant LSP1 proteins will also allow us to accurately map the Ca^{2+}-binding sites in LSP1 and to determine its site of interaction with p43.

ACKNOWLEDGMENTS

This work was supported by grants from the National Cancer Institute of Canada and the Medical Research Council of Canada to J.J., who is a Scholar of the McLaughlin Foundation. J.J.-B. is the recipient of a postdoctoral fellowship award from the Leukemia Society of America, Inc.

REFERENCES

Albrecht DL, Noelle RJ (1988) Membrane Ig-cytoskeletal interactions. I. Flow cytofluorometric and biochemical analysis of membrane IgM-cytoskeletal interactions. J Immunol 141:3915-3922

Babu YS, Sack JS, Greenhough TJ, Bugg CE, Means AR, Cook WJ (1985) Three-dimensional structure of calmodulin. Nature 315:37-40

Baudier J, Gerard D (1983) Ions binding to S-100 proteins: structural changes induced by calcium and zinc on S-100a and S-100b proteins. Biochem 22:3360-3369

Bijsterbosch MK, Meade CJ, Turner GA, Klaus GGB (1985) B-lymphocyte receptors and polyphosphoinositide degradation. Cell 41:999-1006

Berg J (1986) Potential metal-binding domains in nucleic acid binding proteins. Science 232:485

Braun J, Unanue ER (1980) B lymphocyte biology studied with anti-Ig antibodies. Immunol Rev 52:3-28

Braun J, Hochman PS, Unanue ER (1982) Ligand-induced association of surface immunoglobulin with the detergent-insoluble cytoskeletal matrix of the B lymphocyte. J Immunol 128:1198-1204

Coggeshall KM, Cambier JC (1984) B cell activation. VIII. Membrane immunoglobulins transduce signals via activation of phosphatidylinositol hydrolysis. J Immunol 133:3382-3386

Cohen DI, Hedrick SM, D'Eustachio P, Ruddle F, Steinberg AD, Paul WE, Davis MM (1985) Isolation of a cDNA clone corresponding to an X-linked gene family (XLR) closely linked to the murine immunodeficiency disorder, xid. Nature 314:369-372

Collins JH, Greaser ML, Potter JD, Horn MJ (1977) Determination of the amino acid sequence of troponin c from rabbit skeletal muscle. J Biol Chem 252:6356-6362

DeFranco AL, Raveche ER, Asofsky RA, Paul WE (1982) Frequency of B lymphocytes responsive to anti-immunoglobulin. J Exp Med 155:1523-1536

Ellis L, Clausner E, Morgan DO, Edery M, Roth RA, Rutter WJ (1986) Replacement of insulin receptor tyrosine residues 1162 and 1163 compromises insulin-stimulated kinase activity and uptake of 2-deoxyglucose. Cell 45:721-723

Frank G, Weeds AG (1974) The amino-acid sequence of the alkali light chains of rabbit skeletal-muscle myosin. Eur J Biochem 44:317-334

Fullmer CS, Wasserman RH (1981) The amino acid sequence of bovine intestinal calcium-binding protein. J Biol Chem 256:5669-5674

Gershenfeld HK, Weissmann IL (1986) Cloning of a cDNA for a T cell-specific serine protease from a cytotoxic T lymphocyte. Science 232:854

Goroff DK, Stall A, Mond JJ, Finkelman FD (1986) In vitro and in vivo B lymphocyte-activating properties of monoclonal anti-delta antibodies. J Immunol 136:2382-2392

Hedrick SM, Cohen DI, Nielsen EA, Davis MM (1984) Isolation of cDNA clones encoding T cell-specific membrane-associated proteins. Nature 308:149-153

Hornbeck P, Paul WE (1986) Anti-immunoglobulin and phorbol ester induce phosphorylation of proteins associated with the plasma membrane and cytoskeleton in murine B lymphocytes. J Biol Chem 261:14817-14824

Howard M, Farrar J, Hilfiker M, Johnson B, Takatsu K, Hamaoka T, Paul WE (1982) Identification of a T-cell-derived B cell growth factor distinct from interleukin 2. J Exp Med 155:914-923

Jongstra J, Davis MM (1988) Molecular genetic analysis of mouse B lymphocyte differentiation. In: Minna JD, Kuehl M (eds) Cellular and molecular biology of tumors and potential clinical applications. UCLA Symposia 56:261-268

Jongstra J, Schall TJ, Dyer BJ, Clayberger C, Jorgensen J, Davis MM, Krensky AM (1987) The isolation and sequence of a novel gene from a human functional T cell line. J Exp Med 165:601-614

Jongstra J, Tidmarsh GF, Jongstra-Bilen J, Davis MM (1988) A new lymphocyte-specific gene which encodes a putative Ca^{2+}-binding protein is not expressed in transformed T lymphocyte lines. J Immunol 141:3999-4004

Jongstra J, Jongstra-Bilen J, Tidmarsh GF, Davis MM (1988a) The in vitro translation product of the murine lambda 5 gene contains a functional signal peptide. Mol Immunol 25:687-693

Jongstra-Bilen J, Young AJ, Chong R, Jongstra J (1990) Human and mouse LSP1 genes code for highly conserved phosphoproteins. J Immunol 144:1104-1110

Klein DP, Jongstra-Bilen J, Ogryzlo K, Chong R, Jongstra J (1989) Lymphocyte-specific Ca^{2+}-binding protein LSP1 is associated with the cytoplasmic face of the plasma membrane. Mol Cell Biol 9:3043-3048

Kretsinger RH (1979) The informational role of calcium in the cytosol. Adv Cycl Nucleotide Res 11:1-26

Kretsinger RH, Nockolds CE (1973) Carp muscle calcium-binding protein. II. Structure determination and general description. J Biol Chem 248:3313-3326

Maruyama K, Mikawa T, Ebashi S (1984) Detection of calcium binding proteins by ^{45}Ca autoradiography on nitrocellulose membrane after sodium dodecyl sulfate gel electrophoresis. J Biochem 95:511-519

Mizuguchi J, Utsunomiya N, Nakanishi M, Arata Y (1988) Phorbol myristate acetate inhibits increases in membrane fluidity induced by anti-IgM in B cells. J Immunol 140:2495-2499

Monroe JG (1988) Up-regulation of c-fos expression is a component of the mIg signal transduction mechanism but is not indicative of competence for proliferation. J Immunol 140:1454-1460

Myers CD, Kritz MK, Sullivan TJ, Vitetta ES (1987) Antigen-induced changes in phospholipid metabolism in antigen-binding B lymphocytes. J Immunol 138:1705-1711

Nakanishi K, Cohen DI, Blackman M, Nielsen E, Ohara J, Hamaoka T, Koshland ME, Paul WE (1984) Ig RNA expression in normal B cells stimulated with anti-IgM antibody and T-cell-derived growth and differentiation factors. J Exp Med 160:877-892

Osborn M, Weber K (1977) The detergent-resistant cytoskeleton of tissue culture cells includes the nucleus and the microfilament bundles. Exp Cell Res 106:339-349

Pozzan T, Arslan P, Tsien RY, Rink TJ (1982) Anti-immunoglobulin, cytoplasmic free calcium, and capping in B lymphocytes. J Cell Biol 94:335-340

Quigley JP (1976) Association of a protease (plasminogen activator) with a specific membrane fraction isolated from transformed cells. J Cell Biol 71:472-486

Ransom JT, Harris LK, Cambier JC (1986) Anti-Ig induces release of inositol 1,4,5-triphosphate, which mediates mobilization of intracellular Ca^{++} stores in B lymphocytes. J Immunol 137:708-714

Rosenberg AH, Lade BN, Chui D-S, Lin S-W, Dunn JJ, Studier FW (1987) Vectors for selective expression of cloned cDNAs by T7 RNA polymerase. Gene 56:125-135

Schall TJ, Jongstra J, Dyer BJ, Jorgensen J, Clayberger C, Davis MM, Krensky AM (1988) A human T cell-specific molecule is a member of a new gene family. J Immunol 141:1018-1025

von Heyne G (1983) Patterns of amino acids near signal sequence cleavage sites. Eur J Biochem 133:17-21

Weiss A, Imboden J, Shoback D, Stobo J (1984) Role of T3 surface molecules in human T-cell activation: T3-dependent activation results in an increase in cytoplasmic free calcium. Proc Natl Acad Sci USA 81:4169-4173

Weiss MJ, Daley JF, Hodgdon JC, Reinherz EL (1984) Calcium dependency of antigen-specific (T3-Ti) and alternative (T11) pathways of human T-cell activation. Proc Natl Acad Sci USA 81:6836-6840

Wilkinson JM (1980) Troponin c from rabbit slow skeletal and cardiac muscle is the product of a single gene. Eur J Biochem 103:179-188

NONMUSCLE α-ACTININ IS AN EF-HAND PROTEIN

Walter Witke, Angelika A. Noegel, and Michael Schleicher

INTRODUCTION

In 1965, Ebashi and Ebashi isolated an F-actin crosslinking protein from muscle cells which they called α-actinin. In addition to actin, myosin and tropomyosin it was the fourth component so far found in muscle cells. Immunofluorescence studies on striated muscle showed that α-actinin is localized in the Z-lines which separate the sarcomers and anchor the actin filaments. Therefore, a role for α-actinin in linking actin filaments to the Z-line has been discussed. It is generally assumed that α-actinin strengthens the structure and helps the sarcomeric unit to withstand the force generated by the actin-myosin contraction. Whether the in vitro-enhanced Mg-ATPase activity of actomyosin in the presence of α-actinin (Maruyama and Ebashi 1965) plays a significant role in vivo is not clear. Ten years later Lazarides and Burridge (1975) identified α-actinin also in nonmuscle cells. Like muscle α-actinin, its main activity in vitro is the gelation of an F-actin solution by cross linking actin filaments under certain conditions. The finding that α-actinin can be incorporated into artificial membranes and could form trimeric complexes with actin in the presence of diacylglycerol and palmitic acid (Burn et al. 1985) makes it conceivable that α-actinin might anchor actin filaments to the plasma membrane of nonmuscle cells. Since then several isoforms of α-actinin have been isolated from skeletal and smooth muscle (Feramisco and Burridge 1980; Endo and Masaki 1984) or from vertebrate nonmuscle cells like brain (Duhaiman and Bamburg 1984), macrophages (Bennett et al. 1984), platelets (Landon et al. 1985) and fibroblasts (Burridge and Feramisco 1981).

Also from the lower eukaryotes *Dictyostelium discoideum* and *Acanthamoeba castellanii*, α-actinins could be isolated (Condeelis and Vahey, 1982; Fechheimer et al. 1982; Pollard 1981). Molecular size, biochemical characteristics, the elongated structure, and the cross reaction of polyclonal antibodies indicated the close relationship of the α-actinins from different organisms and tissues. This was finally confirmed by sequencing the genes that code for α-actinin from *D. discoideum* (Witke et al. 1986; Noegel et al. 1987), chicken smooth muscle (Arimura et al.,1988), and chicken fibroblast (Baron et al. 1987). One previously described difference between muscle and nonmuscle α-actinin was the strict regulation of the latter by submicromolar Ca^{2+}-concentrations (Burridge and Feramisco 1981). The primary sequence of *D. discoideum* α-actinin revealed the structural basis for the Ca^{2+}-sensitivity of nonmuscle α-actinin. Two complete EF-hand domains closely related to the Ca^{2+}-binding sites in calmodulin were identified (Noegel et al. 1987). The importance of these EF-hand structures for the Ca^{2+}-binding of α-actinin as well as for the cross linking activity was demonstrated by mutation of the EF-hand domains (Witke unpubl.). *D. discoideum* mutants and transformants are available that lack α-actinin either after mutation with nitrosoguanidine (Wallraff et al. 1986; Schleicher et al.; 1988) or after gene disruption via homologous recombination (Witke et al. 1987; Noegel and Witke 1988).

NONMUSCLE α-ACTININ FROM *D.DISCOIDEUM*

D. discoideum is a highly motile amoeba which feeds on bacteria as a free living cell and develops into a multicellular organism as soon as the cells begin to starve. *D. discoideum* α-actinin is a representative of nonmuscle α-actinin with respect to its Ca^{2+}-regulation (Condeelis and Vahey 1982; Fechheimer et al. 1982). Detailed characterization revealed that Ca^{2+} concentrations higher than 0.1 μmol completely inhibit the F-actin cross linking activity. A significant loss of activity could also be observed at pH values above 7. The native α-actinin molecule is a homodimer whose subunits of 95 kDa are arranged in an antiparallel fashion. After rotary shadowing, α-actinin is visualized in the electron microscope as a rod-shaped molecule with a length of about 35 nm (Wallraff et al. 1986). In *D. discoideum* the cellular distribution of α-actinin is diffuse throughout the cell, though some enrichment can be observed in the cortical region where it is collocated with actin filaments and actin bundles.

α-ACTININ CONSISTS OF THREE HIGHLY CONSERVED DOMAINS

D. discoideum α-actinin is encoded by a single copy gene which is separated into three exons (Witke et al. 1986; Witke and Noegel 1990). Only the first 26 amino acids are located in exon I and do not show a striking homology to other α-actinins. On the second exon the putative F-actin-binding site and the region responsible for dimerization are encoded. With the exception of the first four amino acids of the E-helix in the first EF-hand, exon III includes the two EF-hand domains (Figure 1). The intron interrupts the triplet that codes for amino acid five of the first EF-hand. This resembles the intron structure of other calcium-binding proteins like chicken calmodulin, sea urchin specI protein, and mouse myosin light chain 3 (Berchtold et al. 1987). Whether a similar organization of functional domains in distinct exons is also true for other α-actinins has to await the analysis of the genomic sequences of these genes.

Taken together the analysis of the deduced amino acid sequence suggests that α-actinin consists of at least three domains distinct in structure and function:
1. an aminoterminal putative F-actin binding domain,
2. four internal repeats of about 120 amino acids which might be involved in dimerization,
3. a carboxyterminal domain with two EF-hand structures.

THE F-ACTIN BINDING DOMAIN

The amino terminal sequence of *D. discoideum* α-actinin contains a stretch of approximately 250 amino acids which is highly conserved in other muscle and nonmuscle α-actinins (Noegel et al. 1987; Baron et al. 1987). This region shows about 80% similarity as compared to chicken smooth muscle α-actinin. Moreover, the conserved region is not only found in α-actinins but also present in other cytoskeletal proteins like the *D. discoideum* 120 kDa gelation factor (Noegel et al. 1989), *Drosophila* ß-spectrin (Byers et al. 1989) and dystrophin (Hammonds 1987; Koenig et al. 1988). In dystrophin, exon 5 comprises the putative actin-binding site and it is tempting to speculate that proteins belonging to the α-actinin superfamily have received their common actin-binding site by exon shuffling (Figure 2).

Because of the strict conservation of the amino terminal sequence in various α-actinins and other cytoskeletal proteins, we tried to prove that this region represents the binding site for F-actin. Therefore, we expressed a cDNA fragment containing this conserved region (Figure 2) in *E. coli* and tested the partially purified protein in an F-actin sedimentation assay for activity. The expressed protein bound indeed to F-actin (Witke et al. submitted). Recently, we were able to show that the actin-binding site resides in a polypeptide of 76 amino acids (Witke unpubl.).

THE DOMAIN INVOLVED IN DIMERIZATION

The rod-like structure of α-actinin is formed by a fourfold repeat structure which represents the central portion of the molecule (Baron et al. 1987). In analogy to the repeats found in spectrin and dystrophin, α-actinin contains four repeats of about 120 amino acids with a high α-helical potential. In chicken fibroblast α-actinin the described four repeats are fairly similar to spectrin, whereas in *D. discoideum* α-actinin the third and fourth repeat, although present, are not as well pronounced as the first and second. It is questionable whether the similarity of the repetitive regions in α-actinin and spectrin reflect a real homology or if a common secondary structure motif causes some analogy between the central domains of both proteins. Recently Cross et al. (1990) predicted a triple coiled-coil structure for the central domains of dystrophin. This model differs from the Speicher-Marchesi model for spectrin (1984), where internal triple helix repeats are arranged in phase. Cross and co-workers propose that the repeating structure motif in dystrophin is a bead of triple coiled-coil which is out of phase. This structure could confer elasticity to the molecule.

By rotary shadowing and electron microscopy it could be shown (Wallraff et al. 1986) that a monoclonal antibody directed against the carboxyterminal part of α-actinin (Schleicher et al. 1988) bound to both ends of the native dimer. This observation could only be explained by an antiparallel arrangement of the subunits. However, the exact interaction between the subunits is not known. Imamura and co-workers (1988) carried out cross linking experiments with a dimeric 55 kDa proteolytic fragment containing all four repeats.

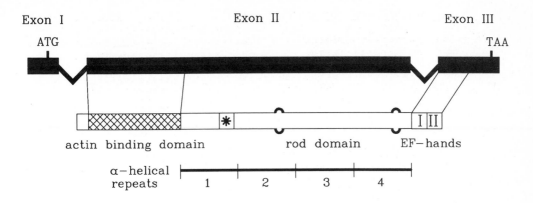

Fig. 1. Exon localization of functional domains of *D. discoideum* α-actinin. The exons are indicated by *filled bars*, introns are drawn as *interruptions*. The *cross-hatched region* represents the actin-binding domain, whereas the rod domain is shown as an *open box*. A proline-rich sequence which is conserved among different α-actinins, ß-spectrin and dystrophin (-Lys-Pro-Pro-Lys-), is indicated by an *asterisk*. The predicted flexible regions are drawn as *arcs*

After further cleavage of the cross linked material they identified an aminoterminal 30 kDa fragment crosslinked to a 16 kDa carboxy terminal fragment. From these data one would suggest that a site in repeat 1 or 2 of one subunit interacts with repeat 4 of the other subunit.

α-ACTININ CONTAINS TWO *EF*-HAND STRUCTURES CLOSE TO THE CARBOXY TERMINUS

Close to the carboxy terminus of *D. discoideum* α-actinin we identified two EF-hand structures similar to the Ca^{2+}-binding domains in calmodulin. In Figure 3 the two EF-hands from *D. discoideum* α-actinin are compared to the Ca^{2+}-binding domains of known EF-hand proteins.

```
           1                                                        50
DDA    TFTAWLNSHLRKLGSS-IEQIDTDFTDGIKLAQLLEVISNDPVFKVNKTP
DDG    ***G*A*NY*-*ERILK**DLA*SLE**VL*IN**I**SKKIL*Y**A*
CSM    *****C******A*TQ-**N*EE**R**L**ML******GERLA*PERG-
CSK    *****C******A*TQ-**N*EE**RN*L**ML******GERL**PDRG-
CFB    -****C******A*TQ-**N*EE**R**L**ML******GERLA*PERG-
DBS    ***K*V****CRVNCR-*ADLYV*MR**KH*IK****L*GERL**PT*GK
HDY    ***K*V*AQFS*F*KQH**NLFS*LQ**RR*LD***GLTGQKL**-E*GS
ABP    ***R*C*E**KCVSKR-*ANLQ**LS**LR*IA****L*QKKMR*H*QR*

           51                                                       100
DDA    -KLRRIHN--IQNVGLCLKHIESHGVKLVGIGAEELVDKNLKMTLGMIWT
DDG    -*I*MQ-KIE--*NNMAVNF*K*E*L*******DI**SQ**LI**L***
CSM    -*M*V-*K--S**NKA*DF*A*K*****S*****I*G*V*********
CSK    -*M*F-*K--*A**NKA*DY*A*K*****S*****I*G*V*********
CFA    -*M*-V*K--S**NKA*DF*A*K*****S*****I*G*V*********
DBS    --M***CL---E**DKA*QFLREQR*H*EN**SHDI**G*ASLN**L***
HDY    --T*V-*A--LN**NKA*RVLQNNN*D**N**STDI**G*H*L***L**N
ABP    -TF*QM-QLE--**SVA*EFLDRESI***S*DSKAI**G***LI**L***

           101                                                      150
DDA    IILRFAIQDI-------SIEEL------SAKEA-----LLL-WCQRK--T
DDG    L***YQ**MS------E*DN--------*P*A*-----**-E*V-**QV-
CSM    **********-------*V**T------***G-----**-*****--*
CSK    **********-------*V**T------***G-----**-*****--*
CFA    **********-------*V**T------***G-----**-*****--*
DBS    *****Q***-------T***VDNKETK***D*-----*LL-***M*--*
HDY    ***HWQVKNVMKNIMAGLQQTN------*E*-------I**S*V-*-QS*
ABP    L**HYS*SM--PMWDEEED*--------E**KQTPKQR**-G*I*

           151                                                      200
DDA    EGYDRVKVGNFHTSFQDGLAFCALIHKH----RPDLINFDSLNKDDKAGN
DDG    AP*KV*-*N**TD*WC**RVLS**----TDSLK*GVREMSTLTG*AVQDI
CSM    AP*KN*NIQ***I*WK***G*****R-----**E**DYGK*R***PLT*
CSK    AP*RN*NIQ***L*WK***GL****R-----*****DYSK****PI**
CFA    AP*KN*NIQ***I*WK***G*****R-----**E**DYGK*R***PLT*
DBS    A**HN*N*R**T**WR****VN*I***----***VQ*EK*S*TNAIH*
HDY    RN*PQ*N*I**T**WS****LN****S-----****FDWN*VVCQQS*TQ

           201                                                      242
DDA    -LQLAFDIA-EK-ELDIPKMLDVSDMLDVVRPDERSVMTYVA
DDG    DRSM--***L*EY--E***IM*AN**-NSL-***L**I***S
CSM    -*NT***V-**Y-*****-**AE*IVGTA****KAI****S
CSK    -IN**ME**-**-H********AE*IVNTPK****AI****S
CFA    -*NT***V-**Y-********AE*IVGTA****KAI****S
DBS    -*NN***V*-*D-K*GLA*L**AE*VF-*EH***K*II***V
HDY    R*EH**N**R--YQ*G*E*L**PE*-V*TTY**KK*ILM*IT
```

Fig. 2. Alignment of the actin-binding domains of different cytoskeletal proteins. The amino acid sequence of the actin-binding domain of *D. discoideum* α-actinin (DDA, Noegel et al. 1987) is compared to the sequence of the *D. discoideum* gelation factor (DDG, Noegel et al. 1989), chicken smooth muscle α-actinin (CSM, Baron et al. 1987), chicken skeletal muscle α-actinin (CSK, Arimura et al. 1988), chicken fibroblast α-actinin (CFB, Arimura et al. 1988), human dystrophin (HDY, Koenig et al. 1988), and human endothelial actin-binding protein (ABP, Gorlin et al. 1990). Identical amino acids are indicated by an *asterisk*

According to the model of Kretsinger (1980), the α-actinin EF-hands are complete, i.e., 12 or more out of 16 amino acids of the consensus sequence for an EF-hand domain are present (Noegel et al. 1987). In *D. discoideum* α-actinin both EF-hands should be functional by these criteria, the first EF-hand scores 12, the second EF-hand 14 amino acids.For the nonmuscle macrophage α-actinin a stoichiometry of two bound Ca^{2+}-ions per subunit were determined (Bennett et al. 1984). Whether both EF-hands in *D. discoideum* α-actinin bind Ca^{2+} is not known so far. Currently, we are studying the function of the two EF-hand structures by altering the Ca^{2+}-liganding loop region. First results indicate that only the second EF-hand contains a high affinity Ca^{2+} binding site because an insertion into the second EF-hand reduces the Ca^{2+}-binding dramatically (Figure 6).

MODEL OF THE *D. DISCOIDEUM* α-ACTININ

Combining the results on the structural domains with biochemical and electron microscopical data, a picture emerges how α-actinin might function and how it is regulated by Ca^{2+}. An investigation of the α-actinin primary sequence with the method of Garnier-Osguthorpe-Robson (1978) revealed two highly flexible regions.

The first one is located 150 amino acids away from the F-actin binding site within the second repeat, the second region of enhanced flexibility resides in the fourth repeat immediately in front of the two EF-hand domains. If the flexible regions of the neighboring antiparallel subunits are opposite to each other, then this arrangement has three consequences for the structure of α-actinin:

1. The flexible regions in the dimer form a hinge at each end of the molecule, separating the actin binding sites and Ca^{2+}-binding domains as units from the central rod.

2. The two subunits would be staggered by about 1.5 repeats.

3. Therefore, the Ca^{2+}-binding domains would not be directly opposite to the actin binding sites but still very close.

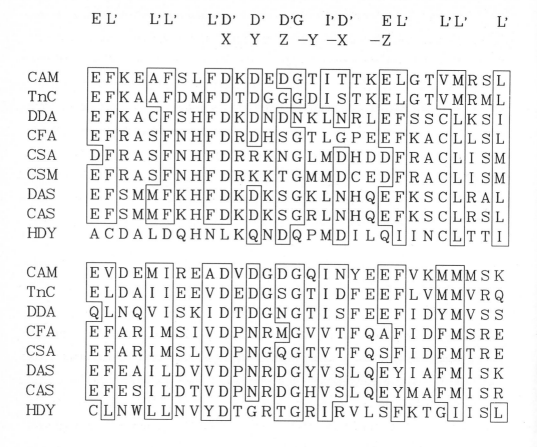

Fig. 3. Comparison of EF-hand domains from known and putative Ca²⁺-binding proteins. According to the EF-hand model for Ca²⁺-modulated proteins there are 16 characteristic positions building an EF-hand structure (Kretsinger 1980). The functionally important amino acids are listed in the *first line* as L' (hydrophobic residues), as D' (hydrophilic amino acids), as E, G and I' (I or V). Important for Ca²⁺ binding are the liganding oxygens that correspond to the octahedral vertices (*second line*). The *upper* and the *lower panel* align respectively calmodulin (CAM, *Trypanosoma brucei gambiense*, residues 11-39 and 120-148: Tschudi et al. 1985), troponin C (TnC, *Rana esculenta*, residues 21-49 and 57-85: Van Erd et al. 1978), *D. discoideum* α-actinin (DDA, residues 734-762 and 770-798: Noegel et al. 1987), chicken fibroblast α-actinin (CFA, residues 714-742 and 755-783: Arimura et al. 1988), chicken smooth muscle α-actinin (CSA, residues 750-778: Baron et al. 1987), chicken skeletal muscle α-actinin (CSM, residues 760-788 and 796-824: Arimura et al. 1988), *Drosophila* α-spectrin (DAS, residues 2269-2297 and 2312-2340: Dubreuil et al. 1989), chicken brain α-spectrin (CAS, residues 2332-2360 and 2375-2403: Wasenius et al. 1989), and human dystrophin (HDY, residues 3129-3157 and 3178-3206, Koenig et al. 1987).

From our point of view a certain flexibility in the α-actinin is of importance for several reasons. First, the actin filaments can easily slide apart and even weak forces at the actin filaments would disrupt the cross linking by α-actinin if it were a rigid molecule. Second, the Ca^{2+}-regulation needs a direct interaction of the EF-hands from one subunit with the neighboring actin-binding site of the other subunit.The dimerization itself is not affected by Ca^{2+}. Therefore dissociation of the dimer is not the reason for inhibition of the cross linking activity. In the Ca^{2+}-saturated state the EF-hands presumably change their conformation, thereby forcing the actin binding sites into a position where they can no longer crosslink actin filaments (Figure 4). In troponin C an "opening" of the EF-hands from a central helix has been observed after Ca^{2+} binding (Herzberg and James 1985). We suggest a similar conformational change in the α-actinin EF-hands which pulls the actin-binding site into a "closed" conformation. A certain flexibility in the head regions of the α-actinin molecule is a prerequisite for this mechanism. A proline-rich sequence in the first repeat of the rod domain probably forms a sharp bend between the actin-binding site and the flexible hinge region which seems to be an important structure for the function of the molecule. All so far known F-actin cross linking proteins belonging to the α-actinin superfamily contain the proline-rich motif "-Lys-Pro-Pro-Lys-". In the heterodimeric spectrin molecule this motif can be found in the ß-subunit in a similar position as in α-actinin. Even in human dystrophin this motif is conserved with exception of the first "Lys" which is changed into an "Arg". In the more detailed model drawn in Figure 4, we propose an explanation for the Ca^{2+}-regulation of nonmuscle α-actinin fitting the current information into the original suggestions made by Noegel et al. (1987).

MOLECULAR GENETICS AS A TOOL TO ELUCIDATE THE IN VIVO ROLE OF α-ACTININ

D. discoideum is for several reasons a good model system for studying the role of actin-binding proteins at the molecular basis. As a haploid organism it is suitable for genetic manipulations. Since a reliable transformation system has been established (Nellen et al. 1984), gene disruption (DeLozanne and Spudich 1987; Witke et al. 1987; Noegel and Witke 1988), gene replacement (Manstein et al. 1989) and antisense RNA inactivation (Knecht and Loomis 1987) can be used as tools for investigating the function of a certain protein.

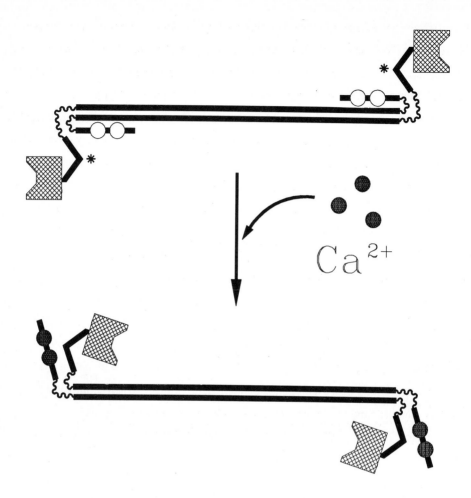

Fig. 4. Model of *D. discoideum* α-actinin. In the α-actinin homodimer, the subunits are arranged in an antiparallel fashion. The carboxy terminal EF-hands shown as *open circles* are folded back in the absence of Ca^{2+} and allow binding of the actin-binding sites (*cross hatched*) to actin filaments. *Asterisks* indicate the position of the proline-rich sequence "-Lys-Pro-Pro-Lys-", which could induce a sharp bend. After Ca^{2+} binding, the EF-hand domains change their conformation which forces the actin-binding sites of the other subunit into a position where they can no longer bind to F-actin. This conformational change might be facilitated by flexible regions

We used *D. discoideum* as an expression system for an altered α-actinin that has been modified in the second Ca^{2+}-binding domain by insertion of a reporter oligonucleotide coding for 13 amino acids from a Sendai virus protein (Figure 5). A monoclonal antibody against the 13 amino acid epitope was available (Einberger et al. 1990). We used two strategies to express the modified α-actinin in *D. discoideum*.First, the construct was expressed in an α-actinin deficient mutant under the control of the actin-6 promoter. Second, we introduced the mutated gene into wild-type cells. By homologous recombination it inserted into the endogenous copy, thereby inactivating the resident gene. The modified α-actinin was purified from transformants and biochemically characterized. As shown in Figure 6, only minute Ca^{2+}-binding could be detected in the mutant protein while the wild type α-actinin bound Ca^{2+} as expected. The properties of the mutant α-actinin on gel filtration columns was indistinguishable from wild type α-actinin indicating that the mutant α-actinin exists as a homodimer. However, the mutant α-actinin did not show an F-actin crosslinking activity. We concluded from these data that the insertion of 13 amino acids into the EF-hand mimics Ca^{2+}-binding by inducing a similar conformational change as the Ca^{2+}-binding does.

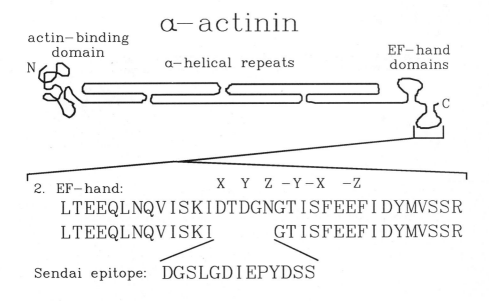

Fig. 5. Mutation of the α-actinin Ca^{2+}-binding site. In the model the two EF-hands of a-actinin are drawn as loops. An oligonucleotide coding for 13 amino acids of a Sendai virus protein was inserted into the cDNA sequence of the second EF-hand. The liganding amino acids are interrupted by the Sendai sequence

332

With *D. discoideum* as a model system we can now ask questions how point mutations in the EF-hand alter the activity of α-actinin or whether one or both EF-hands can bind Ca²⁺. The expression of mutated or truncated α-actinin will be helpful to prove the predicted α-actinin model.

HOW DO Ca²⁺-REGULATED ACTIN BINDING PROTEINS TRANSFER SIGNALS TO THE CYTOSKELETON?

Cellular processes like motility, cytokinesis, phagocytosis, secretion etc. are accompanied by a reorganization of the cytoskeleton. External signals trigger the chemotactic response of motile cells like, for example, macrophages, granulocytes, fibroblasts, and also *D. discoideum*, which shares several features with motile cells of the immune system. The mobilization of Ca²⁺ from internal stores or the influx of external Ca²⁺ is one major event in the second messenger cascade.

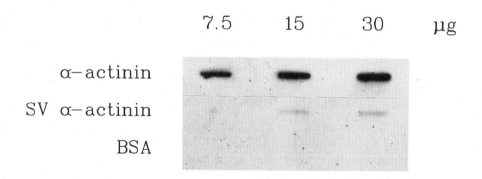

Fig. 6. Ca²⁺-binding of wild-type α-actinin and mutated α-actinin. The mutated α-actinin (*SV α-actinin*) was generated by inserting an oligonucleotide coding for 13 amino acid from a Sendai virus protein into the second EF-hand. Different amounts of proteins produced in *D. discoideum* were bound to nitrocellulose and the filter incubated with ⁴⁵Ca. α-Actinin binds ⁴⁵Ca whereas the mutated SV α-actinin binds only minor amounts. As a control BSA is loaded in the lower panel

Since during stimulation the intracellular Ca^{2+} concentration increases by two or three orders of magnitude, most Ca^{2+}-binding proteins are affected in their activity. Besides α-actinin several other actin binding proteins respond to increased Ca^{2+} concentrations (Pollard and Cooper 1986; Stossel et al. 1985). Severin from *D. discoideum* is activated by micromolar concentrations of Ca^{2+} (Brown et al. 1982), it is closely related to gelsolin (André et al. 1988) and can fragment actin filaments, cap the fast growing ends of actin filaments, nucleate actin polymerization and bind to monomeric actin. Interestingly this protein has no obvious Ca^{2+} binding domain like an EF-hand. The different functions of this protein correspond to domains which have been investigated in detail (Eichinger et al. submitted).

The activity of both severin and α-actinin is regulated by Ca^{2+} in such a way that together they could favor the formation of pseudopods. After an increase of the Ca^{2+}-concentration, the inhibition of α-actinin releases actin filaments from actin bundles and the activation of severin leads to a fragmentation of the actin filaments into short pieces, capped by a severin molecule. Contraction in any other part of the cell would push the cytoplasm into this region of weakest resistance. Different myosins were found in *D. discoideum* which could be responsible for a contraction after stimulation of the cell (Korn and Hammer 1988). As soon as the pressure of the cytoplasm expulses the plasma membrane, actin polymerization in this region could be initiated by the nucleation activity of severin. Severin itself can accelerate the rate-limiting step of actin polymerization, but an even more powerful nucleation occurs at the plasma membrane: the fragmented actin filaments capped by severin at the fast growing end would rapidly diffuse to the plasma membrane and interact with membrane bound actin binding proteins (Wuestehube and Luna 1987; Scheel et al. 1989) or with the membrane itself. PIP2 is able to inhibit severin/actin-interaction (Yin et al. 1990) thereby creating free fast growing ends available for the addition of actin monomers. An enormous number of actin filaments will emanate from the plasma membrane while the Ca^{2+} concentration decreases again due to redistribution into intracellular stores. α-Actinin is re-activated and bundles actin filaments by its cross linking activity. At this point, the function of α-actinin would be the arrangement of actin filaments into an ordered structure which stabilizes the newly formed pseudopod. Rigidity of the pseudopod and the contact of the pseudopod to the surface now would allow the cell to follow in this direction.

This outlined mechanism of pseudopod extension is an oversimplification with respect to the number of involved regulatory proteins and the second messengers influencing this process. Although the model gives an idea how Ca^{2+} regulated actin-binding proteins could play together in modulating the actin cytoskeleton, it is certainly more complicated. Since *D. discoideum* mutants were isolated which lack either severin (André et al. 1989), α-actinin (Wallraff et al. 1986; Witke et al. 1987; Schleicher et al. 1988) or the 120 kDa gelation factor (Brink et al. submitted) but were not strongly affected in motility and chemotactic response it seems that the regulatory actin binding proteins form a redundant network where other proteins with similar functions can substitute the deleted ones (Bray and Vasiliev 1989). There are several other cross linking proteins known in *D. discoideum*: A 30 kDa bundling protein found in *filopodia* (Fechheimer 1987), a 50 kDa bundling protein called ABP 50 (Demma et al. 1990) and proteins similar to filamin and fodrin (Hook and Condeelis 1987; Bennet and Condeelis 1988). Capping proteins which inhibit the elongation of actin filaments and nucleate actin polymerization have also been described in *D. discoideum* (Schleicher et al. 1984; Hartmann et al. 1989). Contrary to the well-defined Ca^{2+}-regulation of α-actinin, only little information is available as to how the activity of these proteins is modulated.

REFERENCES

André E, Lottspeich F, Schleicher M, Noegel A (1988) Severin, gelsolin and villin share a homologous sequence in regions presumed to contain F-actin severing domains. J Biol Chem 263:722-727

André E, Brink M, Gerisch G, Isenberg G, Noegel A, Schleicher M, Segall JE, Wallraff E (1989) A Dictyostelium mutant deficient in severin, an F-actin fragmenting protein, shows normal motility and chemotaxis. J Cell Biol 108:985-995

Arimura C, Suzuki T, Yanagisawa M, Immamura M, Hamada Y, Masaki T (1988) Primary structure of chicken skeletal muscle and fibroblast α-actinin deduced from cDNA sequences. Eur J Biochem 177:649-655

Baron MD, Davison MD, Jones P, Critchley DR (1987) The sequence of chick α-actinin reveals homologies to spectrin and calmodulin. J Biol Chem 262:17623-17629

Bennet H, Condeelis J (1988) Isolation of an immunoreactive analogue of brain fodrin that is associated with the cell cortex of *Dictyostelium amoebae*. Cell Motil Cytoskeleton 11:303-317

Bennett JP, Zaner KS, Stossel TP (1984) Isolation and some properties of macrophage a-actinin: evidence that it is not an actin gelling protein. Biochemistry 23:5081-5086

Berchtold MW, Epstein P, Beaudet AL, Payne ME, Heizmann CW, Means AR (1987) Structural organization and chromosomal assignment of the parvalbumin gene. J Biol Chem 262:8696-8701

Bray D, Vasiliev J (1989) Networks from mutants. Nature 338:203-204

Brink M, Gerisch G, Isenberg G, Noegel AA, Segall JE, Wallraff E, Schleicher M (1990) A *Dictyostelium* mutant lacking an F-actin crosslinking protein, the 120 kD gelation factor. J Cell Biol (in press)

Brown S, Yamamoto K, Spudich JA (1982) A 40.000 dalton protein from *Dictyostelium discoideum* affects assembly properties of actin in a Ca^{2+}-dependent manner. J Cell Biol 93:205-210

Burn P, Rotman A, Mayer RK, Burger MM (1985) Diacylglycerol in large α-actinin/actin complexes in the cytoskeleton of activated platelets. Nature 314:469-472

Burridge K, Feramisco JR (1981) Non-muscle α-actinins are calcium sensitive actin binding proteins. Nature 294:565-567

Byers TJ, Husain-Chishti A, Dubreuil RR, Branton D, Goldstein LSB (1989) Sequence similarity of the amino-terminal domain of *Drosophila* beta-spectrin to α-actinin and dystrophin. J Cell Biol 109:1633-1641

Condeelis J, Vahey M (1982) A calcium and pH-regulated protein from *Dictyostelium discoideum* that cross-links actin filaments. J Cell Biol 94:466-471

Cross RA, Stewart M, Kendrick-Jones J (1990) Structural predictions for the central domain of dystrophin. FEBS Lett 262:87-92

DeLozanne A, Spudich JA (1987) Disruption of the *Dictyostelium* myosin heavy chain gene by homologous recombination. Science 236:1086-1091

Demma M, Warren V, Hock R, Dharmawardhane S, Condeelis J (1990) Isolation of an abundant 50.000-dalton actin filament bundling protein from *Dictyostelium amoebae*. J Biol Chem 265:2286-2291

Dubreuil RR, Byers TJ, Sillman AL, Bar-Zvi D, Goldstein LSB,, Branton D (1989) The complete sequence of *Drosophila* alpha-spectrin: Conservation of structural domains between alpha-spectrin and alpha-actinin. J Cell Biol 109:2197-2205

Duhaiman AS, Bamburg JR (1984) Isolation of brain α-actinin. Its characterization and comparison of its properties with those of muscle α-actinins. Biochemistry 23:1600-1608

Ebashi S, Ebashi F (1965) α-Actinin, a new structural protein from striated muscle. J Biochem 58:7-12

Eichinger L, Noegel AA, Schleicher M (1990) Domain structure in actin-binding proteins: Expression and functional characterization of truncated severin. (submitted)

Einberger H, Mertz R, Hofschneider PH, Neubert WJ (1990) Purification, renaturation and reconstituted protein kinase activity of the Sendai virus large (L) protein: L protein phosphorylates the NP and P proteins in vitro. J Virol (in press)

Endo T, Masaki T (1984) Molecular properties and functions in vitro of chicken smooth muscle α-actinin in comparison with those of striated muscle α-actinins. J Biochem 92:1457-1468

Fechheimer M (1987) The *Dictyostelium discoideum* 30.000-dalton protein is an actin filament-bundling protein that is selectively present in filopodia. J Cell Biol 104:1539-1551

Fechheimer M, Brier J, Rockwell M, Luna EJ, Taylor DL (1982) A calcium and pH-regulated actin binding protein from *D. discoideum*. Cell Motil 2:287-308

Feramisco JR, Burridge K (1980) A rapid purification of α-actinin, filamin and 130000 dalton protein from smooth muscle. J. Biol. Chem 225:1194-1199

Garnier J, Osguthorpe DJ, Robson B (1978) Analysis of the accuracy and implications of simple methods for predicting the secondary structure of globular proteins. J Mol Biol 120:97-120

Gorlin JB, Yamin R, Egan S, Stewart M, Stossel TP, Kwiatkowski DJ, Hartwig JH (1990) Human endothelial actin-binding protein (ABP, nonmuscle filamin): A molecular leaf spring. (submitted)

Hammonds RG (1987) Protein sequence of DMD gene is related to actin binding domain of α-actinin. Cell 51:1

Hartmann H, Noegel AA, Eckerskorn C, Rapp S, Schleicher M (1989) Ca^{2+}-independent F-actin capping proteins: Cap 32/34, a capping protein from *Dictyostelium discoideum*, does not share sequence homologies with known actin-binding proteins. J Biol Chem 264:12639-12647

Herzberg O, James MNG (1985) Structure of the calcium regulatory muscle protein troponin C at 2.8 Å resolution. Nature 313:653-659

Hock RS, Condeelis JS (1987) Isolation of a 240-kilodalton actin-binding protein from *Dictyostelium discoideum*. J Biol Chem 262:394-400

Imamura M, Endo T, Kuroda M, Tanaka T, Masaki T (1988) Substructure and higher structure of chicken smooth muscle α-actinin molecule. J Biol Chem 263:7800-7805

336

Knecht DA, Loomis WF (1987) Antisense RNA inactivation of myosin heavy chain gene expression in *Dictyostelium discoideum*. Science 236:1081-1086

Koenig M, Monaco AP, Kunkel LM (1988) The complete sequence of dystrophin predicts a rodshaped cytoskeletal protein. Cell 53:219-228

Korn ED, Hammer III JA (1988) Myosins of nonmuscle cells. Ann Rev Biophys Chem 17:23-45

Kretsinger RH (1980) Structure and evolution of calcium modulated proteins. CRC Crit Rev Biochem 8:119-174

Landon F, Gache Y, Touitou H, Olomucki A (1985) Properties of two isoforms of human blood platelet α-actinins. Eur J Biochem 153:231-237

Lazarides E, Burridge K (1975) α-Actinin: immunofluorescent localization of a muscle structural protein in non-muscle cells. Cell 6:289-298

Manstein DJ, Titus MA, DeLozanne A, Spudich JA (1989) Gene replacement in Dictyostelium: generation of myosin null mutants. EMBO J 8:923-932

Maruyama K, Ebashi S (1965) α-Actinin, a new structural protein from striated muscle. J Biochem 58:13-19

Nellen W, Silan C, Firtel R (1984) DNA mediated transformation in *Dictyostelium discoideum*: Regulated expression of an actin gene fusion. Mol Cell Biol 4:2890-2898

Noegel AA, Witke W (1988) Inactivation of the α-actinin gene in *Dictyostelium*. Dev Genet 9:531-538

Noegel A, Witke W, Schleicher M (1987) Calcium sensitive non muscle α-actinin contains EF-hand structures and highly conserved regions. FEBS Lett 221:391-396

Noegel AA, Rapp S, Lottspeich F, Schleicher M, Stewart M (1989) The *Dictyostelium* gelation factor shares a putative actin-binding site with α-actinins and dystrophin and also has a rod domain containing six 100-residue motifs that appear to have a cross-beta conformation. J Cell Biol 109:607-618

Pollard TD (1981) Purification of a calcium-sensitive actin gelation protein from *Acanthamoeba*. J Biol Chem 256:7666-7670

Pollard TD, Cooper JA (1986) Actin and actin-binding proteins. A critical evaluation of mechanisms and functions. Ann Rev Biochem 55:987-1035

Scheel J, Ziegelbauer K, Kupke T, Humbel BM, Noegel AA, Gerisch G, Schleicher M (1989) Hisactophilin, a histidine-rich actin-binding protein from *Dictyostelium discoideum*. J Biol Chem 264:2832-2839

Schleicher M, Gerisch G, Isenberg G (1984) New actin binding proteins from *Dictyostelium discoideum*. EMBO J 3:2095-2100

Schleicher M, Noegel A, Schwarz T, Wallraff E, Brink M, Faix J, Gerisch G, Isenberg G (1988) A *Dictyostelium* mutant with severe defects in α-actinin: its characterization using cDNA probes and monoclonal antibodies. J Cell Sci 90:59-66

Speicher DW, Marchesi VT (1984) Erythrocyte spectrin is comprised of many homologous triple helical segments. Nature 311:177-180

Stossel TP, Chaponnier C, Ezzel RM, Hartwig JH, Janmey PA, Kwiatkowski DJ, Lind SE, Smith DB, Southwick FS, Yin HL, Zaner KS (1985) Nonmuscle actin-binding proteins. Ann Rev Cell Biol 1:353-402

Tschudi C, Young AS, Ruben L, Patton CL, Richards FF (1985) Calmodulin genes in trypanosomes are tandemly repeated and produce multiple mRNAs with a common 5' leader sequence. Proc Natl Acad Sci USA 82:3998-4002

van Erd JP, Capony JP, Ferraz Cand Pechere JF (1978) The amino-acid sequence of troponin C from frog skeletal muscle. Eur J Biochem 91:231-242

Wallraff E, Schleicher M, Modersitzki M, Rieger D, Isenberg G, Gerisch G (1986) Selection of *Dictyostelium* mutants defective in cytoskeletal proteins: Use of an antibody that binds to the ends of α-actinin rods. EMBO J 5:61-65

Wasenius V-M, Saraste M, Salven P, Eramaa M, Holm L, Lehto V-P (1989) Primary structure of the brain α-spectrin. J Cell Biol 108:79-93

Witke W, Noegel AA (1990) A single base exchange in an intron of the *Dictyostelium discoideum* α-actinin gene inhibits correct splicing of the RNA but allows transport to the cytoplasm and translation. J Biol Chem 265:34-39

Witke W, Schleicher M, Lottspeich F, Noegel A (1986) Studies on transcription, translation, and structure of α-actinin in *Dictyostelium discoideum*. J Cell Biol 103:969-975

Witke W, Nellen W, Noegel A (1987) Homologous recombination in the *Dictyostelium* α-actinin gene leads to an altered mRNA and lack of the protein. EMBO J 6:4143-4148

Witke W, Gurniak CB, Humbel BM, Einberger H, Neubert J, Noegel AA, Schleicher M Expression of an α-actinin fragment in *Dictyostelium* that carries an actin-binding site. (submitted)

Wuestehube LJ, Luna EJ (1987) F-Actin binds to the cytoplasmic surface of ponticulin, a 17 kD integral glycoprotein from *Dictyostelium discoideum*. J Cell Biol 105:1741-1751

Yin HL, Janmey PA, Schleicher M (1990) Severin is a gelsolin prototype. FEBS Lett 264:78-80

EF-HANDS CALCIUM BINDING REGULATES THE THIOREDOXIN REDUCTASE/THIOREDOXIN ELECTRON TRANSFER IN HUMAN KERATINOCYTES AND MELANOMA

Karin U. Schallreuter and John M. Wood

INTRODUCTION

Human skin, with 1.80 m^2, represents the second largest organ in the body. As a consequence, human epidermis holds a unique position in terms of extracellular influence on its metabolism, cell growth, and mortality. Physical, chemical, and biological factors are constantly changing to challenge homeostasis. Keratinocytes are the major component of the epidermis and the ratio of keratinocytes to the pigment cells (melanocytes) is 36:1 representing the epidermal unit, with Langerhans and Merkel cells of less frequency (Fitzpatrick et al. 1979). In its defense against ultra-violet light generated free radicals, the human skin has evolved a substantial anti-oxidant capacity involving small molecules such as ascorbic acid, enzymes, and the pigment's eumelanin and pheomelanin (Schallreuter and Wood 1989a).

Flesch and Rothman (1948) showed that pigmentation in human skin is regulated by its sulfhydryl content. Two years later, Fitzpatrick and Becker (1950) discovered that the enzyme tyrosinase, the key enzyme responsible for melanin biosynthesis, exists in an inactive form due to its inhibition by thiol-containing molecules. This inhibition could be reversed by ultra-violet light yielding an increase in melanin biosynthesis. In these early studies, the importance of the redox status of the epidermis, vis a vis the dithiol/disulfide equilibrium, was acknowledged, as well as the significance of cell-cell interactions between keratinocytes and melanocytes.

Recently, the important role played by thioproteins in the human epidermis has been confirmed with the thioredoxin reductase (TR)/thioredoxin (T) electron transfer system as the major antioxidant catalysts, where the glutathione reductase (GR)/glutathione (G) system is barely detectable (Schallreuter and Wood 1986; Schallreuter and Wood 1988; Schallreuter et al. 1989a,b,c). The TR/T system is under allosteric regulation by calcium and therefore, this fast exchange ion can regulate the important processes of pigmentation, free radical defense, and deoxyribonucleotide synthesis in the skin (Schallreuter et al. 1986; Schallreuter and Pittelkow 1987, 1988; Schallreuter and Wood 1988, 1989a,b,c).

BIOCHEMISTRY OF THE THIOREDOXIN REDUCTASE/THIOREDOXIN SYSTEM

Moore et al. (1964) first described the TR/T system as electron donors for the biosynthesis of deoxyribonucleotides from ribonucleotides in *Escherichia coli (E. coli)*. TR from *E. coli* was isolated as a dimer (66000 daltons) composed of identical subunits (33000 daltons). Each subunit contains FAD and a dithiol active site with the primary sequence CATC. TR is fully reduced by NADPH and as a consequence, it contains four available electrons, two on $FADH_2$, and two on the dithiolate active site. In this respect, TR is very similar to other well-characterized flavoproteins such as glutathione reductase, lipoamide dehydrogenase, and mercuric ion reductase (Holmgren et al. 1985). However, TR differs from the latter enzymes in two fundamental respects: (1) The dithiol active site structure of TR contains CATC, meanwhile, the other enzymes have C L N V G C sequence homology (Greer and Perham 1985); and (2) based on its physical/chemical properties, it has been suggested that TR has an essential arginine residue in its active site to facilitate the thiol to thiolate acid/base dissociation, whereas the other enzymes have a histidine residue as the active site base (O'Donnell and Williams 1985).

These differences in structure between TR and the other flavoproteins with dithiol active sites are very important to the nucleophilicity of TR at pH 7.0 compared to the other enzymes. TR has a thiolate standard reduction potential of -271 millivolts, and has the capability to reduce superoxide anion radicals (O_2^-) directly i.e., $O_2/O_2^- = -160$ millivolts). Together with its electron acceptor thioredoxin, TR can reduce peroxide $(O_2^=)$. In contrast, the glutathione reductase system (GR) needs glutathione and glutathione peroxidase in order to reduce $O_2^=$ to water (Schallreuter and Wood 1986).

Schallreuter et al. (1990) have used 360 MH$_2$ ^{31}P NMR to show that cell cultures of human keratinocytes have an intracellular pH of 7.05. Since the pK_A for the dissociation of the thiolate active site of TR has been determined by O'Donnell and Williams (1985) as 6.98; two units lower than that for GR (pK - 9.0), a pH-dependent evolution for TR over GR in the epidermis can be rationalized (Schallreuter et al. 1990).

TR from mammalian sources was first purified by Luthman and Holmgren (1982, 1985) in rat liver. The enzyme is a dimer (M_r 116000 daltons). The minimum molecular weight of mammalian TR is 58000 daltons containing one bound FAD and one dithiol active site. Schallreuter et al. (1986, 1988, 1989) have purified TR from human keratinocytes and from human metastatic melanoma tissues confirming a minimum molecular weight of 58000 daltons.

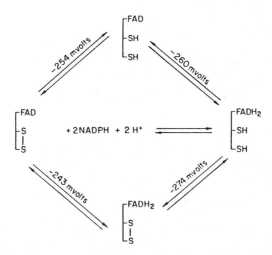

FOUR MICROSCOPIC OXIDATION REDUCTION COUPLES
FOR THIOREDOXIN REDUCTASE AT pH 7.0

Fig. 1. Redox couples in the reversible oxidation/reduction of NADPH and thioredoxin by thioredoxin reductase. (O'Donnell and Williams 1985)

Although the primary sequence of TR from *E. coli* has been determined by Russel and Model (1988), the human/mammalian enzymes have not been sequenced so far. However, human thioredoxin, the natural electron acceptor for TR, has been sequenced and NMR studies revealed structural homology to the same protein from *E. coli* (Wollman et al. 1988). This thioprotein (11400 daltons) has the same active site sequence as the *E. coli* protein (i.e., CAPC). Using the fluorescent probe, 1-deaza-FAD, O'Donnell and Williams (1985) have shown that the active site of TR has a large cleft capable of accepting the small molecular weight natural substrate oxidized thioredoxin. Table 1 summarizes the properties of TRs purified from *E. coli* and human metastatic melanoma, and Figure 1 shows the standard reduction potentials for electron transfer to the four sites in this enzyme.

Table 1. A comparison of the properties of thioredoxin reductases from *Escherichia coli* and human metastatic melanoma

	$TR_{E\ coli}$	$TR_{Melanoma}$
Minimum molecular weight	33000	58000
Active sites	FAD dithiol CATC	FAD dithiol (unknown)
Electron donor	NADPH	NADPH
Natural electron acceptor (substrate)	Thioredoxin (M_r 12500)	Thioredoxin (M_r 11400)
Active site (thioredoxin)	CAPC	CAPC
Allosteric EF-hands	Ca^{2+}	Ca^{2+}
Regulation by calcium	Inhibitor	Inhibitor
Reversible inhibitors	Azelaic acid nitrosoureas	Azelaic acid nitrosoureas
Suicide inhibitor (stereospecific)	13-*cis*-retinoic acid	13-*cis*-retinoic acid

Due to its specific disulfide/interchange reduction properties, it is not surprising that TR/T has already been implicated in nine important reduction reactions in metabolism:
1. The reduction of ribonucleotides to deoxyribonucleotides via the ribonucleotide reductases (Holmgren 1985).

2. The reduction of methionine sulfoxide residues in proteins to methionine (Holmgren 1985).

3. The reduction of sulfite to sulfide (Holmgren 1985).

4. The reduction of insulin (Holmgren 1985).

5. The reduction of nitroxide and superoxide anion radicals (Schallreuter and Wood 1986, 1989).

6. The reduction of vitamin K epoxide via vitamin K epoxide reductase (Silverman et al. 1988).

7. The regulation of melanin biosynthesis through the inhibition of tyrosinase by reduced thioredoxin (Wood and Schallreuter 1988).

8. The reactivation of hydrogen peroxide inactivated 3-phosphoglyceraldehyde dehydrogenase by reduced thioredoxin (Spector et al. 1988).

9. Activation of methionine sulfoxide peptide reductase (Spector et al. 1988).

Considering this broad spectrum of anti-oxidant properties documented for the TR/T system and its abundance in the free radical-rich domain of the epidermis, it seems feasible that such a regulator on intracellular redox conditions suggests control at both the gene and enzyme levels. The expression of TR/T in human keratinocytes (i.e., 5% of the total acidic protein in the cytosol) could be indicative of substantial gene amplification for these two thioproteins in these cells (Schallreuter et al. 1989). However, the most striking discovery has been the recognition that the TR/T electron transfer is regulated by calcium.

ALLOSTERIC REGULATION OF *TR* BY CALCIUM

O'Donnel and Williams (1984) were the first to discover that ^{14}C-labeled N-ethylmaleimide reacts with both active site thiolate groups of TR. This property of the enzyme allowed the use of 4-maleimido-TEMPO to spin-label both active site thiolate groups on TR from *E. coli* and on plasma membrane associated TRs at the surfaces of human keratinocytes, human and guinea pig epidermis (Schallreuter et al. 1986). Electron spin resonance studies of this labeled TR yielded different nitroxide radical spectra with one labeled thiolate being more immobilized than the other (Figure 2).

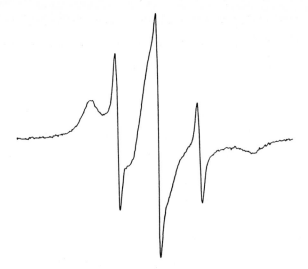

Fig. 2. EPR spectra of 4-maleimido-tempo bound to thioredoxin reductase at the surface of guinea-pig skin. Two spectra are apparent, one 34 G wide (strongly immobilized) and the second 22 G wide (weakly immobilized)

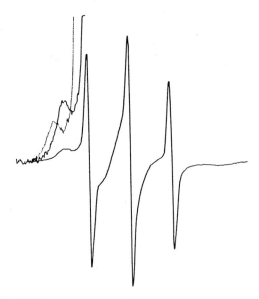

Fig. 3. Shift and reduction in the amplitude of the strongly immobilized nitroxide radical signal upon the addition of 2 mM Ca^{2+} (. . .)

As a consequence, the two active site thiolate groups can be distinguished from each other by this method. When 4-maleimido-TEMPO-labeled TR has been preincubated with calcium (2 x 10^{-3}mol) selective electron transfer from NADPH to enzyme bound FAD was not affected. However, FADH2 did transfer electrons selectively to the more immobilized spin-labeled thiolate group revealing a slow reduction of the nitroxide radical to the hydroxylamine with a decrease in the EPR signal (Figure 3). These results indicated that the electron transfer from FADH2 to the nitroxide radical group was calcium dependent and that a large conformational change occurs in TR due to calcium binding at some distance away from the active site of this enzyme (Figures 4 and 5).

THE REACTION OF TWO SPIN-LABELLED MALEIMIDE
MOLECULES WITH THIOREDOXIN REDUCTASE

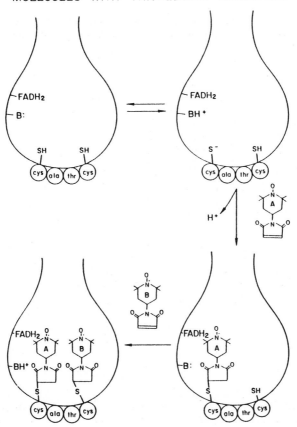

Fig. 4. The reaction of 4-maleimido-tempo with the thiolate active sites of thioredoxin reductase. B+ is a basic amino acid residue in the active site which lowers the pKas for both thiol groups to slightly below 7.0

CLOSURE OF THE ACTIVE SITE OF THIOREDOXIN REDUCTASE BY Ca⁺⁺

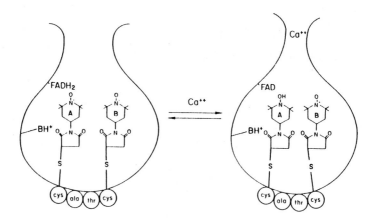

Fig. 5. Proposed closure of the active site of thioredoxin reductase with Ca²⁺ causing electron transfer to the bound nitroxide radical on the most immobilized spin labeled thiol group in the active site. One of the sites is reduced from the paramagnetic nitroxyl radical to the diamagnetic hydroxylamine by the flavoprotein

The preincubation of tissue biopsies from guinea pig skin, human skin, and human primary and metastatic melanomas with calcium showed a marked reduction in membrane-associated TR activities (Schallreuter et al. 1986, 1988, 1990). In addition, the determination of TR activities on pure enzymes from *E. coli*, human keratinocytes, and human melanoma tissues, as well as membrane-associated TR on human keratinocytes, melanocytes, and melanoma cells correlated to extracellular calcium concentration. Calcium has been shown to be a potent inhibitor of electron transfer from the thiolate active site of TR (Schallreuter et al. 1986).

The experimental data suggested that calcium binding closes down the active site cleft in the enzyme implicating an allosteric mechanism for the calcium regulation of TR (Figure 5). This allosteric regulation has been confirmed by a kinetic analysis of calcium-free TR purified from human amelanotic melanoma tissue (Schallreuter and Wood 1989). Calcium-free TR showed normal Michaelis-Menten kinetics, whereas calcium-bound TR yielded classical allosteric inhibition with sigmoidal kinetics (Figure 6).

⁴⁵Calcium has been used to determine the number of binding sites on calcium-free TR. ⁴⁵Calcium-bound enzyme was stable to EDTA and EGTA and could be dialyzed for 24 h

without loss of radioactivity, yielding one bound calcium ion per TR molecule (M_r 58000 daltons) (Schallreuter and Wood 1989; Schallreuter et al. 1989). Calcium-free TR resulted in 70% inhibition upon preincubation with 3 x 10^{-5} M calcium. An even more pronounced inhibition with 88% has been observed when this TR was preincubated with 3 x 10^{-5} M lanthanides such as lanthanum and gadolinium (Figure 7).

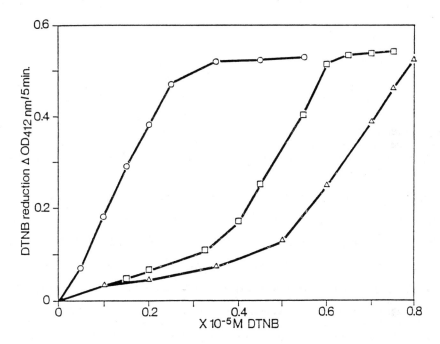

Fig. 6. V vs. [S] plot for DTNB with amelanotic melanoma thioredoxin reductase in the absence of calcium (o) and after preincubation with 6 x 10^{-3} M CaCl$_2$ for 10 min prior to assay (▲). Melanotic melanoma thioredoxin reductase was already isolated with allosteric inhibition (□)

Calcium could be exchanged from its single binding site on TR by the addition of a saturating concentration of its natural substrate oxidized thioredoxin, whereas the lanthanides showed irreversible inhibition due to stronger binding (Figure 8). Lanthanide binding did not influence electron transfer from NADPH to bound FAD, but both calcium and the lanthanides increased the rate of reduction of FAD by NADPH, as determined by fluorescence spectroscopy for the FAD/FADH$_2$ reaction (Figure 9). This result is consistent with the

348

calcium-dependent electron transfer from FADH$_2$ to 4-maleimido-TEMPO spin labeled TR as described earlier (Figure 4, 5). When [45]calcium-bound TR was preincubated with oxidized thioredoxin, the isotope was exchanged from the enzyme, proving that this substrate can reverse allosteric inhibition by calcium (Schallreuter and Wood 1989; Schallreuter et al. 1989).

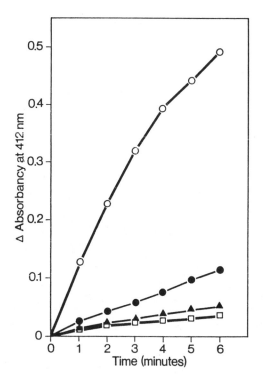

Fig. 7. The inhibition of calcium-free human melanoma thioredoxin reductase by calcium (3 x 10^{-5} M) (• - •) lanthanum (3 x 10^{-5} M) (▲ - ▲) gadolinium (3 x 10^{-5} M) (□ - □) and control (o - o)

Since the sequence of TR from *E. coli* is known, the calcium inhibition/oxidized thioredoxin activation experiments were repeated with this enzyme (Figure 10). The bacterial enzyme responded in a similar way to the TR from amelanotic melanoma in terms of calcium inhibition and exchange (Schallreuter and Wood 1989). An intelligenetics computer analysis of the TR$_{E.\ coli}$ sequence revealed the presence of a single EF-hands calcium-binding site with homology to the first EF-hands site of calmodulin (Schallreuter and Wood 1989; Schallreuter et al. 1989) (Figure 11).

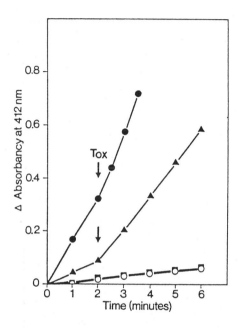

Fig. 8. The reactivation of calcium-bound human melanoma thioredoxin reductase by oxidized thioredoxin (25 x 10^{-6} M) (▲ - ▲) compared to a control (• - •) lanthanum and gadolinium inhibited thioredoxin reductase was not reactivated under the same experimental conditions (i.e. La^{3+}) (○ - ○) and Gd^{3+} (□ - □)

THE REGULATION OF INTRACELLULAR REDOX CONDITIONS IN HUMAN KERATINOCYTES BY EXTRACELLULAR CALCIUM IN THE CULTURE MEDIUM

Earlier studies showed that membrane-associated TR activity at the surface of keratinocytes was determined by the concentration of calcium in the culture medium (Schallreuter et al.

350

1986). Cells grown in medium containing 2×10^{-3} M calcium were approximately half as active in TR compared to cells with 0.1×10^{-3} M calcium.

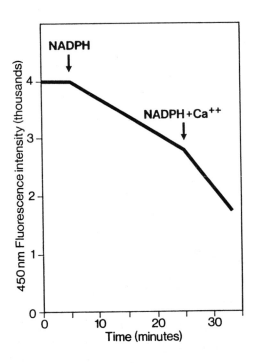

Fig. 9. Quenching of the fluorescence excitation of bound FAD in human melanoma thioredoxin reductase (1.67 mg/ml) by NADPH reduction to $FADH_2$, calcium increased the rate of electron transfer from NADPH to FAD

The influence of extracellular calcium on the intracellular reduction of thioredoxin was determined by labeling the cells with ^{14}C-methylmercury chloride followed by FPLC analysis of cell extracts. This technique allows to distinguish between calcium-bound and calcium-free TR. Cells grown in 0.1×10^{-3} M calcium primarily contained calcium-free enzyme. However, cells established in medium with 2×10^{-3} M calcium TR separated as two peaks, with more calcium-bound than calcium-free enzyme (Schallreuter et al. 1989) (Figure 12).

The different chromatographic properties of calcium-bound versus -free TR indicate a major change in the surface charge caused by calcium binding. These results strengthen even more the major conformational change on the enzyme as predicted from the spin labeling experiments.

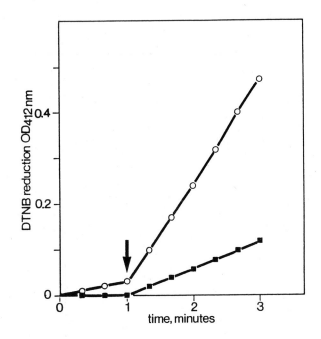

Fig. 10. The effect of oxidized thioredoxin (T) on the rate of DTNB reduction by *E. coli* thioredoxin reductase (TR) in the presence (■) and absence (○) of 6.10^{-3} M $CaCl_2$.0.1 mg of oxidized thioredoxin was added to each reaction at 1.3 min. Reactions contained 1.0 ml of DTNB reaction mixture plus 20 μl of *E. coli* (TR) (4.0 mg/ml), and 25 μmol of oxidized thioredoxin were added to start each reaction at 1.3 min

```
275TQTSIPGVFAAGDVMDHIYRQAITS     thioredoxin reductase

  6TEEQIAEFKEAFALFDKDNNGSISS     calmodulin
```

Fig. 11. Intelligenetics computer determination of sequence homology between thioredoxin reductase (*E. coli*) and the first EF-hands calcium-binding site of calmodulin

Examination of thioredoxin levels in these cell extracts yielded 38% less reduced thioredoxin for cells grown on high calcium concentration compared to cells established in medium containing low calcium.

352

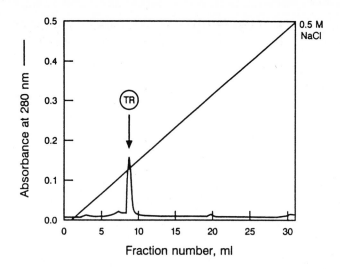

Fig. 12a. Purification of melanoma thioredoxin reductase by FPLC (A_{280nm}; full scale 1.0A units) on a mono Q column HR 5/5 in 0.05 M Tris buffer pH 7.5 with a 0-0.5 M NaCl gradient. This calcium-free enzyme eluted at 1.12 M NaCl

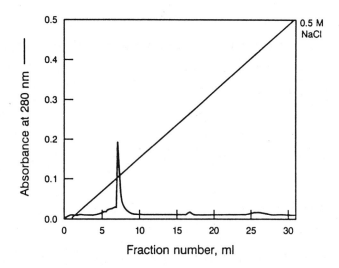

Fig. 12b. Purification of [45]calcium-labeled melanoma thioredoxin reductase (1 [45]calcium/TR molecule) on a mono Q column in 0.05 M tris buffer pH 7.5. Thioredoxin reductase eluted calcium-bound at 0.105 M NaCl

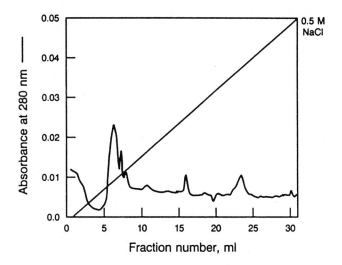

Fig. 12c. FPLC chromatography of $^{14}CH_3HgCl$ labeled extracts of human keratinocytes grown in 2.0 x 10^{-3} M calcium. Thioredoxin reductase eluted as two peaks, one calcium-bound at 0.105 M NaCl, and one calcium-free at 0.12 M NaCl. Thioredoxin eluted at 0.3 M NaCl and contained (8212 cpm $^{14}CH_3Hg$/0.1 OD_{280} unit). These cells contained 34% less reduced thioredoxin than cells grown in medium with 0.1 x 10^{-3} M calcium

These experiments support even further that the extracellular calcium concentration determines the intracellular redox conditions in cell cultures of human keratinocytes with high intracellular calcium causing a more oxidized situation.

CALCIUM UPTAKE IN HYPOPIGMENTATION DISORDERS

The effect of calcium-regulated redox properties on pigmentation were examined in patients with vitiligo and with the tyrosinase positive albinism Hermansky-Pudlak Syndrome (HPS). Vitiligo has been selected, because these patients have depigmented areas of skin surrounded by normally pigmented and sometimes even hyperpigmented epidermis. Subsequently, each patient provides experimental and control samples of skin upon biopsy. HPS albinism was selected primarily because the key enzyme responsible for melanin biosynthesis is present, but tyrosinase activity is not significantly expressed compared to family members lacking this autosomal recessive disorder. Skin biopsies taken from patients with vitiligo (n = 10)

354

revealed TR activities on depigmented areas 30-80% less than found in pigmented control skin of these patients (Schallreuter et al. 1987).

Biopsies from 19 homozygotes with HPS revealed significantly decreased levels of membrane-associated TR compared to normal healthy family members in seven families with the disease. Keratinocytes were established from HPS patients (n = 4). TR has been especially sensitive to low (i.e., 0.1 x 10^{-3} M) calcium concentrations in the culture medium compared to controls (Schallreuter and Witkop 1988). By contrast for vitiligo, a similar calcium sensitivity could only be demonstrated with high calcium (i.e., 2 x 10^{-3} M). [45]Calcium uptake experiments with keratinoctyes from HPS homozygotes and from vitiliginous skin showed that defective calcium uptake was responsible for a decrease in optimum intracellular calcium concentrations. In HPS cells, a slow leakage of calcium has been verified (Schallreuter and Pittelkow 1988, 1989). Analysis of reduced thioredoxin in these cells, using the ^{14}C methylmercury labeling technique, showed that both vitiliginous and HPS cells contained significantly more reduced thioredoxin than controls (Table 2) (Schallreuter et al. 1989, 1990). These results suggest that depigmentation in both HPS and vitiligo may be influenced by defects in calcium transport leading to a more reducing environment for keratinocytes, and as a consequence, for melanocytes in the depigmented epidermis of patients with these hypopigmentation disorders.

Table 2. [14]CH_3HgCl-labeling of reduced thioredoxin (100 μg of protein) in human keratinocytes established from different depigmentation disorders[a]

Keratinocytes established in 0.1 x 10^{-3} M calcium	[14]CH_3Hg^+ incorporated into reduced thioredoxin (cpm)
Type IV Skin (control)	1105 cpm
Type II Skin (control)	2712 cpm
HPS (homozygotes)	3152 cpm
Vitiligo (involved)	2947 cpm
Vitiligo (uninvolved)	973 cpm

[a] [14]CH_3HgCl-labeling method according to Schallreuter et al. (1989).

Based on these results, a general conclusion can be formulated. Keratinocytes with a low redox potential in the epidermis can control tyrosinase activity resulting in inhibition of melanin biosynthesis, and oxidizing conditions would lead to pigmentation. This conclusion gains further support from experiments with pure proteins where reduced thioredoxin inhibits tyrosinase by forming a stable bis-cysteinate complex with one of the copper atoms in the binuclear copper active site of this enzyme (Wood and Schallreuter 1988).

It has been shown in vitro that thioredoxin readily penetrates plasma membranes in lens epithelial cells (Spector et al. 1988). Subsequently, keratinocyte/melanocyte interchange of reduced thioredoxin would explain how the redox status of keratinocytes can regulate melanin biosynthesis in the pigment-producing melanocytes.

CALCIUM REGULATION OF MELANIN BIOSYNTHESIS IN HUMAN MELANOMA

The influence of calcium on pigmentation in human metastatic melanoma the TR/T systems was verified when it was discovered that amelanotic tumors contained calcium-free TR, whereas melanotic tumors mostly contained calcium bound TR (Figure 5) (Schallreuter and Wood 1988, 1989). Considerable heterogeneity for TR activities in melanotic tumors could be demonstrated in five tumors excised from the same patient (Schallreuter and Wood 1988). This biochemical heterogeneity can be explained by the calcium/redox status of these metastases (Schallreuter and Wood 1988, 1990). Examination of 29 primary melanomas and their surrounding skin yielded two groups: (1) tumors with high TR activities compared to normal skin with lower than normal TR in the immediate surrounding epidermis (n = 19), and (2) tumors with low activity with higher than normal activities on surrounding skin (n = 10) (Schallreuter et al. 1990). These gradients of TR activity from tumor to surrounding skin, and vice versa, have been shown to be due to the calcium flux between surrounding keratinocytes and the primary melanoma. A follow up of these 29 patients over a 3-year period indicated that patients with high TR activities on their tumors showed a poorer

356

prognosis where four patients metastasized. However, in the group with low TR activities, there has been no progression so far.

Since the TR/T system represents one alternate pathway for the synthesis of deoxyribonucleotides (Figure 13), then the efflux of calcium from primary melanomas to the surrounding epidermis could activate DNA synthesis, cell division, and as a consequence, tumor growth and invasion. Melanoma cells with high intracellular calcium may be expected to remain in the Go-phase of the growth cycle.

Recently, a very important aspect of the allosteric regulation of TR by calcium has been verified in relation to the treatment of melanoma with nitrosourea compounds. Calcium-free TR, GR, and ribonucleotide reductase are inhibited reversibly by the nitrosourea (Fotemustine), a new drug in phase III clinical trials for metastatic melanoma. It has been shown that nitrosoureas inhibit these thioproteins by the transfer of a chloroethyl carbonium ion to the thiolate active sites. This enzyme deactivation can be reversed by nucleophilic displacement of the chloroethyl-group by a ß-elimination mechanism producing ethylene and chloride ion as the products (Schallreuter and Wood 1989, 1990).

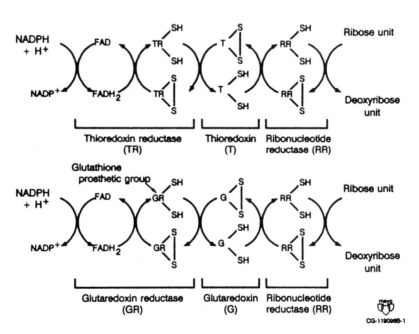

Fig. 13. Alternate pathways for the reduction of ribonucleotides to deoxyribonucleotides by ribonucleotide reductase

Calcium-bound TR has been shown to be considerably more resistant to inactivation by Fotemustine as well as to other inhibitors such as 13-*cis* retinoic acid and azelaic acid (Schallreuter and Wood 1987, 1989, 1990). As described above, calcium binding to TR closes down the active site cleft and as a consequence, both substrates and inhibitors are less accessible. Calcium-bound TR requires 100-fold higher concentration of Fotemustine to yield similar inhibitor kinetics to calcium-free TR.

Preliminary results on patients with metastatic melanomas after treatment with this drug (n = 8) showed a clear correlation between TR levels in these tumors, their pigmentation, and resistance to therapy. Fotemustine-resistant tumors contained higher levels of GR than TR; meanwhile, drug-sensitive tumors were clinically amelanotic and TR was higher than GR (Schallreuter and Wood 1990). It should be noted here that calcium-bound TR can be activated by its natural substrate oxidized thioredoxin which causes calcium exchange to yield calcium-free enzyme. Therefore, the redox status of metastatic melanomas may determine drug sensitivity/resistance as well as controlling pigmentation in these tumors. The individual redox status of different melanoma metastases in one person certainly could explain the heterogeneity of the tumors.

CONCLUSIONS

The TR/T electron transfer system represents the most abundant anti-oxidant catalyst in the human epidermis. High levels of TR/T are present on both plasma membranes and in the cytosol of human keratinocytes, melanocytes, and in some melanoma cells. Since keratinocytes have a finite lifetime in the skin of 9 to 14 days, it seems that the major activity for TR/T is the reduction of active oxygen species such as O_2^- and $O_2^=$ (i.e., UV-generated radicals) rather than as electron donors for ribonucleotide reduction. Clearly, plasma membrane-associated TR, with its activity at the outer cell surface, would be unlikely to function in ribonucleotide reduction (Hansson et al. 1985). The regulation of plasma membrane-associated TR by extracellular calcium, and cytosol TR by intracellular calcium, highlights the importance of this fast exchange ion in the processes of free radical defense and as regulator of intracellular redox conditions/pigmentation in the human epidermis.

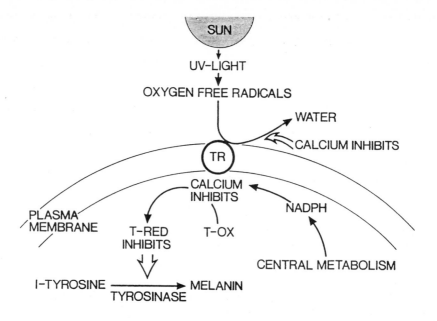

Fig. 14. Scheme for the regulation of free radical defense and melanin biosynthesis in the human epidermis. Extracellular calcium regulates free radical reduction by membrane-associated TR. Intracellular calcium regulates the reduction of thioredoxin by cytosol TR. Reduced thioredoxin controls melanin biosynthesis by complexation to the active site of tyrosinase. Oxidized thioredoxin does not inhibit tyrosinase

In metastatic melanoma, the TR/T system, with its control by calcium, assumes importance as one electron donating pathway for ribonucleotide reductase regulating DNA synthesis and, as a consequence, cell division. The selection of TR/T versus GR/G appears to gain significance in the processes of drug resistance/sensitivity cell division, and pigmentation in these tumors.

Perhaps the best biochemical and clinical example for the importance of calcium in free radical defense and pigmentation resides in the study of tyrosinase-positive albinos with HPS. In these patients, a defective calcium transport system appears to influence free radical defense and pigmentation by membrane-associated TR while lowering the redox status of the epidermis through low intracellular calcium (Schallreuter and Witkop 1988; Schallreuter 1989; Schallreuter and Pittelkow 1989) (Figure 14).

The inability of HPS keratinocytes to retain calcium appears to be a general genetic problem for HPS patients since their platelets do not contain any dense bodies which represent the storage pools for insoluble calcium, ATP, ADP, and serotonin.

REFERENCES

Fitzpatrick TB, Becker SW (1950) Tyrosinase in human skin: demonstration of its presence and role in human melanin formation. Science 113:223-225

Fitzpatrick TB, Eisen AZ, Wolff K, Freeberg IM, Austien KF (eds) (1987) In: Dermatology in general medicine. 3rd ed, McGraw Hill, New York

Flesch P, Rothman S (1948) Role of sulfhydryl compounds in pigmentation. Science 108:505-506

Greer S, Perham RN (1985) DNA sequence of the *Escherichia coli* gene for glutathione reductase: comparison with other flavoprotein disulfide bridge oxidoreductases. In: Thioredoxin and glutaredoxin systems: structure and function. Raven Press, New York, pp 121-130

Hansson HA, Holmgren A, Rozell B, Stemme S (1985) Localization of thioredoxin, thioredoxin reductase and ribonucleotide reductase in cells: immunohistochemical aspects. In: Thioredoxin and glutaredoxin systems: structure and function. Raven Press, New York, pp 177-187

Holmgren A (1985) Thioredoxin. Ann Rev Biochem 54:237-271

Holmgren A, Branden C-I, Jornvall H, Sjoberg B-M (1985) Thioredoxin and glutaredoxin systems: structure and function. Raven Press, New York

Luthman M, Holmgren A (1982) Rat liver thioredoxin and thioredoxin reductase: purification and characterization. Biochemistry 21:6628-6633

Moore EC, Reichard P, Thelander L (1964) Enzymatic synthesis of deoxyribonucleotides purification and properties of thioredoxin reductase from *Escherichia coli B*. J Biol Chem 239:3445-3452

O'Donnell ME, Williams CH Jr (1985) Mechanism of *Escherichia coli* thioredoxin reductase. In: Thioredoxin and glutaredoxin systems: structure and function. Raven Press, New York, pp 131-140

Russel M, Model P (1988) Sequence of thioredoxin reductase from *Escherichia coli*. J Biol Chem 263:9015-9019

Schallreuter KU (1989) Das Hermansky-Pudlak Syndrome: Ein Überblick. Der Hausarzt 40:130-133

Schallreuter KU, Pittelkow MR (1988) Defective calcium uptake system in vitiligo. Arch Dermatol Res 280:137-139

Schallreuter KU, Pittelkow MR (1989) The regulation of thioredoxin reductase by calcium in Hermansky-Pudlak syndrome. Arch Dermatol Res 281:40-44

Schallreuter KU, Witkop CJ Jr (1988) Thioredoxin reductase activity in Hermansky-Pudlak syndrome: a method for identification of putative heterozygotes. J Invest Dermatol 90:372-377

Schallreuter KU, Wood JM (1986) The role of thioredoxin reductase in the reduction of free radicals at the surface of the epidermis. Biochem Biophys Res Commun 136:630-637

Schallreuter KU, Wood JM (1987) Azelaic acid as a competitive inhibitor of thioredoxin reductase in human melanoma cells. Cancer Lett 36:297-305

Schallreuter KU, Wood JM (1988) The activity and purification of membrane-associated thioredoxin reductase from human metastatic melanotic melanoma. Biochim Biophys Acta 967:103-109

Schallreuter KU, Wood JM (1989) Free radical reduction in the human epidermis. Free Rad Biol Med 6:519-532

Schallreuter KU, Wood JM (1989) Thioredoxin reductase in control of the pigmentary system. In: Clinics in dermatology: disorders of pigmentation. J. B. Lippincott, Philadelphia 7:2 92-105

Schallreuter KU, Wood JM (1989) Calcium regulates thioredoxin reductase in human metastatic melanoma. Biochem Biophys Acta 997:242-247

Schallreuter KU, Wood JM (1989) The stereospecific suicide inhibition of human melanoma thioredoxin reductase by 13 *cis* retinoic acid. Biochem Biophys Res Commun 160:2 573-579

Schallreuter KU, Wood JM (to be published) Thioredoxin reductase activity at the surface of human primary cutaneous melanomas and their surrounding skin. Cancer Res

Schallreuter KU, Pittelkow MR, Gleason FK, Wood JM (1986) The role of calcium in the regulation of free radical reduction by thioredoxin reductase at the surface of the skin. Inorg Biochem 28:227-238

Schallreuter KU, Pittelkow MR, Wood JM (1986) Free radical reduction by thioredoxin reductase at the surface of normal and vitiliginous keratinocytes. J Invest Dermatol 87:728-732

Schallreuter KU, Pittelkow MR, Wood JM (1989) EF-hands calcium binding regulates the thioredoxin reductase/thioredoxin electron transfer in human keratinocytes.
Biochem Biophys Res Commun 162:3 1311-1316

Schallreuter KU, Pittelkow MR, Wood JM (to be published) A comparative study of the thioproteins thioredoxin reductase and glutathione reductase in human keratinocytes. J Invest Dermatol

Silverman RB, Nandi DL (1988) Reduced thioredoxin a possible physiological cofactor for Vitamin K epoxide reductase. Biochim Biophys Res Commun 155:1248-1254

Spector A, Goo-Zai Y, Huang Ruey-Ruey, McDermott MJ, Gascoyne PRC, Piguet V (1988) The effect of H_2O_2 upon thioredoxin enriched lens epithelial cells. J Biol Chem 263:4984-4990

Witkop CJ Jr (1989) Albinism, In: Clinics in dermatology: disorders of pigmentation. JB Lippincott, Philadelphia, 7:2 80-91

Wollman EE, d'Auriol L, Rinsky L, Shaw A, Jacquot J-P, Wigfield P, Graber P, Descarps F, Robin P, Galibert F, Berloglio J, Fradlizi D (1988) Cloning and expression of a cDNA for human thioredoxin. J Biol Chem 263:30 15506-15512

Wood JM, Schallreuter KU (1988) Reduced thioredoxin inhibits melanin biosynthesis: Evidence for the formation of a stable bis-cysteinate complex with tyrosinase. Inorganica Chim Acta 151:1

CALCIUM-BINDING CRYSTALLINS

Doraivajan Balasubramanian and Yogendra Sharma

INTRODUCTION

Cataract is an affliction that involves the irreversible opacification of the eye lens and consequent impairment of vision. The etiology of the disease is manifold radiation, free radical reactions, diabetic and other metabolic disorders, metal ion accumulation in the lens and so on. The possibility that calcium may have a role to play in the physiology of cataract has been known for a long time (Burge 1909; Adams 1929). The calcium ion levels in cataractous human lenses are often different from those in normal lenses (Duncan and Bushell 1975; Duncan and van Heyningen 1977; Hightower and Reddy 1982a). Elevated cytosolic calcium concentration of the lens has been correlated with cataract formation in both human and animal model systems (Lohmann et al. 1986; Shearer et al. 1987). The aqueous humor calcium ion concentration in a human eye with a normal transparent lens ranges from 0.5 to 2.0 mM, while it is much greater (0.1-64 mM) in subjects with cataractous lens. Various studies regarding the effect of calcium on the individual crystallins as well as on intact lenses have been performed. Incubation of a normal lens in a solution of $CaCl_2$ results in the lens losing its transparency (Hightower and Reddy 1982b; Hightower 1985) as shown in Figure 1.

The calcium-induced opacification takes place when the lens pH remains slightly acidic (Hightower et al. 1987). Clark et al. (1988) reported the stronger effect of Ca^{2+} on transparency of cortical lens homogenate than nuclear homogenate. Patmore and Maraini (1986) related the increased lens calcium level with membrane permeability and suggested that the loss of membrane integrity may be responsible for the entry of an excessive amount of calcium into the lens. Selenite cataract was also reported to be associated with an elevation of lens calcium and induction of proteolysis (David and Shearer 1984). The

362

transparent structure of the lens is thus maintained, in part, by a careful control of the physiological concentration of calcium (Duncan and Jacob 1984).

The effects of endogenous calcium on lens transparency involve several factors: alteration in the permeability of substances across membrane channels (Clark et al. 1980), change in the organization of membrane proteins in the gap junction (Bernardini and Perrachia 1981), precipitation of calcium oxalate and phosphate (van Heyningen 1972), and aggregation of the lens proteins crystallins at high calcium concentrations (Jedziniak et al. 1972; Spector and Rothschild 1973).

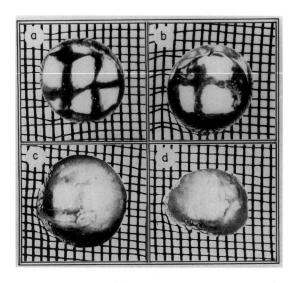

Fig. 1a-d. Photographs of rat lens incubated for 4 h at room temperature in: **a** 10 mM Tris-HCl pH 7 buffer; **b** the above buffer containing 140 mM NaCl; **c** the above buffer containing 10 mM $CaCl_2$; **d** the above buffer containing 5 mM $TbCl_3$

In addition, several other physiological and biochemical processes in the eye lens, e.g., protein biosynthesis and metabolism (Piatigorsky 1980; Hightower 1983), membrane transport (Hamilton et al. 1979; Hightower and Hind 1983), Ca^{2+}-ATPase activity (Borchman et al. 1989; Galvin and Louis 1988), proteinase-mediated hydrolysis of lens filaments and of α-crystallin (Ireland and Maisel 1984; Yoshida et al. 1984), and Ca^{2+}-activated transglutaminase mediated dimerization of ß-crystallin (Lorand et al. 1985) is also dependent on calcium ion concentration. The reviews by Hightower (1985, 1986) summarize these aspects in a comprehensive fashion.

Duncan and van Heyningen (1976) postulated that the internal calcium binding sites in cataractous lenses are different from those of the normal lenses. In the normal rat lens only about 1% of the total calcium is in free state. The concentration of free calcium, however, goes on increasing with age, largely in the perinuclear region of the lens (Duncan and Jacob 1984). In this paper, we concentrate on the chemical aspects of calcium-induced cataractogenesis of the lens, and in particular, on the interaction of Ca^{2+} with the lens crystallins. Calcium ion appears to display preferential interaction with ß-crystallin, and with δ-crystallin.

THE EYE LENS AND ITS PROTEINS

Vertebrate lenses are composed of regularly arranged fiber cells packed with diverse water-soluble proteins called crystallins that form about 90% of the lens (Piatigorsky 1984a). These are classified into four major genetically distinct classes of multimeric proteins, so-called α-, ß-, γ-, and δ-crystallins, which determine the transparency and refractive properties of the lens. Of these, the first three are present in all vertebrate lenses, though in varying proportions depending on the species, while δ-crystallin is the core protein of avian and reptilian lenses (for review see Wistow and Piatigorsky 1988). There are several other noncrystalline proteins, membrane and cytoskeleton proteins, several enzymes and neutral proteases and enzyme inhibitors in the lens (Bloemendal 1981). In this review, we focus our attention on the calcium-binding properties of the various crystallins.

Our analysis of the various crystallins for their calcium-binding properties, has suggested that only ß- and δ-crystallins possess significant calcium-binding properties (Sharma et al. 1989a). We therefore classify these two proteins as calcium-binding crystallins.

ß-CRYSTALLIN

ß-crystallins are the most heterogeneous class of the crystallins and comprise at least 40% of the water-soluble proteins. There are at least six genetically distinct subunits organized as basic ß-crystallin subunits ($ßB_1$, $ßB_2$, $ßB_3$) and acidic ß-crystallin subunits ($ßA_1$, $ßA_2$, $ßA_3$). They all have blocked N-termini. There are some other subunits, which are the products of post-translational modifications of either of these subunits. These subunits occur in aggregates that are resolved into two size classes, $ß_H$ or high molecular weight (160 kDa) oligomer, and $ß_L$ or low molecular weight (60-80 kDa) dimer or trimer. Detailed characterization of various crystallins is described elsewhere (Bloemendal 1981; Wistow and Piatigorsky 1988).

ß-crystallin is a unique mammalian crystallin in that it binds calcium with a capacity of 3-4 mol of calcium per mol of protein and an affinity of about 0.38 mM. $TbCl_3$ also interacts with ß-crystallin, as evident by its fluorescence at 488 and 545 nm via resonance energy transfer (Figure 2a). As is well known, Tb^{3+} and other lanthanide ions are excellent mimics of the calcium ion and offer the advantage of being spectroscopic probes since (a) the fluorescence of several of these ions enhances upon ligand binding, and (b) many of these are paramagnetic in nature.

Where does the calcium ion bind in the ß-crystallin chain? The topology of ß- and γ-crystallins is similar (Slingsby et al. 1988); the latter are built as bilobal molecules with each domain composed of two "Greek key" motifs, similar to those predicted in Γ-crystallin II (Blundell et al. 1981), which associate about an approximate twofold axis to form the ß-sheet (Wistow et al. 1981). To localize the calcium-binding regions in this crystallin, we analyzed the amino acid sequences of various ß-crystallin subunits but didn't detect any obvious calcium-binding motif. However, the sequences in the middle region of the various subunits of ß-crystallin, basic in particular, are notably rich in anionic and polar amino acids that can bind Ca^{2+} as shown in Table 1.

It is important to note that the principal ß subunit, $ßB_2$, forms a stable crystal in presence of calcium, whose diffraction pattern is also different from that crystallized without calcium (Bax and Slingsby 1989).

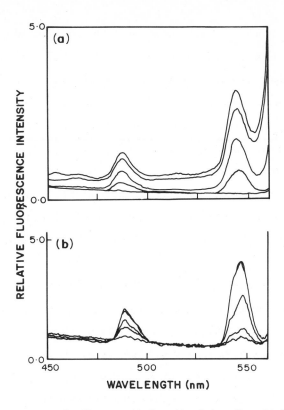

Fig. 2a,b. Fluorescence spectra of terbium upon binding with **a** ß-crystallin and **b** δ-crystallin

Table 1. Putative calcium-binding regions in various subunits of ß-crystallin

ßB₁	CRYSTALLIN	118	Y P R W D T W T S S Y R S D	131
ßB₂	CRYSTALLIN	78	Y P R W D S W T S S R R T D	91
ßB₃	CRYSTALLIN	86	Y P A W D A W S S S R R S D	99

ßB₁ CRYSTALLIN
182 S I T V V S S G T W V G Y Q Y P G Y R G Y Q Y L L E 207
ßB₂ CRYSTALLIN
141 S S V R V Q S G T W V G Y Q Y P G Y R G L Q Y L L E 166
ßB₃ CRYSTALLIN
149 A S I R V I N G T W V G Y E F P G Y R G R Q Y V F E 174

A combination of protein sequence and X-ray crystallographic data has shown a distant but clear structural relationship between ß- and γ-crystallins which has lead to the suggestion of an evolutionary relationship between these two crystallin classes and the existence of a protein structural ßγ-crystallin superfamily (Driessen et al. 1981; den Dunnen et al. 1985).

The protein structure of the ß-crystallin differs markedly from γ-crystallin in the presence of a N- and sometimes a C-terminal extension (Berbers et al. 1983). It is interesting that in spite of a proposed relationship with calmodulin, the protein S actually exhibits much greater similarity with the ßγ-crystallins superfamily (Wistow et al. 1985) but similarity is not very strong in the region where calcium binds (Teintze et al. 1988). Protein S of *Mixococcus xanthus* has two single calcium-binding sites with a sequence that is homologous to those of calmodulin sites that form the Ca^{2+}-binding loop but without the helices on each side (Inouye et al. 1983). It is not known why γ-crystallins, in spite of their similarity with ß-crystallin, do not bind calcium ions. These observations led us to classify ß-crystallin as a single class of unique calcium-binding protein superfamily. The low affinity and moderate capacity of calcium binding and other relevant questions related to ß-crystallins are yet to be answered.

Upon binding calcium, the ß-crystallin molecule undergoes conformational changes in both its secondary and tertiary structures, as evident by circular-dichroism and fluorescence spectroscopy (Sharma et al. 1989a). The rate of photodamage of Trp is also influenced and increased in presence of calcium ions. This may be due to the presence of several Trp and Tyr residues in and around the calcium-binding sites which become accessible to irradiation. It is more meaningful in the sense that the lens is always exposed to radiation.

The carbocyanine dye Stains-all has been used as a probe for calcium-binding proteins. Caday and Steiner (1985) has studied the interaction of this dye with various Ca^{2+}-binding proteins and found that different proteins interact with Stains-all differently, and yield different spectral bands. ß-crystallin induces the J band at 600-630 nm upon interacting with Stains-all (Sharma et al. 1989b). This behavior is similar to parvalbumin but is different from calmodulin and troponin C, which induce the additional γ band (500 nm) in the dye (Caday et al. 1986). We discuss these aspects in chapter 3 in this volume.

The lens-specific distribution of ß-crystallins, its evolutionary conserved structure, its high abundance in lens, and finally its Ca^{2+}-binding properties suggest that this molecule is an easy target in cataract. Thus, any alterations in the ß-crystallin may have broad physiological effects in aging and in cataract. We discuss below some specific examples of how this protein and its calcium-binding properties appear to be affected in lens disorders such as aging and cataract.

δ-CRYSTALLIN

The core protein in avian and reptilian lenses is a molecule that is distinctly different from α, ß, and γ-crystallins and is called δ-crystallin. It is also remarkable in being the only crystallin with an appreciable α-helical chain folding (Piatigorsky 1984b). δ-Crystallin is the principal protein of the embryonic chicken lens and constitutes 40-50% of the soluble protein of mature chicken lens (Piatigorsky 1981). The amount of δ-crystallin, however, varies greatly in different birds and reptiles. This molecule is usually composed of two related primary gene products of slightly different sizes, 48 and 50 kDa, which are differentially synthesized under different ionic conditions. This crystallin is found in the lens as a tetrameric aggregate. The sequence of δ-crystallin suggests a rough twofold repeat (Nickerson et al. 1985).

Incubation of chick lenses in $CaCl_2$ solution also leads to opacification, just as with mammalian lenses. Results of equilibrium dialysis experiments using radioactive $^{45}CaCl_2$, showed that Ca^{2+} ions binds to the δ-crystallin tetramer with a dissociation constant of about 0.23 mM (Sharma et al. 1989a). δ-Crystallin also binds to and increases the Tb^{3+} fluorescence intensity at 488 and 545 nm in much the same way as ß-crystallin does (Figure 2b).

When we analyzed the sequence of δ-crystallin for the presence of any calcium binding sites, we found that the region comprising residues 300-350 in the chicken $δ_2$-crystallin might have the calmodulin type EF-hand helix-loop-helix motif (Kretsinger and Nockolds 1973). In Figure 3, we show the helical wheel or Edmundson projection diagram (Schiffer and Edmundson 1967) of this region. It is worthy of note that the hydrophilic residues are naturally clustered together, while hydrophobic residues are segregated into another cluster, making the helix an amphipathic one. That the residues in the sequence 300-317 adopt the α-helical fold has been surmised (Nickerson and Piatigorsky 1984). In the sequence of chicken $δ_1$-crystallin, a closely related variant protein, the amino acid replacements in this region are conformationally and polarity wise conservative ones of Ile for Val in position 310 and again for Leu-317 of the sequence (Nickerson et al. 1986).

The residues Pro in position 318 and Ser in 319 are both nonhelicogenic and would form the start of the loop region. The 13 residue sequence 318-330 contains several nonhelicogenic residues and interestingly enough, contains in alternating positions residues with side chains that are capable of cation binding, namely Thr, Asn, Asp, Gln, Asp, and Glu. Here again, the replacement of Tyr-321 by Phe and Asn-322 by Ser in δ_1- crystallin (Nickerson et al. 1986) is innocuous in terms of ion-binding ability. The sequence 331-350 contains many helicogenic residues and is predicted to be in the α-helical conformation (Nickerson and Piatigorsky 1984). As Figure 3 reveals, the striking feature here too, is the segregation of the hydrophobic and hydrophilic residues to produce an amphipathic helix. Again, residue replacement seen in δ_1-crystallin (Leu for Phe-333 and Ala for Val 345) are functionally acceptable ones.

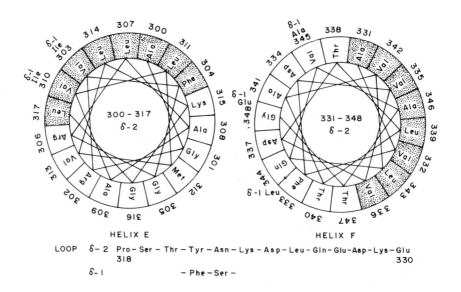

Fig. 3. The α-helical wheel projection residues 300-317 (*left*), and residues 331-350 (*right*), and the loop region sequence 318-330 of chick δ_2-crystallin. The residue changes seen in chick lens δ_1-crystallin are also shown at the respective positions

We have attempted to compare and align the calcium-binding loop in δ-crystallin with that of parvalbumin, calmodulin, and intestinal calcium-binding protein (ICaBP). Table 2 offers such a comparison, along the lines of Tufty and Kretsinger (1975) and Szebenyi and Moffat (1986). In case of δ-crystallin the first residue (x) is Ser or if aligned for maximal similarity, Thr (Table 2). The first Ca-ligand at first position is Asp in most if not all CaM family sequence. The nature of this ligand is critical in promoting the proper fold of the N-terminal half of the binding loop. A loop missing this X-ligand (Asp) does not bind Ca^{2+} with high affinity as suggested in EF hand test rule (Szebenyi and Moffat 1986). This may be a reason for the low affinity of δ-crystallin towards Ca^{2+}, since it is Thr (or Ser) in this protein rather than Asp. In sum, the region 300-350 in δ-crystallin possesses all the features described for the EF-hand motif of the calmodulin family of calcium-binding proteins.

It may be added here that Ca^{2+}-dependent activation of δ-crystallin to bind to avian eye lens cell membranes has been reported (Alcala et al. 1982). In this context, Turnell et al. (1986) have compared the amino acid sequence of δ-crystallin with that of human amyloid protein A and predicted a neutral stretch of four amino acids (Gly 200-Ser 201-Gly 202-Ala 203) in a loop region between presumed α-helices which acts as putative calcium- and lipid-binding sites in δ-crystallin.

The conformational changes induced by the addition of calcium ions to solutions of proteins was seen to be different for ß- and δ-crystallins. In the case of ß-crystallin the fluorescence spectral band is red shifted about 10 nm, and the intensity increased about 10%, upon addition of $CaCl_2$ at a mol ratio of 4 or more. δ-crystallin, however, behaves peculiarly in the sense that, unlike other crystallins, it shows an unusually blue shifted emission band at 315 nm and other at 325 nm. Titration with calcium ions does not cause any band shift in this case but reduces the intensity of emission. The difference absorption spectrum of δ-crystallin, in the presence and absence of Ca^{2+}, displays perturbations in the tertiary structural environment around the aromatic residues in the 260-300 nm region. Such calcium-induced structural changes probably might account for the differences in the crystal stability and geometry of δ-crystallin in the presence and absence of calcium (Mylvaganum et al. 1987).

Like other calcium-binding proteins, δ-crystallin also interacts with the carbocyanine dye, Stains-all, and induces the γ band of the dye at 500 nm both in absorption and CD that disappears upon the addition of calcium. It does not, however, induce the J band (600-650 nm) of the dye, a point that we have discussed elsewhere (Sharma et al. 1989b; see chapt. 3 in this Vol.). It is of interest to note here that δ-crystallin is actually the enzyme argininosuccinate lyase (ASL), that has been recruited for use as the structural component in the lens, either as itself or slightly modified (Piatigorsky et al. 1988). The enzyme ASL is also a tetramer of 50 kDa subunits and acidic pI, and has been studied in some detail by Ratner (1973) but its calcium-binding properties have not been investigated to date.

Table 2. Sequence alignment of calcium binding sites in δ-crystallin with EF-hand loop

	1	2	3	4	5	6
Parvalbumin (90-101)	Asp Ser Asp Gly	Asp Gly	Lys Ile Gly	Val	Asp	Glu
Calmodulin I (20-31)	Asp Lys Asp Gly	Asn Gly	Thr Ile Thr	Thr	lys	Glu
II (56-67)	Asp Ala Asp Gly	Asn Gly	Thr Ile Asp	Phe	Pro	Glu
III (93-104)	Asp Lys Asp Gly	Asn Gly	Tyr Ile Ser	Ala	Ala	Glu
IV (129-140)	Asn Ile Asp Gly	Asp Gly	Glu Val Asn	Tyr	Glu	Glu
ICaBP III (54-65)	Asp Lys Asn Gly	Asp Gly	Glu Val Ser	Phe	Glu	Glu
δ₁	Ser Thr Tyr Asp	Lys Asp	Leu Gln Glu	Asp	Lys	Glu
δ₂	Ser Thr Phe Ser	Lys Asp	Leu Gln Glu	Asp	Lys	Glu
δ₁ (aligned)	Thr Tyr Asp Lys	Asp Leu	Gln Glu Asp	Lys	*	Glu
δ₂ (aligned)	Thr Phe Ser Lys	Asp Leu	Gln Glu Asp	Lys	*	Glu

PHYSIOLOGICAL ASPECTS OF CALCIUM BINDING

Several studies have examined the quantitative and qualitative changes that occur in various crystallins upon aging and cataract (Bloemendal 1981). The best-known age-related processes are the formation of high molecular weight protein aggregate, insolubilization, decrease of water content, deamidation of asparagine and glutamine residues, and of course alteration in calcium level. Besides free radical reactions as intermediate steps in photooxidation and peroxidation are put forward as possible causes of the aging of lens proteins and in cataract. One of the important modifications that occurs in lens crystallins is the oxidation of Trp, His, Met, and Cys residues. Analysis of calcium-binding properties of ß-crystallin modified by irradiation using Stains-all as a probe suggests the loss of its ability to bind calcium ions, as shown in Figure 4 (Sharma et al. to be published).

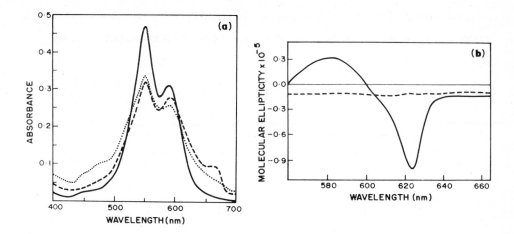

Fig. 4a,b. a Absorption spectra of Stains-all bound to native ($\cdots$), and to irradiated (----) ß-crystallin. *Solid line* represents the spectrum of dye only. **b** Circular-dichroism spectra of Stains-all bound to native (——), and to irradiated ($\cdots$) ß-crystallin

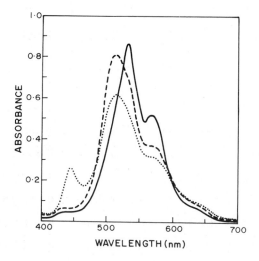

Fig. 5. Absorption spectra of Stains-all (——); dye bound to δ-crystallin (-----); dye bound to irradiated δ-crystallin ($\cdots$)

Almost identical changes are observed in case of its principal subunit (βB_2). These results agree with the data obtained by Duncan and Jacob (1984) in which they postulated that the binding affinity of Ca^{2+}-binding species of lens decreases with age and in cataract (it was not known earlier that it is ß-crystallin that binds calcium). δ-crystallin, however, retains its Ca^{2+}-binding ability even upon prolonged irradiation (Figure 5), which may be one reason why birds are thought to be less prone to cataract.

REFERENCES

Adams DR (1929) The role of calcium in senile cataract. Biochem J 23:902-912

Alcala J, Maisel H, Katar M, Ellis M (1982) δ-crystallin is a chick lens fibre cell membrane extrinsic protein. Exp Eye Res 35:379-383

Bax B, Slingsby C (1989) Crystallization of a new form of the eye lens protein beta B2-crystallin. J Mol Biol 208:715-717

Berbers GAM, Hoekman WA, Bloemendal H, de Jong WW, Kleinschmitt T, Branitzer G (1983) Proline and alanine rich N-terminal extension of the basic bovine ß-cristallin B1 chains. FEBS Lett 161:225-229

Bernardini G, Perrachia C (1981) Gap junction crystallization in lens fibers after an increase in cell calcium. Invest Ophthalmol Vis Sci 21:291-299

Bloemendal H (1981) Crystallins. In: Bloemendal H (ed) Molecular and cellular biology of the eye lens. John Wiley, New York, Chichester, Brisbane, Toronto, pp 2-47

Blundell T, Lindeley P, Miller L, Moss D, Slingsby C, Tickle I, Turnell B, Wistow G (1981) The molecular structure and stability of the eye lens: X-ray analysis of γ-cristallin II. Nature 289:771-777

Borchman D, Paterson CA, Delamere NA (1989) Ca^{2+}-ATPase activity in the human lens. Curr Eye Res 8:1049-1054

Burge WE (1909) Analysis of the ash of the normal and cataractous lens. Arch Ophthalmol 38:447-453

Caday CG, Steiner RF (1985) The interaction of calmodulin with carbocyanine dye Stains-all. J Biol Chem 260:5985-5990

Caday CG, Lambooy PK, Steiner RF (1986) The interaction of Ca^{2+}-binding proteins with the carbocyanine dye Stains-all. Biopolymers 25:1579-1595

Clark JI, Mengel L, Bagg A, Benedek GB (1980) Cortical opacity, calcium concentration and fibre membrane structure in the calf lens. Exp Eye Res 31:399-410

Clark JI, Danford-Kaplan ME, Delaye M (1988) Calcium decreases transparency of homogenate from lens cortex and has no effect on nucleus. Exp Eye Res 47:447-455

David LL, Shearer TR (1984) Ca^{2+} activated proteolysis in the lens nucleus during selenite cataractogenesis. Invest Ophthalmol Vis Sci 25:1275-1283

den Dunnen JT, Moormann RJM, and Schoenmakers JGG, (1985). Rat lens ß-cristallins are internally duplicated and homologous to γ-cristallins. Biochim Biophys Acta 824:295-303

Driessen HPC, Herbrink P, Bloemendal H, de Jong WW (1981) Primary structure of the bovine ß-cristallin B_p chain: internal duplication and homology with γ-cristallin. Eur J Biochem 121:83-91

Duncan G, Bushell AR (1975) Ion analysis of human cataractous lenses. Exp Eye Res 20:223-230

Duncan G, Jacob TLC (1984) Calcium and the physiology of cataract. CIBA Foundation Symp 106:132-152

Duncan G, van Heyningen R (1976) Differences in the calcium binding capacity of normal and cataractous lenses. Documenta Ophthalmol 8:229-232

Duncan G, van Heyningen R (1977) Distribution of non-diffusable calcium and sodium in normal and cataractous human lenses. Exp Eye Res 25:183-193

Galvin A, Louis CF (1988) Calcium regulation by lens plasma membrane vesicles.
    Arch Biochem Biophys 264:472-481
Hamilton PM, Delamare NA, Paterson CA (1979) The influence of calcium on lens ATPase activity.
    Invest Ophthalmol Vis Sci 18:433-436
Hightower KR (1983) The influence of calcium on protein synthesis in the rabbit lens.
    Invest Ophthalmol Vis Sci 24:1422-1426
Hightower KR (1985) Cytotoxic effects of internal calcium on lens physiology: a review.
    Curr Eye Res 4:453-459
Hightower KR (1986) The role of intracellular and bound calcium in the process of opacification. In: Duncan G
    (ed) The lens: Transparency and cataract. Eurage, Rijswijk, The Netherlands, pp 87-95
Hightower KR, Hind D (1983) Cytotoxic effects of calcium on sodium-potassium transport in the mammalian
    cells. Curr Eye Res 2:239 246
Hightower KR, Reddy VN (1982a) Calcium induced cataract. Invest Ophthalmol Vis Sci 22:263-267
Hightower KR, Reddy VN (1982b) Calcium content and distribution in human cataractous lens.
    Exp Eye Res 34:413-421
Hightower KR, McCready JP, Goudsmit EM (1987) Calcium induced opacification is dependent upon lens pH.
    Curr Eye Res 6:1415-1420
Inouye S, Franceschini T, Inouye M (1983) Structural similarities between the development-specific protein S
    from a gram negative bacterium, *Myxococcus xanthus*, and calmodulin.
    Proc Natl Acad Sci USA 80:6829-6833
Ireland M, Maisel H (1984) Evidence for a calcium activated protease specific for lens intermediate filaments.
    Curr Eye Res 3:423-429
Jedziniak JA, Kinoshita JH, Yates E, Hocker L, and Benedek GB (1972) Calcium induced aggregation of
    bovine lens α-cristallin. Invest Ophthalmol Vis Sci 11:905-915
Kretsinger RH, Nockolds CE (1973) Carp muscle calcium binding protein. J Biol Chem 248:3313-3326
Lohmann W, Schmehl W, Strobel J (1986) Nuclear cataract-oxidative damage to the lens.
    Exp Eye Res 43:859-862
Lorand L, Conrad SM, Velasco PT (1985) Formation of a 55000 weight cross-linked dimer in the Ca^{2+}-treated
    lens. A model for cataract. Biochemistry 24:1525-1531
Mylvaganum SE, Slingsby C, Lindely PF, Blundell TL (1987) Preliminary X-ray studies of adult δ-cristallin:
    evidence of a space-group transition. Acta Cryst B 34:580-582
Nickerson JM, Piatigorsky J (1984) Sequence of a complete chicken δ-cristallin cDNA.
    Proc Natl Acad Sci USA 81:2611-26
Nickerson JM, Wawrousek EF, Hawkins JW, Wakil AS, Wistow GJ, Thomas G, Norman BL, Piatigorsky J
    (1985). The complete sequence of the chick δ1-crystallin gene and its 5'-flanking region.
    J Biol Chem 260:9100-9105
Nickerson JM, Wawrousek EF, Borras T, Hawkins JW, Norman BL, Filpula DR, Nagle JW, Ally AH,
    Piatigorsky J (1986) Sequence of the chicken δ2-crystallin gene and its intergenic spacer: extreme
    homology with the δ1-crystallin gene. J Biol Chem 261:552-557
Patmore L, Maraini G (1986) A comparison of membrane potentials, sodium and calcium levels in normal and
    cataractous human lenses. Exp Eye Res 43:1127-1130
Piatigorsky J (1980) Intracellular ions, protein metabolism and cataract formation. Curr Topic Eye Res 3:1-39
Piatigorsky J (1981) Lens differentiation in vertebrates: A review of cellular and molecular features.
    Differentiation 19:134-153
Piatigorsky J (1984a) Lens crystallins and their gene family. Cell 38:620-621
Piatigorsky J (1984b) Delta crystallins and their nucleic acids. Mol Cell Biochem 59:33-56
Piatigorsky J, O'Brien WE, Norman BL, Kalumuck K, Wistow GJ, Borras T, Nickerson JM, Wawrousek EF
    (1988) Gene sharing by δ-crystallin and argininosuccinate lyase.
    Proc Natl Acad Sci USA 85:3479-3483
Ratner S (1973) Enzymes of arginine and urea synthesis. Adv Enzymol 39:1-90
Schiffer M, Edmundson AB (1967) Use of helical wheels to represent the structures of proteins and to identify
    segments with helical potential. Biophys J 7:121-134

Sharma Y, Rao CM, Narasu ML, Rao SC, Somasundaram T, Gopalakrishna A, Balasubramanian D (1989a) Calcium ion binding to ß- and δ-crystallins. The presence of the "EF-hand" motif in δ-crystallin that aids in calcium ion binding. J Biol Chem 264:12794-12799

Sharma Y, Rao CM, Rao SC, Gopalakrishna A, Somasundaram T, Balasubramanian D (1989b) Binding site confirmation dictates the color of the dye Stains-all. A study of the binding of this dye to the eye lens proteins crystallins. J Biol Chem 264:20923-20927

Shearer TR, David LL, Anderson RS (1987) Selenite cataract, a review. Curr Eye Res 6:289-300

Slingsby C, Driessen HPC, Mahadevan D, Bax B, Blundell TL (1988) Evolutionary and functional relationships between basic and acidic ß-crystallins. Exp Eye Res 46:375-403

Spector A, Rothschild C (1973) The effect of calcium upon the reaggregation of bovine α-crystallin. Invest Ophthalmol Vis Sci 12:225-231

Szebenyi DME, Moffat K (1986) The refined structure of vitamin D-dependent calcium-binding protein from bovine intestine: molecular details, ion binding and implications from the structure of other calcium-binding proteins. J Biol Chem 261:8761-8777

Teintze M, Inouye M , Inouye S (1988) Characterization of calcium-binding sites in development-specific protein S *Myxococcus xanthus* using site-specific mutagenesis. J Biol Chem 263:1199-1203

Tufty RM, Kretsinger RH (1975) Troponin C and parvalbumin calcium binding regions predicted in myosin light chain and T_4 Lysozyme. Science 187:167-169

Turnell W, Sarra R, Glover ID, Baum JO, Caspi D, Baltz ML, Pepys MB (1986) Secondary structure prediction of human SAA1. Presumptive identification of calcium and lipid binding sites. Mol Biol Med 3:387-407

van Heyningen R (1972) The human lens. I. A comparison of cataracts extracted in Oxford (England) and Shikarpur (W. Pakistan). Exp Eye Res 13:136-147

Wistow G, Slingsby C, Blundell T, Driessen H, de Jong W, Bloemendahl H (1981) Eye lens proteins: the three dimensional structure of ß-crystallin predicted from monomeric γ-crystallin. FEBS Lett 133:9-16

Wistow G, Summers L, and Blundell T (1985) Myxococcus xanthus spore coat protein S may have a similar structure to vertebrate lens ßγ-crystallins. Nature 315:771-773

Wistow GJ, Piatigorsky J (1988) Lens crystallin: the evolution and expression of proteins for a highly specialized tissue. Ann Rev Biochem 57:479-504

Yoshida H, Murachi T, Tsukahara I (1984) Limited proteolysis of bovine lens α-crystallin by calpain, a Ca^{2+}-dependent cysteine protease, isolated from the same tissue. Biochem Biophys Acta 798:252-259

CLONING AND SEQUENCING OF A NEW CALMODULIN RELATED CALCIUM-BINDING PROTEIN CONTROLLED BY CYCLIC *AMP* DEPENDENT PHOSPHORYLATION IN THE THYROID: CALCYPHOSINE

Anne Lefort, Raymond Lecocq, Frédérick Libert, Christophe Erneux, Françoise Lamy, and Jacques Dumont

INTRODUCTION

The thyroid cell, as many other cells, is subjected to controls through the cyclic AMP, the calcium-phosphatidylinositol and the growth factor receptors tyrosine protein kinases cascades. In the study of the proteins whose synthesis and postranslational covalent modifications are modulated by these cascades, we identified on two-dimensional gel electrophoresis a protein belonging to the calmodulin superfamily, which is phosphorylated in response to cyclic AMP, which binds calcium and which is synthesized in differentiating conditions: calcyphosine (Lefort et al. 1989). In this short review, we summarize recent data on the identification, cloning, sequencing, and properties of calcyphosine.

CALCYPHOSINE: A PROTEIN INVOLVED IN THE REGULATION OF DOG THYROID CELL FUNCTION AND DIFFERENTIATION

Dog thyroid function is mainly regulated by thyrotropin (TSH) through adenylate cyclase activation. Calcyphosine (M_r 24000 and pI 5.4) was initially identified as being a major phosphorylated substrate for cyclic AMP-dependent protein kinase following stimulation by TSH. Indeed, the use of two-dimensional gel electrophoresis allowed us to identify 13 specific proteins whose phosphorylation is increased in response to TSH in dog thyroid slices

376

(Figure 1). This effect, already observed after 10 min incubation with TSH, persists as long as the hormone is present in the incubation medium and is reproduced by all agents which increase the level of cyclic AMP (dibutyryl cyclic AMP and cholera toxin) (Lecocq et al. 1979; Lamy et al. 1984). Most of these phosphorylated proteins are low-abundant proteins as revealed by silver staining of the proteins except for calcyphosine (spot number 5 in Figure 1) which is also the most heavily phosphorylated one. Silver staining of the proteins also reveals two partially overlapping spots at the level of calcyphosine, which is consistent with the fact that the phosphorylated protein contains additional negative charges and which suggests that a sizeable proportion of calcyphosine is phosphorylated .

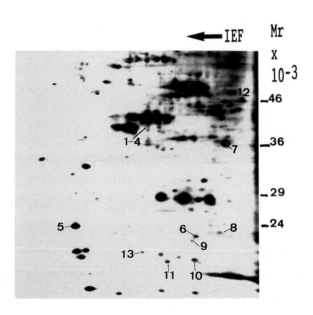

Fig.1. Autoradiography of two-dimensional electrophoretic separation of [^{32}P] phosphate-labeled polypeptides from dog thyroid slices incubated with TSH at 10 mU/ml and (^{32}P)phosphate (0.5 mCi/ml) for 1 h. *Each spot corresponding to a modified protein phosphorylation compared to control is designated by an arbitrary number*

In fact, calcyphosine is subjected to multiple regulations in the dog thyroid cell: (1) It is rapidly phosphorylated during the full functional activation of the cell through the cyclic

AMP cascade; it is not phosphorylated in response to acetylcholine which activates the Ca^{2+}-phosphatidylinositol cascade and has some similar but mostly opposite effects to TSH (Lecocq et al. 1979; Lamy et al. 1984). (2) Its synthesis is enhanced by TSH and cyclic AMP analogs which trigger cell proliferation and maintain expression of the differentiated thyrocyte phenotype. It is decreased by epidermal growth factor (EGF) and 12-O-tetra-decanoyl-phorbol-13-acetate (TPA) which also trigger cell proliferation but repress expression of differentiation (Lamy et al. 1986; Lamy et al. 1989; Lefort et al. 1989). Antiserum against calcyphosine extracted from two-dimensional gels was raised in a rabbit. It proved to be monospecific, detecting only this protein in immunoblots of total dog thyroid proteins separated on two-dimensional gels (Lefort et al. 1989).

STRUCTURE OF CALCYPHOSINE

Calcyphosine cDNA was cloned from a lambda gt11 cDNA library constructed with poly(a) mRNA from dog thyroids. The library was screened using a 1/250 dilution of the anti-calcyphosine serum and positive clones were isolated and sequenced (Sanger et al. 1977). An open reading frame of 189 codons would encode a polypeptide of M_r 21104 (calculated value based on cDNA sequence) which is close to the apparent M_r 24000 of calcyphosine on SDS-polyacrylamide gel electrophoresis (SDS-PAGE). There is an excess of acidic amino acids (35 glutamic and aspartic acids) in comparison to basic amino acids (25 lysine and arginine) which could account for the acidic properties of calcyphosine. The stimulation of (^{32}P) phosphate incorporation into calcyphosine by TSH and by agents which increase the level of cyclic AMP in dog thyroid slices (Lecocq et al. 1979) suggests that this protein could be a substrate for cyclic AMP-dependent protein kinase. Since most physiological substrates for this enzyme contain at least two basic residues N-terminal to the phosphorylatable serine or threonine (Soderling et al. 1990) it is suggested that calcyphosine might be phosphorylated at serine 40. A computer search of the Protein Identification Resource databank (release 15.0, december 1987) using the FASTP program (Lipman and Pearson 1985) revealed significant similarities between calcyphosine and calmodulin. The calcium-regulated proteins of the calmodulin superfamily all share a common structural unit represented by the EF-hand domain described by Kretsinger (1980). This large family comprises proteins presenting two, three, four, or six such calcium-binding domains (Heizmann and Berchtold 1987).

Figure 2 shows the best alignment between calcyphosine and calmodulin. An insertion of 17 residues is present in the fourth putative Ca^{2+}-binding domain of calcyphosine which disrupts its EF-hand structure and, most probably, makes it nonfunctional. Interestingly, this insertion is located precisely at a place corresponding to the position of a conserved intron in the calmodulin gene family (Wilson et al. 1988).

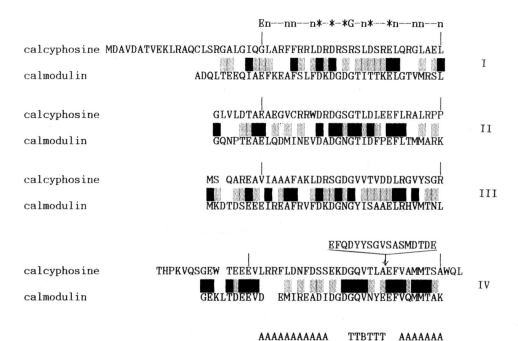

Fig.2. Comparison of the amino acid sequences of bovine brain calmodulin and dog thyroid calcyphosine. *Black symbols* identical residues; *stippled symbols* conservative replacements (positive score in PAM 250 matrix) (Dayhoff 1978). The consensus structure of the EF-hand domain is represented on top (*E* Glutamine; *n* hydrophobic residue; *G* Glycine; * oxygen containing residue; - any residue). *Roman numerals to the right* denote the four calcium-binding domains in calmodulin. The amino acids that constitute the EF-hand calcium-binding domains lie between the residues identified with an *arrow*. The *arrowhead* indicates the insertion site of fragment 161-177. The predicted secondary structure following Garnier (1978) is represented below the sequence (*A* α-helix; *B* ß sheet; *T* turn; *blank* random coil)

An analysis of the calcyphosine sequence reveals the expected pattern of residues conserved in EF-hand structures. The flanking α-helices in calcium-modulated proteins have been proposed to be amphipathic (Kretsinger 1980). In calcyphosine these residues predicted to be on the hydrophobic side of the helices (labeled *n* on Figure 2) agree with the consensus. In the potential calcium-binding loops, the residues found at the positions predicted to bind calcium are all amino acids with oxygen-containing side chains. The glycine at position 15, which is believed to cause a sharp bend in the calcium-binding loop, is conserved in calcyphosine except in domain 1, where an arginine residue is found. It is noteworthy that this arginine is part of the putative phosphorylation site at serine 40. It is tempting to speculate that this characteristic reflects the existence of some kind of cross-signaling between regulation of calcyphosine by the cyclic AMP and the calcium-phosphatidylinositol cascades respectively. Such sequence and structural similarities clearly make calcyphosine a member of the calmodulin superfamily. They indicate strongly that calcyphosine would be capable of binding one or several calcium ions. Calcyphosine ability to bind calcium has been directly confirmed by incubating a Western blot of separated dog thyroid proteins in a buffer containing ^{45}calcium according to Maruyama (1984) (Lefort et al. 1989). The hydropathy profile of calcyphosine, calculated according to the method of Kyte and Doolittle (1982), shows a highly hydrophilic character with no indication of transmembrane(s) segment(s) (Figure 3). The distribution of calcyphosine was tested in various dog tissues using immunodetection by anti-calcyphosine serum on Western blots of SDS-PAGE separated proteins (Lefort et al. 1989). These results were confirmed by hybridization of a Northern blot of dog tissues mRNAs with calcyphosine cDNA used as an hybridization probe (Figure 4). Calcyphosine is present at least in the thyroid, the salivary gland, the kidney, the lung and the brain.

The role of calcyphosine is, as yet, unknown. As calcyphosine is found in several tissues (Figure 4), it is probably not a protein involved in the specialized iodine metabolism of the thyrocyte. Its synthesis is not increased by mitogenic agents (EGF, TPA) other than TSH; it is therefore not an obligatory signal in the proliferation regulating cascades. The wide range of tissues containing this protein suggests a rather general target system such as ion transport or cell motility (Figure 5).

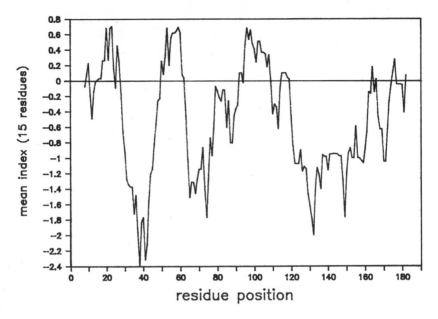

Fig. 3. Hydropathy profile of dog thyroid calcyphosine. The method of Kyte and Doolittle (1982) was used with a window of 15 residues

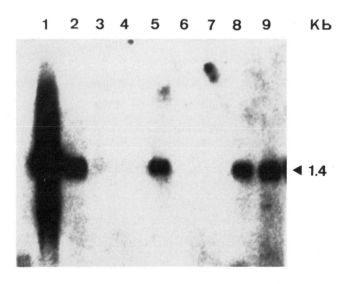

Fig. 4. Tissue distribution of calcyphosine revealed by Northern blot and hybridization with calcyphosine cDNA of 5 µg mRNA from dog. *1* Thyroid; *2* salivary gland; *3* stomach; *4* spleen; *5* kidney; *6* liver; *7* heart; *8* lung; *9* brain

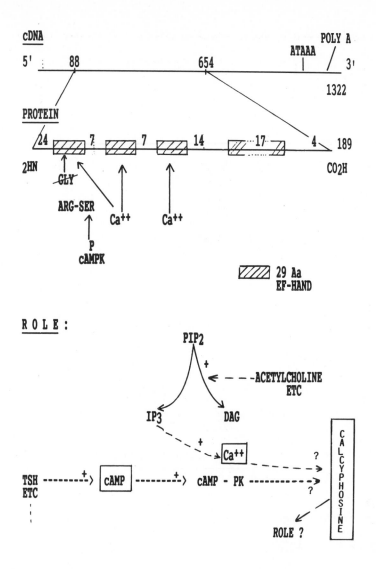

Fig. 5. Structure and regulations of dog thyroid calcyphosine

POTENTIAL BINDING OF CALCYPHOSINE TO TARGET PROTEINS

Primary sequence of calcyphosine indicated that calcyphosine clearly belongs to the calmodulin superfamily of proteins binding calcium ions through EF-hand domains. When

382

calmodulin binds calcium it exposes hydrophobic sites and can be adsorbed to a hydrophobic matrix such as phenyl-sepharose (Gopalakrishna and Anderson 1982).

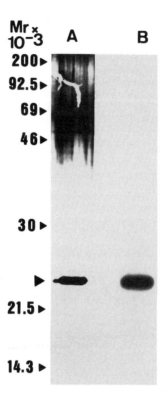

Fig. 6A,B. 15% SDS-PAGE separation of purified calcyphosine from dog thyroid tissue. A silver staining of calcyphosine purified by phenyl-sepharose CL4B. B autoradiography of purified ^{125}I-calcyphosine labeled by lactoperoxidase method. (Van Eldik 1988)

This type of interaction is fully reversible. Our results suggest that calcyphosine may be purified according to the same method. In the presence of calcium, calcyphosine binds to the hydrophobic phenyl-sepharose column and can be eluted by the addition of EGTA (Figure 6A). One of the characteristics of calmodulin is its ability to associate to multiple proteins and/or key enzymes in a calcium-dependent manner (Burgess et al. 1984). To test if calcyphosine could bind other proteins, purified dog thyroid calcyphosine was iodinated by the lactoperoxidase method (Van Eldik 1988) (Figure 6B) and used as a probe in a nitrocellulose overlay of dog thyroid proteins. Preliminary results show that calcyphosine may bind to its own set of protein substrates. This interaction is calcium-dependent.

CONCLUSIONS

Calcyphosine shares common properties with calmodulin:

1. A EF-hand structure based on primary sequence determination after cDNA cloning.

2. Calcium-binding properties.

3. Calcium-dependent interaction to a hydrophobic column which indicates identical conformation changes induced by calcium.

4. A large distribution in dog tissues as shown by both Western and Northern blots.

Nevertheless, dog thyroid calcyphosine presents some important difference with calmodulin: it is phosphorylated in intact cells and slices in response to TSH by a cyclic AMP-dependent phosphorylation process. Its modulation by calcium and cyclic AMP suggests a role at the intersection of the calcium-phosphatidylinositol and cyclic AMP cascades.

To investigate the role of calcyphosine, we have tentatively identified a few proteins of M_r 40500, 43500, 46000 in dog thyroid which may bind to calcyphosine. These last data may suggest that calcyphosine triggers calcium effects and/or changes of cyclic AMP accumulation through the activation of specific target regulatory proteins.

REFERENCES

Burgess WH, Watterson DM, van Eldik LJ (1984) Identification of calmodulin-binding proteins in chicken embryo fibroblasts. J Cell Biol 99:550-557

Dayhoff MO, Schwartz RM, Orcutt BC (1978) A model of evolutionary change in proteins. Atlas of protein sequence and structure. National Biomedical Research Foundation, Silver Spring, Washington, Dayhoff MO (ed), Vol.5, (Suppl. 3):345-352

Garnier J, Osguthorpe DJ, Robson B (1978) Analysis of the accuracy and implications of simple methods for predicting the secondary structure of globular proteins. J Mol Biol 120:97-120

Gopalakrishna R, Anderson WB (1982) Application for purification of calmodulin by phenyl-sepharose affinity chromatography. Biochem Biophys Res Commun 104:830-836

Heizmann CW, Berchtold MW (1987) Expression of parvalbumin and other Ca^{2+}-binding proteins in normal and tumor cells: a topical review. Cell calcium 8:1-41

Kretsinger RH (1980) Structure and evolution of calcium-modulated proteins. CRC Crit Rev Biochem 8:119-174

Kyte J, Doolittle RF (1982) A simple method for displaying the hydropathic character of a protein. J Mol Biol 157:105-132

Lamy FM, Lecocq RE, Contor LS, Dumont JE (1984) Carbamylcholine and thyrotropin activate distinctive pathways of protein phosphorylation in dog thyroid. Biochimica et Biophysica Acta 802:301-305

Lamy FM, Roger PP, Lecocq RE, Dumont JE (1986) Differential protein synthesis in the induction of thyroid cell proliferation by thyrotropin, epidermal growth factor or serum. Eur J Biochem 155:265-272

Lamy FM, Roger PP, Lecocq RE, Dumont JE (1989) Protein synthesis during induction of DNA replication in thyroid epithelial cells: evidence for late markers of distinct mitogenic pathways. J Cell Phys 138:568-578

Lecocq RE, Lamy FM, Dumont JE (1979) Pattern of protein phosphorylation in intact stimulated cells: thyrotropin and dog thyroid. Eur J Biochem 102:147-152

Lefort A, Lecocq R, Libert F, Lamy F, Swillens S, Vassart G, Dumont JE (1989) Cloning and sequencing of a calcium-binding protein regulated by cyclic AMP in the thyroid. EMBO J 8:111-116

Lipman DJ, Pearson WR (1985) Rapid and sensitive protein similarity research. Science 227:1435-1441

Maruyama K, Mikawa T, Ebashi S (1984) Detection of calcium binding proteins by ^{45}Ca autoradiography on nitrocellulose membranes after sodium dodecyl sulfate gel electrophoresis. J Biochem 95:511-519

Sanger F, Nicklen S, Coulson AR (1977) DNA sequencing with chain-terminating inhibitors. Proc Natl Acad Sci USA 74:5463-5467

Soderling TR (1990) Protein kinases. J Biol Chem 265:1823-1826

van Eldik LJ (1988) Preparation and use of iodinated calmodulin for studies of calmodulin-binding proteins. Methods in Enzymology 159:667-675

Wilson PW, Rogers J, Harding M, Pohl V, Pattyn G, Lawson DEM (1988) Structure of chick chromosomal genes for calbindin and calretinin. J Mol Biol 200:615-625

A 22 kDa CALCIUM-BINDING PROTEIN, SORCIN, IS ENCODED BY AMPLIFIED GENES IN MULTIDRUG-RESISTANT CELLS

Marian B. Meyers

INTRODUCTION

A multidrug-resistant (mdr) cell is one which is selected for resistance to a single agent and is simultaneously cross-resistant to a variety of other drugs. The cells described in this overview were selected for resistance to such natural product cancer drugs as vincristine, adriamycin, or actinomycin D (Biedler and Peterson 1981). Cellular cross-resistance to agents with such diverse molecular structures and such varied cellular mechanisms of toxic action, as characterize the drugs in this category, is both interesting and perplexing. Development of multidrug resistance frequently occurs in cancer patients being treated with chemotherapy and is a major factor in the failure of drug treatment. The need to overcome or circumvent clinical drug resistance is the motivation for the vigorous studies of multidrug resistance being conducted in many laboratories all over the world (Bradley et al. 1988; Tsuruo 1988; Biedler and Meyers 1989; Ling 1989). A number of biochemical and phenotypic characteristics have been shown to be associated with resistance development in cultured cells. The most common characteristic is overproduction of a plasma membrane protein called P-glycoprotein (Pgp). Various molecular weights of Pgp have been reported, the most common one being 170000 (Endicott and Ling 1989). The deduced amino acid sequence and hydropathy plot information for Pgp suggests that the glycoprotein has an ion channel-type structure (Ling 1989). Pgp can associate with or bind to a number of drugs in the multidrug resistance category as well as to other substances including progesterone and calcium channel blockers (Safa et al. 1986; Yang et al. 1989; Safa et al. 1987). This binding capability may be part of the putative function of Pgp as an efflux pump for these substances. Effective removal of drugs from intracellular compartments may be a major mechanism of resistance in cells with increased production of Pgp. Many cell lines have elevated levels of Pgp protein as a result of amplification of Pgp genes (Riordan et al. 1985). There are two human Pgp genes and three each in hamster and mouse cells (Ng et al. 1989). The Pgp genes are part of an amplifiable domain, or amplicon, which includes five other genes. One of these other genes encodes a protein named sorcin (soluble resistance-related calcium-binding protein) (Van der Bliek et al. 1986a; Meyers et al. 1987; Van der Bliek et al. 1988).

IDENTIFICATION OF SORCIN

Sorcin was first identified in cells of a vincristine-resistant Chinese hamster lung line (DC-3F/VCRd-5L). These cells were found to have marker chromosomes containing homogeneously staining regions (HSRs) (Meyers and Biedler 1981). Normally, chromosomes have specific striation patterns after trypsin-Giemsa staining which are characteristic for each numbered chromosome and which are seen along the entire length of each chromosome. Regions of chromosomes which are not banded (have uniform or homogeneously staining patterns) have been shown to contain one or more reiterated or amplified genes. Speculation about a protein product encoded by the presumptive amplified genes in the HSRs in DC-3F/VCRd-5L cells led to analysis of proteins in those cells by two-dimensional (2D) gel electrophoresis. Radioautograms of gels on which ^{35}S-methionine-labeled proteins from DC-3F/VCRd-5L and parental drug-sensitive DC-3F cells were separated revealed that while most of the labeled proteins in the two cell lines were quantitatively and qualitatively very similar, DC-3F/VCRd-5L cells synthesized a great deal more of a protein with an approximate molecular weight of 20000 and an isoelectric point of 5.7 as compared to DC-3F cells (Meyers et al. 1985; Meyers et al. 1987). The new protein was first given the designation V19 because it was originally found in vincristine-resistant cells and had an apparent molecular weight of 19000 as determined by rate of migration on sodium dodecyl sulfate gels. The designation was soon found to be inappropriate. An extensive screen of mdr cells in this and other laboratories, first by 2D gel analysis and then by immunoblot analysis with a polyclonal antibody raised in chickens against sorcin purified by preparative gel electrophoresis (Meyers et al. 1987), showed that mdr cells selected for resistance to a wide variety of drugs including colchicine, actinomycin D, taxol, vinblastine, etoposide, teniposide, and adriamycin could overproduce the protein that would eventually be called sorcin (Meyers and Biedler 1981; Polotskaya et al. 1983; Meyers et al. 1985; Koch et al. 1986; Shen et al. 1986; Meyers et al. 1987; Hamada et al. 1988; Van der Bliek et al. 1988; Roberts et al. 1989; Sugawara et al. 1989). The resistant lines were selected from cells of human, hamster, and mouse origin suggesting that sorcin is a highly conserved protein. While not all mdr cells produce sorcin in abundance, at least half of all mdr cells examined do overproduce the protein and the synthesis is constitutive, i.e., as long as the cell remains an mdr type, sorcin level usually remains high (Table 1). An example of the seeming randomness of drug selection resulting in sorcin overproduction is a study in this laboratory of seven human neuroblastoma [BE(2)-C] sublines selected for resistance to four different drugs, actinomycin D, adriamycin, vincristine, and colchicine. Of the seven, three overproduce sorcin (two actinomycin D-resistant lines and one of the three colchicine-resistant lines).

AMPLIFICATION OF SORCIN GENES

All of the cells which overproduce sorcin show some evidence (usually cytogenetic) of amplified genes, and this fueled the notion that increased synthesis of the protein was the result of increased gene copy number (Meyers and Biedler 1981). This idea was reinforced by a study of reversion of DC-3F/VCRd-5L cells. The cells were grown in the absence of drug for more than a year. During that course of time, the length of the HSRs on the marker chromosomes decreased while the amount of sorcin also decreased, suggesting that there were sorcin genes within the HSRs which were lost during the time off drug (Biedler et al. 1983). Further, the level of resistance to vincristine decreased during that time as did the amount of Pgp. In vitro translation of total poly (A$^+$)mRNA from DC-3F, DC-3F/VCRd-5L, and the revertant of the vincristine-resistant cells, followed by 2D electrophoresis of the products, showed that sorcin mRNA was increased in abundance in DC-3F/VCRd-5L as compared to DC-3F and that the amount of sorcin mRNA was much decreased in the revertant population (Meyers et al. 1985). Sucrose density fractionation of mRNA from DC-3F/VCRd-5L cells followed by 2D analysis of in vitro translation products from each fraction allowed estimation of the size of sorcin mRNA at about 1000 nucleotides, sufficient material to encode a 20 kDa protein (Meyers and Biedler in press).

Table 1. Multidrug-resistant cells examined for presence of sorcin

Cell line	Selective agent	Relative resistance	Sorcin[a]	Comments
Hamster[b]				
DC-3F	–	1	–	
DC-3F/VCRd-5L	VCR	2750	+	Amplified genes, increase in protein/mRna
DC-3F/AD X	ACT	10000	–	
DC-3F/DM XX	DM	880	–	
AUXB1	–	1	–	
CH[R]C5	CHC	160	+	Amplified genes, increase in protein/mRNA
Mouse[c]				
MAZ	–	1	–	
MAZ/VCR	VCR	3900	+	Increase in protein/mRNA
QUA	–	1	–	
QUA/ADj	ACT	9300	+	
QUA/ADsx	ACT	1420	–	1% of soluble protein is sorcin
L1210	–	1	–	
LIA5μm	VM26	1300	+	
L1210/VM	VM26	610	+	
L1210/VP	VP16	84		
Human[d]				
BE(2)-C	–	1	–	Neuroblastoma
BE(2)-C/ADR	ADR	60	–	Sorcin increase lost with time in culture
BE(2)-C/ACT.2	ACT	183	+	Sorcin increase constitutive
BE(2)-C/CHCg.2	CHC	48	+	
BE(2)-C/CHCb.2	CHC	54	–	
K562/ADM	ADR	95	+	
A2780	–	1	–	Ovarian carcinoma line
2780AD	ADR	160	+	Amplified sorcin genes

[a] – Indicates no increase in sorcin/mRNA or protein or amplification of sorcin genes
+ Indicates increase in protein and/or mRNA or amplification of sorcin genes
[b] References: van der Bliek et al. 1986a,b
de Bruijn et al. 1985
[c] References: Meyers et al. 1985
Roberts et al. 1989
[d] References: Meyers et al. 1985
Meyers and Biedler (in press)
Suguwara et al. 1989
van der Bliek et al. 1988
Abbreviations are: ACT, actinomycin D; ADR, adriamycin; CHC, colchicine; DM, daunorubicin; VCR, vincristine; VM26, teniposide; VP16, etoposide.

At about the time the information about sorcin mRNA was being determined, Dr. Piet Borst of the Netherlands Cancer Institute and his colleagues identified five gene classes amplified in a colchine-resistant Chinese hamster ovary line (CH[R]C5) and prepared a panel of cDNAs representing the various classes (de Bruijn et al. 1986). The use of these probes, in Southern and Northern analyses of DNA and RNA from DC-3F, DC-3F/VCRd-5L, and two other mdr sublines of DC-3F, one selected for resistance to actinomycin D (DC-3F/AD X) and one for resistance to daunorubicin (DC-3F/DM XX), revealed differential gene amplification and expression of the five gene classes among the three resistant lines (Van der Bliek et al. 1986a,b). DC-3F/VCRd-5L is the only one of these three mdr lines to overproduce the sorcin protein, but all three lines overproduce Pgp. Table 2 records ratios of DNA amounts in the resistant lines compared to control DC-3F cells. There are three class 2 genes designated 2a, 2b, and 2c (Van der Bliek et al. 1986a,b). A sixth gene class was discovered after this analysis was completed and is not included here (Van der Bliek et al. 1988). The levels of RNA expressed in these cells is commensurate with the approximate gene copy number except that class 3 gene expression in DC-3F/DM XX is disproportionately lower than expected and class 2 expression in DC-3F/AD X is disproportionately higher. These inconsistencies are being explored at the present time. Class 2 genes have been shown to encode Pgp.

Table 2. Gene amplification (fold increase) in multidrug-resistant DC-3F sublines

Cell line	Gene class						
	1	2a	2b	2c	3	4	5
DC-3F/VCRd-5L	1	1	1	150	40	30	30
DC-3F/AD X	4	4	4	6	1	1	1
DC-3F/DM XX	6	8	80	80	30	0.5	0.5

Class 4 and 5 genes are amplified in DC-3F/VCRd-5L but not in the other two lines. Either one of these was presumed to be a good candidate for an amplified sorcin gene. Class 4 cDNAs hybridize to two transcript sizes of 1000 and 2500 nucleotides. Class 5 cDNAs hybridize to a 3600 nucleotide transcript. The size coincidence between the 1000 nucleotide transcript recognized by class 4 cDNAs and the 1000 nucleotide mRNA encoding sorcin, previously determined in the sucrose density fractionation experiments, compelled the determination of the sequence of the class 4 cDNAs. The amino acid sequence of sorcin, deduced from the nucleic acid sequences of sorcin cDNAs encompassing the total 1000 nucleotide transcript, was compared to the amino acid sequences of tryptic peptides obtained from sorcin protein, purified by HPLC from an actinomycin D-resistant mouse subline. The cells of this subline, QUA/ADj, contain 1% by weight of sorcin in total soluble protein (Meyers et al. 1987). The deduced and actual sequences matched, demonstrating that (1) class 4 genes encode sorcin, (2) sorcin genes are amplified in DC-3F/VCRd-5L cells, and (3) sorcin genes are not amplified in the other two mdr lines which do not overproduce the sorcin protein. Thus, sorcin can be co-amplified with Pgp genes (Van der Bliek et al. 1986a). This co-amplification has been shown to occur in a number of other cell lines.

Co-amplification of gene classes 1,3,5, and 6 with Pgp genes occurs quite frequently as well, although the protein products of these gene are unknown (Van der Bliek et al. 1988). The amplicon containing sorcin and Pgp genes consists of at least six linked genes which map to the 1q26 chromosome in hamster (Jongsma et al. 1987; Biedler and Melera in press) and to chromosome 7q21 in human cells (Callen et al. 1987). The array of genes surrounding the Pgp genes in rodents is conserved in humans, perhaps suggesting some relationship among the genes or proteins.

Sorcin cDNA probes hybridize to four bands in Eco R1-digested genomic DNA from hamster cells (Van der Bliek et al. 1986b). These probes hybridize to two bands in Bam HI-digested DNA from human cells, a prominent 15 kb band and a weak 8 kb band. The 8 kb band is more sensitive to hybridization stringency than is the 15 kb band and it is this 8 kb band which corresponds to the human sorcin gene. The sorcin-related gene represented by the 15 kb band segregates to human chromosome 4. The nature of the gene is unknown (Van der Bliek et al. 1988).

As previously mentioned, sorcin cDNAs hybridize to transcripts of 1000 and 2500 nucleotides. These are transcribed from one gene and processed at multiple adenylation sites with overlapping sequences at the 5' end. The 1000 nucleotide transcript actually consists of a nested set of transcripts differing at their 3' ends. The purpose or function of the alter native polyadenylation sites is not known, although they may have regulatory significance (Van der Bliek et al. 1986a).

The 2500 nucleotide transcript is not a precursor of the shorter RNAs. There is no evidence that the larger RNA is translated into a sorcin-like protein larger than 20 kDa. How ever, in light of the fact that a sorcin-related gene has been detected in human cells (see above), the possibility of the existence of a sorcin-related protein (possibly a subunit associated with sorcin in a heterodimeric protein) remains. Indeed, the 2D gels used to detect sorcin in DC-3F/VCRd-5L also contained spots representing a 61 kDa and a 90 kDa protein overproduced in those cells as compared to DC-3F. These proteins have yet to be identified and their relationship to sorcin, if any, is unknown (Meyers et al. 1985).

AMINO ACID SEQUENCE OF SORCIN

The sorcin polypeptide contains 198 amino acids for a calculated molecular weight of about 22000 (Figure 1). There is no evidence that the protein is glycosylated although there is one putative N-glycosylation site within the sequence. There are four putative calcium-binding domains.

```
MAYPGHPGAGGGYYPGGYGG
APGGPSFPGQTQDPLYGYFA
SVAGQDGQIDADELQRCLTQ
SGIAGGYKPFNLETCRLMVS
MLDRDMSGTMGFNEFWELWA
VLNGWRQHFISFDSDRSGTV
DPQELQKALTTMGFRLNPQT
VNSIAKRYSTSGKITFDDYI
ACCVKLRALTDSFRRRDSAQ
QGMVNFSYDDFIQCVMTV
```

Fig. 1. Amino acid sequence of sorcin. (Van der Bliek et al. 1986a)

The sites proximal to the NH-terminal are EF-hand domains with homology to calmodulin calcium-binding sites. The second two sites are not typical EF-hand motifs although they also manifest considerable homology with calmodulin domains. The linker sequences between the calcium-binding sites are shorter in sorcin than in other calmodulin-like proteins and the ratio of aspartate to glutamate residues in sorcin is 3:1 instead of nearly 1:1 as in most well-characterized calcium-binding proteins. Each of the second two calcium-binding domains is immediately preceded by a putative cAMP-dependent protein kinase (A kinase) recognition site (-KRYS and -RRRDS-). The NH_2-terminal region of about 30 amino acids contains a high percentage of glycine, proline, and tyrosine residues, suggestive of potential for interaction of sorcin with cellular membranes (Van der Bliek et al. 1986a).

Direct binding assays with $^{45}Ca^{2+}$ have shown that sorcin protein binds calcium. This type of assay involves separation of proteins from drug-sensitive and -resistant cells by gel electrophoresis, transfer to nitrocellulose, and incubation with radiolabeled calcium. In many such studies, sorcin was found to be among the most, and in some cases the most, abundant calcium-binding protein in the resistant sorcin-overproducing cells (Koch et al. 1986; Meyers et al. 1987).

PHOSPHORYLATION STUDIES

In vitro phosphorylation studies with total soluble proteins from DC-3F, DC-3F/VCRd-5L, and other cells conducted in the presence and absence of various kinase activators including cAMP suggest that sorcin is phosphorylated in cells and that the level of phosphorylation is increased in the presence of cAMP, but not in the presence of calcium (Meyers 1989a). Furthermore, the catalytic subunit of cAMP-dependent protein kinase (A kinase) catalyzes phosphorylation of purified sorcin (Meyers 1989b). That sorcin is phosphorylated in the intact cell was determined by immunoprecipitation of sorcin metabolically labeled with ^{32}P-orthophosphate. The level of phosphosorcin detected under these conditions is enhanced when calcium is added to the buffers in which the radiolabeled cells are lysed. Under any conditions used, the level of detectable phosphosorcin in intact cells is low compared to total amount of sorcin. Assumptions to be drawn from this information are that sorcin is phosphorylated in the living cell, probably as a result of A kinase activity, and that calcium plays a role in the phosphorylation function. Extent of binding of calcium to sorcin may also affect location of sorcin within a cell. Examination of 2000 g, 10000 g, and 100000 g particulate fractions from homogenates of VM26-resistant L1210 cells lysed in the presence and absence of calcium showed that sorcin was present in the sedimented fractions when calcium was present in the lysis buffers. Inclusion of EGTA in lysis buffers causes sorcin to be present only in supernatant material (Roberts et al. 1987). Sorcin is, probably, not principally a calcium storage protein but one whose function and site of action is controlled by level of bound calcium and possibly by state of phosphorylation of the protein.

The overall level of in vitro protein phosphorylation in multidrug-resistant cells is increased and altered as compared to drug-sensitive controls. However, it is not clear what role, if any, sorcin plays in this general change in protein phosphorylation. Resistant cells which do not overproduce sorcin also manifest this altered protein phosphorylation profile. One clue suggesting a role for sorcin in overall level of protein phosphorylation comes from a study of phosphatase activity in DC-3F/VCRd-5L (sorcin overproducer) and DC- 3F/AD X (sorcin nonoverproducer) cells. The level of this activity in the vincristine-resistant line is 33% lower than in DC-3F/AD X or the parental DC-3F cells (Meyers 1989a).

The general phosphorylation change in mdr cells may be associated with Pgp phosphorylation requirements and/or with a general modulation of the state of differentiation in mdr cells (Meyers 1989a). With reference to the first possibility, it is known that Pgp is phosphorylated (Center 1985; Hamada et al. 1987; Meyers 1989a) and that the phosphorylation may be calcium-mediated (Meyers 1989a) and/or activated by cAMP (Mellado and Horwitz 1987). Pgp phosphorylation may be controlled in part by calcium-mediated phosphatase activity. The level of phosphorylation of Pgp in isolated plasma membranes from DC-3F/VCRd-5L and DC-3F/AD X cells is decreased by 50% when phosphorylated membranes are prepared for electrophoresis in calcium-containing buffers as compared to the samples pre pared in buffers containing no calcium or containing EDTA (Meyers unpubl.). Speculatively, several different kinases and phosphatases catalyze Pgp phosphorylation and dephosphorylation in a complex system of regulation of the membrane protein (Hamada et al. 1987; Meyers 1989). Some studies have suggested a relationship between level of Pgp phosphorylation and ability of the protein to efflux intracellular drug. In one such study, treatment of adriamycin-resistant human leukemia cells with verapamil, a calcium channel blocker known to override drug resistance, i.e., produce a cytotoxic effect in mdr cells when given with concentrations of drug to which the cells are normally resistant, will produce hyperphosphorylated Pgp and decreased level of intracellular drug (Hamada et al. 1987).

The possibility that phosphoprotein changes are associated with modulation of differentiation state of mdr cells refers to an hypothesis that stems from the facts that (1) drugs in the multidrug resistance category are differentiating agents (Honma et al. 1986; Sartorelli et al. 1987; Biedler et al. in press); (2) some mdr cells have altered morphology, tumorigenic potential, and express proteins indicative of a more differentiated phenotype as compared to drug-sensitive controls (Biedler and Peterson 1981; LaQuaglia et al. 1989); and (3) treatment of certain cells with retinoic acid will produce an increase in Pgp (although there is no evidence that the induced cells are mdr cells) (Mickley et al. 1989). The hypothesis states that treatment of tumor cells with drugs in the mdr category, e.g., vincristine and adriamycin, causes some cells to enter into a differentiative pathway in the short term, and chronic exposure and selection with the drugs renders that differentiative change constitutive. Thus, mdr cells may be simultaneously more differentiated than the parental untreated tumor cells. Sorcin's participation in these two events, Pgp phosphorylation and alteration in state of differentiation, in mdr cells will be discussed shortly.

HOMOLOGY OF SORCIN AND CALPAIN

Sorcin is homologous to the light chain of the thiol protease, calpain (Van der Bliek et al. 1986a). Calpain is a heterodimeric protein (there are actually several forms of the protease) with a large catalytic subunit of about 80 kDa and a small regulatory subunit of about 30 kDa. The calpain heterodimer of about 110 kDa is readily purified by HPLC (Pontremoli and Melloni 1986). Sorcin does not co-purify with any subunit under any conditions used so far. There is no evidence at this time that sorcin has protease activity. Therefore, there is no known functional relationship between sorcin and calpain to provide a basis for the observed sequence homology. That a homology with calpain does exist prompts a search for a subunit associated with sorcin. Information already described supports the possibility that such a subunit exists, although it is certain that any interaction between sorcin and a putative subunit partner will not be the same as the interaction between the two calpain subunits.

SORCIN IS A COMPONENT OF NORMAL CELLS

Sorcin polyclonal antibody (Meyers et al. 1987) was used in immunoblot analysis to determine whether sorcin could be detected in normal tissues and, if so, to determine the tissue distribution of the protein. Sixteen different tissue types from mice were solubilized by detergent and sonication for the study. Sorcin was found to be most abundant in heart (more abundant in adult than in newborn heart) but also readily detectable in kidney, brain, and skeletal muscle. Sorcin was not detectable in mouse adrenal, small intestine, pancreas, or liver (Meyers 1989b). The sorcin antibody may react with a conserved calcium-binding domain in a specific set of calcium-binding proteins (or there may be a number of forms of sorcin) because the antibody recognizes a small number of proteins in normal cells with molecular weights other than 22 kDa, e.g., a 50 kDa species in peripheral blood and a 37 kDa form in heart.

Sorcin has been purified from normal human heart and skeletal muscle by the same HPLC techniques used to isolate the protein from multidrug-resistant cultured cells. Sorcin represents less than 0.1% of total solubilized heart protein. A 37 kDa protein, cross-reactive with the polyclonal sorcin antibody, elutes with sorcin during ion exchange chromatography in the first stage of purification (suggesting that the 22 kDa sorcin and the 37 kDa protein have the same or similar isoelectric point) but is separable from sorcin by molecular sizing gel filtration chromatography. The 22 kDa sorcin from heart binds calcium and its phosphorylation is catalyzed by the catalytic subunit of A kinase with the same kinetics as is sorcin from cultured cells.

SPECULATION ABOUT SORCIN'S ROLE IN MULTIDRUG RESISTANCE
SORCIN AND P-GLYCOPROTEIN PHOSPHORYLATION

A role for sorcin in the development of mdr cells has not been determined. Co-amplification of sorcin and Pgp may be fortuitous; a result of the known proximity of the two genes. Consequently, sorcin may not play a direct role in the development of multidrug resistance. However, the possibility exists, that sorcin is part of a Pgp operational unit and that there is a functional relationship between sorcin and Pgp. (The unit may include the other, as yet unknown, proteins encoded by the other genes in the Pgp amplicon.) Sorcin may then influence the nature of the resistant phenotype by direct or indirect interaction with Pgp. As previously mentioned, the level of phosphorylation of Pgp is modulated, at least in part, by calcium. Sorcin, as a calcium-binding protein, may participate in the control of calcium levels in the cell or in the micro environment around Pgp and, thus, affect the part of Pgp function regulated by phosphorylation. This notion is particularly intriguing considering the fact that calcium binding may affect the localization of sorcin within the cell (Roberts et al. 1987).

OTHER FUNCTIONS FOR SORCIN IN *MDR* CELLS

While sorcin may or may not play a role in the development of multidrug resistance, overproduction of the protein almost certainly modifies the phenotype of an mdr cell. The binding of sorcin to calcium, possibly considerable amounts of calcium, will, in itself, alter signal transduction and calcium-mediated events in the overproducing cell. This change could affect cross resistance profiles and other related facets of multidrug resistance. Sorcin does not bind vincristine, and, therefore, does not act as a means for sequestering drug away from cytotoxic target sites, and the protein does not increase rate of drug metabolism in the cultured cells studied (Meyers et al. 1983). Parenthetically, sorcin binds phenothiazine (as does calmodulin). However, calmodulin-phenothiazine binding is disrupted by EGTA, and sorcin bound to phenothiazine can only be removed by urea treatment (Meyers 1989b). This difference is suggestive of different structural features between the two calcium-binding proteins. How the tight binding of sorcin to phenothiazine might affect mdr cells is not known.

SORCIN AND DIFFERENTIATION IN *MDR* CELLS

The fact that sorcin's distribution in normal tissue is quite specific, i.e., is abundant primarily in heart, kidney, skeletal muscle, and brain, suggests that turn-on of expression of this protein is part of the program of differentiation for these tissues. That mdr cells derived from tissues whose origins are diverse, not necessarily from heart, kidney, or brain tissue, can manifest increased production of sorcin may suggest that an unscheduled differentiative pathway was perturbed in those cells during development of multidrug resistance. That is, overproduction of sorcin may be a marker for altered differentiation state in mdr cells. This is suggested in light of the previously stated hypothesis that mdr cells are more differentiated or have an altered differentiation pathway as compared to sensitive cell counterparts. Does overproduction of sorcin in mdr cells provide those cells with some characteristics typical of cardiac or kidney function? This kind of question will be addressed in the process of testing the hypothesis concerning differentiation and multidrug resistance.

FUTURE PLANS

Monoclonal antibodies directed against purified "native" sorcin are being prepared and will be used in several lines of investigation including an examination of subcellular localization of sorcin in normal and mdr tumor cells and studies of the interrelationship between calcium binding and level of phosphorylation of sorcin. Transfection experiments with a sorcin construct are underway to help define sorcin's role in multidrug resistance. Studies of sorcin in heart cells to probe sorcin's normal role are planned. These studies of sorcin provide an important opportunity to investigate a protein whose function depends on two second messengers, calcium and cAMP.

ACKNOWLEDGMENTS

The author's research is supported by NIH grant CA 28595 and a Bristol-Myers Award.

REFERENCES

Biedler JL, Melera PW (in press) In: Roninson I (ed) Molecular and cellular biology of multidrug resistance in tumor cells. Plenum Press, New York

Biedler JL, Meyers MB (1989) Multidrug resistance (Vinca alkaloids, actinomycin D, and athracycline antibiotics). In: Gupta RS (ed) Drug resistance in mammalian cells, Vol II, CRC Press, Inc., Boca Raton, Florida, pp 57-88

Biedler JL, Peterson RHF (1981) Altered plasma membrane glycoconjugates of Chinese hamster cells with acquired resistance to actinomycin D, daunorubicin, and vincristine. In: Sartorelli AC, Lazo JS, Bertino JR (eds) Molecular actions and targets for cancer chemotherapeutic agents. Academic Press, New York, pp 453-482

Biedler JL, Spengler BA, Ross RA (in press) Cellular maturation and oncogene expression during drug-induced differentiation in vitro: A brief overview. In: Ragaz J (ed) International Symposium on Effects of Therapy on Biology and Kinetics of Surviving Tumor, Alan R. Liss, New York

Biedler JL, Chang T-D, Meyers MB, Peterson RHF, Spengler BA (1983) Gene amplification and phenotypic instability in drug-resistant and revertant cells. In: Chabner BA (ed) Rational basis for chemotherapy. Alan R. Liss, New York, pp 71-92

Bradley G, Juranka PF, Ling V (1988) Mechanism of multidrug resistance. Biochim Biophys Acta 948: 87-128

Callen DF, Baker E, Simmers RN, Seshadri R, Roninson IB (1987) Localization of the human multidrug resistance gene, MDR 1 to 7q21.1. Hum Genet 77:142-144

Center MS (1985) Mechanisms regulating resistance to adriamycin. Evidence that drug accumulation in resistent cells is modulated by phosphorylation of a plasma membrane glycoprotein. Biochem Pharmacol 34:1471-1476 de Bruijn MHL, van der Bliek AM, Biedler JL, Borst B (1986) Differential amplification and disproportionate expression of five genes in three multidrug-resistant Chinese hamster lung cell lines. Mol Cell Biol 6:4717-4722

Endicott JA, Ling V (1989) The biochemistry of P-glycoprotein-mediated multidrug resistance. Annu Rev Biochem 58:137-171

Hamada H, Hagiwara K-I, Nakajima T, Tsuruo T (1987) Phosphorylation of the M_r 170,000 to 180,000 glycoprotein specific to multidrug-resistant tumor cells: effects of verapamil, trifluoperazin and phorbol esters. Cancer Res 47:2860-2865

Hamada H, Okochi E, Oh-Hara T, Tsuruo T (1988) Purification of the M_r 22,000 calcium-binding protein (sorcin) associated with multidrug resistance and its detection with monoclonal antibodies. Cancer Res 48:3173-3178

Honma Y, Honma C, Bloch A (1986) Mechanism of interaction between antineoplastic agents and natural differentiation factors in the induction of human leukemic cell maturation. Cancer Res 46:6311-6315

Jongsma APM, Spengler BA, van der Bliek AM, Borst P, Biedler JL (1987) Chromosomal localization of three genes coamplified in the multidrug-resistant CH^RC5 Chinese hamster ovary cell line. Cancer Res 47:2875-2878

Koch G, Smith M, Twentyman P, Wright K (1986) Identification of a novel calcium-binding protein (CP_{22}) in multidrug-resistant murine and hamster cells. FEBS Lett 195:275-279

LaQuaglia MP, Kopp EB, Spengler BA, Meyers MB, Biedler JL (1989) Altered state of transformation and/or differentiation in multidrug-resistant human neuroblastoma cells. Proc Am Assoc Cancer Res 30:429

Ling V (1989) Colchicine and colcemid. In: Gupta RS (ed) Drug resistance in mammalian cells, Vol II, CRC Press, Inc., Boca Raton, Florida, pp 139-154

Mellado W, Hurwitz SB (1987) Phosphorylation of the multidrug-resistance associated glycoprotein. Biochem 26:6900-6904

Meyers MB (1989a) Protein phosphorylation in multidrug resistant Chinese hamster cells. Cancer Commun 1:233-241

Meyers MB (1989b) Sorcin is a cardiac calcium-binding protein. Proc Am Assoc Cancer Res 30:505

Meyers MB and Biedler JL (1981) Increased synthesis of a low molecular weight protein in vincristine-resistant cells. Biochem Biophys Res Commun 99:228-235

Meyers MB and Biedler JL (1990) Protein changes in multidrug-resistant cells. In: Roninson, I (ed) Molecular and cellular biology of multidrug resistance in tumor cells. Plenum Press, New York (in press)

Meyers MB, Kreis W, Degnan TJ, Biedler JL (1983) Mechanisms of resistance to vincristine (VCR) in Chinese hamster lung cells. Proc Am Assoc Cancer Res 24:42

Meyers MB, Spengler BA, Chang TD, Melera PW, Biedler JL (1985) Gene amplification-associated cytogenetic aberrations and protein changes in vincristine-resistant Chinese hamster, mouse, and human cells. J Cell Biol 100:588-597

Meyers MB, Schneider KA, Spengler BA, Chang T-d, Biedler JL(1987) Sorcin (V19), a soluble acidic calcium-binding protein overproduced in multidrug-resistant cells. Biochem Pharm 36:2373-2380

Mickley LA, Bates SE, Richert ND, Currier S, Tanaka S, Foss F, Rosen N, Fojo AT (1989) Modulation of the expression of a multidrug resistance gene (mdr-1/P-glycoprotein) by differentiating agents. J Biol Chem 264:18031-18040

Ng WF, Sarangi F, Zastawny RL, Veinot-Drebot L, Ling V (1989) Identification of members of the P-glycoprotein multigene family. Mol Cell Biol 9:1224-1232

Polotskaya AV, Gudkov AV, Kopnin BP (1983) Overproduction of specific polypeptides in Djungarian hamster and mouse cells resistant to colchicine and adriablastin. Bull Exp Biol Med 9:95-96

Riordan JR, Deuchars K, Kartner N, Alon N, Trent J, Ling V (1985) Amplification of P-glycoprotein genes in multidrug-resistant mammalian cell lines. Nature (London) 316:817-819

Roberts D, Lee T, Parganas E, Wiggins L, Yalowich J, Ashmun R (1987) Expression of resistance and cross-resistance in teniposide-resistant L1210 cells. Cancer Chemother Pharmacol 19:123-130

Roberts D, Meyers MB, Biedler JL, Wiggins LG (1989) Association of sorcin with drug resistance in L1210 cells. Cancer Chemother Pharmacol 23:19-25

Safa AR, Glover CJ, Meyers MB, Biedler JL, Felsted RL (1986) Vinblastine photoaffinity labeling of high molecular weight surface membrane glycoprotein specific for multidrug-resistant cells. J Biol Chem 261:6137-6140

Safa AR, Glover CJ, Sewell JL, Meyers MB, Biedler JL, Felsted RL (1987) Identification of the multidrug resistance-related membrane glycoprotein (gp150-180) as an acceptor for calcium channel blockers. J Biol Chem 262:7884-7888

Sartorelli AC, Morin MJ, Ishiguro K (1987) Cancer chemotherapeutic agents as inducers of leukemic cell differentiation. In: Harrap KR, Connors TA (eds) New avenues in develop mental cancer chemotherapy. Bristol-Myers Cancer Symposia. Academic Press, Inc., New York, pp 229-244

Shen D-W, Cardarelli C, Hwang J, Cornwell M, Richert N, Ishii S, Pastan I, Gottesman MM (1986) Multiple drug-resistant human KB carcinoma cells independently selected for high-level resistance to colchicine, adriamycin, or vinblastine show changes in expression of specific proteins. J Biol Chem 261:7762-7777

Suguwara I, Mizumoto K, Ohkochi E, Hamada H, Tsuruo T, Mori S (1989) Immunocytochemical identification of the M_r 22,000 calcium-binding protein (sorcin) in an adriamycin-resistant myelogenous leukemia cell line. Jpn J Cancer Res 80:469-474

Tsuruo T (1988) Mechanisms of multidrug resistance and implication for therapy. Jpn J Cancer Res (Gann) 79:285-296

van der Bliek AM, Meyers MB, Biedler JL, Hes E, Borst P (1989a) A 22-kD (sorcin/V19) encoded by an amplified gene in multidrug-resistance cells, is homologous to the calcium-binding light chain of calpain. EMBO J 5:3201-3208

van der Bliek AM, van der Velde-Koerts T, Ling V, Borst P (1986b) Overexpression and amplification of five genes in a multidrug-resistant Chinese hamster ovary cell line. Mol Cell Biol 6:1671-1678

van der Bliek AM, Baas F, van der Velde-Koerts T, Biedler JL, Meyers MB, Ozols RF, Hamilton TC, Joenje H, Borst P (1988) Genes amplified and overexpressed in human multidrug-resistant cell lines. Cancer Res 48:5927-5932

Yang C-PH, DePinho SG, Greenberger LM, Arceci RJ, Horwitz SB (1988) Progesterone interacts with P-glycoprotein in multidrug-resistant cells and the endometrium of gravid uterus. J Biol Chem 264:782-788

Section IV

Calcium-Binding Proteins in Bacteria, Fungi, and Invertebrates. Structure, Function, and Evolution

COORDINATION AND STABILITY OF CALCIUM-BINDING SITE: D-GALACTOSE/D-GLUCOSE-BINDING PROTEIN OF BACTERIAL ACTIVE TRANSPORT AND CHEMOTAXIS SYSTEM

Nand K. Vyas, Meenakshi N. Vyas, and Florante A. Quiocho

INTRODUCTION

In the course of refining crystallographically the crystal structure of the D-galactose/D-glucose-binding protein (Gal/GlcBP)[1] at 1.9 Å resolution, we discovered a bound calcium (Vyas et al. 1987). The Gal/GlcBP is a member of a family of proteins (collectively called "binding proteins" which are located in the periplasmic space of Gram-negative bacteria. All binding proteins (~2 dozen) serve as initial receptors for the high-affinity active transport systems for a variety of amino acids, dipeptides, carbohydrates, and oxyanions (for recent reviews see Furlong 1987). Moreover, four of these proteins also act as receptors for bacterial chemotaxis (Macnab 1987). Each process further requires protein components lodged in the cytoplasmic membrane. The membrane components for active transport are the ones actually responsible in translocating nutrients from the periplasm to the cytoplasm. Translocation is triggered presumably by the interaction of the substrate-loaded binding proteins and each corresponding membrane component. On the other hand, the interaction between binding protein and transmembrane signal transducer proteins, the membrane components for chemotaxis, trigger taxis toward chemical attractants.

Binding proteins consist of a single polypeptide chain, ranging in size from 23500 to 52000 daltons, with a single high affinity substrate-binding site (K_d values in the micromolar range).

---

[1]Abbreviations used: Gal/GlcBP, D-galactose/D-glucose-binding protein.

In recent years our laboratory has been engaged in the determination of the X-ray structures of seven binding proteins, L-arabinose-binding protein, Gal/GlcBP, leucine/isoleucine/ valine-binding protein, leucine-specific binding protein, sulfate-binding protein, and phosphate-binding protein (Quiocho 1990; Luecke and Quiocho 1990). The structures of all these proteins have been determined and refined extensively at high resolutions.

The Gal/GlcBP has the distinction of being the first binding protein discovered to participate in both active transport (Anraku 1968) and chemotaxis (Hazelbauer and Adler 1971). It has 309 amino acids and a molecular weight of 33306 daltons. As noted, the binding protein is able to recognize D-galactose and D-glucose, epimers of each other, at the C-4 sugar position. The K_d for galactose is 0.4 x 10^{-6} and K_d for glucose shows slightly tighter binding (Miller et al. 1980, 1983).

The discovery of a bound cation in the structure of Gal/GlcBP was totally unexpected as there were no prior indications of its existence. Also completely unexpected is the revelation of the similarity of the calcium-binding site of Gal/GlcBP with the calcium-binding sites found in the so-called EF-hand of calcium-modulating proteins. In this chapter we describe the structure of the calcium-binding site of Gal/GlcBP, compare it with the structure of the site in the EF-hand site, and discuss the stability of the site.

STRUCTURE OF THE D-GALACTOSE/D-GLUCOSE-BINDING PROTEIN

The C-α trace of Gal/GlcBP backbone, obtained from the well-refined structure at 1.9 Å resolution (R-factor =14.7%), is shown in Figure 1. This structure (Vyas et al. 1987, 1988), like all the binding protein structures (Quiocho 1990), has the following prominent features: (1) It has the shape of a prolate ellipsoid with dimensions of about 70 Å x 35 Å x 35 Å. (2) It is composed of two distinct globular domains, designated N-domain and C-domain as the former contains the amino or N-terminus and the later the carboxyl or C-terminus. (3) Approximately the first third of the chain constitutes the major part of the N-domain, the second third makes up the C-domain and the final segment meanders between the two domains. (4) Although neither of the domains is folded from one continuous segment, both domains exhibit similar packing of secondary structure, a central core of a pleated sheet flanked on both sides by two or three helices. (5) The two domains are connected by three separate short peptide segments. Although these segments are widely separated along the polypeptide chain, they are spatially close together, providing a boundary or base for the cleft between the two domains and a flexible hinge between the two domains. The Gal/GlcBP, as purified and crystallized, contains tenaciously bound D-glucose (Miller et al. 1980). Structure determination showed the glucose to be bound and completely engulfed in the cleft between the two domains (Figure 1).

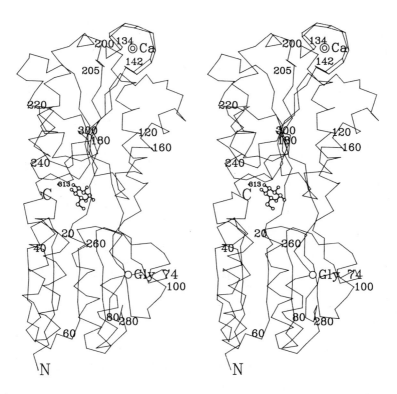

Fig. 1. The stereoscopic view of the C-α trace of Gal/GlcBP. The amino and carboxy terminals are labeled as N and C and every 20th residue is numbered. The Gal/GlcBP molecule has 309 residues that form a bilobate structure. Both domains are structurally very similar, despite lack of any significant sequence homology between the two. The sugar-binding site is in the deep cleft between the two domains and the bound ß-D-glucose molecule is shown here with a bound solvent molecule Wat 313. The natural mutant defective in chemotaxis has only one change from Gly to Asp at residue 74 in Gal/GlcBP. The mutation site is about 18 Å from the nearest hydroxyl of the bound sugar. The calcium-binding loop is located between the first α-helix and the first ß-strand of the C-domain. The bound metal ion is approx. 45 Å from the sugar site. The loop residues (134-142) and a dipeptide (204-205) are drawn as *thick ribbon* and the position of the Ca^{2+} is marked as *two concentric circles*

It also revealed in great detail the atomic interactions between the glucose and Gal/GlcBP (Vyas et al. 1988). The crystal structure of the complex between Gal/GlcBP and D-galactose which has also been recently determined (Vyas et al. unpubl.) confirms the proposed simple mechanism which enables the protein to recognize both sugar epimers (Vyas et al. 1988).Determining the structure of a Gal/GlcBP mutant isolated from a strain defective in chemotaxis, but fully competent in active transport, has also revealed the location of the site involved in the interaction with the Trg transmembrane signal transducer (Vyas et al. 1988). This interaction is thought to trigger chemotaxis toward the sugars. The site is located in the N-domain, 18 Å from the nearest hydroxyl (6-OH) of the bound galactose (Figure 1).

CALCIUM-BINDING SITE OF Gal/Glc*BP*

As there were no prior data indicating the presence of a bound calcium in Gal/GlcBP, its discovery was a complete surprise. Finding of a bound calcium came about in an indirect manner. While carefully analyzing the ordered water molecules bound to Gal/GlcBP during structure refinement at 1.9 Å resolution, we noted that consistently one water molecule was exhibiting abnormal features: it had significantly lower isotropic temperature factor than the average of the values of all water molecules or the average of the values of protein atoms in the vicinity, and it was making very short contacts (around 2.4 Å) with an extraordinarily large number of atoms (total of seven) of the protein. These features strongly indicated the presence of a bound metal in the native protein. Moreover, it was apparent that the metal coordination resembles those found in EF-hand sites (discussed below).

Further structure refinement, with the calcium in place of the water molecule, brought the isotropic temperature factor of the cation to a level comparable to the average of the values of the atoms of the residues coordinating the metal. Independent determination of the metal content by X-ray fluorescence technique confirmed that Gal/GlcBP contained one molecule of bound calcium (Vyas et al. 1987). Also the cadmium used as a heavy atom derivative in the multiple isomorphous replacement was bound at the same site where calcium ion binds (Vyas et al. 1983).

Table 1. The sequence alignment of Ca²⁺-binding loops from known tertiary structures of proteins

Protein	Seq.	1 (X)	2	3 (Y)	4	5 (Z)	6	7 (-Y)	8	9 (-X)	10	11	12 (-Z)	Seq.	Total charge on ligands
Gal/GlcBP	134	Asp	Leu	Asn	Lys	Asp	Gly	Gln	Ile	Gln	-	Ile[a]	Glu	205	-3
Parv	51	Asp	Gln	Asp	Lys	Ser	Gly	Phe	Ile	Glu	Glu	Asp	Glu	62	-4
	90	Asp	Ser	Asp	Gly	Asp	Gly	Lys	Ile	Gly	Val	Asp	Glu	101	-4
TnC	30	Asp	Ala	Asp	Gly	Gly	Gly	Asp	Ile	Ser	Thr	Lys	Glu	41	-3
	66	Asp	Glu	Asp	Gly	Ser	Gly	Thr	Ile	Asp	Phe	Glu	Glu	77	-3
	106	Asp	Lys	Asn	Ala	Asp	Gly	Phe	Ile	Asp	Ile	Glu	Glu	117	-3
	142	Asp	Lys	Asn	Asn	Asp	Gly	Arg	Ile	Asp	Phe	Asp	Glu	153	-3
ICaBP	54	Asp	Lys	Asn	Gly	Asp	Gly	Glu	Val	Ser	Phe	Glu	Glu	65	-3
	20	Asp	Lys	Asp	Gly	Asn	Gly	Thr	Ile	Thr	Thr	Lys	Glu	31	-3
CaM	56	Asp	Ala	Asp	Gly	Asn	Gly	Thr	Ile	Asp	Phe	Pro	Glu	67	-3
	93	Asp	Lys	Asp	Gly	Asn	Gly	Tyr	Ile	Ser	Ala	Ala	Glu	104	-3
	129	Asp[b]	Ile	Asp	Gly	Asp	Gly	Glu	Val	Asn	Tyr	Glu	Glu	140	-4

The sequence of the calcium binding site in Gal/GlcBP (Mahoney et al. 1981) is aligned with EF-hand loops of Parv (parvalbumin; Coffee and Bradshaw 1973), TnC (troponin C; Wilkinson 1976), ICaBP (vitamin-D-dependent calcium-binding protein; Fullmer and Wasserman 1981) and CaM (calmodulin; Watterson et al. 1980). Octahedral vertices (X, Y, Z, -Y, -X, -Z) are shown in Fig. 4. The ligand at position 7 is provided by the peptide carbonyl oxygen while the others are from side chain oxygens (Vyas et al. 1987).

[a] Dipeptide segment 204, 205. [b] In the original sequence of CaM (from bovine brain) had Asn at X position (residue 129) (Watterson et al. 1980). This assignment has appeared in many published sources, including the one of the structure deposited in the Brookhaven protein data bank (pdb3cln.ent by Babu et al. 1988). An Asn residue at this position would be the only exception to those listed in this table. Meanwhile, Marshak et al. (1984) showed that residue 129 is an Asp residue. Although electron densities of Asn and Asp are often indistinguishable in protein structures, especially those refined at resolution no better than 2.5 Å, the hydrogen-bonding pattern of these two residues can be used to distinguish between Asn and Asp. (With well-refined structures at resolutions better than 2 Å, it should be possible to distinguish between the two residues by their thermal parameters). In the calmodulin structure deposited in the data bank, the NH₂ group of the side chain of Asn129 is in close contact (2.64 Å) with the main chain NH of Gly134. This contact is not only unfavorable but also lead to a repulsion of like pairs. However, an Asp129 residue at the position creates a good hydrogen bond between OD2 and the peptide NH of Gly134.

The refinement of cadmium ion bound GBP at 2.5 Å further established the fact that cadmium and calcium ions not only bind at the same site but display very similar coordination (Vyas et al. 1989).

As shown in Figure 1, the calcium-binding site is located in the C-domain, at one end of the elongated protein. The site is approximately 30 Å from the sugar-binding site and 45 Å from the signal transducer binding site. The site is composed mainly of a nine residue loop (residues 134-142) and a dipeptide segment (204-205). The amino acid sequence of the loop and dipeptide segment is shown in Table 1.

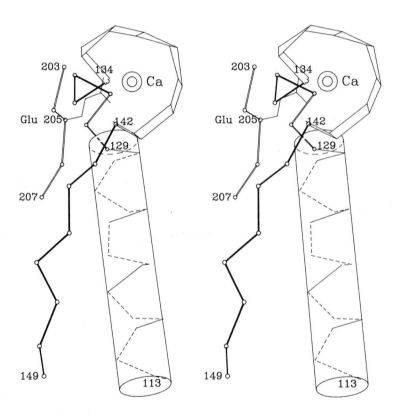

Fig. 2. The secondary structure associated with the calcium-binding loop in Gal/GlcBP. The calcium-binding loop (residues 134-142) provides the first five ligand and remaining two are from the residue Glu205. This 9 residue loop is similar to the first 9 residues of the 12 residues of calcium binding loop in the EF-hand. The calcium-binding loop of Gal/GlcBP is tethered to a ß-turn (residues 131-134) at amino terminal and its carboxy terminal to a ß-strand (residues 142-149). Glu205 is structurally equivalent to the 12th residue of the EF-hand

The nine residue loop is preceded by a type I reverse turn and followed by the first ß-strand (residues 142-147) of the C-domain, forming a turn-loop-strand motif (Figure 2). The reverse turn (residues 131-134) itself is preceded by the first a-helix (residues 113-128) of the C-domain (see Fig 2).

Table 2. Ca^{2+} coordination distances

Residue	Group	Atom	Distance (Å)
Asp134	Carboxylate	OD1	2.4
Asn136	Carboxylate	OD1	2.4
Asp138	Carboxylate	OD1	2.3
Gln140	Carbonyl	O	2.3
Gln142	Carboxylate	OE1	2.4
Glu205	Carboxylate	OE1	2.5
Glu205	Carboxylate	OE1	2.6

COORDINATION

Figure 3 shows the atomic structure around the calcium which is coordinated by seven ligands, all protein oxygen atoms from six amino acid residues. The nine residue peptide loop deploys five ligands from every alternate residue, Asp134 OD1, Asn136 OD1,Asp138 OD1, Gln140 O, and Gln142 OE1. The last two calcium ligands are supplied by Glu205 OE1 and OE2. The Ca^{2+} coordination is an almost pentagonal bipyramid, with apices occupied by oxygen atoms from the side chains of the first (Asp134 OD1) and ninth (Gln142 OE1) positions of the loop. The range of calcium-oxygen distances is 2.3 to 2.6 Å (Table 2), in good agreement with the range found in small molecules (Einsphar and Bugg 1984).

STABILITY OF THE CALCIUM-BINDING SITE

The calcium-binding site structure is maintained by the calcium-oxygen bonds. It is further maintained by fixing the side chains coordinating the calcium through several hydrogen bonds (Figures 3 and 5 and Table 3): (1) The OD2 of the carboxyl group of Asp134 (X) is

410

hydrogen-bonded to the peptide NH group of Gly139. (2) The ND2 of Asn136 (Y) is hydrogen-bonded to OD2 of Asp138 (Z).

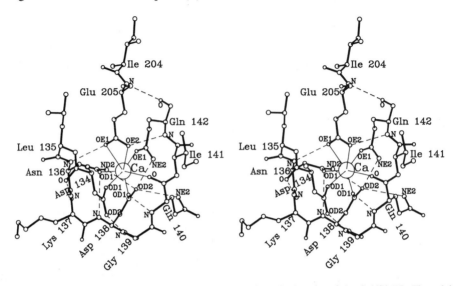

Fig. 3. Stereoscopic view of the atomic structure of the calcium binding site of the Gal/GlcBP. The calcium binding site is composed of two segments (residues 134-142 and 204-205). The coordination of calcium to oxygen ligands are shown by *thin lines* and hydrogen bonding between the atoms in the vicinity of the calcium binding are shown by *dashed lines*. Calcium coordination distances are listed in Table 2 and hydrogen bonding distances are listed in Table 3. The overall conformation of the calcium-binding 9 residue loop and the residue Glu205 is comparable to the EF-hand loop. Although the calcium-binding segments of GBP are separated by 72 residues, they are linked by two hydrogen bonds involving residues Gln142 and Glu205. All residue names, calcium coordinating atoms, and atoms participating in hydrogen bonds are labeled. It is to be noted that the hydrogen-bonding capacity of the each coordinating group is further utilized in stabilizing the ion-binding site. The calcium coordination in Gal/GlcBP, like standard EF-hand protein ligands (Kretsinger 1987), is depicted as an octahedron, as shown in Figure 4, with protein oxygen atoms at the X (1), Y (3), Z (5), -Y (7), -X (9), and -Z which is occupied by a bidentate carboxyl group of Glu205. This depiction will facilitate comparison of calcium-binding sites in the following sections

(3) Asp138 (Z) is the most heavily involved in hydrogen-bonding interaction: OD1 accepts a hydrogen bond from the peptide NH of Gln140 (-Y) and OD2 accepts a total of four hydrogen bonds, one each from Gln140, NE2, Gln142 NE2 (-X), a water molecule, and, as already indicated in (2), from Asn136 ND2(Y). (4) Gln142 (-X) NE2 donates a hydrogen bond to the side-chain carbonyl of Gln174. (5) The OE1 and OE2 atoms of Glu205 (-Z) are the recipient of hydrogen bonds from the peptide backbone NH of Asn136 (Y) and Gln142 (-X), respectively.

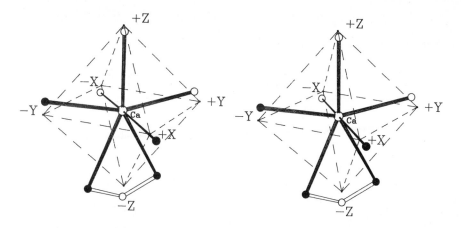

Fig. 4. The pentagonal bipyramid coordination of the bound calcium ion in Gal/GlcBP is shown to be very similar to the octahedral coordination where last two ligands from residue Glu205 share a common octahedral vertex -Z. All the calcium-binding sites in EF-hand form pentagonal bipyramid as observed in Gal/GlcBP. For the comparison purpose the octahedral vertices can be divided in two parts. The first part is the conserved part. It is composed of vertices +X, -Y and -Z. All oxygen atoms at these vertices are shown in *solid circles*. The residue at -Y vertex is variable but ligand is conserved since it is always a main chain carbonyl oxygen. The net charge of the conserved ligands is sufficient to balance the positive divalent calcium ion. Remaining vertices +Y, +Z and -X form the variable part of the octahedron and ligands at these vertices are provided by variable residues. The net charge of ligands at the variable vertices is usually negative

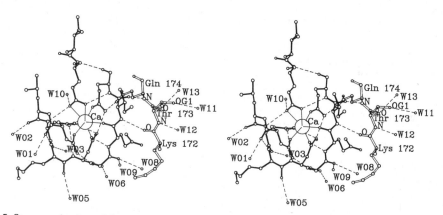

Fig. 5. Stereoscopic view of the calcium-binding site of the Gal/GlcBP. The calcium binding site shown here has very similar orientation as shown in Figure 3. For the purpose of clarity, labels of the calcium-binding site common to Figure 3 are not shown. Calcium-binding site of Gal/GlcBP forms a part of a short parallel ß-sheet involving loop residues 140, 141, and a nearby ß-strand (residues 172-174). Such a ß-sheet is found in other calcium-binding loops of the EF-hand but strands are antiparallel (see Figure 6). Residue at the 8th position is usually a isoleucine that is shielded from the bulk solvent in all the EF-hand loops. In Gal/GlcBP, a threonine residue (Thr173) is appropriately located to shield Ile141. Also it is to be noted that the net excess charge of the calcium binding can be dissipated by forming hydrogen-bonded arrays originating from the bound calcium coordination sphere to the bulk solvent. Almost all the carbonyl oxygens exposed to the solvent have bound water molecules

There are other hydrogen-bonding interactions that further stabilize the calcium-binding site. These involve atoms of the main chain (Figure 5 and Table 3).

Table 3. Hydrogen bonds of the Ca²⁺ binding

Atom X		Atom Y		X...Y (Å)	Comment
ß-sheet					
Ile141	N	Lys172	O	2.9	
Glu174	N	Ile141	O	2.6	
Main chain		Main chain			
Asp134	N	Gln131	O	3.2	Turn(type I)
Leu135	N	Trp133	O	2.8	γ-turn
Lys137	N	Asp134	O	3.0	Turn(type I)
Glu205	N	Gln142	O	3.0	
Val206	N	Gln142	O	3.2	
Main chain		Side chain			
Asn136	N	Glu205	OE1	3.3	
Asp138	N	Asp138	OD1	2.7	
Gly139	N	Asp134	OD2	2.7	
Gln140	N	Asp138	OD1	2.9	Asx turn
Gln142	N	Glu205	OE2	3.2	
Side chain		Side chain			
Asn136	ND2	Asp38	OD2	2.9	
Gln140	NE2	Asp38	OD2	2.8	
Gln142	NE2	Asp38	OD2	2.8	
Gln142	NE2	Gln74	OE2	2.8	
Protein		Water			
Asp34	O	WatWO1	O	2.8	
Leu35	O	WatWO2	O	2.8	
Asn36	O	WatWO3	O	3.5	
Asn36	ND2	WatWO4	O	3.3	
Lys37	O	WatWO5	O	2.9	
Asp38	O	WatWO6	O	2.9	
Asp38	OD2	WatWO7	O	2.6	
Gly39	O	WatWO8	O	2.9	
Gly39	O	WatWO9	O	3.0	
Glu42	OE1	WatW1O	O	2.8	

Of interest is the parallel ß-sheet-like hydrogen bonding between residues 141 of the calcium-binding loop and residues 172 and 174 of a strand (Figure 5).

METAL-BINDING STUDIES

We have recently determined the relative affinities of the binding of different cations to the single site in Gal/GlcBP in solution (Vyas et al. 1989). In order of affinity, the metals are: $Ca^{+2} \sim Tb^{+3} \sim Pb^{+2} > Cd^{+2} > Sr^{+2} > Mg^{+2} >> Ba^{+2}$. The K_d for calcium was determined to be 2.0 mM. We have also crystallographically refined the structure of Gal/GlcBP with the calcium substituted by cadmium, compared it with the native structure with bound calcium (Vyas et al. 1987), and found them to have identical structures (Vyas et al. 1989). The results of these structural and solution studies support the hypothesis that for a given metal-binding site, cation hydration energy, size, and charge are major factors contributing to metal binding affinity. Neither hydration energy alone nor size alone can account for the experimental binding data indicated above.

Differences in calcium- and cadmium-binding affinities to proteins have been attributed to cadmium's preference for sixfold coordination (Szebenyi and Moffat 1986). This attribution is inconsistent with the finding that the structure of the metal binding site of Gal/GlcBP (7-coordination) remains the same when the calcium is replaced by cadmium (Vyas et al. 1989; Swain et al. 1989). The NMR experiments of other proteins predict identical environments for bound calcium and cadmium (Cave et al. 1979; Vogel et al. 1985).

The bound calcium in Gal/GlcBP is 5.8 Å from the center of the six-membered ring of Trp127 and 11.4 Å from the center of the Trp133. This offered an opportunity to exploit the unusual spectral properties of terbium when bound to proteins. Replacement of the calcium by terbium resulted in considerable enhancement of its characteristic green fluorescence due to an energy transfer from the tryptophans, primarily from Trp127 (Vyas et al. 1989). One of the advantages of using terbium as a calcium probe is its sensitivity to its surrounding environment. The dependence of transfer efficiency on the distance between the donor tryptophan residue and terbium to the sixth power makes it particularly responsive to small conformational changes about the calcium-binding site. We have explored the possibility that sugar binding could alter the environment of the bound terbium. Neither D-galactose nor D-glucose binding to sugar-free, terbium-loaded Gal/GlcBP causes a change in terbium fluorescence. This is not surprising since the sugar-binding site is approximately

30 Å from the calcium-binding site (Figure 1) and the tryptophan nearest the bound calcium (Vyas et al. 1987).

CALCIUM FUNCTION

There is no definitive evidence for the function of the calcium bound to Gal/GlcBP. Thus far, the Gal/GlcBP is the only one of about two dozen binding proteins in which a bound calcium has been unambiguously determined. The binding proteins are located in the periplasmic space where presumably free calcium concentrations are similar to extracellular concentrations (~0.001 M). Therefore, Gal/GlcBP, despite its relatively low K_d for calcium, will be saturated with calcium under physiological conditions and not subject to regulation. The Gal/GlcBP calcium-binding site and sugar-binding site are fairly distant from each other and functionally independent. In addition, regulatory sites, such as those in calmodulin, release calcium rapidly (k_{off} ~10 s^{-1}), enabling them to respond quickly to changes in calcium concentration. In contrast to calmodulin and consistent with a structural role, terbium, which binds with a K_d similar to calcium, is released slowly (k_{off}=0.003 s^{-1}), from the metal site of Gal/GlcBP. Thus, indirect results suggest a structural rather than a regulatory role as has been observed in calcium-modulating proteins (see Vyas et al. 1989).

COMPARISON OF CALCIUM-BINDING SITES

From a structural standpoint, proteins with calcium-binding sites may be grouped into two: one group consists of intracellular proteins which contain two or more copies of highly homologous and symmetry-related-calcium-binding sites formed from a 12-residue peptide segment (for example see Figure 6 for sites III and IV of the troponin C). The Ca^{2+}-binding sites in such structures are popularly known as EF-hands or calmodulin folds (see sequence comparison in Table 1). After the first structure determination of parvalbumin, an intracellular protein, Kretsinger (1987) has named such sites as EF-hands or calmodulin fold. The 12 residue Ca^{2+}-binding loop (or EF loop) is preceded by an E helix and followed by an F helix. The last four residues of this EF loop are part of the F-helix. In the 12-residue EF-hand calcium-binding loop or EF loop, residues 1 (X), 3 (Y), 5 (Z), 7 (-Y), 9 (-X) and 12 (-Z) provide seven oxygen ligands at six vertices of the octahedron (see Figure 4 and Table 1). The 12th residue, always a glutamate, of the EF loop provides both carboxylate oxygens as ligands and share a site at vertex -Z.

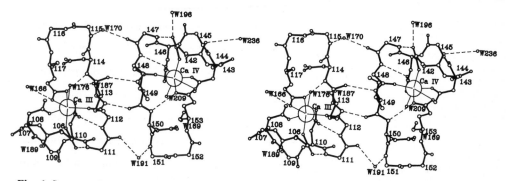

Fig. 6. Stereoscopic view of the pair of calcium binding loops III and IV of the troponin C from chicken (Satyshur et al. 1988; coordinates from the protein data bank of Brookhaven National Laboratory). The view of this calcium binding site III is almost in the similar orientation as of the Gal/GlcBP shown in Fig 3 and 5. In troponin C and in other EF-hand sites the calcium binding loops are paired and are joined by hydrogen bonds of the antiparallel ß-strands (as shown in Figure 7b)

The second group consists of extracellular proteins with metal-binding sites which are not only different from one another or from the calmodulin fold site but are also formed from at least two separate peptide segments. The site in each of these proteins normally exist as one site or without a pair. Since Gal/GlcBP is located in the periplasmic space, it may be considered as an extracellular protein. Nevertheless, the structure of the calcium-binding site in the Gal/GlcBP very much resembles the site in the calmodulin fold.

The nine-residue loop in the Gal/GlcBP metal site adapts a conformation and provides five metal ligands very similar to the highly conserved first 9 residues of the 12-residue EF-hand loop (Table 1). Moreover, Glu205, which provides bidentate oxygen ligands to the metal, occupies a position equivalent to the invariant Glu residue at coordinating position -Z (12th position) of the EF loop, and the backbone atoms of residue 204 are partly equivalent with the residue at position 11 in EF-hand loop. Superimpositioning of the main-chain coordinates of the Gal/GlcBP calcium-binding site [the nine residue loop (134-142) and the dipeptide segment (204-205)] on the parvalbumin EF loop show a root-mean-squares discrepancy of

0.60 Å which is comparable with values obtained from comparisons of several EF-hand proteins (Herzberg and James 1985; Vyas et al. 1987).

Although Glu205 of Gal/GlcBP provides two oxygen atoms from its carboxylate group at -Z vertex, it is not the 12th residue but the 72th residue from the first residue of the loop at the vertex X (Asp134). Kretsinger (1987) has noted, rather remarkably, that of all the possible combination to wrap the 12-residue loop of EF-hand from X to -Z of the octahedral vertices, only X, Y, Z, -Y, -X, -Z have been observed in all the known EF-hand structures. In Gal/GlcBP, everything is similar from X to -Z but the segment intervening vertices -X and -Z is 62 residues long in contrast to two in EF-loops. The extra 60-residues segment joining the -X and -Z vertices is involved in the globular folding of the C-domain of Gal/GlcBP (see Figure 1)

Several notable similarities between the sequences of Gal/GlcBP calcium-binding loop and EF-loop exist (Table 1). These key residues include those that contain side-chain ligands at positions 1 (X), 3 (Y) and 5 (Z). There is also well-conserved Gly at position 6 and Val/Ile at position 8, which provides flexibility and stability to the loop. In both the Gal/GlcBP and EF-loop, position 7 (-Y) is occupied by the peptide carbonyl oxygen. In this regard the Gly residue at position 6 found in the calcium-binding site of all EF loops and Gal/GlcBP as well is crucial as it adopts a main-chain conformation (f=70,y=5 in Gal/GlcBP) which insures that chain direction changes appropriately to allow the peptide carbonyl oxygen of the residue 7 (-Y) to coordinate the calcium. The conformation of Gly139 is stabilized by a hydrogen bond between its peptide NH and Asp134 OD2 (Figure 3). The branched aliphatic side chain at position 8 is directed towards the interior of the domain, thereby contributing to stability of the loop. We have noted previously that the invariant glutamate residue at position 12 (-Z) in EF loop is structurally equivalent to Glu205 of Gal/GlcBP.

Despite the fact that the first nine residue loop of Ca^{2+}-binding site is homologous in sequence and structure with the first nine residues of the EF loop, the Gal/GlcBP loop is preceded by a reverse turn and followed by a ß-strand (Figure 2). All intracellular Ca^{2+}-binding proteins have paired EF-hands, whereas in Gal/GlcBP there is only one Ca^{2+}-binding site (Table 1, Figures 2 and 6). Coffee and Salano (1976) have shown that the pairing of EF-hand is essential for the stability of the calcium-binding sites, according to their study, tryptic digestion of parvalbumin (residues 1-75 and 76-108) results in a complete loss of Ca^{2+}-binding activity. In this regard, can understanding of the unpaired calcium-binding site of Gal/GlcBP provide some clues to the evolutionary/ancient lone calcium-binding site? All known calcium-binding sites in the EF-hands are related by a local twofold symmetry through the center of two antiparallel ß-strands segments of each 12 residue loop (Figure 6 and 7b). Residues at positions 7 to 9 are involved in the antiparallel strand interactions (Herzberg and James 1985). Residues at positions 7-9 of the lone calcium-binding loop in Gal/GlcBP are also involved in a short ß-sheet type of hydrogen bonding (Figure 5 and Table 3). However, in this case, the hydrogen bonding is between parallel strands (Figures 5 and 7a). In Gal/GlcBP, the backbone of Ile141 (position 8) donates and accepts two hydrogen bonds from residues 172 and 174 of the neighboring parallel strand (Figure 5). Moreover, the aliphatic side chain at the 8th position in both Gal/GlcBP (Figure 5) and EF-hands (Figure 6) are involved in hydrophobic interactions (Herzberg and James 1985; Vyas et al. 1987). In the case of Gal/GlcBP, the Ile at position 8 is in contact with a Thr173 in the adjacent strand. In contrast, because of the local twofold symmetry of the pair of EF-hands, two isoleucines (each at position 8 of the antiparallel strands) form an apolar interaction (Figure 6).

CHARGE STABILIZATION OF CALCIUM-LOADED BINDING SITES

In the course of analyzing the geometry of the calcium binding sites in proteins of known three-dimensional structures we found, to our surprise, that all but two of 22 sites examined (bound calcium included) do not have a zero net charge (unpublished data). Some of the structures examined are listed in Table 1. Net negative charge arises from excess carboxylate side chains which are coordinated to the calcium. A net positive charge on the bound calcium is observed in sites containing predominantly main-chain peptide carbonyl oxygens coordinated to the cation (ICaBP domain I-II for example and unpublished data). Many of the peptide units associated in these sites are in turn further coupled to hydrogen bond arrays consisting of hydrogen-bonded peptide units. While this may be, to our knowledge, the first time that this general observation has been made, nevertheless, the important question is: "How is the excess charge(s) neutralized or stabilized?" The answer to this question as

418

amplified below is related to the dipolar interactions which the residues coordinated to the calcium further make.

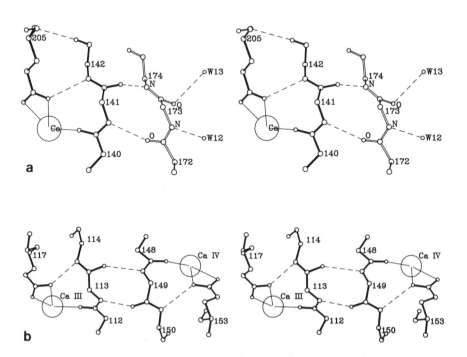

Fig. 7a,b. a Stereoscopic view of the pair of parallel ß-stands (residues 140-142 and 172-174) associated with the calcium-binding loop of the Gal/GlcBP. The orientation of these strands is very similar to those shown in Figure 5. In troponin C and in other EF-hand binding sites the calcium-binding loops are paired and are joined by hydrogen bonds of the antiparallel ß-strands as shown in Figure 7b. Since Gal/GlcBP has only one calcium-binding site, the main-chain nitrogen and oxygen atoms of residue 173 are hydrogen bonded to solvent molecules. **b** Stereoscopic view of the pair of anti-parallel ß-strands formed between the calcium-binding loops III and IV of the troponin C protein. The orientation of these strands is very similar to that shown in Figure 6. It is to be noted that the size and location of these antiparallel ß-strands are similar to the parallel strands for the Gal/GlcBP shown in Figure 7a

Two key features of these calcium-binding sites have important bearing on the question. First, carboxylate side chains of residues which coordinate the calcium are further engaged in hydrogen bonding with main-chain peptide NH groups. Often the CO group of the same peptide unit does in turn participate in a hydrogen bond array consisting of other peptide units or is pointing towards the bulk solvent. Examples of these carboxylate and main-chain interactions in the binding sites of the Gal/Glc-binding protein and of the

EF-hand loop include the highly conserved Asp and Glu at X and at -Z, respectively, and those frequently located at positions 3 (or Y) and 5 (or Z). For instance, in the Gal/Glc-binding protein, the OD2 of Asp134 at position 1 (X) accepts a hydrogen bond from the main-chain peptide unit NH of residue 139 and the CO of the same peptide unit is hydrogen bonded with water molecule WO6 which is sticking out into the solvent (Figures 3 and 5). The OD1 of Asp 138 at position 5 makes the same array of hydrogen bonds. Each OE1 and OE2 of the Glu205(-Z) is hydrogen-bonded to backbone NH group. Moreover, the hydrogen-bond array involving OE2 is one of the most extensive; OE2 is hydrogen bonded to peptide NH of Gln142 and the CO (Ile141) of the same peptide accepts a hydrogen bond from the peptide NH of residue 174, which in turn has the CO of the same peptide hydrogen bonding with a water molecule (Figure 5). Similar examples of hydrogen-bonding arrays involving the EF-loop charged groups are shown in Figure 6.

The second feature is that the main-chain peptide units which provide the peptide carbonyl oxygens coordinated to the calcium (Table 1) are further linked, via their NH groups, in arrays of hydrogen bonds. The main chain peptide unit CO group is found at position 7 (or -Y) in EF-hand loops and in the loop of the Gal/Glc-binding protein. The NH group associated with the peptide unit (8th residue) residue donates a hydrogen bond to a peptide carbonyl oxygen of the neighboring strand in the ß-sheet. The difference in Gal/GlcBP and EF-hand structures is that in the former the ß-sheet is parallel (Figure 5) whereas in the latter it is anti-parallel (Figure 6).

On the basis of the two common key features, we propose a mechanism whereby the dipole interactions with main chain peptide units which are associated with the carboxylate side chains or directly with the calcium are mainly responsible in neutralizing or stabilizing the net negative or positive charge of the calcium-binding site. The stabilization via this mechanism could be further facilitated by the involvement of these peptide units in hydrogen bond arrays. In the absence of calcium, where the net negative charge of the site is increased, the proposed mechanism plays an even more important role in charge stabilization. The proximity of the negatively charged residues to the bulk solvent will also help stabilize the charges. In this regard Ca^{2+}-binding sites I and II that are apo in the crystal structure of Troponin C (Satyshur et al. 1988; Herzberg and James 1988) provide a good example. First the twofold related ß-strands of the Ca^{2+}-binding loop form two more hydrogen bonds compared to holo sites III and VI of the same protein. Second, several of the coordinating carboxylate groups form similar hydrogen bond arrays involving peptide -NH and -CO groups.

The stabilization of the net positive charge on the calcium by dipoles of the main-chain CO group is reminiscent of the stabilization of cations bound to neutral ionophores or by uncharged chelating agents.

While the carboxylate side-chain to main-chain interactions found in calcium-binding sites are known to be crucial in maintaining the structural stability and tertiary structure of the calcium-binding site, their importance in charge stabilization has, heretofore, not been recognized. Interestingly, the first feature is observed in all calcium-binding sites with net negative charge (unpubl. data). The second feature is generally observed in calcium-binding sites with net positive charge (unpubl. data).

The electrostatic interaction described herein is one example of the growing importance of peptide unit mediated hydrogen bond dipoles in charge stabilization. The mechanism of charge stabilization was originally proposed in the stabilization of isolated charged ligands and residues (e.g., sulfate dianion, leucine zwitterion, and positively charged arginine residue) sequestered in proteins (Quiocho et al. 1987a,b). This mechanism has wide application in other biological systems.

CONSERVED AND VARIABLE PARTS OF EF-HAND OR SIMILAR CALCIUM-BINDING SITE

One can propose on the basis of structural features discussed above that the calcium-binding sites of the EF loop and Gal/GlcBP can be considered as being formed by a conserved and a variable part. Residues Asp, Gly, and Glu at 1 (X), 6 and 12 (-Z), respectively, form the conserved part and the rest forms the somewhat variable part. The conserved part could provide minimum stability and conformation necessary to form an EF-loop. The importance of these three conserved residues can be appreciated when one considers the fact that these residues cannot be replaced without affecting the loop structure. Asp at position 1 (X) is essential because both of its side-chain oxygen atoms (OD1 and OD2) are involved in providing a ligand and in accepting a hydrogen bond from NH of Gly at position 6. This hydrogen-bonding interaction helps maintain the conformation of the Gly residue. As has already been noted, the conformation of the Gly residue at position 6 is crucial in enabling the peptide carbonyl oxygen of the variable seventh residue to coordinate the calcium. Glu at 12 (-Z) position, besides providing two oxygen ligands, accepts hydrogen bonds from the peptide NH of the residues at position 3 (Y) and 9 (-X).

The interaction of the bound Ca^{2+} with residue at 7 (-Y) in Gal/GlcBP and EF-hands has been also conserved despite its variable side chain (Table 1). The variable side chain of the 7th residue does not directly contribute to the Ca^{2+} binding as its main-chain carbonyl oxygen atom is the ligand. Residue at the 8th position is well conserved and it plays important role in stabilizing the site through apolar interactions with the rest of the protein stricture and particularly stabilizing the pair of EF-hands that are related by twofold rotation symmetry (see Figure 6).

Thus, in calcium-binding loops of EF-hand and Gal/GlcBP, the roles of residues at 1 (+X), 6, 7 (-Y), 8, and 12 (+Z) are well conserved and can stabilize the charge on Ca^{2+} whereas ligand providing residues at 3 (+Y), 5 (+Z) and 9 (-X) are variable and contribute to increase the negative charge on the site. A charge relay mechanism in Ca^{2+}-binding sites has been discussed above.

One of the major goals in structural studies of Ca^{2+}-binding proteins is understanding what determines low or high affinity binding of the calcium ion. In this regard, it is possible to propose that the variable half of the octahedron governs the specificity and affinity of the Ca^{2+}-binding site. Also it is important to see how charges of the metal ion and the ligand providing groups are stabilized.

ACKNOWLEDGMENTS

This work was supported by the Howard Hughes Medical Institute and a grant from NIH. We thank Mrs. Connie Wallace for her assistance in the preparation of this manuscript.

REFERENCES

Anraku Y (1968) Transport of sugars and amino acids in bacteria. I. Purification and specificity of the galactose- and leucine-binding proteins. J Biol Chem 243:3116-3122

Babu YS, Bugg CE, Cook WJ (1988) Structure of calmodulin refined at 2.2 Å resolution. J Mol Biol 204:191-204

Cave A, Daures MF, Parello J, Saint-Yves A, Sempere R (1979) NMR studies of primary and secondary sites of parvalbumins using the two paramagnetic probes Gd(III) and Mn(II). Biochimie (Paris) 61:755-765

Coffee CJ, Bradshaw RA (1973) Carp muscle calcium-binding protein. J Biol Chem 248:3305-3312

Coffee C, Solano C (1976) Preparation and properties of carp muscle parvalbumin fragments A (residues 1-75) and B (residues 76-108). Biochem Biophys Acta 453:67-80

Einspahr H, Bugg CE (1984) Crystal structure studies of calcium complexes and implications for biological systems. In: Sigel H (ed) Calcium and its role in biology (metal ions in biological systems). Vol. 17, Marcel Dekker, New York, pp 51-97

Fullmer CS, Wasserman RH (1977) Calcium binding proteins and calcium function. In: RA Wasserman, Corradino RA, Carafoli E, Kretsinger RH, MacLennan DH, Siegel FL (eds), North Holland Press, p 303

Furlong CE (1987) Osmotic-shock-sensitive transport system. In: Neidhardt FC (ed in chief) Escherichia coli and Salmonella typhimurium. American Society for Microbiology, Washington D.C., Vol. I, p 768

Hazelbauer GL, Adler J (1971) Role of the galactose binding protein in chemotaxis of Escherichia coli toward galactose. Nature New Biol 230:101-104

Herzberg O, James MNG (1985) Common structural framework of the two Ca^{2+}/Mg^{2+} binding loops of troponin C and other Ca^{2+} binding proteins. Biochemistry 24:5298-5302

Herzberg O, James MNG (1988) Refined crystal structure of troponin C from turkey skeletal muscle at 2.0 Å resolution. J Mol Biol 203:761-779

Kretsinger RH (1987) Calcium coordination and calmodulin fold: divergent versus convergent evolution. In: Cold Spring Harbor Symposia on Quantitative Biology, Vol LII, Cold Spring Harbor Laboratory, p 499

Luecke H, Quiocho FA (1990) The high specificity of a phosphate transport protein is determined by hydrogen bonds. Nature (submitted)

Macnab RM (1987) Motility and chemotaxis. In: Neidhardt FC (ed in chief) Escherichia coli and Salmonella typhimurium. American Society for Microbiology, Washington DC, Vol I, p 732

Mahoney WC, Hogg RW, Hermodson MA (1981) The amino sequence of the D-galactose-binding protein from Escherichia coli. Biol Chem 256:4350-4356

Marshak DR, Clarke M, Roberts DM, Watterson DM (1984) Structural and functional properties of calmodulin from the eukaryotic microorganism Dictyostelium discoideum. Biochemistry 23:2891-2899

Miller DM III, Olson JS, Quiocho FA (1980) The Mechanism of sugar binding to the periplasmic receptor for galactose chemotaxis and transport in Escherichia coli. J Biol Chem 255:2465-2471

Miller DM III, Olson JS, Pflugrath JW, Quiocho FA (1983) Rates of ligand binding to periplasmic proteins involved in bacterial transport and chemotaxis. J Biol Chem 258:13665-13672

Quiocho FA, Sack JS, Vyas NK (1987a) Stabilization of charges on isolated ionic groups sequestered in proteins by polarized peptide units. Nature 329:561-564

Quiocho FA, Vyas NK, Sack JS, Vyas MN (1987b) Atomic protein structures reveal basic features of binding of sugars and ionic substrates and calcium cation. In: Cold Spring Harbor Symposium on Quantitative Biology LII, p 453

Quiocho FA (1990) Atomic structures of periplasmic binding proteins and the high-affinity active transport system in bacteria. Phil Trans R Soc Lond B 326:341-351

Satyshur KA, Rao ST, Pyzalska D, Drendel W, Greaser M, Sundaralingam M (1988) Refined structure of chicken skeletal muscle troponin C in the two-calcium state at 2-Å resolution. J Biol Chem 263:1628-1647

Swain AL, Kretsinger RH, Amma EL (1989) Restrained least squares refinement of native (calcium) and cadmium-substituted carp parvalbumin using X-ray crystallographic data at 1.6 Å resolution. J Biol Chem 264:16620-16628

Szebenyi DME, Moffat K (1986) The refined structure of vitamin D-dependent calcium-binding protein from bovine intestine. J Biol Chem 261:8761-8777

Szebenyi DME, Obendorf SK, Moffat K (1981) Structure of vitamin D-dependent calcium-binding protein from bovine intestine. Nature 294:327-332

Vogel HJ, Drakenberg T, Forsen S, O'Neil JD Jr, Hoffmann T (1985) Structural differences in the two calcium binding sites of the porcine intestinal calcium binding protein: a multinuclear NMR. Biochemistry 24:3870-3876

Vyas NK, Vyas MN, Quiocho FA (1983) The 3 Å resolution structure of a D-galactose-binding protein for transport and chemotaxis in *Escherichia coli*. Proc Natl Acad Sci USA 80:1792-1796

Vyas NK, Vyas MN, Quiocho FA (1987) A novel calcium binding site in the galactose binding protein of bacterial transport and chemotaxis. Nature 327:635-638

Vyas NK, Vyas MN, Quiocho FA (1988) Sugar and signal-transducer binding sites of the *Escherichia coli* galactose chemoreceptor protein. Science 242:1290-1295

Vyas MN, Jacobson BL, Quiocho FA (1989) The calcium-binding site in the galactose chemoreceptor protein. J Biol Chem 264:20817-20821

Watterson DM, Sharief F, Vanaman TC (1980) The complete amino acid sequence of the Ca^{2+}-dependent modulator protein (calmodulin) of bovine brain. J Biol Chem 255:962-975

Wilkinson JM (1976) The amino acid sequence of troponin C from chicken skeletal muscle. FEBS Lett 70:254

CALCIUM-BINDING SITES IN MYELOPEROXIDASE AND LACTOPEROXIDASE

Karla S. Booth, Winslow S. Caughey, Shioko Kimura, and Masao Ikeda-Saito

INTRODUCTION

Myeloperoxidase and lactoperoxidase are important enzymes in the microbicidal defense of the host. Myeloperoxidase, involved in the microbicidal and inflammatory action of neutrophils, catalyzes the formation of hypochlorous acid from hydrogen peroxide and chloride ion. Hypochlorous acid causes rapid degradation of biological compounds and may be the ultimate toxin in the neutrophil antimicrobial system. Myeloperoxidase is a tetrameric 150 kDa enzyme consisting of two light chains (M_r 15 kDa) and two heavy chains (M_r 60 kDa); each heavy chain has a single iron prosthetic group (Andrews and Krinsky 1981). Lactoperoxidase, an enzyme found in milk, catalyzes the oxidation of thiocyanate in the presence of peroxide. The enzyme consists of a single polypeptide chain of unknown sequence with M_r 78.5 kDa, 10% carbohydrate content and one heme prosthetic group (Paul and Ohlsson 1985).

Calcium has long been recognized as being involved in stimulus-effector coupling of many neutrophil events (Simchowitz and Spilberg 1979; Matsumoto et al. 1979; Takeshige et al. 1980). A variety of effectors such as fMet-Leu-Phe, immune complexes, complement components, and the lectin concanavalin A can interact through receptors and trigger numerous responses in neutrophils including aggregation, degranulation, and the respiratory burst (Rossi 1986). Calcium mobilization has been proposed to act in signaling processes in both bacterial killing and mediation of the inflammatory response (Korchak et al. 1984). On the other hand, calcium has been shown to serve a structural role in several metalloproteins (Kretsinger 1976). The importance of calcium in maintaining protein structure particularly in the heme environment has been suggested for horseradish peroxidase (Haschke and Friedhoff 1978; Ogawa et al. 1979).

Bovine myeloperoxidase and lactoperoxidase were shown by direct metal analyses using inductively coupled plasma atomic emission spectrometry to contain equimolar calcium and iron with no other metal detected in significant amount (Booth et al. 1989). Calcium is bound with high affinity and could be removed upon exposure to 6 M guanidine hydrochloride/ EGTA which resulted in precipitation of proteins. Two plausible binding sites for calcium in human myeloperoxidase have been found based on computer analyses of amino acid sequences and secondary structures of known calcium binding proteins.

PREPARATION OF MYELOPEROXIDASE AND LACTOPEROXIDASE

Myeloperoxidase was purified from fresh bovine blood. Granulocytes were isolated from whole blood, sonicated and centrifuged to remove cellular debris (Pember et al. 1984). Subsequent detergent-extraction with Triton X-114 removed the membranous proteins (Pember et al. 1984). Myeloperoxidase was purified from the aqueous phase by ion exchange and size exclusion chromatography as described previously (Ikeda-Saito et al. 1989). The enzyme preparations had an A_{430nm} / A_{280nm} greater than 0.92. Horseradish peroxidase (Type VI) and lactoperoxidase were purchased from Sigma. Lactoperoxidase was purified further by ion exchange chromatography on CM Sepharose CL-6B followed by gel filtration with Sephacryl S-200. Chemicals were of reagent grade and used without further purification. All solutions were prepared from double glass-distilled water.

REMOVAL OF CALCIUM FROM MYELOPEROXIDASE AND LACTOPEROXIDASE

Calcium removal from myeloperoxidase and lactoperoxidase was attempted by exposing protein samples to the metal chelators diethylenetriaminepentaacetic acid (DTPA) or [ethylenebis(oxyethylenenitrilo)]tetraacetic acid (EGTA). Approximately 35 μl of protein sample were added to 2 ml of 10 mM DTPA (or 10 mM EGTA) in 10 mM potassium phosphate, pH 7.0. The sample was centrifuged in an Amicon Centricon 10 concentrator to a volume of ca. 25 μl. This procedure was repeated once and the sample resuspended in 10 mM potassium phosphate, pH 7.0, for metal analysis.

Calcium removal was achieved from horseradish peroxidase, myeloperoxidase, and lactoperoxidase by a modification of the procedure of Haschke and Friedhoff (1978). Enzyme preparations were mixed with 6 M guanidine hydrochloride, 10 mM EGTA, pH 7.5, and allowed to stand at room temperature for approximately 2 h. This protein solution (1 ml of 40 μM enzyme) was reduced to ca. 25 μl in an Amicon Centricon 10 concentrator at room temperature. The filtrate, solution which passed through the 10000 M_r cutoff membrane of the Centricon 10, was saved and analyzed for metal content. The protein sample was resuspended in 1 ml of an appropriate buffer for metal analysis. If a precipitate had formed, the sample was centrifuged to separate the protein solution from protein precipitate; both the supernatant and precipitate were analyzed for metal content.

ANALYSIS OF METAL CONTENT

Metal contents were determined by inductively coupled plasma atomic emission spectrometry (ICP-AES) with a Jarrell-Ash Model 975 ICP AtomComp Spectrometer which allows accurate and simultaneous determination of 21 metals (Wolnik 1988). A set of standard solutions of metal ions in an appropriate matrix were used to improve accuracy; solutions of iron and calcium in the concentration range expected for the protein solution were prepared in the same buffer as the protein. Metal concentrations for the protein solution were determined based on a standard curve of the standard iron and calcium solutions.

COMPUTER ANALYSES OF AMINO ACID SEQUENCES AND SECONDARY STRUCTURES OF MYELOPEROXIDASE AND CALCIUM-BINDING PROTEINS

Human myeloperoxidase and calcium-binding protein amino acid sequences were compared using the FASTA program (Pearson and Lipman 1988). Calcium-binding protein sequences used for comparison included rat calmodulin, rabbit troponin-C, mouse myosin alkali light chain, rat myosin regulatory light chain, rat parvalbumin, rat oncomodulin, rat 9 kDa calcium-binding protein, and human calcium protease small subunit (Perret et al. 1988a). Secondary structure was predicted using the algorithm of Garnier et al. (1978).

MYELOPEROXIDASE AND LACTOPEROXIDASE BIND CALCIUM WITH HIGH AFFINITY

Metal analyses of bovine myeloperoxidase and lactoperoxidase by ICP-AES detected the presence of calcium and iron in equimolar amounts. Iron was present in the same concentration as iron determined by UV/visible spectrometry using the extinction coefficients of 89 mM^{-1}cm^{-1} at 430 nm (Ikeda-Saito et al. 1989) for myeloperoxidase and 114 mM^{-1}cm^{-1} at 413 nm (Kimura and Yamazaki 1978) for lactoperoxidase. Trace amounts of copper and zinc (<1% molar amount of Fe) were essentially removed upon dialysis of the enzymes against 10 mM DTPA or 10 mM EGTA. However, neither of these treatments removed calcium from the enzymes, indicating that the calcium was intrinsic to both myeloperoxidase and lactoperoxidase (Table 1).

Attempted removal of calcium from myeloperoxidase by the guanidine hydrochloride/EDTA method reported for horseradish peroxidase (Haschke and Friedhoff 1978) was not successful. However, a modified procedure using EGTA instead of EDTA (as described in the previous section on removal of calcium) did remove calcium.

Table 1. Effect of DTPA and EGTA on calcium to iron atom ratios of bovine myeloperoxidase and lactoperoxidase

	Ca/Fe	
Treatment	Myeloperoxidase	Lactoperoxidase
Untreated	0.99 ± 0.09	0.83 ± 0.08
10 mM DTPA	0.96 ± 0.01	0.92 ± 0.06
10 mM EGTA	0.96	1.07 ± 0.03

Ratios are averages of at least two experiments except for MPO in 10 mM EGTA where only one experiment was conducted.

The efficacy of the procedure was verified by removing calcium from commercial horseradish peroxidase with a calcium to iron ratio of 2.0. After exposure to guanidine hydrochloride/EGTA, horseradish peroxidase remained in solution; ICP-AES analysis showed that calcium had been removed.Exposure of myeloperoxidase or lactoperoxidase to guanidine hydrochloride /EGTA in the same way caused precipitation of a portion of the enzymes. Metal analysis of myeloperoxidase showed a Ca to Fe ratio of 0.1 in the precipitate, whereas a ratio of 1.3 was found for the protein remaining in solution (Table 2). Metal analysis of lactoperoxidase showed a calcium to iron ratio of 0.8 in the solution and a ratio of 0.1 in the precipitate. The calcium to iron ratio of the filtrates revealed that calcium had been removed from both myeloperoxidase and lactoperoxidase.

Table 2. Calcium to iron atom ratios of bovine myeloperoxidase and lactoperoxidase treated with guanidine hydrochloride and EGTA

	Ca/Fe	
	Myeloperoxidase	Lactoperoxidase
Supernatant	1.28 ± 0.18	0.83 ± 0.35
Precipitate	0.14 ± 0.13	0.10 ± 0.14
Filtrate	21.5 ± 2.0	20.0 ± 2.3

Ratios are averages of at least two experiments.

These findings indicate that myeloperoxidase and lactoperoxidase bind calcium with high affinity and that calcium was removed only with denaturation of the enzyme. Although the role(s) of calcium in these peroxidases cannot be ascertained by these observations alone, important structural or catalytic roles can be suggested. Stabilization of enzyme structure by calcium has been shown in several metalloproteins (Kretsinger 1976), including horseradish peroxidase (Haschke and Friedhoff 1978; Ogawa et al. 1979). Calcium stabilizes the high spin state of the heme iron in horseradish peroxidase (Shiro et al. 1986). Recent NMR evidence suggests that the heme crevice of myeloperoxidase is very similar to the one for lactoperoxidase (Dugad et al. 1990). Evidence from NMR studies had previously suggested that the heme crevice in lactoperoxidase was very similar to both horseradish peroxidase and cytochrome c peroxidase (Thanabal and LaMar 1989).

MYELOPEROXIDASE CONTAINS PLAUSIBLE CALCIUM BINDING SITES BASED ON AMINO ACID SEQUENCE AND SECONDARY STRUCTURE

Human myeloperoxidase cDNA deduced amino acid sequence (Morishita et al. 1987; Johnson et al. 1987) was searched for a calcium-binding site by comparing it with several sequences belonging to the calcium-binding protein gene superfamily (Perret et al. 1988a) using the FASTA program (Pearson and Lipman 1988). This is based on the assumption that the calcium-binding domain of myeloperoxidase has sequences similar to those of known calcium-binding proteins. We believe that the human myeloperoxidase primary sequence is very similar to that of bovine myeloperoxidase in those regions related to basic characteristics (Ikeda-Saito et al. 1989). Calcium-binding proteins exhibit a common structural motif of helix-loop-helix with the central calcium-binding loop containing six amino acids as potential calcium ligands (Perret et al. 1988a; Shiro et al. 1986; Figure 1). In most of the calcium-binding proteins, this calcium binding motif repeats four times and the binding domains I and II are more homologous to III and IV, respectively, than to each other.

The FASTA program revealed significant similarities in two portions of myeloperoxidase with one of four calcium-binding domains of calmodulin and one of two binding domains of the 9 kDa calcium-binding protein, respectively (Figure 1). The region similar to the calmodulin calcium-binding sequence, designated MPO-Ca-I, is found in the myeloperoxidase small subunit, just before the beginning of the large subunit. The other region similar to the 9 kDa calcium-binding protein domain, designated MPO-Ca-II, was found in the large subunit close to the predicted proximal histidine residue of myeloperoxidase (Kimura and Ikeda-Saito 1988). This finding is significant in relation to the results reported for horseradish peroxidase, in which calcium is important in maintaining protein structure in the heme environment (Haschke and Friedhoff 1978; Ogawa et al. 1979).

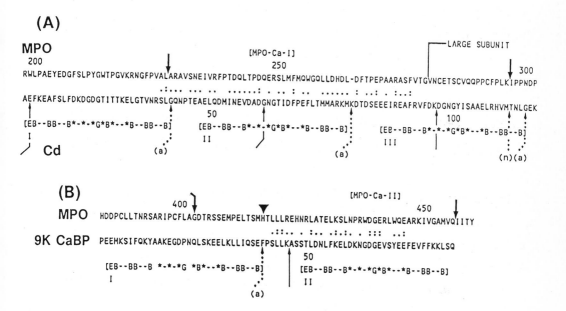

Fig. 1A,B. Comparison of the amino acid sequences of human myeloperoxidase with **A** rat calmodulin (*Cd*; Perret et al. 1988a) and **B** rat 9 kDa calcium-binding protein (*9K CaBP*; Perret et al. 1988a,b). Regions which the FASTA program identified as having significant similarities to each other are shown by a *colon* for an identity and a *dot* for a conservative replacement (Pearson et al. 1988). [*EB--BB--*--*G--B*] represents consensus sequence of calcium-binding regions with *E* glutamic acid; *B* hydrophobic residue with isoleucine preferred; * potential calcium ligands; and *G* glycine. Two other calcium-binding sites of calmodulin (I and III) and another site of the 9K CaBP (I) are also shown below the sequences (Perret et al. 1988a). Introns are indicated by *arrows*: ↓ introns interrupt exons between codons; ↓ introns interrupt codons between the first and second nucleotides: bold solid arrows (myeloperoxidase); *solid arrows* (Cd or 9K kCaBP); *dotted arrows (a)* introns of the four-site ancestral calcium binding gene; *dotted arrows (n)* a intron found in nematode calmodulin (Perret et al. 1988a). The predicted proximal histidine residue of myeloperoxidase is marked by a *closed triangle*. (Kimura and Ikeda-Saito 1988)

432

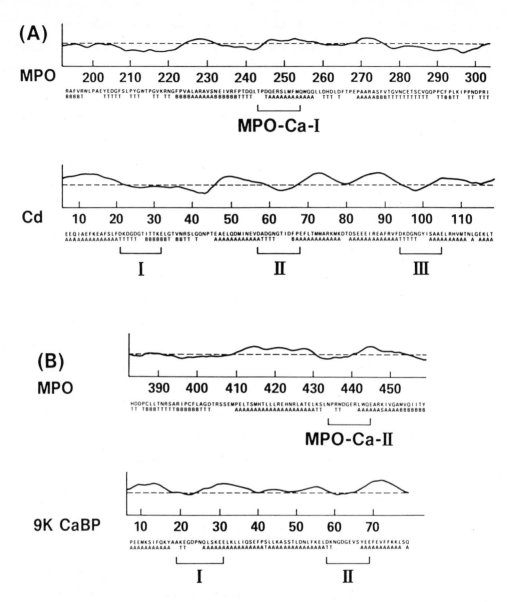

Fig. 2A,B. Secondary structure comparison of human myeloperoxidase with **A** rat calmodulin (*Cd*; Perret et al. 1988a) and **B** rat 9 kDa calcium-binding protein (*9K CaBP*; Perret et al. 1988a,b). Secondary structures are compared in the same amino acid sequence regions as Figure 1. Using the algorithm of Garnier et al. (1978), the relative probabilities of four secondary structures of a protein are calculated; alpha-helicity is depicted graphically on an arbitrary scale in the figure along with amino acid sequence. Protein structures are predicted based on those giving the highest probability values among four structures, and are shown below the amino acid sequences: *A* alpha-helix; *B* beta-sheet; *T* turn; and *blank* random coil. Calcium-binding sites of calmodulin (*I, II,* and *III*) and of 9 kDa calcium-binding protein (*I* and *II*), and predicted calcium-binding sites of myeloperoxidase (*MPO-Ca-I* and *MPO-Ca-II*) are indicated

In addition, secondary structure analyses demonstrated that the MPO-Ca-II region is surrounded by alpha helices as is seen in known calcium-binding domains (Figure 2). These results suggest that the MPO-Ca-II region is the more likely calcium-binding site in myeloperoxidase.

All the presently known calcium-binding proteins are believed to have diverged from a common ancestral gene (Perret et al. 1988a,b). Although unique intron sites are found in different calcium-binding proteins, some introns have been conserved (Perret et al. 1988a). Based on comparisons of these intron positions, evolutionary relationships among calcium-binding proteins have been proposed (Perret et al. 1988a). A one-site primordial ancestral gene which contained only one calcium-binding site, underwent two successive duplications, giving rise to a four-site ancestral gene. If a part of the myeloperoxidase gene which shares homology to a calcium-binding domain was derived from a stage of the calcium-binding ancestral gene by exon shuffling (Gilbert et al. 1986; Davie et al. 1986; Doolittle et al. 1986), it is expected to share introns with known calcium binding proteins. Therefore, we compared intron positions of myeloperoxidase with those of calmodulin and the 9 kDa calcium-binding protein (Figure 1). Although rat calmodulin did not share any common intron sites with myeloperoxidase, one of the introns of the proposed four-site ancestral genes and that of the nematode calmodulin apparently shared a site with myeloperoxidase (Figure 1; Perret et al. 1988a). On the other hand, the MPO-Ca-II region did not seem to share any intron sites with the 9 kDa calcium-binding protein or the proposed ancestral gene. Note that the 9 kDa calcium-binding protein is proposed to have diverged from the third and fourth domains of the four-site ancestral gene which does not have any introns after the fourth domain (Perret et al. 1988a). These results may favor the MPO-Ca-I region as the more likely calcium-binding site in myeloperoxidase. However, it is interesting to note that if the 9 kDa calcium-binding protein diverged from the first and second domains of the four-site ancestor, the myeloperoxidase intron between amino acid residues 455 and 456 almost coincides with the location of an intron in the ancestral gene (Figure 1; see the intron position in the calmodulin sequence that is present after the second domain). In addition, the MPO-Ca-II region is close to the proposed proximal histidine residue, both of which are on one continuous exon and are surrounded by conserved sequences among all known peroxidases (Kimura and Ikeda-Saito 1988). These results suggest that a calcium-binding sequence may have been inserted into an ancestral peroxidase gene. Further, this would suggest that a calcium-binding sequence in lactoperoxidase can also be found near the heme-binding site if lactoperoxidase shares the same ancestral peroxidase gene as myeloperoxidase and thyroid peroxidase (Kimura and Ikeda-Saito 1988). A shared ancestry seems more likely with the recent finding of the similarity of the heme crevice in lactoperoxidase and myeloperoxidase (Dugad et al. 1990).

Our work was the first report of equimolar amounts of intrinsic calcium and iron in myeloperoxidase and lactoperoxidase (Booth et al. 1989). Two plausible calcium-binding sites in myeloperoxidase were identified based on computer analyses of amino acid sequences and secondary structures of various calcium-binding proteins. However, further studies are necessary to determine the exact binding site for calcium and its possible structural or catalytic role in these enzymes.

ACKNOWLEDGMENT

This work was supported by U.S. Public Health Service Grants No. HL-15980 and GM 39492.

REFERENCES

Andrews PC, Krinsky NI (1981) The reductive cleavage of myeloperoxidase in half, producing enzymically active hemimyeloperoxidase. J Biol Chem 256:4211-4218

Booth KS, Kimura S, Lee HC, Ikeda-Saito M, Caughey WS (1989) Bovine myeloperoxidase and lactoperoxidase each contain a high affinity site for calcium. Biochem Biophys Res Commun 160:897-902

Davie EW, Ichinose A, Leytus SP (1986) Structural features of the proteins participating in blood coagulation and fibrinolysis. Cold Spring Harbor Symp Quant Biol 51:509-514

Doolittle RE, Feng DF, Johnson MS, McClure MA (1986) Relationships of human protein sequences to those of other organisms. Cold Spring Harbor Symp Quant Biol 51:447-455

Dugad LB, LaMar GN, Lee HC, Ikeda-Saito M, Booth KS, Caughey WS (1990) A nuclear Overhauser effect study of the active site of myeloperoxidase: structural similarity of the prosthetic group to that on lactoperoxidase. J Biol Chem (in press)

Garnier J, Osguthorpe DJ, Robson B (1978) Analysis of the accuracy and implications of simple methods for predicting the secondary structure of globular proteins. J Mol Biol 120:97-120

Gilbert W, Marchionni M, McKnight G (1986) On the antiquity of introns. Cell 46:151-154

Haschke RH, Friedhoff JM (1978) Calcium-related properties of horseradish peroxidase. Biochem Biophys Res Commun 80:1039-1042

Ikeda-Saito M, Lee H C, Adachi K, Eck HS, Prince RC, Booth KS, Caughey WS, Kimura S (1989) Demonstration that spleen green hemeprotein is identical to granulocyte myeloperoxidase. J Biol Chem 264:4559-4563

Johnson KR, Nauseef WM, Care A, Weelock MJ, Shane S, Hudson S, Koeffler HP, Selsted M, Miller C, Rovear G (1987) Characterization of cDNA clones for human myeloperoxidase: predicted amino acid sequence and evidence for multiple mRNA species. Nucleic Acids Res 15:2013-2028

Kimura S, Ikeda-Saito M (1988) Human myeloperoxidase and thyroid peroxidase, two enzymes with separate and distinct physiological functions, are evolutionarily related members of the same gene family. Proteins: Struct Funct Genet 3:113-120

Kimura S, Yamazaki I (1978) Heme-linked ionization and chloride binding in intestinal peroxidase and lactoperoxidase. Arch Biochem Biophys 189:14-19

Korchak HM, Vienne K, Rutherford LE, Weissmann G (1984) Neutrophil stimulation: receptor membrane and metabolic events. Fed Proc 43:2749-2754

Kretsinger RH (1976) Calcium-binding proteins. In: Snell EE, Boyer PD, Meister A, Richardson CC (eds) Ann Rev Biochem. Annual Reviews Inc, Palo Alto, California, p 239

Matsumoto T, Takeshige K, Minakami S (1979) Inhibition of phagocytic metabolic changes of leukocytes by an intracellular calcium-antagonist 8-(N,N-diethylamino)-octyl-3,4,5-trimethoxybenzoate. Biochem Biophys Res Commun 88:974-979

Morishita K, Kubota N, Asano S, Kaziro Y, Nagata S (1987) Molecular cloning and characterization of cDNA for human myeloperoxidase. J Biol Chem 262:3844-3851

Ogawa S, Yoshitsugu S, Morishima I (1979) Calcium binding by horseradish peroxidase c and the heme environmental structure. Biochem Biophys Res Commun 90:674-678

Paul KG, Ohlsson PI (1985) The chemical structure of lactoperoxidase. In: Pruitt KM, Tenovuo JO (eds) The lactoperoxidase system: chemistry and biological significance. Marcel Dekker, New York, p 15

Pearson WR, Lipman DJ (1988) Improved tools for biological sequence comparison. Proc Natl Acad Sci USA 85:2444-2448

Pember SO, Heyl BL, Kinkade JM, Lambeth JD (1984) Cytochrome b 558 from (bovine) granulocytes. Partial purification from Triton X-114 extracts and properties of the isolated cytochrome. J Biol Chem 259:10590-10595

Perret C, Lomri N, Thomasset M (1988a) Evolution of the EF-hand calcium-binding protein family: evidence for exon shuffling and intron insertion. J Mol Evol 27:351-364

Perret C, Lomri N, Gouhier N, Auffray C, Thomasset M (1988b) The rat vitamin D-dependent calcium-binding protein (9K CaBP) gene. Complete nucleotide sequence and structural organization. Eur J Biochem 172:43-51

Rossi F (1986) The superoxide-forming NADPH oxidase of the phagocytes: nature, mechanisms of activation and function. Biochim Biophys Acta 853:65-89

Shiro Y, Kurono M, Morishima I (1986) Presence of endogenous calcium ion and its functional and structural regulation in horseradish peroxidase. J Biol Chem 261:9382-9390

Simchowitz L, Spilberg I (1979) Generation of superoxide radicals by human peripheral neutrophils activated by chemotactic factor. Evidence for the role of calcium. J Lab Clin Med 93:583-593

Takeshige K, Nabi ZF, Tatschek B, Minakami S (1980) Release of calcium from membranes and its relation to phagocytic metabolic changes. Biochem Biophys Res Commun 95:410-415

Thanabal V, LaMar GN (1989) A nuclear overhauser effect investigation of the molecular and electronic structure of the heme crevice in lactoperoxidase. Biochemistry 28:7038-7044

Wolnik KS (1988) Inductively coupled plasma emission spectrometry. Methods Enzymol 158:190-205

A DEVELOPMENT-SPECIFIC CA²⁺-BINDING PROTEIN FROM *MYXOCOCCUS XANTHUS*

Martin Teintze, Masayori Inouye, and Sumiko Inouye

INTRODUCTION

Myxococcus xanthus is a Gram-negative soil bacterium that undergoes a unique developmental cycle (for reviews see Rosenberg 1984). When starved for nutrients on a solid surface, the cells utilize their gliding motility to aggregate and form mounds called fruiting bodies. Some of the rod-shaped cells then differentiate into round or ovoid myxospores, which are resistant to heat, desiccation, UV irradiation, and sonication. During development there are many changes in the pattern of protein synthesis, the most striking of which is the appearance of protein S (Inouye et al. 1979a,b). Synthesis of protein S is induced early in development and increases until it reaches 15% of total protein synthesis at the stage of mound formation. It accumulates in the cytoplasm as a soluble protein until the onset of sporulation, after which most of it is found assembled on the surface of the myxospores (Inouye et al. 1979b) and a smaller amount inside the spores (Teintze et al. 1985a,b). Protein S can be removed from the surface of the spores by extraction with 1 N NaCl or 10 mM EDTA or 10 mM EGTA and can be reassembled on the spores by adding a 10 mM excess of Ca²⁺ ion or removing the NaCl by dialysis (Inouye et al. 1979b). Protein S is an acidic (pI 4.5), heat-stable protein of 173 amino acid residues and a molecular weight of 18792, with a very high content of β-structure (Inouye et al. 1981, 1983a); it crystallizes differently in the presence and absence of Ca²⁺ (Inouye et al. 1980).

THE GENES FOR PROTEIN S

When the gene for protein S was cloned and sequenced, it was found that there are two very homologous genes (designated *ops* and *tps*) in the same orientation, separated by a short spacer region (Inouye et al. 1983a,b). Only the downstream (*tps*) gene codes for protein S (Inouye et al. 1983a); the *ops* gene is expressed later in development, after the onset of sporulation, and its product (designated protein S1) is found only inside the spores (Teintze et al. 1985a,b). In addition to the 88% homology between the two genes, there is also striking internal homology in the sequences of the proteins (see Figure 1); they can be divided into four homologous domains, with domains 1 and 3 as well as domains 2 and 4 having particularly extensive homologies (Inouye et al. 1983a). This pattern of homologies, probably resulting from two successive gene duplication events, is also found in calmodulins and crystallins. In fact, protein S has homologies to both bovine brain calmodulin (Inouye et al. 1983a) and γ-crystallin II (Wistow et al. 1985), which will be discussed below.

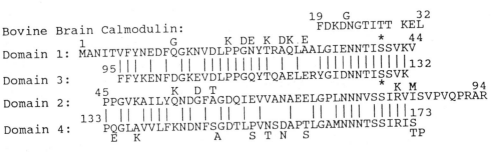

Fig. 1. The sequence of protein S, showing the homologies in the four-domain structure. The sequence of one of the calmodulin Ca²⁺-binding sites is shown above. *Asterisks* mark the residues that were changed to Arg in the studies described below. The amino acid residues that are different in protein S1 are shown *above* or *below* the corresponding position in the sequence of protein S

The protein S mRNA has a very long half-life and is found only in developing cells (Nelson and Zusman 1983; Inouye 1984). The major site for initiation of transcription is 51 bases upstream of the *tps* gene initiation codon and the sequence of the development-specific promoter has been identified (Inouye 1984).

The gene for a developmentally expressed sigma factor has been identified and sequenced (Apelian et al. 1990); this gene (*sigB*) is expressed from the middle to late stages of development, when sporulation occurs. Deletion of *sigB* blocks the expression of the *ops* gene coding for protein S1; mound formation and sporulation proceed normally, but the stability and viability of the spores is reduced (Apelian et al. 1990).

Deletion mutants of *M. xanthus* have been constructed in which either the *ops* or the *tps* gene or the entire region containing both genes has been removed (Komano et al. 1984; Teintze et al. 1985a). Deletion of the *tps* gene has no effect on aggregation, differentiation, sporulation or the yield, stability, and viability of the spores (Komano et al. 1984; Teintze et al. 1985a). Protein S is therefore not required for development in *M. xanthus* in spite of the large quantities that are synthesized specifically during this time when the cells are starved for nutrients. Deletion of the *ops* gene delays fruiting body formation slightly, while the mutant with both genes deleted exhibits a more significant delay in fruiting body formation and a 69% decrease in the yield of myxospores; the two very homologous proteins may therefore be able to partially substitute for each other in whatever role they may have inside the developing cells (Komano et al. 1984; Teintze et al. 1985a; Furuichi et al. 1985).

Mutants containing an *ops* gene coding region fused to the *tps* gene promoter develop normally and have both protein S and protein S1 on their spores; the *ops* gene can also be expressed in *E. coli*, and the resulting protein S1 can be assembled on the surface of protein S-deficient myxospores (either from a deletion mutant or stripped with EDTA) in a Ca^{2+}-dependent manner just like protein S (Teintze et al. 1985a,b). In *M. xanthus* protein S1 is made so late in development that it must remain inside the cell, which by then is surrounded by the impenetrable spore coat. Protein S, on the other hand, is made early in development and accumulates inside the cells until they begin to differentiate into spores (Inouye et al. 1979b); at that point the protein must be released so that it can assemble on the outside of the myxospores. Since protein S is very hydrophilic and has no signal sequence, it is hard to imagine it being suddenly secreted, although this has been claimed (Nelson and Zusman 1983). However, up to 90% of the developing cells lyse when the others differentiate into spores (Wireman and Dworkin 1977), so it is more likely that the protein S is released by the lysing cells and then assembles on the surface of the remaining spores. When *tps* deletion mutants are allowed to develop close to wild-type cells but separated from them by a membrane, the mutant and wild-type spores end up with normal quantities of protein S on their spore surfaces, indicating that protein S is made in excess and diffuses freely after being released to assemble on any nearby spore (Teintze et al. 1985a).

CALCIUM-BINDING SITES

Protein S binds 2 mol Ca^{2+}/mol protein with K_d values of 2.7 x 10^{-5} M and 7.6 x 10^{-5} M (Teintze et al. 1988). The sequence of protein S has four regions with homology to the Ca^{2+}-binding sites of bovine brain calmodulin (Inouye et al. 1983a; Watterson et al. 1980); of these, the ones at the ends of domains 1 and 3 (see Figure 1), are the most homologous, suggesting that these form the two calcium-binding sites in protein S (Teintze et al. 1988). This has been confirmed by site-specific mutagenesis studies in which residues predicted to be involved in binding to the Ca^{2+} ions from the homology to calmodulin were replaced in protein S and the effect on Ca^{2+} binding was measured (Teintze et al. 1988).

Ser^{40} and Ser^{129} were selected as the sites for mutagenesis because of the alignment with the calmodulin Ca^{2+}-binding sequence (see Figure 1) and their location in a proposed three-dimensional structure based on homologies to γ-crystallin (Wistow et al. 1985), which predicts these residues to be facing possible Ca^{2+}-binding sites on the surface of the molecule, whereas Ser^{41} and Ser^{130}, for example, have their side chains buried. Ser^{40} and Ser^{129} were changed to arginine residues because the presence of a positive charge at those locations was likely to interfere with the binding of the Ca^{2+} ions. Both mutations were made in the cloned *tps* gene using a synthetic oligonucleotide, and a double mutant was also constructed (Teintze et al. 1988). The mutant genes were each transduced into the chromosome of a *M. xanthus* strain from which the protein S gene had previously been deleted.

The mutant strains grow and develop normally and express the mutant protein S genes at the proper time in development. The mutant forms of protein S can each be immuno-precipitated from lysates of developing cells of the respective strains using rabbit antiserum to the wild-type protein S. All three mutants produce the same amount of protein, which is comparable to the wild-type protein S production at that stage of development. However, while [Arg^{129}]protein S assembles on the spores in the same quantities as the wild-type protein, [Arg^{40}]protein S is present in much smaller amounts on the spores of its mutant and [Arg^{40}, Arg^{129}]protein S does not assemble on spores at all (Teintze et al. 1988).

The [Arg^{40}]protein S and [Arg^{40}, Arg^{129}]protein S do not bind measurable amounts of Ca^{2+}; [Arg^{129}]protein S binds Ca^{2+}, but with a lower affinity than the wild-type protein. The number of binding sites in [Arg^{129}]protein S is probably one, but could not be determined unequivocally because of the low affinity and the problem of nonspecific binding to such an acidic protein at high Ca^{2+} concentrations (Teintze et al. 1988). These results are consistent with the hypothesis that the sequences in domains 1 and 3 of protein S which are homologous to the calmodulin Ca^{2+}-binding sites are in fact involved in Ca^{2+}binding by protein S. The decreased Ca^{2+} binding by [Arg^{129}]protein S probably reflects a greatly reduced affinity at the binding site in domain 3; this may also reduce the affinity of the other site due to cooperative interactions. The fact that this mutant protein continues to assemble normally on the spores is consistent with the observation that a "tryptic core fragment" lacking domains 3 and 4 is also able to assemble on the spores (Inouye et al. 1981, 1983a). Thus the binding of a Ca^{2+} ion to domain 1 is probably sufficient to induce a conformational change in the protein that results in its assembly on the spores. Unlike calmodulin, however, the binding of Ca^{2+} by protein S does not result in a conformational change that is detectable by a change in electrophoretic mobility (Teintze et al. 1988).

The inability of [Arg^{40}]protein S to bind Ca^{2+} and its reduced ability to assemble on the myxospores has two possible explanations: either the mutated site in domain 1 is the high-affinity site and the binding of Ca^{2+} to the remaining site in domain 3 in the absence of cooperativity is too weak to detect and does not change the conformation of the protein sufficiently to permit proper assembly on the spores, or the mutation in domain 1 changes the conformation of the protein in such a way that no Ca^{2+} binding to either site is possible. However, the fact that this protein is still recognized by antibodies to the wild-type protein S, is not degraded in the developing cells, and still assembles on the surface of the spores to a certain degree indicates that [Arg^{40}]protein S still resembles the wild-type protein.

HOMOLOGIES TO OTHER PROTEINS

In addition to the four-domain structure and the sequence homology at the Ca^{2+} binding sites described above, protein S has other similarities to calmodulin. Both are small, hydrophilic, heat-stable proteins; both are acidic, with a pI of 4.5 for protein S versus 4.2 for calmodulin, and both lack cysteine residues; both have a trypsin-sensitive site between domains 2 and 3 (Inouye et al. 1983a; Newton et al. 1984). However, apart from the Ca^{2+}-binding sites there is no significant sequence homology between the protein S and calmodulin; in fact, whereas calmodulin is primarily α-helical, protein S is mostly β sheet and lacks the characteristic helix-loop-helix or EF-hand structures found in calmodulins and related proteins (Inouye et al. 1981; Babu et al. 1985; Teintze et al. 1988). In this respect protein S resembles *E. coli* periplasmic D-galactose-binding protein, whose structure has a single Ca^{2+}-binding loop with a sequence very homologous to the calmodulin sites, but without the helices on each side (Vyas et al. 1987). Unlike calmodulin, protein S does not stimulate 3',5'cyclic-nucleotide phosphodiesterase and calmodulin antagonists have no effect on the assembly of protein S on spores (Teintze et al. 1984).

Protein S has homologies throughout its sequence to the eye lens proteins of the $\beta\gamma$-crystallin family, and a three-dimensional structure for protein S has been proposed based on the X-ray crystallographic data for γ-crystallin (Wistow et al. 1985). The γ-crystallin sequence also contains four internally homologous domains and is made up almost entirely of β-structure like protein S; in addition, certain key residues conserved in the $\beta\gamma$-crystallins because of their importance to the folding of the protein appear to be present in protein S also (Wistow et al. 1985). Although γ-crystallins do not bind Ca^{2+}, β-crystallin has recently been shown to bind four Ca^{2+} ions/aggregate unit of 160 kDa and δ-crystallin has a calmodulin-type helix-loop-helix or "EF-hand" calcium ion-binding sequence (Sharma et al. 1989). The evolutionary relationships between protein S and the calmodulin and crystallin protein families is therefore still a matter for speculation.

ANOTHER CALMODULIN-LIKE PROTEIN IN *M. XANTHUS*?

Growth and development of *M. xanthus* are very sensitive to the known calmodulin antagonists trifluoperazine (TFP), N-(6-aminohexyl)-5-chloro-1-napthalenesulfonamide (W7), and N-(4-aminobutyl)-5-chloro-2-napthalenesulfonamide (W13); as is the case with calmodulin, the nonchlorinated analogs of W7 and W13 (W5 and W12, respectively) have no effect (Teintze et al. 1984). As shown in Figure 2, even very low concentrations of TFP and W-7 inhibited the growth of *M. xanthus* cells in rich liquid media, whereas W-5 did not; in control experiments on *E. coli*, no effect was observed with any of the drugs, even at the higher concentrations that were bacteriocidal for *M. xanthus*. When W-7 was added to *M. xanthus* cells on agar plates, it inhibited motility and development, as shown in Figure 3.

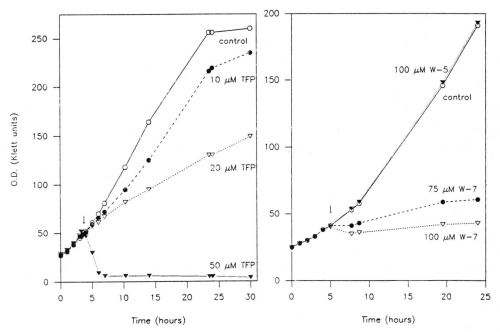

Fig. 2. The effect of calmodulin antagonists on vegetative growth of *M. xanthus*. Cultures were shaken at 30 °C and drug was added at the time indicated by the *arrows*

Panel 1a shows the cells on the periphery of the spot on rich agar swarming out using their ability to glide on solid surfaces; panels 1b, 1c, and 1d show the effect of increasing concentrations of W-7 in the agar; panel 1e shows that the nonchlorinated analog W-5 had no effect. Panel 2a shows the mature fruiting bodies containing refractile myxospores that form when cells are spotted on clone fruiting agar (Hagen et al. 1978); again, W-7 had dramatic effects (b-d), whereas W-5 (e) did not. Similar results were obtained using TFP atone half the concentrations used for W-7, and using W-13 and W-12 at twice the concentrations used for W-7 and W-5, respectively.

In addition, a partially purified extract from *M. xanthus* cells was able to activate calmodulin-dependent bovine heart 3',5'cyclic nucleotide phosphodiesterase in the presence of Ca^{2+} but not in the presence of EGTA; when this extract was passed over phenothiazine-agarose, an affinity column for calmodulins, a protein bound to the column in the presence of Ca^{2+} and was eluted with EGTA (Teintze et al. 1984).

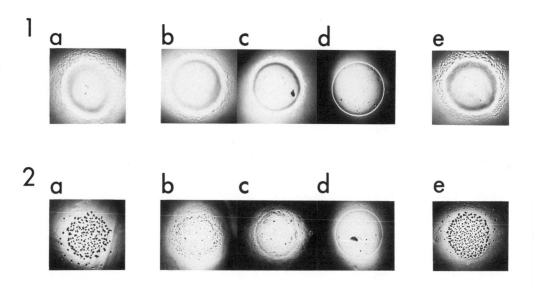

Fig. 3. The effect of calmodulin antagonists on *M. xanthus* growth (**1**) and development (**2**). *M. xanthus* cells were spotted on rich agar (**1**) and starvation agar (**2**) containing: *a* control; *b* 50 µM W-7; *c* 75 µM W-7; *d* 100 µM W-7; *e* 100 µM W-5. Plates were incubated 3 days at 30 °C

Protein S does not bind to this column and has no homology to calmodulin in the region where the latter is thought to bind the above-mentioned drugs. Furthermore, protein S is not present in the vegetative cells from which the extract that activated the phosphodiesterase was made and the self-assembly of protein S on myxospores is not affected by TFP or W7 at the concentrations that inhibit growth and development. These data make it highly unlikely that the observed effects of the calmodulin antagonists are the result of interaction with protein S, but instead suggest the presence of another calmodulin-like protein in *M. xanthus* with as yet unspecified functions in growth and development.

ACKNOWLEDGMENT

S. Inouye was supported by grant GM 26843 from the National Institutes of Health.

REFERENCES

Apelian D, Inouye M, Inouye S (1990) Development-specific sigma factor essential for late-stage differentiation of *Myxococcus xanthus*. Genes Dev (in press)

Babu YS, Sack JS, Greenhough TJ, Bugg CE, Means AR, Cook WJ (1985) Three-dimensional structure of calmodulin. Nature 315:37-40

Furuichi T, Komano T, Inouye M, Inouye S (1985) Functional complementation between the two homologous genes, *ops* and *tps*, during differentiation of *Myxococcus xanthus*. Mol. Gen. Genet. 199:434-439

Hagen DC, Bretscher AP, Kaiser D (1978) Synergism between morphogenic mutants of *Myxococcus xanthus*. Dev Biol 64:284-296

Inouye M, Inouye S, Zusman D (1979a) Gene expression during development of *Myxococcus xanthus*: pattern of protein synthesis. Dev Biol 68:579-591

Inouye M, Inouye S, Zusman D (1979b) Biosynthesis and self-assembly of protein S, a development-specific protein of *Myxococcus xanthus*. Proc Natl Acad Sci USA 76:209-213

Inouye S (1984) Identification of a development-specific promoter of *Myxococcus xanthus*. J Mol Biol 174:113-120

Inouye S, Inouye M, McKeever B, Sarma R (1980) Preliminary crystallographic data for protein S, a development-specific protein of *Myxococcus xanthus*. J Biol Chem 255:3713-3714

Inouye S, Harada W, Zusman D, Inouye M (1981) Development-specific protein S of *Myxococcus xanthus*: purification and characterization. J Bacteriol 148:678-683

Inouye S, Franceschini T, Inouye M (1983a) Structural similarities between the development-specific protein S from a Gram-negative bacterium, *Myxococcus xanthus*, and calmodulin. Proc Natl Acad Sci USA 80:6829-6833

Inouye S, Ike Y, Inouye M (1983b) Tandem repeat of the genes for protein S, a development-specific protein of *Myxococcus xanthus*. J Biol Chem 258:38-40

Komano T, Furuichi T, Teintze M, Inouye M, Inouye S (1984) Effects of deletion of the gene for the development-specific protein S on differentiation in *Myxococcus xanthus*. J Bacteriol 158:1195-1197

Nelson DR, Zusman D (1983) Evidence for long-lived mRNA during fruiting body formation in *Myxococcus xanthus*. Proc Natl Acad Sci USA 80:1467-1471

Newton DL, Oldewurtel MD, Krinks MH, Shiloach J, Klee CB (1984) Agonist and antagonist properties of calmodulin fragments. J Biol Chem 259:4419-4426

Rosenberg E (ed) (1983) Myxobacteria: development and cell interactions. Springer, New York, Berlin, Heidelberg

Sharma Y, Rao CM, Narasu ML, Rao SC, Somasundaram T, Gopalakrishna A, Balasubramanian D (1989) Calcium ion binding to δ- and to β-crystallins. J Biol Chem 264:12794-12799

Teintze M, Thomas R, Inouye M, Inouye S (1984) A calmodulin-like protein in *Myxococcus xanthus*. Fed Proc 43:1519

Teintze M, Furuichi T, Thomas R, Inouye M, Inouye S (1985a) Differential expression of two homologous genes coding for spore-specific proteins in *Myxococcus xanthus*. In: Hoch J, Setlow P (eds) Spores IX: the molecular biology of microbial differentiation. American Society for Microbiology, Washington, D.C., pp 253-260

Teintze M, Thomas R, Furuichi T, Inouye M, Inouye S (1985b) Two homologous genes coding for spore-specific proteins are expressed at different times during development of *Myxococcus xanthus*. J Bacteriol 163:121-125

Teintze M, Inouye M, Inouye S (1988) Characterization of calcium-binding sites in development-specific protein S of *Myxococcus xanthus* using site-specific mutagenesis. J Biol Chem 263:1199-1203

Vyas NK, Vyas MN, Quiocho FA (1987) A novel calcium binding site in the galactose-binding protein of bacterial transport and chemotaxis. Nature 327:635-638

Watterson DM, Sharief F, Vanaman TC (1980) The complete amino acid sequence of the Ca^{2+}-dependent modulator protein (calmodulin) of bovine brain. J Biol Chem 255:962-975

Wireman JW, Dworkin M (1977) Developmentally induced autolysis during fruiting body formation by *Myxococcus xanthus*. J Bacteriol 129:796-802

Wistow G, Summers L, Blundell T (1985) *Myxococcus xanthus* spore coat protein S may have a similar structure to vertebrate lens βγ-crystallins. Nature 315:771-773

UNIQUE Ca²⁺-BINDING PROTEINS IN METAZOAN INVERTEBRATES

Jos A. Cox, Ying Luan-Rilliet, and Takashi Takagi

INTRODUCTION

Many external signals, like hormones, growth factors, sensory stimuli, neurotransmitters, or nerve impulses, are translated into intracellular information via the so-called Ca^{2+} signal, i.e., a transient increase (often oscillatory) in the free Ca^{2+} concentration from the resting 0.1 μM to peak values ranging from 0.2 to 2 μM. The Ca^{2+} signal is regulated by many devices at the level of the plasma membrane, of IP3-sensitive vesicular organelles, of cytosolic Ca^{2+} sequestering agents and, exceptionally, of the mitochondria. Also, the Ca^{2+} signal, first confined to the cytosol, is transduced into the mitochondria, where it regulates the production of ATP. The Ca^{2+} signal is useless unless it can be coupled to the cellular response systems, which range from muscle contraction, cell division, secretion, metabolic changes, and perception, to memory storage. Kretsinger proposed that proteins only can be mediators of the Ca^{2+} signal and that most of these Ca^{2+}-binding proteins (CaBP[1]) possess a recurrent structural motif, i.e., the EF-hand Ca^{2+}-binding domain (for recent review, see Kretsinger 1987). A functionally important discrimination must be made between two types of CaBPs: (1) those that convey the Ca^{2+}-signal to a specific response system through conformational changes and activation of the target, and (2) those that buffer the Ca^{2+} and Mg^{2+} concentrations inside the cell. The former type possesses so-called Ca^{2+}-specific sites, the latter $Ca^{2+}+Mg^{2+}$ mixed sites.

---

[1] Abbreviations used are: CaBP, Ca^{2+}-binding protein; SCP, sarcoplasmic Ca^{2+}-binding protein; CaVP, Ca^{2+} vector protein; IgCalvin, Ig-fold-containing Ca^{2+} vector-regulated protein; Spec, S. *purpuratus* ectoderm; Ig-fold, immunoglobulin fold.

This chapter deals with selected CaBPs that are specific to the higher invertebrate animals, i.e., belonging all to the kingdom of the Metazoa, subdivision Eumetazoa (for updated classification, see Lake 1990). Those treated here belong to animals of the following phyla: coelenterates, annelids, molluscs, arthropods, and echinoderms. Their counterpart may exist in vertebrates, but then with very different characteristics or in amounts that have prevented them from being identified until now. The soluble sarcoplasmic calcium-binding proteins (SCPs), functionally resemble vertebrate parvalbumins and most likely are involved in the regulation of Ca^{2+}/Mg^{2+} homeostasis. Two typical invertebrate proteins, amphioxus Ca^{2+} vector protein (CaVP) and squid optic lobe Ca^{2+}-binding protein (squidulin), seem to act as specific Ca^{2+}-signal transducers and in at least one (CaVP) the target (named IgCalvin) is identified. The echinoderm Spec proteins, which are transiently expressed during development, have not been characterized enough to speculate about their role.

SARCOPLASMIC CALCIUM-BINDING PROTEINS
GENERAL CHARACTERISTICS

In 1974-1980, in search of parvalbumins in invertebrate muscle, we and others discovered another type of soluble sarcoplasmic calcium-binding protein in crustacea (Cox et al. 1976), annelids (Cox and Stein 1981; Gerday et al. 1981), molluscs (Collins et al. 1983) and protochordates (Kohler et al. 1978). The general properties of this protein were too different from those of parvalbumins to allow its classification in the same subfamily (for review, see Wnuk et al. 1982; Gerday 1988). A SARC branch was thus introduced in the phylogenetic tree of EF-hand-containing proteins. Recently, sequence studies revealed that a few other EF-hand-containing proteins belonging to the AEQ subfamily, namely the coelenterate proteins aequorin and luciferin-binding protein and the *Streptomyces erythraeus* CaBP, the first sequenced EF-hand protein of a prokaryote, show high homology with SCPs (Cox and Bairoch 1988).

---

[2] Branch lengths represent the minimum number of nucleotide substitutions that could account for amino acid replacement in the bounding nodes for a given connecting branch.

In the phylogenetic tree (Moncrief et al. 1990), these proteins are located on the same branch, with a common insertion point at a distance of 73 branch lengths[2] from the ancestral protein of all EF-hand proteins.

Like parvalbumins, SCPs are characterized by extensive polymorphism in their subunit and/or polypeptide composition. Sequence analyses revealed that in amphioxus the polymorphism (five isoforms) is due to a few amino acid substitutions in a short 17-residue-long segment in the first Ca^{2+}-binding domain (Takagi et al. 1986; Takagi and Cox 1990a). This suggests that in amphioxus the isoforms are generated by alternative splicing of the primary RNA transcript with a mutually exclusive pattern. The existence of isoforms in the shrimp is also due to some point mutations, but spread over the whole length of the sequence (Takagi and Konishi 1984a,b; Takagi et al. 1984), suggesting that here the isoforms originate from multiple genes. The different isoforms in clam, oyster, earthworm, crayfish, and lobster have not been sequenced yet. In crustacea, polymorphism is increased since SCPs of this phylum naturally occur as dimers with the following subunit composition: α_2, $\alpha\beta$, and β_2 (Wnuk and Jauregui-Adell 1983). The monomer molecular weight of all SCPs is 20 to 22 kDa and, except for the case of crustaceans, SCPs are monomeric. The Stokes radii of the SCPs correspond well to those of globular proteins.

SCPs do not interact with hydrophobic matrices such as phenyl-Sepharose or fluphenazine-Sepharose. Similarly, none of the SCP's forms electrophoretically stable complexes with the amphophilic peptides melittin and seminalplasmin, which often serve as models to monitor the interactions of calmodulin with their target proteins (Cox 1988). Neither do immobilized SCPs retain any protein when charged with muscle extracts of different invertebrates. These data suggest that SCPs are incapable of protein-protein interaction. A reservation has to be made for the case of the dimeric crustacean SCPs, which might dissociate and thus uncover a potential site for protein-protein contact.

DISTRIBUTION

The interphylum distribution pattern of proteins belonging to the SARC and AEQ families can be summarized in four points:

1. Parvalbumins and SCPs have not yet been found in comparable amounts in the same animal; the former are restricted to vertebrates.

2. SCPs occur in various muscle types of invertebrate phyla. In some phyla and classes still no SCP could be detected (for details, see Wnuk et al. 1982).

3. No SCP could be detected in coelenterates, but some of these animals possess SCP-related proteins, among them the bioluminescent aequorin (Shimomura et al. 1962) and the luciferin-binding protein (Charbonneau and Cormier 1979).

4. At least one prokaryote, namely *S. erythraeus*, possesses a SCP-related Ca^{2+}-binding protein (Swan et al. 1987). Its sequence homology is highest with annelid SCP. It is not clear if this protein is as old as the bacterium, i.e., 1400 million years, or was later taken over from a eukaryotic host by gene transfer.

From the SCP distribution within each phylum no rule could be postulated. For instance, molluscan gastropods and cephalopods do not contain significant amounts, but bivalves are rich in SCPs. Nevertheless, fast-moving animals have more chance to contain SCP in abundance. Within a given species the concentration of SCPs is, like that of parvalbumins (Celio and Heizmann 1982), higher in fast contracting muscle (Cox et al. 1976). If, as assumed, SCPs act as intracellular Ca^{2+} and Mg^{2+} buffers (see later), their distribution may be as unpredictable as that of parvalbumins (Heizmann and Berchtold 1987) and depend on the individual and specific need for ion buffering of each cell, organ, and animal species.

Recent immunocytochemical studies using polyclonal antibodies against amphioxus SCPs revealed selective and reproducible stain of individual cells. These affinity-purified antibodies are useful for the cellular characterization of the visual system of *D. melanogaster* (Buchner et al. 1988). The staining with anti-amphioxus SCPII is very specific and completely different from anti-parvalbumin and anti-calbindin: only 10 to 20 neurons are densely stained in the entire brain. Anti-amphioxus SCPII has recently also been shown to stain selectively a 20 kDa, pI 4.6 protein in the single giant neuron of Aplysia (Dr. Paul, Salzburg, pers. commun.). SCPs being neuronal markers, we anticipate that these proteins are not vital for cell life. Their involvement in differentiation may have a kinetic incidence on the function of the cell, as has been postulated for the role of parvalbumins in the central nervous system (Heizmann 1984).

STRUCTURE

The complete amino acid sequence of 11 SCPs from four different phyla has recently been reported (Takagi and Konishi 1984a; Takagi and Konishi 1984b; Takagi et al. 1984; Kobayashi et al. 1984; Takagi et al. 1986; Collins et al. 1988; Jauregui-Adell et al. 1989; Takagi and Cox 1990a). As a whole, all SCPs contain four putative Ca^{2+}-binding domains with the EF-hand motif. The overall sequence identity between SCPs is high (above 80%) within each phylum, but low for SCPs of different phyla (15 to 20%). In fact only seven amino acid residues (three in domain I and four in domain III) are conserved in all SCPs. This overall high sequence variability is consistent with observations that SCPs do not interact with proteins. Within the Ca^{2+}-binding domains conservation is dictated by the constraints of the EF-hand motif, though some are still functional while others have degenerated. Domains I and III are functional in all SCPs. In addition, domain II is active in crustacean and protochordate SCPs and domain IV in that of annelids. A particularity in the primary structure of domain I of SCPs warrants a comment: in the overwhelming majority of functional EF-hands, the -Z coordinating ligand in Ca^{2+}-binding domain I is Glu, whereas it is Asp in all SCPs; two residues farther towards the C-terminus, all SCPs possess a Glu residue. The same pattern is found in luciferin-binding protein and in *S. erythraeus* CaBP. The proteolytic fragment corresponding to domain I of *Nereis* SCP binds Ca^{2+} with higher affinity than any other excised or synthesized single EF-hand domain (Luan-Rilliet and Cox unpubl. observ.), except the 34 residue-long analog of skeletal muscle troponin C (Reid et al. 1981). The particular sequence of domain I of SCPs may correspond to an ancient single EF-domain, efficient Ca^{2+}-binding polypeptide.

Given the highly stringent geometric configuration of an EF-hand structure, it is unlikely that the "fingerprint" characteristic of any subfamily will be found inside the Ca^{2+}-binding domains. A useful criterion for the classification in the SCP and AEQ (Charbonneau et al. 1985) subfamily is the length of the peptide segments not directly involved in Ca^{2+}-binding, especially of the linkers between the EF domains. In all SCPs domains I and II are separated by a 18 to 20 residue-long linker, II and III by a 11 to 12, III and IV by one to two residues. This irregular spacing of the Ca^{2+}-binding domains applies also to the CaBP from *S. erythraeus* (Swan et al. 1987), but has not been observed in the other EF-hand subfamilies.

The elucidation of the tridimensional structure of SCP crystals is in progress. Kretsinger et al. (1980) crystallized crayfish SCP a_2 at pH 6.5 in the $P2_12_12_1$ space group with one dimer per asymmetric unit. *Nereis* SCP crystallizes at pH 7.6 in the space group $P2_1$ with two monomers per asymmetric unit (Babu et al. 1987) and X-ray diffraction data have been processed to 3.0 Å (Cook et al. 1990). The distance between Ca^{2+} ions in the paired C-terminal sites is 11.8 Å; the unpaired Ca^{2+} ion is at approximately 30 Å from the Ca^{2+} pair. In the 3D-resolved proteins like calmodulin, troponin C, parvalbumin, and 9 kDa calbindin, pairing of EF-hand domains is common practice because it stabilizes the global structure and attributes functional Ca^{2+}-binding to each or at least one of the two domains. Single EF-hand domains indeed show little or no affinity for Ca^{2+}, but the cation affinity can be increased by use of lanthanides or trifluoroethanol-containing buffers (for more information, see Reid 1987). The unusual sequence of the C-terminal part of domain I in SCPs, luciferin-binding protein and *S. erythraeus* CaBP points to an unpaired, yet stable and high-affinity EF-hand. One can indeed imagine that not only Asp in the -Z position, but also Glu in the -Z(+2) position makes contact with Ca^{2+}. Knowledge of the high-resolution tridimensional structure of *Nereis* SCP allows to check our prediction on the stability of unpaired domain I and will be the basis for the interpretation of the structure of the other SCPs, the bioluminescent proteins and the prokaryotic CaBP.

INTERACTION WITH Ca^{2+} AND Mg^{2+}

Ion-binding properties of the SCP and AEQ subfamilies are in general complex and involve positive and negative cooperativity (for review, see Wnuk et al. 1982; Cox 1989). Crustacean SCPs possess one Ca^{2+}-specific and two Ca^{2+}-Mg^{2+} sites per monomer; protochordate SCPs contain two Ca^{2+}-specific and one Ca^{2+}-Mg^{2+} site; annelid SCP's contain three Ca^{2+}-Mg^{2+} sites and mollusc SCPs one Ca^{2+}-specific and one Ca^{2+}-Mg^{2+} site. Annelid SCP displays the "simplest" Ca^{2+}-binding characteristics and binds Mg^{2+} with a marked positive cooperativity. Ca^{2+}-binding in the absence of Mg^{2+} displays no positive cooperativity, but millimolar $[Mg^{2+}]$

induces a strong apparent positive cooperativity in Ca^{2+}-binding. This results from the allosteric Mg^{2+}-dissociation (Engelborghs et al. 1990). No direct Ca^{2+}-binding studies have been performed on aequorin, luciferin-binding protein and the *S. erythraeus* CaBP. Under physiological ionic conditions the mean apparent Ca^{2+} dissociation constants for all the SCP and AEQ proteins are similar (30 to 100 nM) and comparable to those of parvalbumins.

Pronounced conformational changes usually accompany the binding of Ca^{2+} to SCPs. Interestingly and in contrast to the case of parvalbumin (Cox et al. 1979), most cation-dependent conformational changes in SCPs are sequential, i.e., they closely follow the appearance of the species $SCP \cdot Ca_n$, with n being different for different types of conformational probes (for this notion, see Cox et al. 1984): in crayfish SCP, the major structural changes occur when the second (α-helix content), third (tryptophan fluorescence) or fourth (thiol reactivity) Ca^{2+} binds to the dimer (Wnuk et al. 1981). In *Nereis* SCP binding of the first Ca^{2+} or Mg^{2+} induces all the conformational change, monitored by Trp fluorescence. The same pattern of ligand-induced conformational changes prevails during aequorin and obelin luminescence (Moisescu and Ashley 1977; Allen et al. 1977; Stephenson and Sutherland 1981): the binding of three Ca^{2+} per protein is required for the control of the luminescence reaction. It should be noted here that SCPs show no Ca^{2+}-dependent exposure of a hydrophobic patch at their surface, confirming that these proteins are unable to interact with hydrophobic matrices, with model peptides or with target proteins (Cox 1990).

Nereis SCP is the simplest and most interesting member of the SCP*SARC subfamily for its abundance, relative simple cation interaction, resemblance to the prokaryote CaBP and to aequorin, and for the expectation that its 3D structure will soon be resolved. Stopped-flow experiments (Engelborghs et al. 1990) showed that Trp fluorescence changes upon Ca^{2+} binding are instantaneous whereas Mg^{2+} binding involves a fast pre-equilibrium, followed by two slow consecutive conformational changes. Upon dissociation of Ca^{2+} or Mg^{2+} from the protein, the Trp fluorescence change is very slow if the protein converts to the metal-free configuration; it is nearly as fast as cation dissociation if the protein converts to the complementary cation-bound state in Ca^{2+}-Mg^{2+} exchange reactions. Indirectly, this study yielded information on the cation dissociation constants, which range from 18 to 22 s^{-1} for the case of Ca^{2+} and to 130 to 220 s^{-1} for Mg^{2+}.

FUNCTIONS

The high concentrations of SCPs, the presence of Ca^{2+}-Mg^{2+} sites, and absence of protein-protein interaction, incited us to consider these proteins as the functional counterparts of parvalbumins in invertebrates. We thus assume that SCPs only function is to influence the Ca^{2+} and Mg^{2+} concentrations inside the cells. Two different hypotheses have been put forward for the exact role of these proteins in the contraction-relaxation process in muscle and are briefly discussed below.

1. Soluble Relaxing Factors. According to this idea, parvalbumins and SCPs act as shuttles transporting Ca^{2+} from the myofibrils to the sarcoplasmic reticulum during the relaxation phase. Direct arguments were provided by Pechère et al. (1977), and by Somlyo et al. (1981), and Gillis (1985). For the case of SCPs that contain solely Ca^{2+}-Mg^{2+} sites the hypothesis, however, does not withstand objections of a kinetic order (Cox et al. 1979; Robertson et al. 1981). Indeed, the duration of contraction and relaxation in fast skeletal muscle is very short and the free Ca^{2+} gradient rises in a few ms and decays with a $t_{1/2}$ of ca. 50 ms. Therefore, for sarcoplasmic CaBPs to be active during each contraction-relaxation cycle, their kinetics of Ca^{2+} binding and dissociation would have to match precisely these rates and be slightly slower than the kinetics of the Ca^{2+}-specific sites of troponin C, the trigger sites in muscular contraction. At the Ca^{2+}-Mg^{2+} sites, disassociation of Ca^{2+} is very slow ($t_{1/2}$ is ca. 100 ms for *Nereis* SCP, see previous paragraph). For parvalbumins a $t_{1/2}$ of 260 ms has been obtained (Potter et al. 1978). Thus, Ca^{2+}-specific sites respond fast enough to sense the rapid changes of sarcoplasmic free $[Ca^{2+}]$ and can act as a relaxing factor, but Ca^{2+}-Mg^{2+} sites react much too slowly to regulate Ca^{2+} concentrations during each twitch.

2. Protectors Against High Ca^{2+} Levels During Prolonged Contractions + Regulators of Intracellular $[Mg^{2+}]$. Only repeated stimulations of muscle (smooth tetanus) can modify the occupancy of the Ca^{2+}-Mg^{2+} sites. In other words these mixed sites act as a slow responding Ca^{2+} sink and thus protect the muscle against excess cytosolic $[Ca^{2+}]$. Most likely they also accelerate energy provision: indeed the concentrations of mixed sites in fast skeletal muscle amount to up to 2 mM and potentially important amounts of Mg^{2+} can be liberated from these sites during muscle tetanization. The free sarcoplasmic Mg^{2+} concentration is estimated at 0.5 mM (Gupta and Moore 1980), indicating that the Mg^{2+} homeostasis can be noticeably affected by the Ca^{2+}-Mg^{2+} sites. Since Mg^{2+} is a cofactor for different glycolytic enzymes and a co-substrate in most reactions involving phosphorus-containing metabolites, it can be anticipated that prolonged stimulations of fast muscles lead to Ca^{2+} binding to sarcoplasmic CaBPs, to an increase in free Mg^{2+}, and to an acceleration of glycolysis.

AMPHIOXUS CALCIUM VECTOR PROTEIN
GENERAL CHARACTERISTICS, Ca^{2+} BINDING AND DISTRIBUTION

In 1986, a new Ca^{2+}-binding protein, called CaVP, was described (Cox 1986) in the muscle of amphioxus (*Branchiostoma lanceolatum*), the protochordate that is closest in evolution to the vertebrates (Lake 1990). This abundant 18.3 kDa protein binds two Ca^{2+} ions with high affinity, apparently at so-called Ca^{2+}-specific sites. Like calmodulin and troponin C, it is highly asymmetric, since it shows an apparent molecular weight of 28 kDa upon gel filtration. CaVP interacts in a Ca^{2+}-dependent manner with amphophilic model peptides, similarly to other well-known intracellular Ca^{2+} vectors such as calmodulin and troponin C (Comte et al. 1983, 1986). It can, however, not functionally substitute for the two latter regulators. *In vivo* most of the CaVP participates in a 50 kDa complex with an endogenous protein (Cox 1986), which we now call IgCalvin. The ability of CaVP to form Ca^{2+}-dependent protein-protein contacts allows its classification as a calcium vector protein (Cox, 1990), even though its role is not actually known.

Antibodies against CaVP showed no cross-reactivity with SCPs, calmodulin, or troponin C, nor with any component in extracts of different organs from rat and fish (Cox 1986). Nevertheless, in immunolocalization experiments (Muntener unpubl. data) positive staining is observed in rat skeletal muscle. Thus, if this CaVP exists in vertebrates, either its concentration is much lower than in amphioxus, or its immunogenic profile is different. In amphioxus anti-CaVP stains intensively the body wall muscles and the neuronal chord. A weakly, but specific and very particular staining pattern has been observed in other organs such as the liver and the gills of the pharynx, indicating that CaVP is not exclusively a muscular protein.

STRUCTURE

Its amino acid sequence revealed that CaVP cannot be classified in to one of the ten EF-hand subfamilies, i.e., it must be considered as a unique protein (Moncrief et al. 1990) as, for instance, the B subunit of calcineurin and caltractins. Typical of CaVP are that: (1) Only EF-domains III and IV are canonical according to rules established for over 200 EF-hand domains and most likely are "active" since the protein possesses two Ca^{2+}-specific sites. (2) In its N-terminal half native CaVP contains two abortive EF-hands and a disulfide bridge linking the first α-helix of domain I to the second of domain II. (3) Domain III contains two ϵ-trimethyllysin residues in the α-helices flanking the Ca^{2+}-binding loop.

The crystal structure of CaVP has not yet been elucidated, but the extensive sequence homology with calmodulin and troponin C, especially in the C-terminal half (ca. 40% sequence identity and 20% conservative replacements with respect to the C-terminal halves of both proteins), permitted the construction of two different tridimensional models of CaVP, one based on the crystallographic structure of CaM with a "Ca^{2+}-filled sites" configuration in the N-terminal half, and one based on TnC generating a "Ca^{2+}-empty sites" configuration (Cox et al. 1990). Both models predict the presence of a long central a-helix, a distance between the two thiols that is optimal for the formation of a disulfide bridge, and a poor accessibility of the disulfide bond to reducing reagents. The CaM-derived model further predicts two surface-exposed hydrophobic patches of ca. 700 $Å^2$; the TnC model predicts only one such surface. The optical properties of the Trp and Tyr residues of CaVP indicate that the CaM-derived model represents the most plausible prediction. This modelization study comforted the hypothesis that CaVP is a modulator protein.

IgCALVIN, TARGET OF CaVP

Under the physiological conditions of the first steps of the purification of CaVP, all of it is associated with a protein of 28 kDa, which we named IgCalvin. IgCalvin can be purified after dissociation of the complex by 6 M urea and Ca^{2+}-dependent chromatography on calmodulin-Sepharose (Takagi and Cox 1990). During the purification IgCalvin is occasionally proteolyzed to a 26 kDa + 5.7 kDa fragment, the former of which could also be purified. Interestingly, both the 22.0 kDa and the 5.7 kDa fragments no longer bind to CaM-Sepharose, suggesting that the proteolytic cutting took place in the middle of the CaVP-binding domain. Polyclonal antibodies raised against IgCalvin showed immunoreactivity in extracts of rat and bovine brain. Thus, in contrast to CaVP, IgCalvin may be better conserved in different vertebrates.

IgCalvin is composed of 243 residues with a M_r of 27.7 kDa (Takagi and Cox 1990b). The topography of this protein is as follows: the N-terminal 21-residue-long segment is rich in Pro and Ala and resembles the N-terminal segment of skeletal muscle myosin light chain kinase (Guerriero et al. 1986); the next 18-residue long segment (from 33 to 50) is an amphophilic, positively charged α-helix, and represents the CaVP-binding segment. Binding of CaVP to this segment may be modulated by phosphorylation since Ser[43] is a potential substrate of protein kinase C. The rest of the protein contains one canonical (residues 57 to 149) and one ancestral (residues 151 to 242) immunoglobulin-fold (Ig-fold). The basic structure of the Ig-fold is a combination of 2 ß-sheets of respectively four and three anti-parallel ß-strands, which surround a hydrophobic interior (for review, see Williams and Barclay 1988). The interaction between the sheets is stabilized by a conservative disulfide bond. The Ig-fold has been found in many proteins comprising all types of immunoglobulins and immunoglobulin receptors, T cell receptors, MHC antigens, cell adhesion molecules, growth factor receptors, and viral proteins. In these proteins the Ig-fold domains are located extracellularly and belong to three subclasses: V, C1, and C2. Recently, a new subclass, C3, has been described in the chaperone proteins (Holmgren and Bränden 1990), intracellular proteins that stabilize and/or transport newly synthesized polypeptides (Ellis and Hemmingsen 1989). The two Ig-folds of IgCalvin belong to the C2 subclass and particularly resemble those present in the neural cell adhesion molecules. However, they cannot form the stabilizing disulfide bond (which anyway is not a vital element of the Ig-fold, as Cys is occasionally replaced by a hydrophobic residue, see Williams 1987).

IgCalvin displays pronounced sequence homology with segments in two intracellular proteins, namely smooth muscle myosin light chain kinase and telokin (Ito et al. 1989). Telokin, a 18 kDa protein abundant in smooth muscle of unknown function, is the C-terminal part of smooth muscle myosin light chain kinase starting just beyond the calmodulin-binding domain and absent in the enzyme of skeletal muscle. Telokin is expressed independently as the translation product of a 2.7-kilobase mRNA (Guerriero et al. 1986) and is three times more abundant than myosin light chain kinase. The protein contains a 20-residue-long phosphorylatable N-terminal segment, a typical Ig-fold displaying, without any gap, 40% homology with the first Ig-fold of IgCalvin, and a 22-residue-long acidic tail. It is specific of smooth muscles in avians and mammalians (Russo et al. 1987).

FUNCTION OF CaVP AND IgCALVIN

A structural property that functionally discriminates the EF-hand proteins is the Ca^{2+}-dependent exposure of a hydrophobic patch at their surface. It should distinguish between "vectors" that activate response elements, and "buffers," which modulate the ion fluxes. CaVP most likely possesses two hydrophobic patches (Cox et al. 1990) also the other characteristics of "vectors" and its response element very likely uses IgCalvin as intermediate. Now it is felt that for the elucidation of the function of CaVP, the study of IgCalvin, and specially of its most salient structural motifs, the Ig-folds, will constitute the most promising approach.

What is usually the function of the Ig-fold? According to Williams and Barclay (1988), "the Ig-fold can be considered as providing a stable platform for the display of specific determinants for recognition reactions on the faces of ß sheets or at the bends between ß strands. The determinants involved are likely to be proteins". Self-recognition is commonly observed in cell adhesion molecules (Kadmon et al. 1990). Since IgCalvin does not present evidence of an enzymatic activity, it is likely that it is involved in stable interaction with an enzyme or a cytoskeletal component, perhaps in linking proteins together or linking proteins to membranes. Such interaction would be modulated by CaVP and post translational modifications direct the localization of IgCalvin. According to the emerging rule of redundancy (Doolittle 1989), it thus looks that an intracellular protein-scaffold with aptitude for protein binding was selected as basic structure for extracellular recognition in vertebrates.

SQUIDULIN OR *LOLIGO* CaBP

Molluscs not only contain Ca^{2+}-buffering proteins such as SCPs, but also a unique CaBP with vector properties, at least in the class of the cephalopods. This protein, called squidulin, was first found in squid (*Loligo pealei*) optic lobe by Head and collaborators (1983). The protein was isolated by Ca^{2+}-dependent chromatography on phenothiazine-Sepharose, a well-known method for the purification of calmodulin. The 16.9 kDa protein is abundant (6 mg/100 g optic lobe tissue). Squidulin and bovine brain calmodulin share 68% sequence identity with a single residue insertion between domains III and IV in squidulin (Head 1989). Phylogenetic calculations by Moncrief et al. (1990) suggest that squidulin branched from the calmodulin line after the divergence of the troponin C subfamily but before any other known member of the calmodulin family. The most apparent structural divergence with calmodulin seems to be without the continuous long central α-helix. Squidulin possesses four canonical EF-hand structures capable of binding Ca^{2+}. Using the binding data of Sheldon and Head (1988), but a slightly different interpretation based on the obvious symmetry of the isotherms, we calculated a $K_{ass(Ca2+)}$ of 5.3×10^5 M^{-1} for all sites. Mg^{2+} shifts the Ca^{2+}-binding isotherm to the right and induces strong positive cooperativity in Ca^{2+}-binding. The mean $K_{ass(Ca2+)}$ shift to 7.9×10^4 M^{-1} allows deduction of a $K_{ass(Mg2+)}$ value of 1.9×10^3 M^{-1}. As for the case of *Nereis* SCP, it is likely that the apparent cooperativity in Ca^{2+} binding is due to homotropic cooperativity in Mg^{2+} dissociation. In squidulin cation-dependent conformational changes are typically sequential, i.e., concomitant with the appearance of a given species: changes in the Tyr environment occur in two steps, associated with the binding of the first and of the second Ca^{2+}. Squidulin has most of the characteristics of a vector protein since (1) it binds chlorpromazine and model peptides such as melittin, (2) Mg^{2+}-induced conformational changes are very different from those induced by Ca^{2+}. However, squidulin does not substitute for the function of calmodulin or troponin C (perhaps for the lack of the central α-helix) and no target is actually known.

ECHINODERM CaBPs

The review on unique invertebrate Ca^{2+}-binding proteins would not be complete without a brief description of the Spec proteins found in echinoderms. This is a group of ten proteins that accumulate extensively after fertilization in ectoderm cells of the sea urchin *Strongylocentrotus purpuratus* (Hardin et al. 1985). The Spec mRNAs and proteins thus serve as molecular markers of aboral ectoderm cells, and Spec gene expression has become a model for control of gene expression. The major pair of Spec proteins is encoded by a single Spec1 gene, while several minor proteins are encoded by six or seven Spec2 genes (Hardin et al. 1985, 1988). The proteins have molecular masses of 14 to 17 kDa and those genes that have been sequenced all encode proteins having four EF-hand domains. Phylogenetically these proteins must be considered as a subfamily close to the common four-domain ancestral protein (situated at 21 branch lengths, compared to 10 for calmodulin and 19 for troponin C, from the common ancestor) (Moncrief et al. 1990). No information is available on their ion-binding properties and their function also remains unknown. Hosoya et al. (1986) isolated a 15 kDa CaBP from eggs of the sea urchin *Hemicentrotus pulcherrimus*, that is very asymmetric in shape and binds in a Ca^{2+}-dependent way to fluphenazine-Sepharose. If this protein corresponds to one of the Spec proteins, it is plausible that they all are vectors of the Ca^{2+} signal. In the distantly related sea urchin *Lytechinus pictus*, the Spec-related homolog, LSP1, is duplicated and contains eight EF-hands (Xiang et al. 1988). Phylogenetically this protein does not cluster with the Spec proteins (Moncrief et al. 1990).

CONCLUSIONS

EF-hand calcium-binding proteins are a versatile family to handle the Ca^{2+} signal inside the cell. They all evolved, with different rates of diversification, from an ancestral protein containing a pair of Ca^{2+}-binding sites, or even from a single EF-hand domain, of the type reminiscent to domain I in the SARC and AEQ subfamily. This ancestor was merely engaged in simple regulation through Ca^{2+}-dependent association with target enzymes, which many present-day EF-hand proteins still do. The main force of interaction seems hydrophobic.

Some proteins still lost the faculty for hydrophobic protein-protein interaction and became merely involved in fine control of the intracellular free Ca^{2+} concentration. Evolution further endowed proteins with a high diversity in the number, affinity and divalent ion selectivity of Ca^{2+}-binding sites, which further helped the cell to adjust in time and space its many response elements to the Ca^{2+} signal. Ultimate sophistication in the control consists of the differential expression of EF-hand coding genes in normal and diseased cells and tissues.

REFERENCES

Allen DG, Blinks JR, Prendergast FG (1977) Aequorin luminescence: relation of light emission to calcium concentration. A calcium-independent component. Science 195:996-998

Babu YS, Cox JA, Cook WJ (1987) Crystallization and preliminary X-ray investigation of sarcoplasmic calcium-binding protein from *Nereis diversicolor*. J Biol Chem 262:11884-11885

Buchner E, Bader R, Buchner S, Cox JA, Emson PC, Flory E, Heizmann CW, Hemm S, Hofbauer A, Oertel WH (1988) Cell-specific immuno-probes for the brain of normal and mutant *Drosophila melanogaster*. I Wildtype visual system. Cell Tissue Res 253:357-370

Celio MR, Heizmann CW (1982) Calcium-binding protein parvalbumin is associated with fast contracting muscle fibers. Nature 297:504-506

Charbonneau H, Cormier MJ (1979) Ca^{2+}-induced bioluminescence in *Renilla reniformis*. Purification and characterization of a calcium-triggered luciferin- binding protein. J Biol Chem 254:769-780

Charbonneau H, Walsh KA, McCann RO, Prendergast FG, Cormier MJ, Vanaman TC (1985) Amino acid sequence of the calcium-dependent photoprotein aequorin. Biochemistry 24:6762-6771

Collins J, Cox JA, Theibert JL (1988) Amino acid sequence of a sarcoplasmic calcium-binding protein from the sandworm *Nereis diversicolor*. J Biol Chem 263:15378-15385

Comte M, Maulet Y, Cox JA (1983) Calcium-dependent high-affinity complex formation between calmodulin and melittin. Biochem J 209:269-272

Comte M, Malnoë A, Cox JA (1986) Affinity purification of seminalplasmin and characterization of its interaction with calmodulin. Biochem J 240:567-573

Cook WJ, Ealick SE, Babu YS, Cox JA (1990) Three-dimensional structure of a sarcoplasmic calcium-binding protein from the sandworm *Nereis diversicolor*. Seventh international symposium on calcium-binding proteins in health and disease. Banff, Alberta, p 147 (Abstract)

Cox JA (1984) Sequential events in calmodulin on binding with calcium and interaction with target enzymes. Fed Proc 43:3000-3004

Cox JA (1986) Isolation and characterization of a new M_r 18,000 protein with calcium vector properties in amphioxus muscle and identification of its endogenous target protein. J Biol Chem 261:13173-13178

Cox JA (1988) Interactive properties of calmodulin. Biochem J 249:621-629

Cox JA (1990) Calcium vector protein and sarcoplasmic calcium binding proteins from invertebrate muscle. In: Dedman JR, Smith VL (eds) Stimulus-response coupling: the role of intracellular calcium. Telford Press, Caldwell, pp 85-110

Cox JA, Bairoch A (1988) Sequence homologies in prokaryote and invertebrate calcium-binding proteins. Nature 331:491-492

Cox JA, Stein EA (1981) Characterization of a new sarcoplasmic calcium-binding protein with magnesium-induced cooperativity in the binding of calcium. Biochemistry 20:5430-5436

Cox JA, Wnuk W, Stein EA (1976) Isolation and properties of a sarcoplasmic calcium-binding protein from crayfish. Biochemistry 15:2613-2618

Cox JA, Winge DR, Stein EA (1979) Calcium, magnesium and the conformation of parvalbumin during muscular activity. Biochimie 61:601-605

Cox JA, Alard P, Schaad O (1990) A predicted structure of amphioxus calcium vector protein from comparative molecular modeling with calmodulin and troponin C. (submitted)

Doolittle RF (1989) Similar amino acid sequences revisited. Trends Biochem Sci 14:244-245

Ellis RJ, Hemmingsen SM (1989) Molecular chaperons: proteins essential for the biogenesis of some macromolecular structures. Trends Biochem Sci 14:339-342

Engelborghs Y, Mertens K, Willaert K, Luan-Rilliet Y, Cox JA (1990) Kinetics of conformational changes in *Nereis* sarcoplasmic calcium-binding protein upon binding of divalent ions. (submitted)

Gerday Ch (1988) Soluble calcium binding proteins in vertebrate and invertebrate muscles. In: Gerday Ch, Bolis L, Gilles R (eds) Calcium and calcium binding proteins. Springer, Berlin, Heidelberg, pp 23-39

Gerday Ch, Collin S, Gerardin-Otthiers N (1981) The soluble calcium binding protein from sandworm (*Nereis virens*) muscle. J Muscle Res Cell Motil 2:225-238

Gillis JM (1985) Relaxation of vertebrate skeletal muscle. A synthesis of the biochemical and physiological approaches. Biochim Biophys Acta 811:97-145

Guerriero V, Russo MA, Olson NJ, Putkey JA, Means AR (1986) Domain organization of chicken gizzard light chain kinase deduced from a cloned cDNA. Biochemistry 25:8372-8381

Gupta RK, Moore RD (1980) ^{31}P NMR studies of intracellular free Mg^{2+} in intact frog skeletal muscle. J Biol Chem 255:3987-3993

Hardin PE, Angerer LM, Hardin SH, Angerer RC, Klein WH (1988) Spec2 genes of *Strongylocentrotus purpuratus*. Structure and differential expression in embryonic aboral ectoderm cells. J Mol Biol 202:417-431

Hardin SH, Carpenter CD, Hardin PE, Bruskin AM, Klein WH (1985) Structure of the Spec1 gene encoding a major calcium-binding protein in the embryonic ectoderm of the sea urchin *Strongylocentrotus purpuratus*. J Mol Biol 186:243-255

Head JF (1989) Amino acid sequence of a low molecular weight, high affinity calcium-binding protein from the optic lobe of the squid *Loligo pealei*. J Biol Chem 264:7202-7206

Head JF, Spielberg S, Kaminer B (1983) Two low-molecular-weight Ca^{2+}-binding proteins isolated from squid optic lobe by phenothiazine-Sepharose affinity chromatography. Biochem J 209:797-802

Heizmann CW (1984) Parvalbumin, an intracellular calcium binding protein, distribution, properties and possible roles in mammalian cells. Experientia 40:910-921

Heizmann CW, Berchtold MW (1987) Expression of parvalbumin and other Ca^{2+}-binding proteins in normal and tumor cells: a topical review. Cell Calcium 8:1-41

Holmgren A, Bränden C-I (1990) Crystal structure of chaperon protein PapD reveals an immunoglobulin fold. Nature 342:248-251

Hosoya H, Iwasa F, Ohnuma M, Mabuchi I, Mohri H, Sakai H, Hiramoto Y (1986) A novel 15 kDa Ca^{2+}-binding protein present in the eggs of the sea urchin, *Hemicentrotus pulcherrimus*. FEBS Lett 205:121-126

Ito M, Dabrowska R, Guerriero V, Hartshorne DJ (1989) Identification in turkey gizzard of an acidic protein related to the C-terminal portion of smooth muscle myosin light chain kinase. J Biol Chem 264:13971-13974

Jauregui-Adell J, Wnuk W, Cox JA (1989) Complete amino acid sequence of the sarcoplasmic calcium-binding protein (SCP-I) from crayfish (*Astacus leptodactylus*). FEBS Lett 243:209-212

Kadmon G, Kowitz A, Altevogt P, Schachner M (1990) The neural cell adhesion molecule N-CAM enhances L1-dependent cell-cell interaction. J Cell Biol 110:193-208

Kobayashi T, Takasaki Y, Takagi T, Konishi K (1984) The amino acid sequence of sarcoplasmic calcium-binding protein obtained from sandworm, *Perinereis vancaurica tetradentata*. Eur J Biochem 144:401-408

Kobayashi T, Takagi T, Konishi K, Cox JA (1987) The primary structure of a new M_r 18,000 calcium vector protein from Amphioxus. J Biol Chem 262:2613-2623

Kohler L, Cox JA, Stein EA (1978) Sarcoplasmic calcium-binding proteins in protochordate and cyclostome muscle. Mol Cell Biochem 20:85-93

Kretsinger RH (1987) Calcium coordination and the calmodulin fold: divergent versus convergent evolution. Cold Spring Harbor Symp Quant Biol 70:499-510

Kretsinger RH, Rudnick SE, Smeden DA, Schatz VB (1980) Calmodulin, S-100 and crayfish sarcoplasmic calcium binding protein crystals suitable for X-ray diffraction studies. J Biol Chem 255:8154-8156

Lake JA (1990) Origin of the Metazoa. Proc Natl Acad Sci USA 87:763-766

Moncrief ND, Kretsinger RH, Goodman M (1990) Evolution of EF-hand calcium-modulated proteins. I. Relationships based on amino acid sequences. J Mol Evolution (in press)

Moisescu DG, Ashley CC (1977) The effect of physiologically occurring cations upon aequorin light emission. Determination of the binding constants. Biochim Biophys Acta 460:189-205

Pechère J-F, Derancourt J, Haiech J (1977) The participation of parvalbumins in the activation-relaxation cycle of vertebrate fast skeletal muscle. FEBS Lett 75:111-114

Potter JD, Johnson JD, Mandel F (1978) Fluorescence stopped-flow measurements of Ca^{2+} and Mg^{2+} binding to parvalbumin. Fed Proc Am Soc Exp Biol 37:1608

Reid RE (1987) A synthetic 33-residue analogue of bovine brain calmodulin calcium binding site III: synthesis, purification and calcium binding. Biochemistry 26: 6070-6073

Reid RE, Gariépy J, Saund AK, Hodges RS (1981) Calcium-induced protein folding. Structure-affinity relationships in synthetic analogues of the helix-loop-helix calcium binding unit. J Biol Chem 256:2742-2751

Robertson SP, Johnson JD, Potter JD (1981) The time course of Ca^{2+}-exchange with calmodulin, troponin, parvalbumin and myosin in response to transient increases in Ca^{2+}. Biophys J 34:559-569

Russo MA, Guerriero V, Means AR (1987) Hormonal regulation of a chicken oviduct messenger ribonucleic acid that shares a common domain with gizzard myosin light chain kinase. Mol Endocrinol 1:60-67

Sheldon A, Head JF (1988) Calcium-binding properties of two high affinity calcium-binding proteins from squid optic lobe. J Biol Chem 263:14384-14389

Shimomura O, Johnson FH, Saiga Y (1962) Extraction, purification and properties of aequorin, a bioluminescent protein from the luminous hydromedusan, *Aequorea*. J Cell Comp Physiol 59:223-239

Somlyo AV, Gonzales-Serattos H, Schuman H, McCleilan G, Somlyo AP (1981) Calcium release and ionic changes in the sarcoplasmic reticulum. J Cell Biol 90:577-594

Stephenson DG, Sutherland PJ (1981) Studies on the luminescent response of the Ca^{2+}-activated photoprotein, obelin. Biochim Biophys Acta 678:65-75

Swan DG, Hale RS, Dhillon D, Leadlay PF (1987) A bacterial calcium-binding protein homologous to calmodulin. Nature 329:84-85

Takagi T, Cox JA (1990a) Amino acid sequences of four isoforms of amphioxus sarcoplasmic calcium-binding proteins. (submitted)

Takagi T, Cox JA (1990b) Amino acid sequences of IgCalvin, the target of calcium vector protein, submitted

Takagi T, Konishi K (1984a) Amino acid sequence of α chain of sarcoplasmic calcium binding protein obtained from shrimp tail muscle. J Biochem (Tokyo) 95:1603-1615

Takagi T, Konishi K (1984b) Amino acid sequence of the ß chain of sarcoplasmic calcium binding protein (SCP) obtained from shrimp tail muscle. J Biochem (Tokyo) 96:59-67

Takagi T, Kobayashi T, Konishi K (1984) Amino acid sequence of sarcoplasmic calcium-binding protein from scallop (*Patinopecten yessoensis*) adductor striated muscle. Biochim Biophys Acta 787:252-257

Takagi T, Konishi K, Cox JA (1986) The amino acid sequence of two sarcoplasmic calcium-binding proteins from the protochordate amphioxus. Biochemistry 25:3585-3592

Williams AF (1987) A year in the life of the immunoglobulin superfamily. Immunology Today 8:298-303

Williams AF, Barclay AN (1988) The immunoglobulin superfamily. Domains for cell surface recognition. Ann Rev Immunol 6:381-405

Wnuk W, Jauregui-Adell J (1983) Polymorphism in high-affinity calcium-binding proteins from crustacean sarcoplasm. Eur J Biochem 131:177-182

Wnuk W, Cox JA, Stein EA (1981) Structural changes induced by calcium and magnesium in a high affinity protein from crayfish sarcoplasm. J Biol Chem 256:11538-11544

Wnuk W, Cox JA, Stein EA (1982) Parvalbumin and other soluble sarcoplasmic Ca-binding proteins. In: Cheung WY (ed): Calcium and cell function, Vol II. Academic Press, New York, pp 243-278

Xiang M, Bédart P-A, Wessel G, Filion M, Brandhorst BP, Klein WH (1988) Tandem duplication and divergence of a sea urchin protein belonging to the troponin C superfamily. J Biol Chem 263: 17173-17180

SPEC PROTEINS: CALCIUM-BINDING PROTEINS IN THE EMBRYONIC ECTODERM OF SEA URCHINS

William H. Klein, Mengqing Xiang, and Gary M. Wessel

INTRODUCTION

Since their initial characterization several years ago, the Spec genes have served as models for the study of gene activation during sea urchin embryogenesis (Bruskin et al. 1981, 1982). Seven or eight related Spec genes[1] of *Strongylocentrotus purpuratus* are activated a few hours after fertilization and are expressed exclusively in cell lineages giving rise to aboral ectoderm (Hardin et al. 1988; Tomlinson and Klein 1990). This cell type is a squamous epithelium of the dorsal/lateral surfaces of the late-stage embryo and larva. Because they are representative of differentiation in aboral ectoderm, these genes have proven to be valuable markers for examining the ontogeny of this embryonic cell type (Nemer 1986; Hurley et al. 1989; Stephens et al. 1989; Wessel et al. 1989). Thus, it should be possible to trace the origins of Spec gene activation and hence aboral ectoderm specification back to maternal factors present in the egg. To this end, recent experiments have begun to dissect the *cis* elements and *trans* factors responsible for activating the Spec genes (Gan et al. 1990a,b; Tomlinson et al. 1990).

A second interest in the Spec gene family has been the role that the gene products play in the differentiating aboral ectoderm cells. A number of investigations have provided details on the properties of the Spec proteins (Carpenter et al. 1984; Muesing et al. 1984;

---

[1]In this article we refer only to the Spec1-Spec2 gene family encoding the intracellular calcium-binding proteins. Other Spec genes unrelated to the Spec1-Spec2 gene family include Spec3, a gene encoding a protein involved in embryonic ciliogenesis (Eldon et al. 1987; 1990), Spec4, a gene encoding the cytoplasmic actin CyIIIa (Bruskin et al. 1981; Bruskin 1983) and Spec5, a gene encoding an uncharacterized protein (Bruskin et al. 1981). All of these Spec genes were originally identified as genes expressed in *Strongylocentrotus purpuratus ectoderm* (Bruskin et al. 1981).

Klein et al. 1985). While it is clear that the proteins encoded by the Spec genes have EF-hand (helix-loop-helix) motifs, and probably most bind Ca^{2+} with high affinity (Kretsinger et al. 1988; Moncrief et al. 1990), the function of the Spec proteins remains a mystery. The Spec1 protein, the most thoroughly characterized of the group, accumulates to high levels in aboral ectoderm cells and, besides calmodulin (Floyd et al. 1986), represents the major calcium-binding protein at the beginning of the larval stage (Bruskin et al. 1982; Carpenter et al. 1984; Klein et al. 1984). The related Spec2 proteins composing the remainder of the family accumulate to lower levels than Spec1 but are also prominent proteins in aboral ectoderm cells (Bruskin et al. 1982; Klein et al. 1984; 1985). In contrast with the cell type specificity of the Spec proteins, calmodulin appears to be present in all cells of the embryo (Floyd et al. 1986). The Spec proteins are not found in adult tissues of the sea urchin, nor in other organisms (unpubl. results); instead they appear to be specialized proteins for sea urchin embryogenesis. Hence, their restricted location and their relatedness to other intracellular EF-hand calcium-binding proteins are the major facts on which predictions of function must be based.

In this chapter, we review some of the earlier models for Spec protein function and discount them in light of more recent information. We summarize results on the conservation of the Spec proteins within *S. purpuratus* and among other sea urchin species. Contrary to expectation, we find that the Spec proteins are highly divergent, maintaining relatedness only in their calcium-binding loops. We also describe studies that imply a connection, at least in some sea urchin species, between the extracellular environment and Spec gene expression. From these considerations, we propose that the Spec proteins play a role in increasing the concentrations of calcium ions in aboral ectoderm cells, thereby leading to an increased transfer of calcium into the blastocoel for use in the formation of the larval endoskeleton. We hypothesize that the Spec proteins function in a manner quite similar to the mammalian intestinal 9-kDa calbindin.

EARLIER MODELS FOR SPEC PROTEIN FUNCTION

Molecular biologists studying sea urchin development have devoted considerable efforts towards identifying genes that are expressed in restricted temporal and spatial patterns in the embryo (e.g., Davidson 1986). Initial studies on ectoderm differentiation involved characterizing mRNAs that accumulated specifically in these cells. Besides the Spec mRNAs, the CyIIIa and CyIIIb mRNAs, each encoding cytoplasmic actins, were also found associated with aboral ectoderm cells (Cox et al. 1986). The CyIIIa mRNA was particularly relevant since its accumulation appeared to coincide very closely with that of Spec1 mRNA (Angerer and Davidson 1984). It was suggested, therefore, that the Spec1 and CyIIIa genes were coordinately regulated and possibly functionally linked (Carpenter et al. 1984). The latter suggestion was based on the notion that other EF-hand proteins such as calmodulin, myosin light chains, and troponin C, can interact either transiently or permanently with contractile filaments.

Two models were suggested, both predicting an interaction between CyIIIa actin and Spec1 and both based on correlations with changes in cell shape (Carpenter et al. 1984). The gradual flattening of the aboral ectoderm cells between the blastula and larval pluteus stages occurs at approximately the same time as CyIIIa actin and Spec1 accumulate. It was postulated that this flattening, while a passive response in itself, was accompanied by the restructuring of microfilaments in the cortex of the ectodermal cells, which itself resulted in the formation of the rigid ectodermal wall of the larva (Akhurst et al. 1987). Perhaps the Spec1 protein, in response to an influx or efflux of Ca^{2+}, mediated this cell shape change (Carpenter et al. 1984).

A second model predicted a much later role for Spec proteins. After several weeks of larval growth, metamorphosis occurs, which involves a dramatic contraction of the ectodermal epithelium (Cameron and Hinegardner 1978). This contraction is believed to be driven by microfilaments and might be analogous to microfilament-mediated contractions in smooth and other nonmuscle cells involving the calmodulin-mediated activation of myosin light chain kinase and the generation of a phosphorylated regulatory myosin light chain ATPase (Adelstein and Eisenberg 1980). In the ectoderm cells of sea urchin embryos, Spec proteins could act like myosin light chains to mediate the contractile events (Carpenter et al.

1984; Klein et al. 1985). In this regard, the Spec2a and Spec2c proteins and LpS1 [the Spec counterpart in another sea urchin species, *Lytechinus pictus* (see below)] contain some semblance of a myosin light chain kinase phosphorylation site in their amino acid sequences (Xiang et al. 1988). In addition, calcium ionophores can cause premature contraction and metamorphosis, implying that calcium ion is involved in these processes (Klein et al. 1985).

In spite of these correlations, it has been difficult to produce any evidence in support of the idea that Spec proteins mediate cell shape changes or that they are associated with polymerized or unpolymerized actin. Since microfilaments are generally found in the cortical region of cells underlying the plasma membrane, it might be expected that Spec proteins would also be found there. However, antibodies against the Spec1 protein show localization throughout the cytoplasm of aboral ectoderm cells (Carpenter et al. 1984). From indirect immunofluorescence labeling or colloid-gold electron microscopic analysis, there is no convincing evidence that Spec proteins are enriched in cortical regions or are associated with microfilaments (unpubl. results). Cell fractionations have shown that Spec1 fractionates as a cytosolic protein (unpubl. results). Immunoprecipitation experiments, as well as gel overlays using radiolabeled Spec1 or LpS1 protein, failed to demonstrate any reproducible interactions between Spec1 (or LpS1) and other proteins. Moreover, Spec proteins do not appear to be sea urchin versions of myosin light chains; Spec does not copurify with myosin and is not immunologically similar; and Spec1 and LpS1 are not phosphorylated with myosin light chain kinase from chicken skeletal muscle.

Of course it is impossible to draw definitive conclusions with negative data of this sort. Nevertheless, we find that the biological basis for Spec proteins being associated with actin or actin filaments is not compelling. Recently, several other genes have been identified whose products accumulate solely in aboral ectoderm cells. One such gene encodes an arylsulfatase that is believed to be secreted into the blastocoel and to play a role in modifying sulfated proteoglycans in the extracellular matrix (ECM) (Yang et al. 1989a). A second gene encodes a metallothionine that is believed to play a role in protection against metal toxicity (Wilkinson and Nemer 1987). These examples demonstrate that the differentiation of the aboral ectoderm results in a variety of cytodifferentiation products, and the notion that these products would all be functionally coupled is an oversimplification. At present there is just no convincing biological rationale for a functional relationship between the Spec proteins and actin.

The model suggesting that Spec proteins play a role in metamorphosis suffers from the fact that these proteins accumulate to high levels several weeks earlier. Furthermore, not all aboral ectoderm cells are involved in the contraction process. Only a few form processes or connecting bridges filled with microfilaments (Akhurst et al. 1987; Cameron pers. commun.). As far as is known, Spec proteins are equally distributed in all aboral ectoderm cells.

While Spec-actin protein interactions associated with cell shape changes have not been conclusively ruled out, it would seem fruitful at this point to develop other hypotheses.

DIVERSITY OF SPEC PROTEIN SEQUENCES

The sequences of four Spec proteins from *S. purpuratus* are known, Spec1, Spec2a, Spec2c, and Spec2d. These are displayed in Figure 1; all contain four calcium-binding domains with EF-hand (helix-loop-helix) motifs indicated in bold type in Figure 1. Spec2a and Spec2c are closely related proteins (72% amino acid match), presumably the result of a recent gene duplication. The Spec1 sequence is more divergent, although clear matches are also apparent outside the calcium-binding domains. It is not clear what the functional basis of the sequence diversity is, if any. It could be that these proteins play slightly different roles in the aboral ectoderm cells or that, whatever their function is, there is no strict requirement for a high degree of sequence conservation.

Comparison of the Ca^{2+}-binding domains with those of other EF-hand proteins suggests that all three proteins can bind Ca^{2+} with high affinity (Moncrief et al. 1990). Spec1 has four good binding domains and Spec2a and Spec2c have three good domains (I, III, and IV) (Moncrief et al. 1990). We have previously shown, indirectly, that Spec1 can bind Ca^{2+} based on SDS-PAGE mobility shifts in the presence and absence of Ca^{2+} (Muesing et al. 1984) and on its ability to bind to phenyl-sepharose in the presence but not in the absence of Ca^{2+} (unpubl. results). Spec mRNA levels indicate that the Spec1 protein is a fewfold higher in abundance than the Spec2a or Spec2c proteins, assuming there are no drastic differences in translation rates or protein stability (Hardin et al. 1988).

The Spec2d protein is highly diverged from the others (Hardin and Klein 1987; Figure 1). The only similarity observed, and it is weak, is in the calcium-binding domains. Inspection of these domains predicts that none of them can bind Ca^{2+} (see Figure 1 legend). Furthermore, the Spec2d mRNA is present at only 2% of the levels of the Spec1 mRNA in the late stage embryo and larva, suggesting that the Spec2d protein is a very minor component of the Spec proteins (Hardin et al. 1988). Based on its low abundance and the prediction that it cannot bind Ca^{2+}, it may be that the Spec2d protein is nonfunctional and destined to become a nontranscribed pseudogene.

In order to determine whether the Spec proteins were conserved in other sea urchin species, we isolated a cDNA clone representing the Spec counterpart in *L. pictus*, a sea urchin that diverged from an ancestor common with *S. purpuratus* 35 million years ago (Xiang et al. 1988). No *L. pictus* cDNA clones could be obtained when Spec cDNAs were used as probes, so we resorted to using an oligonucleotide probe that represented a consensus calcium-binding domain sequence (Hardin et al. 1987). The cDNA clone we isolated, LpS1, coded for a protein that was strikingly different from the Spec proteins of *S. purpuratus* (Figure 1). The translated sequence predicted a protein twice the molecular weight of the Spec proteins with eight rather than four calcium-binding domains and a high degree of sequence divergence: only the calcium-binding domains were conserved (Figure 1; Xiang et al. 1988). Antibodies against the LpS1 reading frame showed reaction to three prominent, closely migrating 34-kDa proteins as determined by two-dimensional gel-Western analysis. In addition, Spec1 antibodies showed cross-reaction to these same proteins, while no reaction with these antibodies was ever observed with calmodulin, myosin light chains, or troponin C. The LpS1 mRNA accumulated to high levels during embryogenesis and was restricted to the aboral ectoderm cells of *L. pictus*. Thus, we concluded that the major Spec counterpart in *L. pictus* was a highly divergent protein encoded by a gene that presumably underwent an internal duplication of its coding exons sometime after the *Lytechinus* and *Strongylocentrotus* lineages split. We have recently isolated the LpS1 gene and shown that it has twice the number of coding exons and was indeed generated by an internal duplication. Moreover, unlike the Spec gene family in *S. purpuratus*, LpS1 appears to be a single-copy gene in the *L. pictus* genome (unpubl. results).

A major conclusion from these studies is that the Spec proteins are not conserved in phylogeny. Figure 1 shows the LpS1 protein sequence divided in half, each half containing four calcium-binding domains. Inspection of the domains indicates that LpS1 has seven domains capable of binding Ca^{2+}. LpS1 binds strongly to phenyl-sepharose in the presence of Ca^{2+}, further implying it can bind Ca^{2+}. Thus, unlike Spec2d, which is also highly diverged from the major *S. purpuratus* Spec proteins, LpS1 appears to be a bona fide Ca^{2+}-binding protein. Based on its ontogenic properties and on the specificity of cross-reactivity of the Spec1 and LpS1 antibodies, it is reasonable to assume that the LpS1 and Spec genes arose from an ancestral Spec gene and that LpS1 has the same function in *L. pictus* as the Spec proteins have in *S. purpuratus*. In *L. variegatus* the homolog of LpS1, termed LvS1, also represents a duplicated version of the Spec proteins (Figure 1). Even though the two *Lytechinus* species are closely related, the partial sequence of the LvS1 cDNA clone shows substantial divergence from the LpS1 sequence (Figure 1).

A recent cladogram analysis of EF-hand proteins by Kretsinger and his colleagues based on maximum parsimony shows that LpS1 is more distantly related to the Spec proteins than to other members of the group such as caltractin and CDC 31 (Moncrief et al. 1990). This is a puzzling observation and may suggest that LpS1's recruitment to function in aboral ectoderm cells was derived from a different ancestral EF-hand gene than Spec. However, it is more likely that a surprising amount of divergence is allowed in the Spec/LpS1 proteins without affecting their function. In spite of their divergence, all of the proteins, with the exception of Spec2d, have retained their ability to bind calcium ions as judged by their amino acid sequences. A logical extension of these arguments is that the major role for the Spec/LpS1 proteins in aboral ectoderm cells is simply to bind Ca^{2+} with high affinity.

LpS1 GENE EXPRESSION AND THE EMBRYONIC EXTRACELLULAR MATRIX

Many events of embryogenesis appear to be mediated by cell-ECM interactions (Trelstad 1984). Components of the ECM such as laminin, fibronectin, collagen, and heparin sulfate proteoglycan come in contact with cell surfaces that have receptors for them (Buck and Horowitz 1987). In the sea urchins, disruption of the ECM during embryogenesis can be achieved by a number of compounds (Wessel et al. 1989). For instance, β-amino-propionitrile (BAPN) inhibits the enzyme lysyl oxidase, which is responsible for the cross-linking of newly synthesized collagen molecules. A lack of cross-linking in collagens lends to their instability, proteolytic digestion, and rapid removal from the matrix. This altered ECM has a profound and selective effect on embryogenesis in the sea urchin. Early development is unaffected: the embryo undergoes blastulation and mesenchyme cell ingression into the blastocoel. The subsequent events of gastrulation and skeletogenesis, however, are inhibited (Wessel et al. 1989). Other agents that disrupt the ECM in different ways produce the same defects in the developing embryo. These effects are not associated with general toxicity, and the arrested embryos will continue to develop normally once the agent is removed (Wessel et al. 1989).

Because the disruption of the ECM causes such pronounced defects in endoderm and mesoderm differentiation, we were interested in determining whether ectoderm differentiation and the general differentiation of the embryo as monitored by the accumulation of cytodifferentiation markers would also be affected.

Fig. 1A,B. Comparisons of Spec protein sequences. **A** Amino acid sequence alignment of Spec, LpS1, and LvS1 proteins based on nucleotide sequence of their cDNA clones. Because LpS1 and LvS1 contain eight rather than four calcium domains, their sequences are divided into the first four domains (LpS1[1], LvS1[1]) and the last four domains (LpS1[2], LvS1[2]). Only partial sequence is available for LvS1. The *underlined* regions and *Roman numerals* indicate the 22-residue EF-hand calcium-binding domains. In Spec1, all domains are predicted to bind calcium. Spec2a and Spec2c domains I, III, and IV bind calcium but domain II does not due to the change of the 11th residue glycine to lysine or arginine in the domain. Spec2d is predicted not to bind calcium because there are changes of key amino acid residues in all of the four calcium-binding loops. All LpS1 domains except the last domain bind calcium. In the last domain there is a substitution of the sixth residue asparagine with phenylalanine. **B** Percent amino acid sequence match among the Spec protein sequences

A.

```
Spec1         M A A Q L L F T D E E V T E F K R R F K N K D T D K S K S I T A E E L G E F F K
Spec2a        M A V Q L L F T E E E K A L F K S S F K S E D T D G D G K I T S E E L R A A F K
Spec2c        M A V Q L F F T E E Q R K V F K S S F K S I D A D G D G K I T P E E L K A A F K
Spec2d        M A A N L L F S E D Q I K E Y K T K F D A F D R N N D G N F P T M F L G N A M K
LpS1(1) MSDRM A K F K A G M P K D A I E_AL K Q E F K D N_YD T N K D G T V S C A E L V K L M N
LpS1(2)       P I G M G P C K D E E Y R E_YY K N E F E K F D K N G D G S L T T A E M S E F M_SK
LvS1(1) MSDAL A K F T R G V P K S V I E_EM K K E F K D N_YD T N N D G T V S C A E L A K L M D
LvS1(2)       P V R K D S F L N E N L L E_YI K K _____
                                                          I
```

```
Spec1         S T G K S Y T D K Q I D K M I S D V D T D E S G T I D F S E M L M G I A E Q M V
Spec2a        S I E I D L T Q E K I D E M M G M V D K D G S K D M D F S E F L M R K A E Q W -
Spec2c        S I E I E L T Q E K I D E M M S M V D K D G S R P V D F S E I L M K K A E Q M -
Spec2d        S V G H V L T A A - - - - E L E N S R R V R K G T T T F P Q F L A M I L D - - -
LpS1(1)       W T E - E M A Q N - - - - I I A R L D V N S D G H M Q F D E F I L Y M - E G S T
LpS1(2)       S T - K - Y S D K E I - E_YL I S R V D L N D D G R V Q F N E F F M H L - D G V S
LvS1(1)       C P E - E E A Q K - - - - M I A T V D V N C D G R M Q F D
LvS1(2)       _____G R M Q F S E F L L S V - Q N V S
                                             II
```

```
Spec1         K W T W K E E H Y T K A F D D M D K D G N G S L S P Q E L R E A L S A S K P P M
Spec2a        R - G - R E V Q L T K A F V D L D K D H N G S L S P Q E L R T A M S A C T_DP P M
Spec2c        R - G - K G A Q Y F K A F D A L D T D K S G S L S P E E L R T A L S A C T_DP P M
Spec2d        K - - - K C R - - - K V F K A M D K D D K D K L_LS A D E V R Q A M L S F D R Q I
LpS1(1)       K E R L Y S S D E I K_QM F D D L D K D G N G R I S P D E L N K G V R E I Y T K V
LpS1(2)       K - - - - - - D H I K_QQ F M A I D K D K N G K I S P E E M V F G I T K I Y R Q M
LvS1(1)                           R S S P E E L S K G V R E I Y T K L
LvS1(2)       K - - - - - - D D I K_NQ F M A I D K D K N G K I S P E E M V T G I K E I Y A S M
                                        III
```

```
Spec1         K R K K I K A I I Q K A D A N K D G K I D R E E F M K L I K S C
Spec2a        T E K E I D A I I E K A D C N G D G K I C L E E F M K L I H S S
Spec2c        T K E E I D A I I K K A D G N N D G E I R R A E F V R M I Q S S Y
Spec2d        T E D K I K E M I E K A D F P N D G K C S L E E F V K M V M N F C
LpS1(1)       V D G M A N K L I Q E A D K D G D G H V N M E E F F D T L V V K L
LpS1(2)       V D F E V_AK L I K E S S F E D D D G Y I N F N E F V N R F F S N CPYKINSLYWPIYLGCAVSI
LvS1(1)       I E G M A T K L I Q E A D K D D D G H V N M E E F C D T L V E K L
LvS1(2)       V D S E V_AK L I K E - A D F D G G D C V D I W E F
                            IV
```

B.

Spec1	Spec1							
Spec2a	58	Spec2a						
Spec2c	52	72	Spec2c					
Spec2d	32	34	31	Spec2d				
LpS1(1)	27	28	23	21	LpS1(1)			
LpS1(2)	32	27	28	24	33	LpS1(2)		
LvS1(1)	26	26	24	21	74	30	LvS1(1)	
LvS1(2)	32	33	29	28	40	60	36	LvS1(2)

Fig. 1A,B.

In both *S. purpuratus* and *L. variegatus* a wide variety of markers that were not specifically associated with endoderm or mesoderm differentiation accumulated normally in ECM-deficient embryos. These included the Spec1 and Spec2a/Spec2c mRNAs. Thus, although morphogenesis had been dramatically altered in these embryos, the differentiation of aboral ectoderm cells as judged by the accumulation of the Spec mRNAs appeared to proceed normally (Wessel et al. 1989).

To identify genes whose activity might be responsive to the ECM, a differential cDNA screen was performed comparing RNA from control and ECM-deficient embryos. One clone repeatedly isolated was LvS1, the *L. variegatus* homolog of LpS1. Accumulation of LvS1 and LpS1 mRNAs were both shown to be strongly dependent on the presence of a normal ECM. Virtually no mRNA accumulated in embryos cultured in the presence of BAPN or other ECM-disrupting agents. Removal of the embryos from these agents resulted in the disposition of a new ECM and the accumulation of LvS1/LpS1. Nuclear run-on experiments with nuclei from BAPN-treated and untreated embryos clearly showed that the lack of accumulation of LpS1 mRNA was due to a cessation of transcription. This effect was quite specific because the accumulation of another mRNA, LpS3 (also found only in ectoderm cells but unrelated to LpS1 or LvS1) was unaffected (Wessel et al. 1989).

We concluded from these experiments that transcription of the LpS1 and LvS1 genes requires the presence of a normal ECM. Presumably, a signaling event is continuously transmitted from the ECM to the ectodermal cell nucleus to maintain active transcription of these genes and disruption of the ECM interferes with the signal transduction. However, the Spec genes of *S. purpuratus* are not affected to any significant extent by ECM disruption. These data suggest that, in evolution, some event in a signaling cascade was either acquired by *Lytechinus* or lost by *Strongylocentrotus*.

More important, the results point out that at least in some species a tight coupling exists between the accumulation of the ectodermal calcium-binding proteins and the extracellular environment.

POTENTIAL ROLE FOR SPEC PROTEINS IN ASSISTING CALCIUM TRANSPORT

Several interesting features about the Spec proteins and their *Lytechinus* counterparts, LpS1 and LvS1, have emerged. They appear to be proteins capable of binding calcium with high affinity. The large degree of variability in their amino acid sequences both within a species (Spec1 versus Spec2a and Spec2c) and between species (Specs versus LpS1 and LvS1) suggests that they are not involved in extensive protein-protein interactions or part of a multiprotein structure. These types of interactions generally require conservation of large regions of the molecule. Our evidence suggests that the Spec proteins are monomeric cytosolic proteins. Because the only remnants of sequence conservation between the *Strongylocentrotus* and *Lytechinus* proteins are the calcium domains, it may be that the sole function of the Spec proteins is to bind Ca^{2+}.

These considerations support the notion that Spec proteins are closer in their mode of action to parvalbumin and the 9-kDa and 28-kDa calbindins than to calmodulin, myosin light chains, or troponin C. The precise function of parvalbumin and the calbindins is unknown, but all are thought to act via their sole ability to bind Ca^{2+} with high affinity (Pechere et al. 1977; Wasserman et al. 1978). The parvalbumins may act as kinetic buffers in fast-twitch muscle (Gillis et al. 1982), and the calbindins may assist in calcium transport in the kidney and intestine (Christakos et al. 1989). The 9-kDa intestinal calbindins are of particular interest. These proteins have been found in the duodenum of avians and mammals, and calbindin gene expression is regulated in a vitamin D-dependent fashion (Christakos et al. 1989). Calbindin function is clearly related to calcium absorption in the intestine, though their actual role is unclear (Feher et al. 1989). Some investigators believe the proteins assist transport of calcium through cells of the intestinal epithelium from the lumen to the body cavity for use in bone formation and maintenance (Feher et al. 1989; Kretsinger et al. 1982; Bronner et al. 1987). The intestinal wall represents a severe barrier to Ca^{2+} flow: there are millimolar levels of free Ca^{2+} in the lumen and body cavity but only micromolar levels in the intestinal epithelial cells. Kretsinger and his coworkers have calculated that by binding Ca^{2+}, the 9-kDa calbindin would increase Ca^{2+} levels in the intestinal cells such that diffusion rates for calcium across the epithelial boundary would increase correspondingly (Kretsinger et al. 1982). The increase in calculated diffusion rates matches closely with what is observed in experimental situations (Bronner et al. 1987).

The sea urchin embryo faces a similar problem of delivering calcium across an epithelial barrier. In this case, the aboral ectoderm separates the blastocoelic cavity from the surrounding sea water. Calcium ion is required for a variety of functions within the blastocoel, the major one being the formation of the endoskeleton, a calcium carbonate-organic matrix deposited by the primary mesenchyme cells. Skeletogenesis begins at the early gastrula stage and continues throughout embryogenesis and larval growth.

We hypothesize that the Spec proteins function in much the same way as the 9-kDa intestinal calbindins. Their abundance and ability to bind calcium ion elevates Ca^{2+} levels in aboral ectoderm cells in the later stages of embryogenesis. This is presumably a time when free Ca^{2+} is being depleted in the blastocoel because of the deposition of the endoskeleton. The increase in Ca^{2+} levels in the ectodermal cells in turn causes an increase in diffusion rates and assists in transporting Ca^{2+} to the blastocoel. This argument for function puts no restraints on the divergence of Spec protein sequences other than keeping their Ca^{2+}-binding properties. Furthermore, it could be argued that a feedback mechanism, which couples the blastocoelic ECM to the production and maintenance of high levels of Spec proteins, would guarantee the proper flow of Ca^{2+} into the blastocoel. Such a linkage could have evolved in *Lytechinus*, or, alternatively, because it was not absolutely necessary, been lost in *Strongylocentrotus*.

It has recently become clear that the aboral ectoderm may be involved in secreting or transporting a variety of molecules into the blastocoel, including growth factors (Yang et al. 1989b), enzymes (Yang et al. 1989a), and ECM components (G.M. Wessel, unpubl. results). It would not be surprising if another one of its roles was to regulate the flow of calcium ion from sea water to the blastocoel. This would imply that other components of the Ca^{2+} transport system would be present on the apical and basal membranes of the aboral ectoderm cells.

A prediction that arises from this hypothesis is that the Ca^{2+} levels in aboral ectoderm cells and in the blastocoel should be lower when Spec proteins are absent. Preliminary experiments have been directed at this issue. Since BAPN eliminates the expression of the

LpS1 gene in *L. pictus*, Ca^{2+} levels using $^{45}Ca^{2+}$ were monitored in BAPN-treated and control embryos. In control embryos Ca^{2+} uptake increased dramatically in concert with LpS1 protein accumulation and continued to increase throughout development. In BAPN-treated embryos, although a blastocoel was present, no skeleton formed, and $^{45}Ca^{2+}$ levels were severely inhibited. While these experiments do not demonstrate Spec protein function, the results do suggest that Ca^{2+} levels can be modulated coincident with Spec expression, both of which are correlated to embryonic development.

A more specific method of eliminating LpS1 would be the use of an LpS1 antisense RNA. This is a feasible approach since large amounts of RNA can be stably injected into sea urchin eggs (Colin et al. 1986), and in other cases, specific aberrations can be seen (Izant et al. 1985; Cabrera et al. 1987). It should then be possible to monitor decreases in LpS1 protein levels by using LpS1 antibodies. If the antisense approach reduces the levels of LpS1 protein, the effect this has on Ca^{2+} uptake and skeletogenesis could be observed. Defects in these events would provide strong support for our model of Spec function.

ACKNOWLEDGMENTS

The work reported here was supported by an NICHD grant to WHK (HD22619) and a postdoctoral fellowship to GMW (HD06899).

REFERENCES

Adelstein RS, Eisenberg E (1980) Regulation and kinetics of the actin-myosin-ATP interaction. Annu Rev Biochem 49:921-956

Akhurst RJ, Calzone FJ, Lee JJ, Britten RJ, Davidson EH (1987) Structure and organization of the CyIII actin gene subfamily of the sea urchin, *Strongylocentrotus purpuratus*. J Mol Biol 194:193-203

Angerer RC, Davidson EH (1984) Molecular indices of cell lineage specification in the sea urchin embryo. Science 226:1153-1160

Bronner F, Pansu D, Stein WD (1987) The vitamin D-dependent Ca-binding proteins assure active Ca transport in intestine and kidney. In: Norman AW, Vanaman TC, Means AR (eds) Calcium binding proteins in health and disease. Academic Press, San Diego, pp 95-97

Bruskin AM (1983) A family of genes expressed in the embryonic ectoderm of sea urchins. Ph.D. thesis, Indiana University, Bloomington, IN

Bruskin AM, Tyner AL, Wells DE, Showman RM, Klein WH (1981) Accumulation in embryogenesis of five mRNAs enriched in the ectoderm of the sea urchin pluteus. Devel Biol 87:308-318

Bruskin AM, Bedard PA, Tyner AL, Showman RM, Brandhorst BP, Klein WH (1982) A family of proteins accumulating in ectoderm of sea urchin embryos specified by two related cDNA clones. Devel Biol 91:317-324

Buck CA, Horowitz AF (1987) Cell surface receptors for extracellular matrix molecules. Annu Rev Cell Biol 3:179-205

Cabrera CV, Alonso MC, Johnston P, Phillips RG, Lawrence PA (1987) Phenocopies induced with antisense RNA identify the Wingless gene. Cell 50:659-663

Cameron RH, Hinegardner RT (1978) Early events in sea urchin metamorphosis, description and analysis. Morph 157:21-32

Carpenter CD, Bruskin AM, Hardin PE, Keast M, Anstrom J, Tyner AL, Brandhorst BP, Klein WH (1984) Novel proteins belonging to the troponin D superfamily are encoded by a set of mRNAs in sea urchin embryos. Cell 36:663-671

Christakos S, Gabrielides C, Rhoten WB (1989) Vitamin D-dependent calcium binding proteins: Chemistry, distribution, functional considerations and molecular biology. Endocrine Rev 10:3-26

Colin AM, Hille MB (1986) Injected mRNA does not increase protein synthesis in unfertilized, fertilized, or ammonia-activated sea urchin eggs. Devel Biol 115:184-192

Cox KH, Angerer LM, Lee JJ, Davidson EH, Angerer RC (1986) Cell lineage-specific programs of expression of multiple actin genes during sea urchin embryogenesis. J Mol Biol 188:159-172

Davidson EH (1986) Gene activity in early development, 3rd edn. Academic Press, Orlando FL, pp 213-246

Eldon ED, Angerer LM, Angerer RC, Klein WH (1987) Spec3: embryonic expression of a sea urchin gene whose product is involved in ectodermal ciliogenesis. Genes Devel 1:1280-1292

Eldon ED, Montpetit IC, Nguyen T, Decker G, Valdizan MC, Klein WH, Brandhorst BP (1990) Localization of the sea urchin Spec3 protein to cilia and Golgi complexes of embryonic ectoderm cells. Genes Devel 4:111-122

Feher JJ, Fullmer CS, Fritzsch GK (1989) Comparison of the enhanced steady state diffusion of calcium by calbindin-D9K and calmodulin: possible importance in intestinal calcium absorption. Cell Calcium 10:189-203

Floyd EE, Gong L, Brandhorst BP, Klein WH (1986) Calmodulin gene expression during sea urchin development: Persistence of a prevalent maternal protein. Devel Biol 113:501-511

Gan L, Wessel GM, Klein WH (1990a) Spec2- but not Spec1-lac z fusions are correctly expressed in sea urchin embryos. (submitted)

Gan L, Zhang W, Klein W (1990b) Repetitive DNA sequences linked to the sea urchin Spec genes contain transcriptional enhancer-like elements. Devel Biol (in press)

Gillis JM, Thomason DB, LeFevre J, Kretsinger RH (1982) Parvalbumin and muscle relaxation: A computer simulation study. J Muscle Res Cell Motil 3:377-398

Hardin PE, Klein WH (1987) Unusual sequence conservation in the 5' and 3' untranslated regions of the sea urchin Spec mRNAs. J Mol Evol 25:126-133

Hardin PE, Angerer LM, Hardin SH, Angerer RC, Klein WH (1988) Spec2 genes of Strongylocentrotus purpuratus: Structure and differential expression in embryonic aboral ectoderm cells. J Mol Biol 202:417-431

Hardin SH, Keast MJ, Hardin PE, Klein WH (1987) Use of consensus oligonucleotides for detecting and isolating nucleic acids encoding calcium binding domains of the troponin C superfamily. Biochem 26:3518-3523

Hurley DL, Angerer LM, Angerer RC (1989) Altered expression of spatially regulated embryonic genes in the progeny of separated sea urchin blastomeres. Development 106:567-579.

Izant JG, Weintraub HC (1985) Constituitive and conditional suppression of exogenous and endogenous genes by antisense RNA. Science 229:345-352.

Klein WH, Spain LM, Tyner AL, Anstrom J, Showman RM, Carpenter CD, Eldon ED, Bruskin AM (1984) A family of mRNAs expressed in the dorsal ectoderm of sea urchin embryos. In: Malacinski GM, Klein WH (eds) Molecular aspects of early development. Plenum Press, New York, pp 131-140

Klein WH, Carpenter CD, Philpotts LE, Brandhorst BP (1985) The sea urchin Spec family of calcium binding protein: Characterization and consideration of possible role in larval development. In: Sawyer RH, Showman RM (eds) The cellular and molecular biology of invertebrate development. The Belle W. Baruch Library in Marine Science Number 15, University of South Carolina Press, Columbia SC, pp 275-295

Kretsinger RH, Mann JE, Simmons JG (1982) Model of facilitated diffusion of calcium by the intestinal calcium binding protein. In: Norman AW, Schaefer D, Herrath DV, Grigoleit HG (eds) Vitamin D. Chemical, biochemical and clinical endocrinology of calcium metabolism, W. de Gruyter, New York, pp 233-248

Kretsinger RH, Moncrief ND, Goodman M, Czelusniak J (1988) Homology of calcium-modulated proteins: Their evolutionary and functional relationships. In: Morad M, (ed) The calcium channel: structure, function and implications. Springer, Berlin, Heidelberg, pp 16-35

Moncrief, ND, Kretsinger, RH, Goodman, M (1990) Evolution of EF-hand calcium-modulated proteins: (I) Relationships based on amino acid sequences. J Mol Evol 30:522-562

Muesing M, Carpenter CD, Klein WH, Polisky B (1984) High-level expression in *Escherichia coli* of calcium-binding domains of an embryonic sea urchin protein. Gene 31:155-164

Nemer M (1986) An altered series of ectodermal gene expressions accompanying the reversible suspension of differentiation in the zinc-animalized sea urchin embryo. Devel Biol 114:214-224

Pechere J-F, Derancourt J, Haiech J (1977) The participation of parvalbumins in the activation-relaxation cycle of vertebrate fast skeletal-muscle. FEBS Lett 75:111-114

Stephens L, Kitajima T, Wilt F (1989) The role of cell interactions in tissue-specific gene expression of sea urchin embryos. Development 107:299-307

Tomlinson CR, Klein WH (1990) Temporal and spatial regulation of the aboral ectoderm-specific Spec genes during sea urchin embryogenesis. Mol Rep Devel (in press)

Tomlinson CR, Kozlowski M, Klein WH (1990) Ectoderm nuclei from sea urchin embryos contain a Spec DNA binding protein similar to the vertebrate transcription factor USF. (submitted)

Trelstad RL (1984) The role of the extracellular matrix in development. AR Liss, Inc. NY.

Wasserman RH, Fullmer CS, Taylor AN (1978) The vitamin D-dependent calcium-binding proteins. In: Lawson DEM (ed) Vitamin D. Academic Press NY, pp 133-136

Wessel GM, Zhang W, Tomlinson CR, Lennarz WJ, Klein WH (1989) Transcription of the Spec1-like gene of *Lytechinus* is selectively inhibited in response to disruption of the extracellular matrix. Development 106:355-365.

Wilkinson DG, Nemer M (1987) Metallothionine genes MTa and MTb expressed under distinct quantitative and tissue-specific regulation in sea urchin embryos. Mol Cell Biol 7:48-58

Xiang M, Bedard P-A, Wessel G, Filion M, Brandhorst BP, Klein WH (1988) Tandem duplication and divergence of a sea urchin protein belonging to the troponin C superfamily. J Biol Chem 263:17173-17180

Yang Q, Angerer LM, Angerer RC (1989a) Structure and tissue-specific developmental expression of a sea urchin arylsulfatase gene. Devel Biol 135:53-65

Yang Q, Angerer LM, Angerer RC (1989b) Unusual pattern of accumulation of mRNA encoding EGF-related protein in sea urchin embryos. Science 246:806-808

CALCIUM-BINDING PROTEINS FROM *TETRAHYMENA*

Tohru Takemasa, Takashi Takagi, and Yoshio Watanabe

INTRODUCTION

Ciliated protozoan *Tetrahymena* is a useful model organism capable of providing important clues for understanding various cellular and molecular mechanisms in eukaryotes, since it can easily be cultivated on simple axenic media up to concentrations of about 106 cells/ml and since it can be analyzed by using various methods of biochemistry, physiology, genetics, and cell biology.

The following calcium-dependent phenomena are known in ciliates; ciliary reversal (avoiding reaction), contractile ring contraction (cell division), food vacuole formation (endocytosis), contractile vacuole contraction (regulation of osmotic pressure) and cytoproct contraction (exocytosis). Of the above, ciliary reversal has been well studied electro-physiologically (Eckert 1972) or by using detergent extracted cell models (Naitoh and Kaneko 1972 and 1973) or nonreversal mutants (Chang and Kung 1973; Takahashi and Naitoh 1978; Takahashi 1979; Takahashi et al. 1980).

It has been postulated that calcium-binding proteins play important roles as calcium-dependent sensory (regulatory) proteins in these calcium-dependent phenomena. As we have been very interested in the above mentioned calcium-dependent phenomena, we have attempted to detect and isolate calcium-binding proteins and characterize them biochemically in *Tetrahymena*. In order to study these calcium-binding proteins thoroughly, it is useful to clone their cDNAs, because they can be used as templates to produce antisense RNAs or intact proteins in *E. coli* expression systems. Moreover, by making site-specific changes in the nucleotide sequences, they can also be applied to produce selected portions of antisense RNAs or mutagenized proteins.

In this chapter, we discuss the structures of three calcium-binding proteins in *Tetrahymena*, which were recently cloned as cDNAs in our laboratory.

TETRAHYMENA CALMODULIN

Since ciliary reversal is induced by micromolar levels of free calcium (Naitoh and Kaneko 1973), we first focused on the ubiquitous calcium-binding protein, calmodulin. We succeeded in isolating calmodulin for the first time from *Tetrahymena pyriformis* and characterized some of its biochemical properties (Suzuki et al. 1979; Nagao et al. 1979; Suzuki et al. 1981; Watanabe and Nozawa 1982). Primary structural analysis of T. *pyriformis* calmodulin was done by Yazawa et al. (1981) by direct amino acid sequencing, and it was revealed that 14 residues diverged from the sequence of vertebrate calmodulin. It was also found that T. *pyriformis* calmodulin has 147 amino acid residues whereas all the other calmodulins from animal kingdom have 148 residues.

In our laboratory, we have preferably used micronucleated T. *thermophila* from which many useful mutants have been isolated (Takahashi et al. 1980; Tamura et al. 1984; Yasuda et al. 1984), so that we cloned the cDNA for calmodulin from a T. *thermophila* cDNA library (Takemasa et al. in prep.).

The amino acid sequence of T. *thermophila* calmodulin deduced from the cDNA is shown in Figure 1 with calmodulins from other organisms (for sources, see legend of Figure 1). Since the first methionine in calmodulin is known to be removed in the course of processing, the relative molecular mass of the anticipated peptide consisting of 148 amino acid residues is calculated to be 16676 daltons. The strict conservation of primary structure in calmodulin is well known and there is no difference among the amino acid sequences of vertebrate calmodulins (for sources, see legend of Figure 1). It has been proved that the electric eel calmodulin reported by Lagacé et al. (1983) was exceptional (one of the isoforms of calmodulin) and another isoform having identical amino acid sequence with that of vertebrates also exists in this organism (Toda et al. 1985).

```
                     Helix      Ca-binding loop    Helix

VERTEBRATES    ADQLTEEQIA EFKEAFSLF DKDGDGTITTKE LGTVMRSL
Electric eel   ---------- --------- ------------ --------
Ascidian       ---------- --------- ------------ --------
Sea slug-1     ---------- --------- ------------ --------
Sea slug-2     ---------- --------- ------------ --------
Fruit fly      ---------- --------- ------------ --------
P.tetraurelia  (AQE)------ ------A-- ------------ --------
T.pyriformis   ---------- --------- ------------ --------
T.thermophila  ---------- --------- ------------ --------
Trypanosome    ----SN---S --------- ------------ --------


VERTEBRATES    GQNPTEA ELQDMINEV DADGNGTIDFPE FLTMMARK
Electric eel   ------- --------- ------------ ------K-
Ascidian       ------- --------- ----D------- --------
Sea slug-1     ------- --------- ------------ --------
Sea slug-2     ------- --------- ------------ --------
Fruit fly      ------- --------- ------------ --------
P.tetraurelia  ------- --------- ------------ --SL----
T.pyriformis   ------- --------- ----D------- --SL----
T.thermophila  ------- --------- ------------ --SL----
Trypanosome    ------- --------- -Q--S------- ---L----


VERTEBRATES    MKDTDSEE EIREAFRVF DKDGNGYISAAE LRHVMTNL
Electric eel   -------- --------- ------------ --------
Ascidian       -------- --------- ------F----- --------
Sea slug-1     -------- --------- ------------ --------
Sea slug-2     --E----- --------- ------F----- --------
Fruit fly      -------- --------- ------F----- --------
P.tetraurelia  --EQ---- -LI---K-- -R----L----- --------
T.pyriformis   -------- -LI---K-- -R--D-L-T--- --------
T.thermophila  -----T-- -LI---K-- -R----L----- --------
Trypanosome    -Q-S---- --K------ ------F----- ---I----


VERTEBRATES    GEKLTDE EVDEMIREA DIDGDGQVNYEE FVQMMTAK
Electric eel   ------- --------- ------------ --------
Ascidian       ------- --------- ------------ --T---S-
Sea slug-1     ------- --------- ------------ --T---S-
Sea slug-2     ------- --------- ------------ --T---S-
Fruit fly      ------- --------- ------------ --T---S-
P.tetraurelia  ------D --------- ------HI---- --R--VS-
T.pyriformis   ------- --------- ------HI---- --R--()--
T.thermophila  ------- --------- ------HI---- --R--M--
Trypanosome    ------- --------- -V-----I---- --K--MS-
```

Fig. 1. Comparison between calmodulin amino acid sequences of vertebrate calmodulins [human (Sasagawa et al. 1982), cow (Watterson et al. 1980), rat (Nojima and Sokabe 1987), chicken (Putkey et al. 1983), frog (Chien and David 1984) electric eel (Lagacé et al. 1983)], invertebrate calmodulins [sea ascidian (Yazawa 1988), sea slugs (Yazawa 1988), fruit fly (*Drosophila melanogaster*) (Smith et al. 1987)], and protozoan calmodulins [*Paramecium tetraurelia* (Schaefer et al. 1987), *Tetrahymena pyriformis* (Yazawa et al. 1981), *Tetrahymena thermophila* (Takemasa et al. in prep.), *Trypanosoma brucei* (Tschudi et al. 1985)]. The sequence of the first three amino acids (*parenthesis*) of *P. tetraurelia* has not yet been determined. Amino acid residues conserved with respect to those of vertebrate calmodulin are represented by *hyphens*

```
                                    -45     AATTC ATAAAATTCA AAAAACAACA AACAAACAAT   -11

          AAATAAATAA ATG GCT CAA TAC TCT CAA ACT CTC AGA TCT TCT GGT TTC ACT TCC ACT   48
                     Met Ala Gln Tyr Ser Gln Thr Leu Arg Ser Ser Gly Phe Thr Ser Thr
                      1               ↓                   10
GTT GGT CTC ACC GAT ATT GAA GGT GCT AAG ACT GTC GCT AGA AGA ATC TTC GAA AAC TAC   108
Val Gly Leu Thr Asp Ile Glu Gly Ala Lys Thr Val Ala Arg Arg Ile Phe Glu Asn Tyr
             20                              30
GAC AAG GGC AGA AAG GGC AGA ATC GAA AAC ACC GAC TGT GTT CCC ATG ATT ACT GAA GCC   168
Asp Lys Gly Arg Lys Gly Arg Ile Gln Asn Thr Asp Cys Val Pro Met Ile Thr Glu Ala
             40                              50
TAC AAG TCC TTC AAC TCC TTC TTT GCC CCC TCT TCT GAT GAC ATC AAG GCC TAC CAC AGA   228
Tyr Lys Ser Phe Asn Ser Phe Phe Ala Pro Ser Ser Asp Asp Ile Lys Ala Tyr His Arg
             60                              70
GTC CTC GAC AGA AAC GGT GAC GGT ATT GTT ACT TAC TAA GAT ATT GAA GAA CTT TGC ATC   288
Val Leu Asp Arg Asn Gly Asp Gly Ile Val Thr Tyr Gln Asp Ile Glu Glu Leu Cys Ile
             80                              90
AGA TAC CTC ACT GGT ACC ACT GTC TAA AGA ACT ATC GTC ACT GAA GAA AAG GTT AAG AAG   348
Arg Tyr Leu Thr Gly Thr Thr Val Gln Arg Thr Ile Val Thr Glu Glu Lys Val Lys Lys
             100                             110      ↓
TCC AGC AAG CCC AAA TAC AAC AAT GAA GTT GAA GCT AAG CTC GAC GTC GCT AGA AGA CTC   408
Ser Ser Lys Pro Lys Tyr Asn Pro Glu Val Glu Ala Lys Leu Asp Val Ala Arg Arg Leu
             120                             130
TTC AAG AGA TAC GAC AAG GAC GGT TCT GGT TAA TTA CAA GAT GAC GAA ATC GCT GGC TTA   468
Phe Lys Arg Tyr Asp Lys Asp Gly Ser Gly Gln Leu Gln Asp Asp Glu Ile Ala Gly Leu
             140                             150
TTA AAG GAC ACC TAT GCT GAA ATG GGT ATG TCC AAC TTC ACC CCT ACT AAG GAA GAC GTT   528
Leu Lys Asp Thr Tyr Ala Glu Met Gly Met Ser Asn Phr Thr Pro Thr Lys Glu Asp Val
             160                             170
AAG ATC TGG TTA TAA ATG GCT GAC ACC AAC TCT GAT GGT TCA GTC TCC CTT GAA GAA TAC   588
Lys Ile Trp Leu Gln Met Ala Asp Thr Asn Ser Asp Gly Ser Val Ser Leu Glu Glu Tyr
             180                             190
GAA GAC CTC ATT ATC AAG TCT CTC CAA AAG GCT GGT ATT AGA GTC GAA AAG TAA TCC TTA   648
Glu Asp Leu Ile Ile Lys Ser Leu Gln Lys Ala Gly Ile Arg Val Glu Lys Gln Ser Leu
             200                             210
GTT TTC TGA   GCAAATAAT AACCAATTTC TTTAACAAAT CAATACACAC TAATATTAAT TTGTCCTAAC   717
Val Phe ***
     218
          TAATATGAAA ATAACATATA ACACTGTACA GAGAAATTTT ATGTATGCTT TGTATTTGAG ACATTTTTA   787

          ACTTTTCTTT ATAAAATCAT TAAAAAAAAA                                            817
              ●  ●●●●●
```

Fig. 2. Nucleotide sequence of the TCBP-25 cDNA and its deduced amino acid sequence. Amino acid sequence *underlined* corresponds to the degradation product which was determined by direct sequencing. At least two introns are located at the points marked with the *arrows*. Twelve amino acids *in the boxes* represent putative calcium-binding loops homologous to the test sequence for an EF-hand calcium-binding site. A stop codon TGA is marked with *asterisks* and a putative polyadenylation signal, TATAAA, is marked with *closed circles*. (Takemasa et al. 1989)

Even in invertebrate animals, structural conservation of calmodulin is well retained as seen in those of ascidians (Yazawa 1988), see slug (Lagacé et al. 1983) and *Drosophila* (Smith et al. 1987).

As for ciliated protozoa, the primary structures of calmodulins differ greatly from those of vertebrates, but there also exist common amino acids changing positions, so to speak "hot spots of divergence". Among the 12 substitutions of *T. thermophila* calmodulin, 10 changes are conserved in *Paramecium tetraurelia* (Schaefer et al. 1987) and *Tetrahymena pyriformis* (Yazawa et al. 1981), indicating that a certain direction must be present in these changes. Some of them may contribute to specific properties of ciliate calmodulin, such as activation of *Tetrahymena*'s membrane-bound guanylate cyclase, and not adenylate cyclase (Nagao et al. 1979; Kudo et al. 1981). Recently, Nozawa et al. (1989) demonstrated that residue 143 was important for the activation of the guanylate cyclase by using recombinant calmodulins. In the near future, we will intend to determine which amino acid(s) contributes to the property by using gene engineering techniques.

Table 1 shows the percentage homologies among calmodulin amino acid sequences from various animals. It is obvious that protozoan calmodulins are much different from those of other animals. Moreover, it is interesting that protozoan calmodulins cannot be categorized into a single group either; for example, the showing lowest homology with calmodulins from Trypanosomes (Tschudi et al. 1985) are not vertebrate calmodulins but *Tetrahymena* calmodulins. From the genealogical viewpoint, it is worthy to note that, even in the same genus *Tetrahymena*, the amino acid sequences of *T. thermophila* and *T. pyriformis* calmodulins are significantly different from each other. It can be supposed that protozoa have evolved diversely and independently as compared with other animals.

Table 1. Percentage homologies between calmodulin sequences

	Vertebrates	Electric eel	Ascidian	Sea slug-1	Sea slug-2	Fruit fly	P. tetraurelia	T. pyriformis	T. thermophila
Electric eel	99.3% (1)								
Ascidian	97.3% (4)	96.6% (6)							
Sea slug-1	98.6% (2)	98.0% (3)	98.6% (2)						
Sea slug-2	97.3% (4)	96.6% (5)	98.6% (2)	98.6% (2)					
Fruit fly	98.0% (3)	97.3% (4)	99.3% (1)	99.3% (1)	99.3% (1)				
P. tetraurelia	88.5% (17)	87.8% (18)	88.5% (17)	89.2% (16)	89.2% (15)	89.2% (16)			
T. pyriformis	90.5% (14)	89.9% (15)	90.5% (14)	89.9% (15)	89.2% (16)	89.9% (15)	93.2% (10)		
T. thermophila	91.9% (12)	91.2% (13)	90.5% (14)	91.2% (13)	90.5% (14)	91.2% (13)	94.6% (8)	96.6% (5)	
Trypanosoma	88.5% (17)	87.8% (18)	89.9% (15)	89.2% (16)	89.2% (14)	89.9% (15)	84.5% (23)	85.1% (22)	86.5% (20)

Each homology was determined from the same data as shown in Figure 1.

Numbers in parenthesis represent of diverged residues in the amino acid sequences (among 148 residues).

A

	I/III	II/IV	I/II	II/III	III/IV	I/IV
Identical amino acid residues	33.3%	36.7%	10.0%	16.7%	13.3%	10.0%
Identical nucleotides	58.9%	50.0%	33.3%	35.6%	43.3%	40.0%

B

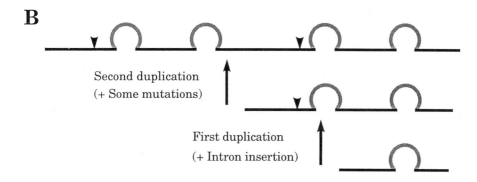

Second duplication
(+ Some mutations)

First duplication
(+ Intron insertion)

Fig. 3A,B. A model for the evolution of TCBP-25. **A** Homologies between calcium-binding domains (helix-loop-helix) at the levels of both amino acid and nucleotide. *I, II, III* and *IV* represent calcium-binding domains I, II, III and IV, respectively. Homologies between domains (for example, I/III = domain I versus domain III) are expressed as percentages. **B** Schematic illustration of two successive gene duplications. *Arrowheads* indicate the positions of introns. *Arrow* indicates the putative evolutionary process

TETRAHYMENA CALCIUM-BINDING PROTEIN OF 25 kDa (*TCBP*-25)

Various calcium-dependent phenomena in *Tetrahymena* do not always occur at micromolar level of calcium, but at concentrations ranging from 10^{-7} to 10^{-3} M. Therefore, we have hypothesized that there must be calcium-binding proteins other than calmodulin in *Tetrahymena* and have attempted to identify them. Consequently, we detected a 25 kDa calcium-binding protein, that we have named TCBP-25. (For detailed description on the detection of this protein, see references; Ohnishi and Watanabe 1983; Kobayashi et al. 1988; Takemasa et al. 1989).

To learn the structure of TCBP-25, we isolated the gene for TCBP-25 (Takemasa et al. 1989). The cDNA sequence and deduced amino acid sequence of TCBP-25 is shown in Figure 2. TCBP-25 is composed of 218 amino acid residues and the relative molecular mass is calculated to be 24702 daltons. This protein possesses four putative EF-hand type calcium-binding domains and therefore it is a member of the calmodulin family. In detail, the Y and Z positions of domain I are replaced by nonconsensus amino acid residues, so that there remains a possibility that loop I cannot bind calcium from the analogy of the defective loop I of bovine heart muscle troponin C (van Eerd and Takahashi 1976).

The most characteristic feature of TCBP-25 is that the regions between calcium-binding loops are somewhat long (30-50 residues) as compared with those of other calcium-binding proteins such as calmodulin, troponin C, myosin light chain, B subunit of calcineurin, etc. (20-29 residues).

Kretsinger and Barry (1975) proposed that EF-hands are arranged in pairs, basing on the structure of troponin. Their data suggest that calcium-binding domain I is genetically close to domain III, and II is close to IV.

This suggestion holds true for TCBP-25 at the amino acid level: the EF-hand regions of domains I and III are very similar to each other as well as those of domains II and IV, while the other combinations share relatively low homology (Figure 3A). Such a tendency holds true for the comparisons at the level of the nucleotide sequence (Figure 3A). The similarities between domains (I/III) and between domains (II/IV) indicates that the genes for these four domains are derived from a pair of a prototype calcium-binding domain gene.

Moreover, the positions of the two introns found in the genomic TCBP-25 gene (arrows in Figure 2 and Figure 3B) at short distances before the domains I and III also strongly support the above idea. Figure 3B shows schematically the gene duplication which may have occurred in the process of generating TCBP-25 gene. A homology search for proteins related to TCBP-25 was carried out , but TCBP-25 showed little homology with all the proteins reported (EMBL-GDB or SWISS-PROT from SDC-GENETIX version 8.0), except for TCBP-23 which will be mentioned in the next section.

```
-39   CAAAAAAAA AAAGTAGAAA GCAAACAAAC ACTAAAAGAA   ATG GAA CAC CAA ATC ATC ACC    21
                                                   Met Glu His Gln Ile Ile Thr
                                                    1

TAA AAC GTC TAC GCT CCT GAT ACT GAA GCC AAG CTT GAT GTT GCT AGA AAG CTT TTT GCT    81
▼▼▼
Gln Asn Val Tyr Ala Pro Asp Thr Glu Ala Lys Leu Asp Val Ala Arg Lys Leu Phe Ala
         10                          20

TAG TTT GAT TCT AAC AAG AAC GGT ACT TTA GAT CCT AGC GAA GTT GCT GGA CTC ATC AAG   141
▼▼▼
Gln Phe Asp Ser Asn Lys Asn Gly Thr Leu Asp Pro Ser Glu Val Ala Gly Leu Ile Lys
             30                          40

ACT ACT TTT GAA AAT ATG GGT GTT AAG GAC TAC AGC GTC ACT GCT GAT GAT GTC AAG CTT   201
Thr Thr Phe Glu Asn Met Gly Val Lys Asp Tyr Ser Val Thr Ala Asp Asp Val Lys Leu
         50                          60

TAC ATG AAG AGT GTT GAT GTT GAT AAC AAC GGT CTT GTT TCC TAC TCT GAA TAC GAA GAA   261
Tyr Met Lys Ser Val Asp Val Asp Asn Asn Gly Leu Val Ser Tyr Ser Glu Tyr Glu Glu
         70                          80

TAC GTC ATT GCT TGC CTC AAG AAA GCT GGC TTC GAC TGT GAA GTT AAG CAA AAG GTA AAA   321
Tyr Val Ile Ala Cys Leu Lys Lys Ala Gly Phe Asp Cys Glu Val Lys Glu Lys Val Lys
         90                          100

AGA TCT GCC AAA AAG AGA GAC GCT GCT ACT GAA ATG AAG TTG GAC GTT GCC AGA AGA CTC   381
Arg Ser Ala Lys Lys Arg Asp Ala Ala Thr Glu Met Lys Leu Asp Val Ala Arg Arg Leu
         110                         120

TTC GCT AAG TAC GAC TCT GAT AAG AGT GGT TAA TTA GAA GAA AAG GAA GTT TAT GGT GTC   441
Phe Ala Lys Tyr Asp Ser Asp Lys Ser Gly Gln Leu Glu Glu Lys Glu Val Tyr Gly Val
         130                         140

ATT ACT GAA ACC TAT AAG CAA ATG GGT ATG GAT TAC AAG CCC ACT GAA GCT GAT GTT AAG   501
Ile Thr Glu Thr Tyr Lys Gln Met Gly Met Asp Tyr Lys Pro Thr Glu Ala Asp Val Lys
         150                         160

CTT TGG ATG TCC ATG ACT GAT ACT GAT AAG AAT GGA ACT GTC TCT ATT GTT GAA TAT GAA   561
Leu Trp Met Ser Met Thr Asp Thr Asp Lys Asn Gly Thr Val Ser Ile Val Glu Tyr Glu
         170                         180

GAT TTC GTC ATT TCT GGT CTT AAG AAG GCT GGT TTC ATG GTC AAG GAA TTC ACT CAA GCT TGA  624
Asp Phe Val Ile Ser Gly Leu Lys Lys Ala Gly Phe Met Val Lys Glu Phe Thr Gln Ala ***
         190                         200                         207

TCTTATTTTA ATAAACATAT TTCAAATAAA AATTGAACTG TATTTGTGTA TTTCATGTCA TTTCTTAAAA         694
         ● ●●●●●              ●●●●●●

TTTTAAATGA TTTTTTTAAT ATCGTTAAAA                                                     724
```

Fig. 4. Nucleotide sequence of the TCBP-23 cDNA and its deduced amino acid sequence. Amino acid sequence *underlined* corresponds to the degradation product, determined by direct sequencing. Twelve amino acids *in the boxes* represent putative calcium-binding loops homologous to the test sequence for an EF-hand calcium-binding site. Triplet codons "TAA" and "TAG" (marked with *open triangles*) are confirmed to code for glutamine from the actual amino acid sequencing of the degradation product (Martindale 1989). Two AATAAAs (marked with *closed circles*) represent putative polyadenylation signals. (Takemasa et al. 1990)

TETRAHYMENA CALCIUM-BINDING PROTEIN OF 23 kDa (*TCBP*-23)

TCBP-23 is the third calmodulin family protein in *Tetrahymena thermophila* (for detailed description on the detection of this protein, see reference Takemasa et al. 1990). In order to learn the primary structure of this polypeptide, we cloned and sequenced the cDNA for TCBP-23. The result of sequencing and the deduced primary structure is shown in Figure 4. This polypeptide is composed of 207 amino acid residues and the relative molecular mass is calculated to be 23413 daltons, so we call this protein TCBP-23. There are four putative calcium-binding domains, all of which completely match Kretsinger's consensus sequence. A computer search was performed, but TCBP-23 also showed little homology with all the proteins in the data base.

Exceptionally, TCBP-23 shows substantial homology (35%) with TCBP-25 at full length (Takemasa et al. 1990). As we mentioned in the previous section, not only TCBP-25 but also TCBP-23 has the special features that residues between calcium-binding domains, especially those between domains II and III, are very long as compared with those of other calmodulin family proteins.

COMPARISON BETWEEN *TCBP*s AND OTHER CALCIUM-BINDING PROTEINS

Representative calcium-binding proteins are schematically aligned with those of two novel calcium-binding proteins found in *Tetrahymena* (Figure 5; for sources, see legend). Although EF-hand calcium-binding proteins, such as calmodulin, troponin C, myosin light chain, and B subunit of calcineurin, are generally known to be small proteins (<20 kDa), some of EF-hand calcium-binding proteins have relatively higher molecular weights (>20 kDa); for example, "calbindins" are mammalian intestinal calcium-binding proteins of 28-30 kDa (Figure 5).

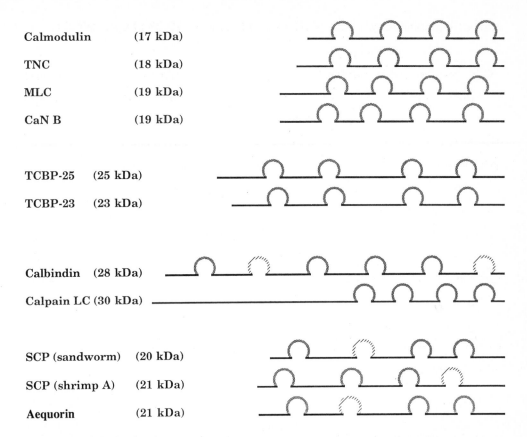

Calmodulin	(17 kDa)
TNC	(18 kDa)
MLC	(19 kDa)
CaN B	(19 kDa)
TCBP-25	(25 kDa)
TCBP-23	(23 kDa)
Calbindin	(28 kDa)
Calpain LC	(30 kDa)
SCP (sandworm)	(20 kDa)
SCP (shrimp A)	(21 kDa)
Aequorin	(21 kDa)

Fig. 5. Schematic illustration of the structures of various calcium-binding proteins. Each *arc* indicates calcium-binding loop. *Arcs with oblique stripes* indicate the loops which are demonstrated or strongly inferred not to bind calcium. The *length of each scheme* indicates the relative molecular mass of each protein. Data sources are; *Tetrahymena* calmodulin, Takemasa et al. prep.; troponin C (TNC), Collins et al. 1977; myosin light chain (MLC), Matsuda et al. 1981; B subunit of calcineurin (CaN B), Aitken et al. 1984; *Tetrahymena* calcium-binding protein of 25 kDa (TCBP-25), Takemasa et al. 1989; *Tetrahymena* calcium-binding protein of 23 kDa (TCBP-23), Takemasa et al. 1990; calbindin, Hunziker 1986; calpain light chain (Calpain LC), Emori et al. 1986; sarcoplasmic calcium-binding proteins (SCPs) from shrimp (aA), Takagi and Konishi 1984 and from sandworm, Kobayashi et al. 1984; aequorin, Inouye et al. 1985

However, these have six putative EF-hand calcium-binding domains, so TCBPs (TCBP-25 and TCBP-23) having four calcium-binding domains are not related to this protein group. On the other hand, calpain light chains (30 kDa), which are the small subunits of a calcium-dependent protease, are categorized as four EF-hand type calcium-binding proteins. But these are fusion peptides and their EF-hand domains are concentrated in their C-terminal regions (Figure 5). Therefore, calpain light chains are dissimilar to TCBPs. Another group sarcoplasmic calcium-binding proteins are four EF-hand type calcium-binding proteins of 20-22 kDa. Aequorin is also categorized in this group. Interestingly, one (or two) of their calcium-binding domains of this protein group is (are) known to be defect in calcium-binding activity. As shown in Figure 5, the differences between sarcoplasmic calcium-binding proteins and TCBPs (the relative molecular masses and the spacing of calcium-binding domains) are evident, so it can be said that TCBPs are very unique in their structures among all calcium-binding proteins reported.

In order to ascertain whether TCBPs have evolved from the same ancestral calcium-binding domain gene, their calcium-binding domains were compared as shown in Figure 6. As postulated by the hypothesis of Goodman et al.(1979), the high homologies in both amino acid and nucleotide sequence of domains (I/III) and domains (II/IV) in TCBP-23 are seen.

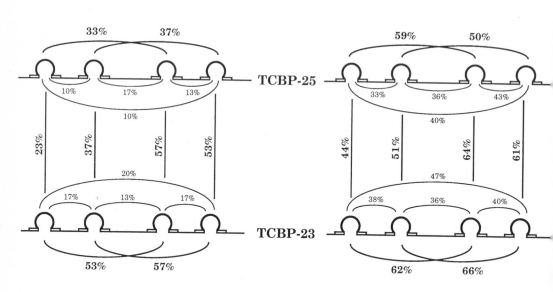

Fig. 6. Homologies between calcium-binding domains from TCBP-25 and TCBP-23. Calcium-binding domains are illustrated as *open box* (helix)-*arc* (loop)-*open box* (helix) structures. Domain homologies of intra- (*bows*) or Inter- (*vertical lines*) molecules are indicated either by percentage identities of amino acid residues or nucleotides

Moreover, homologies of domains (III/III) and domains (IV/IV) between TCBP-25 and TCBP-23 are also very high. These two points of evidence indicate that domains (I-II) and domains (III-IV) in TCBP-23 and domains (III-IV) in TCBP-25 are closely related to one another. This is a corroboration that the genes for TCBPs may have evolved from the same ancestral gene for a prototype peptide having two calcium-binding domains. As to the domains (I-II) in TCBP-25, they may have originated from domains (III-IV), but some recombinations (mutations) might have occurred during the second duplication (from III-IV to I-II-III-IV, ref. Figure 3B). We have no idea why such a recombination would occur, but it must be the reason for the existence of TCBP-25. One possibility is the expected functional difference due to first domain of TCBP-25, where two residues in the consensus sequence for calcium binding are altered (see above). This may cause a difference in sensitivities for free calcium of TCBP-25 and TCBP-23 to work at different physiological states. At any rate, it is interesting that two similar calcium-binding proteins distinct from any other calcium-binding proteins known so far exist in a single cell.

PROSPECT

Here we have shown that *Tetrahymena thermophila* has three kinds of calcium-binding proteins. We are now investigating their functions in *Tetrahymena*, how they share or how they cooperate in various calcium-dependent phenomena. In order to do this, it is necessary to obtain the full length cDNA for them. With these cDNAs, we can make anti-sense RNAs to block translation of independent natural transcripts, or we can produce intact or mutated protein using expression vectors in the living organism. Such methods would be useful to elucidate the functions of novel calcium-binding proteins *in vivo*.

494

REFERENCES

Aitken A, Klee CB, Cohen P (1984) The structure of the B subunit of calcineurin. Eur J Biochem 139:663-671

Chang SY and Kung C (1973) Genetic analysis of heat-sensitive pawn mutant of *Paramecium aurelia*. Genetics 75:49-56

Chien Y-H and David IB (1984) Isolation and characterization of calmodulin genes from *Xenopus laevis*. Mol Cell Biol 4:507-513

Collins JH, Greaser ML, Potter JD, Horn MJ (1977) Determination of the amino acid sequence of troponin C from rabbit skeletal muscle. J Biol Chem 252:6356-6362

Eckert R (1972) Bioelectric control of ciliary activity. Science 176:473-481

Emori Y, Kawasaki H, Imajoh S, Kawashima S, Suzuki K (1986) Isolation and sequence analysis of cDNA clones for the small subunit of rabbit calcium-binding protease. J Biol Chem 261:9472-9476

Goodman M, Pechere J-F, Haiech J, Demaille JG (1979) Evolutionary diversification of structure and function in the family of intracellular calcium-binding proteins. J Mol Evol 13:331-352

Hunziker W (1986) The 28-kDa vitamin D-dependent calcium-binding protein has a six-domain structure. Proc Natl Acad Sci USA 83:7578-7582

Inouye S, Noguchi M, Sakaki Y, Takagi Y, Miyata T, Iwanaga S, Miyata T, Tsuji FI (1985) Cloning and sequence analysis of cDNA for the luminescent protein aequorin. Proc Natl Acad Sci USA 82:3154-3158

Kobayashi T, Takasaki Y, Takagi T, Konishi K (1984) The amino acid sequence of sarcoplasmic calcium-binding protein obtained from sandworm, *Perinereis vancaurica tetradentata*. Eur J Biochem 144:401-408

Kobayashi T, Takagi T, Konishi K, Ohnishi K, Watanabe Y (1988) Amino acid sequence of a calcium-binding protein (TCBP-10) from *Tetrahymena*. Eur J Biochem 174:579-584

Kretsinger RH and Barry CD (1975) The predicted structure of the calcium binding component of troponin. Biochim Biophys Acta 405:40-52

Kudo S, Ohnishi K, Muto Y, Watanabe Y, Nozawa Y (1981) Paramecium calmodulin can stimulate membrane-bound guanylate cyclase in *Tetrahymena*. Biochem Int 3:255-263

Lagacé L, Chandra T, Woo SLC, Means AR (1983) Identification of multiple species of calmodulin messenger RNA using a full length complementary DNA. J Biol Chem 258:1684-1688

Martindale DW (1989) Codon usage in *Tetrahymena* and other ciliates. J Protozool 36:29-34

Matsuda G, Maita T, Umegane T (1981) The primary structure of L-1 light chain of chicken fast skeletal muscle myosin and its genetic implication. FEBS Lett 126:111-113

Nagao S, Suzuki Y, Watanabe Y, Nozawa Y (1979) Activation by a calcium-binding protein of guanylate cyclase in *Tetrahymena pyriformis*. Biochem Biophys Res Commun 90:261-268

Naitoh Y and Kaneko H (1972) Reactivated Triton-extracted models of Paramecium: modification of ciliary movement by calcium ions. Science 176:523-524.

Naitoh Y and Kaneko H (1973) Control of ciliary activities by adenosine triphosphate and divalent cations in Triton-extracted models of *Paramecium caudatum*. J Exp Biol 58:657-676.

Nojima H and Sokabe H (1987) Structure of a gene for rat calmodulin. J Mol Biol 193:439-445

Nozawa Y, Nagao S, Kanoh H, Ozawa T, Matsuki S (1989) Activation of *Tetrahymena* plasma membrane guanylate cyclase by recombinant calmodulin: site-directed mutagenesis of rat brain calmodulin cDNA. In: 8th Int Congr Protozool 10-17 July 1989, Tsukuba, Japan

Ohnishi K and Watanabe Y (1983) Purification and some properties of a new Ca^{2+}-binding protein (TCBP-10) present in *Tetrahymena* cilium. J Biol Chem 258:13978-13985.

Putkey JA, Ts'ui KF, Tanaka T, Lagacé L, Stein JP, Lai EC, Means AR (1983) Chicken calmodulin genes. A species comparison of cDNA sequences and isolation of a genomic clone. J Biol Chem 258:11864-11870

Sasagawa T, Ericsson LH, Walsh KA, Schreiber WE, Fischer EH, Titani K (1982) Complete amino acid sequence of human brain calmodulin. Biochemistry 21:2565-2569

Schaefer WH, Lukas TJ, Blair IA, Schultz JE, Watterson DM (1987) Amino acid sequence of a novel calmodulin from *Paramecium tetraurelia* that contains dimethyllysine in the first domain. J Biol Chem 262:1025-1029

Smith VL, Doyle KE, Maune JF, Munjaal RP, Beckingham K (1987) Structure and sequence of the *Drosophila melanogaster* calmodulin gene. J Mol Biol 196:471-485

Suzuki Y, Hirabayashi T, Watanabe Y (1979) Isolation and electrophoretic properties of a calcium-binding protein from the ciliate *Tetrahymena pyriformis*. Biochem Biophys Res Commun 90:253-260

Suzuki Y, Nagao S, Abe K, Hirabayashi T, Watanabe Y (1981) *Tetrahymena* calcium-binding protein is indeed a calmodulin. J Biochem (Tokyo) 89:333-336

Takagi T and Konishi K (1984) Amino acid sequence of a chain of sarcoplasmic calcium binding protein obtained from shrimp tail muscle. J Biochem (Tokyo) 95:1603-1615

Takahashi M (1979) Behavioral mutant in *Paramecium caudatum*. Genetics 91:393-408

Takahashi M and Naitoh Y (1978) Behavioral mutants of Paramecium caudatum with the defective membrane electrogenesis. Nature (London) 271:656-659

Takahashi M, Onimaru H, Naitoh Y (1980) A mutant of *Tetrahymena* with nonexcitable membrane. Proc Jpn Acad Sci B 56:585-590

Takemasa T, Ohnishi K, Kobayashi T, Takagi T, Konishi K, Watanabe Y (1989) Cloning and sequencing of the genes for *Tetrahymena* calcium-binding 25-kDa protein (TCBP-25). J Biol Chem 264:19293-19301

Takemasa T, Takagi T, Kobayashi T, Konishi K, Watanabe Y (1990) The third calmodulin family protein in *Tetrahymena*: Cloning of the cDNA for *Tetrahymena* calcium-binding protein of 23 kDa (TCBP-23). J Biol Chem 265:2514-2517

Tamura R, Takahashi M, Watanabe Y (1984) Molecular mechanism of cell division in *Tetrahymena thermophila* I. Analysis of execution period of a division-arrest ts-mutant. Zool Sci (Tokyo) 1:50-61

Toda H, Yazawa M, Abe Y, Sakiyama F, Yagi K (1985) Comparison of primary structures of calmodulins. Seikagaku (in Japanese) 57:1037 (4B-a10)

Tschudi C, Young AS, Ruben L, Patton CL, Richards FF (1985) Calmodulin genes in trypanosomes are tandemly repeated and produce multiple mRNAs with a common 5' leader sequence. Proc Natl Acad Sci USA 82:3998-4002

van Eerd J-P and Takahashi K (1976) Determination of the complete amino acid sequence of bovine cardiac troponin C. Biochemistry 15:1171-1180

Watanabe Y and Nozawa Y (1982) Possible roles of calmodulin in a ciliated protozoan *Tetrahymena*. In: Cheung WY (ed), Calcium and cell function, Vol. II, Academic Press Inc., New York, pp 297-323

Watterson DM, Sharief F, Vanaman TC (1980) The complete amino acid sequence of the Ca^{2+}-dependent modulator protein (calmodulin) of bovine brain. J Biol Chem 255:962-975

Yasuda T, Tamura R, Watanabe Y (1984) Molecular mechanism of cell division in *Tetrahymena thermophila* II. Ultrastructural changes found in a division-arrest ts-mutant. Zool Sci (Tokyo) 1:62-73

Yazawa M, Yagi K, Toda H, Kondo K, Narita K, Yamazaki R, Sobue K, Kakiuchi S, Nagao S, Nozawa Y (1981) The amino acid sequence of the *Tetrahymena* calmodulin which specifically interacts with guanylate cyclase. Biochem Biophys Res Commun 99:1051-1057

Yazawa M (1988) Molecular evolution of calmodulin. Protein, nucleic acid, and enzyme. (in Japanese) 33:2091-2101

SEQUENCE AND POSSIBLE FUNCTION OF A NOVEL Ca^{2+}-BINDING PROTEIN ENCODED IN THE *SHAKER*-LOCUS OF *DROSOPHILA*

Olaf Pongs, Imke Krah-Jentgens, Dieter Engelkamp, and Alberto Ferrus

INTRODUCTION

In the 1960s, the multifaceted role of cyclic nucleotides and their associated phosphorylation systems became apparent. Similar second messenger functions for Ca^{2+} were also suggested. In certain favorable instances, such as muscle contraction and neurotransmitter release, these were actually established. The role of Ca^{2+} in the control of neurotransmitter release has been investigated in great detail. In addition, the control of neuronal excitability and plasticity are of special interest. An effect of Ca^{2+} on the gating of voltage-gated ion channels has been suggested, but little is known about the initial Ca^{2+} receptor molecules along the stimulus-response pathway which decodes the information contained in a Ca^{2+} transient and which translates it into an effect on the gating of ion channels.

Shaker (Sh) mutants of *Drosophila melanogaster* are in this context particularly interesting because they are ideally suited to study the molecular basis of the relationship between a behavioral trait and an altered neuronal excitability caused by dysfunctional voltage-gated K$^+$-channels. Action potentials from the cervical giant interneuron (CGF) have abnormal duration in some mutants and abnormal frequencies in others (Tanouye et al. 1981; Tanouye and Ferrus 1985; Ferrus et al. 1990). Some *Sh* mutants have an abnormal electro-physiology because of lacking or abnormal A-type K$^+$-channels (Zagotta et al. 1988; Gisselmann et al. 1989). *Sh* mutants have been mapped within a transcription unit encoding a family of A-type K$^+$-channels (Schwarz et al. 1988; Kamb et al. 1988; Pongs et al. 1988). Yet other *Sh* mutants do not map within the K$^+$-channel transcription unit (Ferrus et al. 1990). For example, the mutant T(X;Y)V7 maps some 110 kb away from the transcription start site of K$^+$-channel mRNA (Krah-Jentgens 1989). The chromosomal breakpoint in this

Sh mutant marks the proximal limit of the *Sh* locus. The chromosomal rearrangement causes leg shaking and CFG action potential abnormalities similar to *Sh* mutants which do not abolish but reduce K$^+$-channel activities. Therefore, we entertained the hypothesis that T(X;Y)V7 affects a gene which encodes a modulator of K$^+$-channel activity and thereby reduces the effectiveness of K$^+$-channel activity in the CNS of *Drosophila melanogaster*. We find that this gene encodes a Ca^{2+}-binding protein with novel sequence and structural properties. This raises the possibility that we have cloned a Ca^{2+}-receptor molecule which modulates the gating of K$^+$-channels in the nervous system.

RESULTS

We have isolated several overlapping cDNA-clones which define a transcription unit linked to the chromosomal breakpoint T(X;Y)V7. This transcription unit encodes a ~6 kb long mRNA (not shown). The longest open reading frame in this RNA is only 561 nucleotides long (Krah-Jentgens 1989). This suggests that only 8% of the transcript sequence is translated into protein. The derived protein sequence is 187 amino acids long with a molecular mass of 21671 (Figure 1). The sequence is very hydrophilic, rich in charged amino acids and does not exhibit any extensive hydrophobic stretches.

```
M G K K S S K I K Q D T I D R L T T D T Y F T E K        25
E I R Q W H K G F L K D C P N G L L T E Q G F I K        50
I Y K Q F F P Q G D P S K F A S L V F R V F D E N        75
N D G S I E F E E F I R A L S V T S K G N L D E K       100
L Q W A F R L Y D V D N D G Y I T R E E M Y N I V       125
D A I Y Q M V G Q Q P Q S E D E N T P Q K R V D K       150
I F D Q M D K N H D G K L T L E E F R E G S K A D       175
P R I V Q A L S L G G G                                 187
```

Fig. 1. Derived amino acids sequence of a calcium-binding protein (CBS) encoded in a transcription unit of the *Shaker* locus. Amino acid residues are numbered beginning with the initial methionine. Numbers of the last residues are given in the *right hand side*. The sequence is given in the standard one letter code. The isolation and characterization of the corresponding cDNA/genomic DNA is described in Krah-Jentgens (1989)

A similarity dot matrix analysis (Staden 1982) of the derived protein sequence (Figure 2A) revealed that it contains a repeated structural motif, which occurs three times between amino acid residues 69 to 85, 105 to 121, and 152 to 168, respectively. The repeated sequence motif corresponds to the Ca^{2+}-binding domains of Ca^{2+}-binding proteins, most notably to those of troponin C (Van Eerd et al. 1978) and calmodulin (Beckingham et al. 1987). The similarities are limited to the Ca^{2+}-binding domains (Figure 2B,C). The Ca^{2+}-binding domains of the derived protein are in spacing and sequence more similar to those of troponin C than to those of calmodulin. Troponin C and calmodulin contain four Ca^{2+}-binding domains, in contrast the derived protein only three. Also, the exon/intron organization of the V7 transcription unit is different from that of the *Drosophila* calmodulin gene (Beckingham et al. 1987). The V7-transcription unit, therefore, does not encode a member of the troponin C/calmodulin gene family.

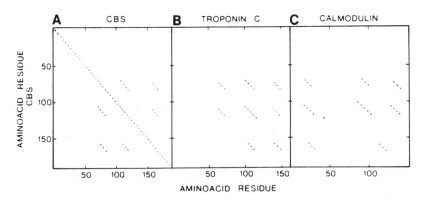

Fig. 2A-D. Similarity matrix analysis comparing the amino acids sequence of CBS protein (ordinate) with itself (A) with troponin C of frog skeletal muscle (Van Eerd et al. 1978) (B) with calmodulin of *Drosophila melanogaster* (C) (Beckingham et al. 1987) using a computer program to generate diagonal lines indicating segments of six residues that show sequence identity >50%. D Primary sequence alignment between Ca^{2+}-binding domains of CBS protein and the ones of troponin C and of calmodulin. *Numbers* on the *left* and the *right hand side* indicate the position of the given first and last amino acid in the respective protein sequence. X,Y,Z mark the amino acid side chains which constitute in the consensus EF-hand the Ca^{2+}-contact sites (Persechini et al. 1989).

Figure 2D aligns the amino acids sequences of the three Ca^{2+}-binding domains of the derived protein with those of the third Ca^{2+}-binding domain of troponin C and calmodulin, respectively (Figure 2D), which gave the best fits in the similarity matrix analysis. Identical amino acid residues are found where the amino acid side chains (marked X,Y, an Z in Figure 2D) of the Ca^{2+} binding domains are directly involved in binding Ca^{2+} as suggested by the EF-hand consensus sequence (Persechini et al. 1989). Giving these compelling similarities we propose to name the protein CBS for calcium-binding protein of the *Shaker* locus. CBS mRNA is preferentially expressed in the nervous system. *In situ* hybridization experiments indicate that CBS mRNA accumulates in the embryo in mature cells showing the characteristic neuromeric pattern (Figure 3A). The RNA expression is maintained in the CNS during development. In the adult (Figure 3B) CBS mRNA is distributed homogeneously in the somata of the CNS, which includes the somata of thoracic ganglia.

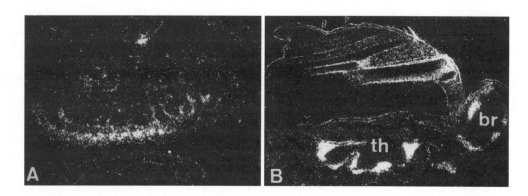

Fig. 3A,B. Distribution of CBS RNA in the nervous system of *Drosophila melanogaster*. **A** Sagittal section of an 18 h *Drosophila* embryo hybridized with an antisense CBS riboprobe. The expression is restricted to the central core of CNS neuromeres. **B** Sagittal section of a normal adult hybridized with the same riboprobe as in **A**. The expression is localized homogeneously throughout the CNS. *br* brain; *th* thoracic ganglion. Details of *in situ* hybridization experiments are described in Pongs et al. (1988)

DISCUSSION

The structure of the derived CBS protein provides some clues for its function. The derived primary sequence of the CBS protein shows three internally repeated sequence motifs which have a strong similarity to Ca^{2+} binding domains of the EF-hand type. Outside of the Ca^{2+}-binding domains, the CBS protein does not have any particular similarity to sequences of proteins which are available in the NBRF or EMBL data bank. Thus, the CBS protein is a novel Ca^{2+}-binding protein specifically functioning in the nervous system of *Drosophila melanogaster*. The predicted structure of the CBS protein is different from the prototypes of the calcium-binding protein superfamily, troponin C and calmodulin. These two proteins have adopted a symmetrical arrangement of two Ca^{2+}-binding domains at the top and the bottom of a central α-helical core (Herzberg and James 1985; Babu et al. 1985), respectively. The CBS protein cannot acquire such a symmetrical structure because it has an odd number of Ca^{2+}-binding domains. Only the first two Ca^{2+}-binding domains have a spacing comparable to the ones observed in troponin C or calmodulin.

We have found that the T(X;Y)V7 chromosomal breakpoint is in the immediate vicinity (within 1 kb) of the polyadenylation signal of the CBS transcription unit. V7 flies have increased levels of CBS mRNA and, therefore, appear to overproduce the CBS protein (Krah-Jentgens, 1989). We entertain the hypothesis that the phenotypic traits of V7 flies are caused by an overproduction of CBS protein. V7 flies have a *Shaker* phenotype. Also, V7 flies have irregular action potentials (Tanouye et al. 1981) in interneurons of the *Drosophila* central nervous system as well as a facilitated neurotransmitter release at the synapse of motoneurons (Mallart, Ferrus and Pongs unpubl.)

The action potential abnormality in V7 flies is very similar to *Shaker* mutants which have a reduced number of *Sh* K^+-channel forming proteins (Baumann et al. 1989). Accordingly, the activity of CBS protein would be to reduce the number of *Sh* K^+-channels in the nervous system of *Drosophila*. We like to speculate that the *Sh* K^+-channels are like other potassium channels a focal point for second messenger-dependent neuromodulation. In Aplysia, for example a cAMP and Ca^{2+}/calmodulin-dependent kinase modulates the effectiveness of K^+-channels (Abrahams and Kandel 1988). It is conceivable that a Ca^{2+}/CBS-dependent kinase phosphorylates *Sh* K^+-channels and thereby closes the ion channel. Then, an overproduction of CBS protein might give rise to an increased Ca^{2+}/CBS-dependent kinase activity, which in turn leads to an abnormally increased number of closed *Sh* K^+-channels and, thus, to the firing of irregular action potentials with a delayed repolarization.

502

SUMMARY

The mutation T(X;Y)V7 defines the right-most breakpoint within the *Shaker* gene complex (*ShC*) of *Drosophila*. This mutation is associated with a dominant perturbation of action potential profiles in the cervical giant interneuron (CGF). The cloning of the entire *ShC* has permitted to locate the site of T(X;Y)V7. This breakpoint is located 130 kb proximal to be transcription unit where the potassium channel subunits are coded. The V7 transcription unit encodes a novel type of Ca^{2+}-binding protein (CBS) which is expressed specifically in the nervous system throughout development. Mutant V7 flies show increased steady-state levels of CBS RNA. We propose that the overexpression of CBS protein causes abnormal action potentials in the central nervous system.

ACKNOWLEDGMENTS

This work was supported by grants from Deutsche Forschungsgemeinschaft (O. Pongs) Volkswagen Stiftung (A. Ferrus and O. Pongs), CAICYT (A. Ferrus).

REFERENCES

Abrams TW and Kandel ER (1988) Learning in *Aplysia* suggests a possible molecular site. TINS 11:128-135

Babu YS, Sack JS, Greenhough TJ, Bugg CE, Means AR, Cook WJ (1985) Three-dimensional structure of calmodulin. Nature 315:37-40

Baumann A, Krah-Jentgens I, Müller R, Müller-Holtkamp F, Canal I, Galceran J, Ferrus A, Pongs O (1989) Molecular basis of biological diversity at the *Shaker* locus of *Drosophila* in molecular biology of neuroreceptors and ion channels (ed A. Maelicke), Springer, Heidelberg, pp 231-244

Beckingham K, Doyle KE, Maune JF (1987) The calmodulin gene of *Drosophila melanogaster*. Meth in Enzymol 139:230-247

Ferrus A, Llamazares S, de la Pompa JL, Tanouye MA, Pongs O (1990) Genetic analysis of the *Shaker* gene complex of *Drosophila melanogaster*. Genetics (in press)

Gisselmann G, Sewing S, Madsen BW, Mallart A, Angaut-Petit D, Müller-Holtkamp F, Ferrus A, Pongs O (1989) The interference of truncated with normal potassium channel subunits leads to abnormal behavior in transgenic *Drosophila melanogaster*. EMBO J 8:2359-2364

Herzberg O, James MNG (1985) Structure of the calcium regulatory muscle protein troponin-C at 2,8 A resolution. Nature 313:653-659

Kamb A, Tseng-Crank J, Tanouye MA (1988) Multiple products of the *Drosophila Shaker* gene contribute to potassium channel diversity. Neuron 1:421-430

Krah-Jentgens I (1989) Klonierung des *Shaker*-Komplexes von *Drosophila melanogaster* und Charakterisierung einer Transkriptionseinheit aus diesem Gen-Komplex. Ph.D. Thesis, Ruhr-Universität Bochum

Persechini A, Moncrief ND, Kretsinger RH (1989) The EF-hand family of calcium-modulated proteins. TINS 12:462-467

Pongs O, Kecskemethy N, Müller R, Krah-Jentgens I, Baumann A, Kiltz HH, Canal I, Llamazares S, Ferrus A (1988) *Shaker* encodes a family of putative potassium channel proteins in the nervous system of *Drosophila*. EMBO 7:1087-1096

Schwarz TL, Tempel BL, Papazian DM, Jan YN, Jan LY (1988) Multiple potassium channel components are produced by alternative splicing at the *Shaker* locus in *Drosophila*. Nature 331:137-142

Staden R (1982) An interactive graphics program for comparing and aligning nucleic acid and amino acid sequences. Nucl Acids Res 10:2951-2961

Tanouye MA, Ferrus A (1985) Action potentials in normal and *Shaker* mutant *Drosophila*. J Neurogenet 2:253-271

Tanouye MA, Ferrus A, Fujita SC (1981) Abnormal action potentials associated with the *Shaker* complex locus of *Drosophila*. Proc Natl Acad Sci USA 78:6548-6552

van Eerd JP, Capony J-P, Ferraz C, Pechere J-F (1978) The amino acid sequence from frog skeletal muscle. Eur J Biochem 91:231-242

Zagotta WN, Brainard MS, Aldrich RW (1988) Single channel analysis of four distinct classes of potassium channels in *Drosophila* muscle. J Neurosci 8:4765-4779

A NOVEL EF-HAND Ca²⁺-BINDING PROTEIN IS EXPRESSED IN DISCRETE SUBSETS OF MUSCLE AND NERVE CELLS OF *DROSOPHILA*

Rita Reifegerste, Claudia Faust, Norbert Lipski, Gertrud Heimbeck, Alois Hofbauer, Gert Pflugfelder, Konrad Zinsmaier, Claus W. Heizmann, Sigrid Buchner, and Erich Buchner

INTRODUCTION

Various Ca²⁺-binding proteins have been identified in vertebrate brain cells but their functions remain largely speculative (review Heizmann and Braun 1990). This is because vertebrate neurons cannot easily be manipulated genetically *in vivo*. In *Drosophila*, however, cloned genes can be mutated and the effects of a lack of the functional gene product on the performance of the otherwise intact neuronal network can be studied.

It thus seemed attractive to search for homologs of vertebrate brain Ca²⁺-binding proteins in this dipteran insect species.

Parvalbumin, calbindin-D-28K, and calretinin are Ca²⁺-binding proteins characterized by domains of conserved amino acids that form a helix-loop-helix conformation and bind Ca²⁺ with high affinity. All three proteins are found in different subpopulations of vertebrate neurons and thus may be assumed to be responsible for specific functional properties of these cells. Starting from the observation that antisera against carp or rat muscle parvalbumin and against chicken or rat calbindin-D-28K all recognize a characteristic subset of neurons in the optic lobes of *Drosophila* (Buchner et al. 1988), we used the carp parvalbumin serum to isolate immunopositive clones from an expression library. One of these clones apparently corresponds to a gene that codes for the antigen recognized by the vertebrate antisera. This clone was selected for further characterization. Here we report the amino acid sequence derived from the 5' incomplete cDNA and the expression pattern of the corresponding gene in the adult fly. The cDNA codes for a protein fragment containing five putative Ca²⁺-binding sites and shows high homology to the vertebrate Ca^{2+}-binding proteins calretinin (Rogers 1987) and calbindin-D-28K (Wilson et al. 1985).

RESULTS AND DISCUSSION

Of a cDNA expression library made in λ gt11 from isolated head poly A^+ mRNA, 200000 plaques (Zinsmaier 1988) were screened immunochemically using a polyclonal antiserum against carp (II) muscle parvalbumin (Gosselin-Rey 1974; Gerday 1988). From the seven strongest of 29 positive signals seven clones were isolated, three of which produced IPTG-induced fusion proteins recognized by the antiserum in Western blots. When ^{35}S-labeled antisense cRNA transcribed from these clones was hybridized to frozen sections of adult flies (for technical details see Zinsmaier et al. 1990), only one clone produced an autoradiographic signal that showed resemblance to the staining pattern of the carp parvalbumin antiserum in the optic lobes (Buchner et al. 1988). In particular, labeling was found over a group of neurons between medulla and lobula plate and over a few small muscles in the thorax. Restriction analysis of the cDNA insert of this clone revealed two EcoRI fragments of 0.2 and 0.7 kb size. In order to obtain higher spatial resolution of the expression pattern and information on the distribution of the gene product in the brain we recloned the 0.7 kb fragment in an expression vector (pGEX, Smith and Johnson 1988) and produced and purified glutathione-S-transferase (GST) fusion protein as described elsewhere (Bürk et al. in prep.). Three mice were injected with 2, 6, and 13 µg of fusion protein, respectively, emulsified in Freunds adjuvans. The mice were boosted after 4 weeks. All three antisera stained fusion protein (45 kDa) and insert protein (18 kDa) in Western blots, only two of them also stained isolated GST carrier protein (26 kDa), indicating that the third antiserum was specific for the insert protein. All three antisera also produced very similar immunohistochemical staining patterns on formaldehyde-fixed frozen sections (Figure 1). In the optic lobes the pattern was compared with Golgi impregnations of Drosophila brain (Fischbach and Dittrich 1989). The stained structures include (1) at least one of the three distal lamina monopolar cells L1, L2, or L3, (2) several columnar neurons of the medulla, (3) a group of columnar neurons of the lobula with cell bodies located between medulla and lobula plate and arborizations in a characteristic layer of the lobula (presumably T2 and/or T3 cells) (Figure 1a), (4) numerous distinct cells of the central brain and the thoracic ganglia (not shown), and (5) a few small thoracic muscles caudally adjacent to the tergo-trochanter muscle (TTM) (Figure 1b). Apart from the lamina signal that may have been too weak to be detected in the *in situ* autoradiographies, this distribution is consistent with the hybridization patterns of the cloned cRNA and also shows significant similarities to the staining of the original carp II parvalbumin antiserum. On Western blots the new antisera produced a single brain-specific signal near 32 kDa where previous $^{45}Ca^{2+}$ blots had identified a protein band with high Ca^{2+} affinity (Lipski 1989). The cloned cDNA was sequenced in both directions after Henikoff subcloning in M13 vector by the Sanger dideoxy chain termination technique. It codes for a single large open reading frame of 238 amino acids and a 159 bp 3' nontranslated trailer terminated by a stretch of 54 adenosine residues. The length of this poly-A tail and the fact that it is preceded 12 bp upstream by the polyadenylation consensus sequence AATAAA suggest that the cDNA is 3' complete. Its 5' end, however, lies within the ORF such that both, cDNA and ORF, are incomplete. Comparison of the encoded amino acid sequence with the NBRF protein data base reveals homologies with various Ca^{2+}-binding proteins of the calmodulin family.

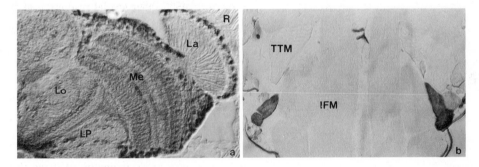

Fig. 1a,b. Distribution of the *Drosophila* Ca^{2+}-binding protein as revealed immunohistochemically by an antiserum against fusion protein expressed from a GST-*Drosophila* cDNA gene construct in *E.coli*. On Western blots this serum stains fusion protein and cleaved cDNA-encoded protein but not GST. The similarity of the distribution of stained perikarya and muscles to *in situ* hybridization signals on autoradiographies from frozen sections incubated with ^{35}S-labeled anti-sense RNA transcribed from the cDNA clone indicates that most or all staining is specific. **a** Horizontal section through optic lobe. **b** Horizontal section through thorax. *R* retina; *La* lamina; *Me* medulla; *Lo* lobula; *LP* lobula plate; *IFM* indirect flight muscles; *TTM* tergo-trochanter muscle

```
                                         E              L*- -L*L*- -L*    D- D- D*G I D*- E   L*- -L*L*- -L*      E

Dros. Ca²⁺-binding prot., dom.II    ₁₅E E L   K S C F M E A Y   D D N Q D G K I D I R E   L A Q L L P M E   E N F ₄₈
Chicken calretinin, dom.II          ₅₉D K M   K E - F M H K Y   D K N A D G K I E M A E   L A Q I L P T E   E N F ₉₁
Dros. Ca²⁺-binding prot., dom.II    ₆₁S V E   F M K I W R E Y   D T D N S G Y I E A D E   L K N F L R D L   L K E ₉₄
Chicken calretinin, dom.III         ₁₀₂S S E   F M E A W R R Y   D T D R S G Y I E A N E   L K G F L S D L   L K K ₁₃₅
Dros. Ca²⁺-binding prot., dom.IV    ₁₀₇I E Y   T D T M L Q V F   D A N K D G R L Q L S E   M A K L L P V K   E N F ₁₄₀
Chicken calretinin, dom.IV          ₁₄₆Q E Y   T Q T I L R M F   D M N G D G K L G L S E   M S R L L P V Q   E M F ₁₇₉
Dros. Ca²⁺-binding prot., dom.V     ₁₅₄K E D   I E K V F S L Y   D R D N S G T I E N E E   L K G F L K D L   L E L ₁₈₇
Dros. Ca²⁺-binding prot., dom.VI    ₁₉₉A A F   E E T I M R G V   G T D K H G K I S R K E   L T M I L L T L   A K I ₂₃₂
Carp muscle parvalbumin, dom.CD     ₄₀A D D   V K K A F A I I   D Q D K S G F I E E D E   L K L F L Q M F   K A D ₇₃
Carp muscle parvalbumin, dom.EF     ₇₉D G E   T K T F L K A G   D S D G D G K I G V D E   F T A L V K A₁₀₈

                                            Helix            X Y Z -Y -X -Z            Helix
                                                               Loop
```

Fig. 2. Comparison of the Ca²⁺-binding domains of chicken calretinin, carp muscle parvalbumin, and the present *Drosophila* Ca²⁺-binding protein. The single-letter amino acid code is used. *Boxed* amino acids correspond to the model of Tufty and Kretsinger (1975) where $L*$ is a hydrophobic amino acid and $D*$ an amino acid with an oxygen atom as Ca²⁺ ligand, D, G, E, and I are highly conserved amino acids in the loop region. X, Y, Z $-Y$, $-X$, $-Z$ indicate the positions of Ca²⁺-binding amino acids

Alignment with the helix-loop-helix model (Tufty and Kretsinger 1975) demonstrates that the present protein fragment contains five EF-hand Ca^{2+}-binding domains in which most or all functionally important amino acids are conserved (Figure 2).With the Ca^{2+}-binding protein of the Shaker complex of *Drosophila* (Pongs et al. chapter 27 in this Vol.) only weak homology is found, restricted to the Ca^{2+}-binding domains.

High homology, on the other hand, exists to calretinin and calbindin-D-28K. Specifically, in a stretch of 150 amino acids for which sequence information is available for chicken calretinin, chicken calbindin-D-28K and the present *Drosophila* Ca^{2+}-binding protein 85 (77) amino acids or 57% (51%) are identical between the latter and calretinin (calbindin), whereas 86 amino acids (57%) are identical between calretinin and calbindin. Significant homologies to both vertebrate proteins are also found outside the helix-loop-helix domains, suggesting that functional similarities of these three cell-specifically expressed brain proteins may extend beyond Ca^{2+} binding. The study of such functional aspects may become possible if mutations in the corresponding gene of *Drosophila* can be isolated, an approach that has been successfully used for the Ca^{2+}-binding protein of the Shaker locus (Pongs et al. this Vol.). The distribution of the present Ca^{2+}-binding protein in the adult visual system (Fig. 1a) suggests that its function in a single identified cell type may be investigated by behavioral analysis of genetic mosaics induced by somatic recombination. The cloning of genes cell-specifically expressed in the brain of *Drosophila* thus may lead to new approaches in neural network analysis.

ACKNOWLEDGMENTS

We would like to thank Dr. M. Heisenberg for his permanent support of this project. Financed by DFG (Bu 566/1-4, Ho 798/3), the State of Bavaria, (PhD Studentship to RR), and by the Swiss National Science Foundation (31-9409.88).

510

REFERENCES

Buchner E, Bader R, Buchner S, Cox J, Emson PC, Flory E, Heizmann CW, Hemm S, Hofbauer A, Oertel WH (1988) Cell-specific immuno-probes for the brain of normal and mutant *Drosophila melanogaster*. Cell Tissue Res 253:357-370

Fischbach K-F, Dittrich APM (1989) The optic lobe of *Drosophila melanogaster*. I. A Golgi analysis of wild-type structure. Cell Tissue Res 258:441-475

Gerday C (1988) Soluble calcium binding proteins in vertebrate and invertebrate muscle. In: Gerday C, Bolis L, Gilles R (eds) Calcium and calcium binding proteins. Molecular and Functional Aspects. Springer, Heidelberg, pp 23-39

Gosselin-Rey C (1974) Fish parvalbumin. In: Drabikowski W, Strzelecka-Golaszewska H (eds) Immuno-chemical reactivity and biological distribution in calcium binding proteins. Polish Publishers, Warszawa, pp 679-701

Heizmann CW, Braun K (1990) Calcium Binding Proteins: Molecular and Functional Aspects. In: The role of calcium in biological systems. LJ Anghileri (ed), CRC-Press, Vol. V, pp 21-66

Lipski N, (1989) Versuche zur immunchemischen Charakterisierung von zellspezifischen Antikörper-Bindungsstellen im *Drosophila*gehirn. Diploma thesis, University of Würzburg

Rogers JH (1987) Calretinin: A gene for a novel calcium-binding protein expressed principally in neurons. J Cell Biol 105:1343-1353

Smith DB, Johnson KS (1988) Single-step purification of polypeptides expressed in *E.coli* as fusions with glutathione S-transferase. Gene 67:31-40

Tufty RM, Kretsinger RH (1975) Troponin and parvalbumin calcium binding regions predicted in myosin light chain and T4 lysozyme. Science 187:167-169

Wilson PW, Harding M, Lawson DEM (1985) Putative amino acid sequence of chick calcium-binding protein deduced from a complementary DNA sequence. Nucleic Acids Res 13:8867-8881

Zinsmaier KE (1988) Konstriktion einer cDNA-Expressionsbibliothek aus Kopf mRNA bei *Drosophila melanogaster*. Diploma thesis, University of Würzburg

Zinsmaier KE, Hofbauer A, Heimbeck G, Pflugfelder GO, Buchner S, Buchner E (1990) A cysteine-string protein is expressed in retina and brain of *Drosophila*. J Neurogenetics (in press)

Section V

Calcium Fluxes, Binding, and Metabolism

PUTATIVE CALCIUM-BINDING SITES IN VOLTAGE-GATED SODIUM AND CALCIUM CHANNEL α_1 SUBUNITS

Joseph A. Babitch

INTRODUCTION

Ion channels are involved in the depolarization and repolarization of excitable membranes, excitation contraction coupling, etc. They can be divided into groups based upon biophysical, electrophysiological, and pharmacological properties (Hille 1984). Recent cloning and sequencing of some ion channel polypeptides has permitted initial attempts to correlate ion channel structure and function. These efforts have revealed that sodium channel α_1 and dihydropyridine-sensitive calcium channel α_1 polypeptides have recognizably similar amino acid sequences, both of which appear to have evolved by two gene duplications from an ancient ion channel probably resembling the *Drosophila Shaker* gene product, a voltage-gated potassium channel (Trimmer and Agnew 1989; Ramaswami and Tanouye 1989).

A computer-assisted analysis of published amino acid sequences revealed that four sodium channel α_1 polypeptides and one dihydropyridine-sensitive, calcium channel α_1 polypeptide had putative EF-hand type calcium-binding sites near their carboxyl terminals (Babitch and Anthony 1987; Babitch 1988). This analysis has been extended to include the many channel amino acid sequences published since the middle of 1987.

SODIUM CHANNELS

Trimmer et al. (1989) have reported the sequence of the rat skeletal muscle sodium α_1 subunit. This polypeptide, like the *Electrophorus* sodium channel sequence, has two putative calcium-binding sites (Figure 1).

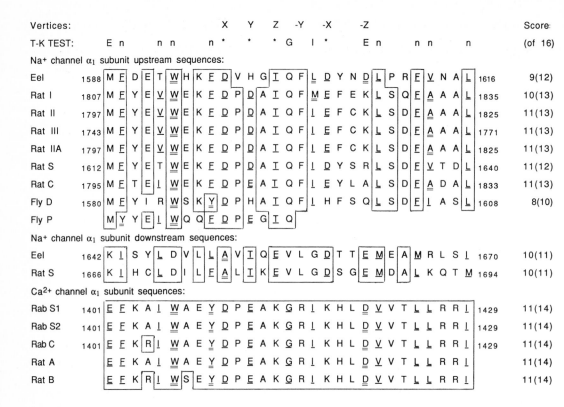

Fig. 1. Amino acid sequences in sodium channel and calcium channel α_1 subunits examined by the Tufty-Kretsinger EF-hand test. Ligand vertices for calcium coordination (X,Y,Z,-Y,-X,-Z) as described by Tufty and Kretsinger (1975) are indicated. Amino acid sequences are depicted using the single letter amino acid code. In the test sequence *E,G* and *I* are specified where indicated. *n* is a hydrophobic residue (*L,I,V,F,M*) and * is an oxygen containing amino acid (*D,N,E,Q,S,T*). Amino acids satisfying the test are *underlined*. Plausible modifications to the test are *double underlined* with the increased scores given in *parentheses*. Origins of these sequences (with references in parentheses) are as follows: Eel, *Electrophorus electricus* electric organ (Noda et al. 1984); Rat I and Rat II, rat brain (Noda et al. 1986); Rat III, rat brain (Kayano et al. 1988); Rat IIA, rat brain (Auld et al. 1988); Rat S, rat skeletal muscle (Trimmer et al. 1989); Rat C, rat cardiac muscle (Rogart et al. 1989); Fly D, *Drosophila* DSC locus (Salkoff et al. 1987a,b); Fly P, *Drosophila para* locus (Ramaswami and Tanouye 1989); Rab Sl, rabbit skeletal muscle (Tanabe et al. 1987); Rab S2, rabbit skeletal muscle (Ellis et al. 1988); Rab C, rabbit cardiac muscle (Mikami et al. 1989; Slish et al. 1989); Rat A, rat aorta (Koch WJ unpubl.); Rat B, rat brain (Hui A unpubl.).

The first of these, amino acids 1612-1640, scores 11(of 16) in the Tufty-Kretsinger EF-hand test (Tufty and Kretsinger 1975) and this can be increased to 12 if tryptophan is considered to be a hydrophobic amino acid. [Parenthetically, this tryptophan is one of only two amino acids in the putative EF-hands, the other being the first ligand in the loop region, which is constant across all twelve sodium and calcium channel α_1 subunit sequences described in the literature (Figure 1)]. The putative calcium-binding loop contains three acidic ligands and only one threonine as a ligand, both prerequisites for calcium binding to an EF-hand type sequence.

In this sequence from skeletal muscle there is a second putative EF-hand sequence at amino acids 1666-1694. This sequence scores 10 in the Tufty-Kretsinger test and could score 11 if alanine acted as a calcium ligand by association through a water molecule, as has been reported for some other calcium-binding proteins. Like the putative EF-hand upstream, this loop contains only one threonine and three acidic ligands. This sequence scores as well as the comparable region in the *Electrophorus* sodium channel sequence and is quite similar to it in other ways (Figure 1). Perhaps this region helps to distinguish a "neuromuscular type" sodium channel α_1 subunit from others such as the *Drosophila* and rat brain sequences which are less likely to be calcium-binding sites, principally due to substitution of arginine for glutamic acid at the Z ligand of the loop. Arguing against this interpretation is the absence of a second putative EF-hand in a tetrodotoxin-resistant sodium channel isoform cloned from rat heart (Rogart et al. 1989; Figure 1). The upstream putative EF-hand in the cardiac sequence is very similar to those of rat skeletal muscle and brain and scores higher than the comparable region in the *Drosophila* DSC sequence (Figure 1).

Auld et al. (1988) have sequenced a second type II rat brain sodium channel α_1 subunit. Compared with the type II rat brain channel sequenced previously (Noda et al. 1986), only six amino acid changes occur, none being in the putative EF-hands. Therefore the scores remain 11(13) and 9(10) respectively.

Part of this region in the *Drosophila para* locus has been sequenced by Ramaswami and Tanouye (1989). Although only half of the putative EF-hand sequence was reported, the changes compared with the DSC locus sequence (Salkoff et al. 1987a) substantially improve chances that the *para* region is a calcium-binding site. At critical positions in the reported

N terminal 15 amino acids of the putative EF hand four differences occur between the DSC and *para* locus gene products (Figure 1). In these 15 amino acids are eight positions considered critical for EF-hand formation, so the variability at critical residues (four of eight) is surprisingly high considering that these two sequences come from the same creature. The first replacement, at position two, is semi conservative (F becomes Y) and slightly decreases the Tufty-Kretsinger score but two other replacements at hydrophobic preferring positions in the first helix of the EF-hand (R and Y being replaced by I and F) significantly improve the score. Another major improvement is the substitution of E for H at the second ligand of the calcium-binding loop. If no other changes occur between the DSC and *para* loci in the rest of the putative EF-hand, these improvements would raise the Tufty-Kretsinger score from 8(10) to 10(12). Still, it is not possible yet to say that this is a likely EF-hand until more of the sequence is revealed because only two acidic ligands are apparent in the sequence reported so far. The sequence reported by Loughney et al. (1989) does not reveal this area of the polypeptide. Similarly, the possible presence of a second EF hand downstream requires additional sequencing. Nevertheless, the preliminary data support the idea of Ramaswami and Tanouye (1989) that the *para* locus is more like vertebrate sodium channels and that the *para* and DSC loci diverged from each other before *para* diverged from vertebrate sodium channels.

When sodium channel α_1 subunit sequences for the upstream putative EF-hand are compared across all sequences eight (of 16) critical positions for EF-hand formation are constant (boxed in Figure 1). Of these, two occur in the first α-helix, three are in the loop, and the final three occur in the second helix. Of the 13 noncritical positions in the putative EF-hand, only one is constant across all sequences and that is the -Y ligand near the center of the loop. At critical positions, replacements are always either conservative, such as the substitutions of methionine or leucine for the preferred isoleucine in the loop, or they are semi-conservative. Even in the nonconserved positions, most changes are conservative or semi conservative with the only radical change being the substitution of arginine for the more prevalent aspartic acid at position 24 (amino acid 1611) of the *Electrophorus* sequence. In the downstream EF-hand region more residues are conserved in the two sequences which do score well enough to be putative EF-hands, but all of the brain sequences have poor scores due to sequence divergence. In the rat skeletal muscle and *Electrophorus* sequences 12 of 16

(75%) of the critical amino acids are identical while only three of 13 (23%) noncritical positions contain the same amino acid in both sequences. Again, as for the upstream putative EF-hands, amino acids are conserved to a much greater degree at positions which are critical for EF-hand formation than are noncritical residues. When critical amino acids change, as for the rat brain downstream sequences, scores tend to decrease.

CALCIUM CHANNELS

Ellis et al. (1988) sequenced the α_1 subunit of the rabbit skeletal muscle dihydropyridine-sensitive calcium channel. They found only three amino acid differences at positions 1808, 1815, and 1835 when compared to the sequences described by Tanabe et al. (1987). None of these changes occurs in the putative calcium binding region (amino acids 1401-1429).

The sequence of the rabbit cardiac muscle α_1 subunit has been reported by Mikami et al. (1989) and Slish et al. (1989). There is one change at a noncritical residue in the first α-helix (alanine to arginine, Figure 1). Therefore, the EF-hand score remains as calculated previously [11(14)]. Overall, 28 of 29 positions are identical in these three sequences. This putative EF-hand region linked to the IVS6 region 25 amino acids upstream is the most highly conserved area in the entire polypeptide: only one amino acid substitution occurs (see above) in 78 consecutive amino acids. Because the overall sequence homology is 66% between the skeletal muscle and cardiac dihydropyridine receptors (Mikami et al. 1989), the much greater homology of the putative EF-hand region suggests that it plays an important role in channel structure and/or function. The group of Dr. Arnold Schwartz (Koch et al. 1989) has continued to sequence the rat brain and aorta calcium channel α_1 cDNAs. The sequence of the putative EF-hand region of the aorta polypeptide is identical to the rabbit skeletal muscle polypeptide while the rat brain sequence has one noncritical change from the rabbit cardiac sequence (Figure 1). Therefore, the calcium channel sequences are very highly conserved, much more so than the sodium channel sequences. It also is worth noting that these putative single EF-hand type calcium-binding sites in the calcium channel sequences have more characteristics of EF-hands than any other channel protein sequences, particularly the specified E at the start of the EF-hand, and the specified G and I in the loop. Perhaps calcium binds with sufficient affinity to these sites to eliminate the need for a second calcium-binding site downstream.

Overall, sequence homology between these sodium and calcium channel subunits is not high. The putative EF-hand regions are also diverse with only two of 29 amino acids (W at position 6 and D at position 10) remaining invariant across all 14 sequences. So it is all the more surprising that the structural features of a particular kind of calcium-binding site should be retained in all of these sequences.

Ellis et al. (1988) also sequenced the α_2 subunit of the rabbit skeletal muscle calcium channel. The internal portion of the polypeptide displays no homologies to an EF-hand, calcium-binding elbow (Stuart et al. 1986) or lockwasher (Vyas et al. 1989). However, there is a region (amino acids 536-549) with a suspicious concentration of acidic residues resembling the second calcium-binding loop of S-100α_1, S-100b and a growth factor-inducible polypeptide described by Calabretta et al. (1986). Considering the importance of negative surface charges for calcium binding (Linse et al. 1988), this region (possibly combined with the acidic region just downstream, residues 556-561) might bind cations as described for canine cardiac calsequestrin "by acting as a charged surface rather than by presenting multiple discrete Ca^{2+}-binding sites" (Scott et al. 1988). Possibly calcium could be bound by more than one region of the polypeptide, as has been described for pig pancreatic α-amylase (Buisson et al. 1987) and proteinase K from *Tritirachium album Limber* (Bajorath et al. 1989), and it is very unlikely that any analysis of protein primary sequences could detect such calcium-binding sites. As described by Ruth et al. (1989), the ß subunit of the dihydropyridine-sensitive calcium channel of rabbit skeletal muscle does not contain EF-hands.

POTASSIUM CHANNELS

Since the previous analysis (Babitch 1988), many potassium channels have been cloned and sequenced (Baumann et al. 1987; Schwarz et al. 1988; Pongs et al. 1988; Tempel et al. 1988; Baumann et al. 1988; Takumi et al. 1988; Butler et al. 1989; Murai et al. 1989; McKinnon

519

1989). None of these sequences has typical calcium-binding regions in agreement with previous observations.

CONCLUDING REMARKS

This analysis indicates that voltage gated sodium and calcium channels, but not potassium channels, can bind calcium with portions of their internal sequences. Because the calcium channel putative EF hands generally score higher than do the sodium channel sequences and the former have additional characteristics, like more traditional EF-hand calcium-binding sites, the calcium channel α_1 polypeptides might bind calcium with greater affinity than the sodium channel α_1 polypeptides. The absence of calcium-binding sites in potassium channel polypeptides suggests that calcium binding is involved in some aspect of depolarization but not repolarization or facilitation of ion permeation. I propose the following scenario: during depolarization of excitable tissues the entering calcium ions bind first to calcium channel and then to sodium channel internal sequences to increase the probability that the passage of ions will be stopped by closing and/or inactivating the channels. One way of testing this proposal, if the putative EF-hand in the *Drosophila para* locus does turn out to have a high Tufty-Kretsinger score, would be to express both the DSC and *para* loci polypeptides in *Xenopus* oocytes and compare their closing and inactivation properties. If, on the other hand, the sodium channel sequences do not bind calcium ions with micromolar affinity, then the calcium channel sequences could be the high affinity internal calcium-binding site which facilitates ion permeation.

ACKNOWLEDGMENTS

This work is supported by NIH (NS-26518) and the TCU Research Foundation. I thank W. J. Koch, A. Hui and A. Schwartz for providing the sequences of the rat aorta and brain calcium channel α_1 subunit putative EF-hands prior to publication. The typing skills of Mrs. Elaine Bozeman and Mrs. Pam Johnson are gratefully acknowledged.

REFERENCES

Auld VJ, Goldin AL, Krafte DS, Marshall J, Dunn JM, Catterall WA, Lester HA, Davidson N, Dunn RJ (1988) A rat brain Na⁺ channel a subunit with novel gating properties. Neuron 1:449-461

Babitch JA (1988) On calcium binding to channel proteins. J theor Biol. 133:525-528

Babitch JA, Anthony FA (1987) Grasping for calcium binding sites in sodium channels with an EF hand. J theor Biol 127:451-459

Bajorath J, Raghunathan S, Hinrichs W, Saenger W (1989) Long-range structural changes in proteinase K triggered by calcium ion removal. Nature 337:481-484

Baumann A, Krah-Jentgens I, Müller R, Müller-Holtkamp F, Seidel R, Kecskemethy N, Casal J, Ferrus A, Pongs O (1987) Molecular organization of the maternal effect region of the *Shaker* complex of *Drosophila*: characterization of an I_A channel transcript with homology to vertebrate Na⁺ channel. EMBO J 6:3419-3429

Baumann A, Grupe A, Ackermann A, Pongs O (1988) Structure of the voltage-dependent potassium channel is highly conserved from *Drosophila* to vertebrate central nervous systems. EMBO J 7:2457-2463

Buisson G, Duee E, Haser R, Payan F (1987) Three dimensional structure of porcine pancreatic α-amylase at 2.9Å resolution. Role of calcium in structure and activity. EMBO J 6:3909-3916

Butler A, Wei A, Baker K, Salkoff L (1989) A family of putative potassium channel genes in *Drosophila*. Science 243:943-947

Calabretta B, Battini R, Kaczmarek L, de Riel JK, Baserga R (1986) Molecular cloning of the cDNA for a growth factor-inducible gene with strong homology to S-100, a calcium-binding protein. J Biol Chem 261:12628-12632

Ellis SB, Williams ME, Ways NR, Brenner R, Sharp AH, Leung AT, Campbell KP, McKenna E, Koch WJ, Hui A, Schwartz A, Harpold MM (1988) Sequence and expression of mRNAs encoding the α_1 and α_2 subunits of a DHP-sensitive calcium channel. Science 241:1661-1664

Hille B (1984) Ionic Channels of Excitable Membranes. Sinauer, Sutherland, MA

Kayano T, Noda M, Flockerzi V, Takahashi H, Numa S (1988) Primary structure of rat brain sodium channel III deduced from the cDNA sequence. FEBS Lett 228:187-194

Koch WJ, Hui A, Shul GE, Ellinor P, Schwartz A(1989) Characterization of cDNA clones encoding two putative isoforms of the α_1 subunit of the dihydropyridine-sensitive voltage-dependent calcium channel isolated from rat brain and rat aorta. FEBS Lett 250:386-388

Linse S, Brodin P, Johansson C, Thulin E, Grundström T, Forsén S (1988) The role of protein surface charges in ion binding. Nature 335:651-652

Loughney K, Kreber R, Ganetzky B (1989) Molecular analysis of the *para* locus, a sodium channel gene in *Drosophila*. Cell 58:1143-1154

McKinnon D (1989) Isolation of a cDNA clone coding for a putative second potassium channel indicates the existence of a gene family. J Biol Chem 264:8230-8236

Mikami A, Imoto K, Tanabe T, Niidome T, Mori Y, Takeshima H, Narumiya S, Numa S (1989) Primary structure and functional expression of the cardiac dihydropyridine-sensitive calcium channel. Nature 340:230-233

Murai T, Kakizuka A, Takumi T, Ohkubo H, Nakanishi S (1989) Molecular cloning and sequence analysis of human genomic DNA encoding a novel membrane protein which exhibits a slowly activating potassium channel activity. Biochem Biophys Res Commun 161:176-181

Noda M, Shimizu S, Tanabe T, Takai T, Kayano T, Ikeda T, Takahashi H, Nakayama H, Kanaoka Y, Minamino N, Kangawa K, Matsuo H, Raftery MA, Hirose T, Inayama S, Hayashida H, Miyata T, Numa S (1984). Primary structure of *Electrophorus electricus* sodium channel deduced from cDNA sequence. Nature 312:121-127

Noda M, Ikeda T, Kayano T, Suzuki H, Takeshima H, Kurasaki M, Takahashi H, Numa S (1986) Existence of distinct sodium channel messenger RNAs in rat brain. Nature 320:188-192

Pongs O, Kecskemethy N, Müller R, Krah-Jentgens I, Baumann A, Kiltz HH, Canal I, Llamazares S, Ferrus A (1988) *Shaker* encodes a family of putative potassium channel proteins in the nervous system of *Drosophila*. EMBO J 7:1087-1096

Ramaswami M, Tanouye MA (1989) Two sodium-channel genes in *Drosophila*: implications for channel diversity. Proc Natl Acad Sci USA 86:2079-2082

Rogart RB, Cribbs LL, Muglia LK, Kephart DD, Kaiser MW (1989) Molecular cloning of a putative tetrodotoxin-resistant rat heart Na$^+$ channel isoform. Proc Natl Acad Sci USA 86:8170-8174

Ruth P, Rohrkasten A, Biel M, Bosse E, Regulla S, Meyer HE, Flockerzi V, Hofmann F (1989) Primary structure of the beta subunit of the DHP-sensitive calcium channel from skeletal muscle. Science 245:1115-1118

Salkoff L, Butler A, Scavarda N, Wei A (1987a) Nucleotide sequence of the putative sodium channel gene from *Drosophila*: the four homologous domains. Nucleic Acids Res 15:8569-8572

Salkoff L, Butler A, Wei A, Scavarda N, Giffen K, Ifune C, Goodman R, Mandel G (1987b) Genomic organization and deduced amino acid sequence of a putative sodium channel gene in *Drosophila*. Science 237:744-749

Schwarz TL, Tempel BL, Papazian DM, Jan YN, Jan LY (1988) Multiple potassium-channel components are produced by alternative splicing at the Shaker locus in *Drosophila*. Nature 331:137-142

Scott BT, Simmerman HKB, Collins JH, Nadal-Ginard B, Jones LR (1988) Complete amino acid sequence of canine cardiac calsequestrin deduced by cDNA cloning. J Biol Chem 263:8958-8964

Slish DF, Engle DB, Varadi G, Lotan I, Singer D, Dascal N, Schwartz A (1989) Evidence for the existence of a cardiac specific isoform of the α_1 subunit of the voltage dependent calcium channel. FEBS Lett 250:509-514

Stuart DI, Acharya KR, Walker NPC, Smith SG, Lewis, M, Phillips DC (1986) α-Lactalbumin possesses a novel calcium binding loop. Nature 324:84-87

Takumi T, Ohkubo H, Nakanishi S (1988) Cloning of a membrane protein that induces a slow voltage-gated potassium current. Science 242:1042-1045

Tanabe T, Takeshima H, Mikami A, Flockerzi V, Takahashi H, Kangawa K, Kojima M, Matsuo H, Hirose T, Numa S (1987) Primary structure of the receptor for calcium channel blockers from skeletal muscle. Nature 328:313-318

Temple BL, Jan YN, Jan LY (1988) Cloning of a probable potassium channel gene from mouse brain. Nature 332:837-839

Trimmer JS, Agnew WS (1989) Molecular diversity of voltage-sensitive Na channels. Annu Rev Physiol 51:401-418

Trimmer JS, Cooperman SS, Tomiko SA, Zhou J, Crean SM, Boyle MB, Kallen RG, Sheng Z, Barchi RL, Sigworth FJ, Goodman RH, Agnew WS, Mandel G (1989) Primary structure and functional expression of a mammalian skeletal muscle sodium channel. Neuron 3:33-49

Tufty RM, Kretsinger RH (1975) Troponin and parvalbumin calcium binding regions predicted in myosin light chain and T4 lysozyme. Science 187:167-169

Vyas MN, Jacobson BL, Quiocho FA (1989) The calcium-binding site in the galactose chemoreceptor protein. Crystallographic and metal binding studies. J Biol Chem 264:20817-20821

ALTERATIONS OF CELLULAR CALCIUM METABOLISM AND CALCIUM BINDING IN HYPERTENSION AND HUMAN CARDIAC HYPERTROPHY

Paul Erne, Maryse Crabos, Dieter Engelkamp, Fritz R. Bühler, and
Claus W. Heizmann

INTRODUCTION

As the central ion in the control of vascular and cardiac contraction, calcium may be expected to be involved in the pathogenesis of cardiovascular disorders. However, while a variety of alterations in cellular calcium metabolism have been described in the pathogenesis of myocardial ischemia (for review see Buja et al. 1988), atherosclerosis (for review see Phair 1988, Kiowski et al. 1989), cardiac hypertrophy and both experimental and human essential hypertension, the causal link remains speculative at this time. This review focusses on alterations of cellular calcium metabolism in hypertension and cardiac hypertrophy and provides evidence for associated abnormalities of cellular calcium binding in these disorders.

CELLULAR CALCIUM ALTERATION IN HYPERTENSION

An increase in systemic vascular resistance characterizes the established phase of both experimental (Hermsmeyer 1987) and human essential (Erne et al. 1987) hypertension. This derangement is due to structural differences between normotensive and hypertensive subjects (Folkow 1971), but also entails functional components (Bolli et al. 1984). Several lines of evidence suggest that abnormalities of calcium metabolism are involved in the pathogenesis of established human hypertension. In particular, low serum concentrations of ionized Ca^{2+} have been observed in at least a fraction of patients with this disease (for review see Young et al. 1990). Furthermore, there is an enhanced vasodilation and antihypertensive effect of calcium antagonists in patients with essential hypertension (Erne et al. 1987, Bühler et al. 1990).

Membrane alterations have repeatedly been suggested as being likely contributors to hypertension (for review see Rinaldi et al. 1988; Hermsmeyer and Erne 1990). Ion channels are the molecular means by which transmembranous calcium influx is linked to intracellular calcium sequestration and hence contraction.The most recent step in progress to understand electrophysiologic alterations in hypertension involved the investigation of two functionally important types of calcium channels in vascular muscle cells. Direct comparison of calcium channels in vascular cells from a neonatal spontaneous hypertensive versus normotensive rat strain showed an enhanced probability for an opening and larger Ca^{2+}-currents of the sustained type of calcium channels in genetic hypertension (Sturek and Hermsmeyer 1986; Rusch and Hermsmeyer 1988; Erne et al. 1989a).

Studies using the platelet as model of the contractile system indicate that alterations of Ca^{2+} handling in human hypertensives may occur at the cellular level. Elevated free calcium concentrations have been reported in platelets from hypertensive subjects (Erne et al. 1984; Bruschi et al. 1985; Le Quang Sang et al. 1985, Lindner et al. 1987) and genetic hypertensive rats (Bruschi et al. 1985; Vasdev et al. 1988) which correlates with the height of the blood pressure (Erne et al. 1984; Vasdev et al.1988; Pritchard et al. 1989). These alterations of calcium metabolism in platelets from essential hypertensive patients have been confirmed by many investigators, and observed for some patients with secondary hypertension. Altered platelet calcium metabolism has recently been used as an early predictor of increased peripheral vascular resistance and preeclampsia in pregnancy-induced hypertension (Zemel et al. 1990). Furthermore, an altered calcium extrusion via the calmodulin-dependent Ca^{2+}-ATPase (Resink et al. 1986) as well as alterations of calcium-dependent phosphorylation of myosin light chain kinase and abnormalities in phosphoinositide turnover have been described in platelets from untreated hypertensive patients (for review see Erne et al. 1987; Resink et al. 1987). However,these studies do not determine whether similar alterations are operative in vascular muscle cells, and, if these changes are causally implicated in elevated blood pressure.

To define vascular muscle mechanisms which are altered in an animal model of hypertension prior to the development of elevated blood pressure, studies were carried out on vascular muscle cells using the fluorescent indicator fura-2 in order to compare Ca^{2+}-concentrations between genetically hypertensive and normotensive animals. While average Ca^{2+}-concentration within vascular muscle cells from hypertensives is not abnormal, the free Ca^{2+}-concentration is elevated even at rest in localized regions of vascular muscle cells from spontaneously hypertensive rats (Erne and Hermsmeyer 1989b). These methods have the potential to monitor rapid and localized calcium transients (Eberhard and Erne 1989) which may allow to characterize the kinetics of calcium binding to proteins (Eberhard et al. unpubl.). In a more recent study, these findings were confirmed by confocal laser scan microscopy indicating a subsarcolemmal increase in intracellular calcium in vascular muscle cells from spontaneously hypertensive rats (Hermsmeyer and Erne submitted).

These alterations of cellular calcium metabolism might be linked to abnormalities of membrane and cytosolic calcium binding sites. Such reduced calcium binding capacity to plasma membrane, which might contribute to the increased calcium permeability of the membrane, has been demonstrated in both experimental and human hypertension (Devynck et al. 1981; Orlov and Postnov 1982; Cirillo et al. 1989), and it has been related to a decreased content of an integral membrane calcium binding protein (IMCAL) (Kowarski et al. 1986). In a preliminary screen for EF-type as well as other calcium binding proteins, we were unable to identify the IMCAL protein by use of an antibody directed to IMCAL from rat mucosa inhuman platelets although an eventual instability of this protein might have hampered this observation (M. Crabos, unpubl. observation). Calmodulin is another putative candidate involved in hypertension and there is some preliminary evidence that one functional Ca^{2+}-binding subdomain caused by a deletion of two amino acids may lead to hypertension (Nojima et al. 1986). Alternatively, a decreased sensitivity of calmodulin target enzymes might be due to an imbalance or abnormal expression of calmodulin activity inhibitors and activators (Huang et al. 1986). Furthermore, an abnormal transmembranous calcium metabolism in human hypertension has recently been related to a calpastatin defect, an inhibitor of calpain (Pontremoli et al. 1988).

CELLULAR CALCIUM ALTERATION IN CARDIAC HYPERTROPHY

Hypertrophy is a common response of terminally differentiated tissues and cells, and cardiac hypertrophy is an adaptive mechanism of the myocardium subsequent to cardiac overload. Overload of cardiac cells is generated locally and can be a consequence of hypertension, ischemia, valvular stenosis, regurgitation or idiopathic cardiomyopathy.

Hypertrophy has an initial energy-sparing effect by the addition of new sarcomeres, thereby unloading the cells of the failing heart. However, other changes induced by hypertrophy at both tissue and cellular level lead to an imbalance of energy production and consumption in the hypertrophied heart. Molecular changes in proteins synthesized by the myocardium play a major part in the adaptation of the hypertrophied heart and may influence the long-term prognosis. In general, adult myocardial cells respond to overload by a transient induction of c-fos and c-myc protooncogenes that encode short-lived nuclear proteins for the regulation of cell proliferation and by acceleration of protein synthesis, in particular of fetal isoforms such as actin, tropomyosin or desmin (for review see Swynghedauw 1989). Furthermore, an altered expression of both myosin light- and heavy chains (Schaub and Hirzel 1987; Kurabayashi et al. 1988) has been observed in response to cardiac overload.

The contraction-relaxation cycle is mediated via intracellular calcium transients. A rise in cytosolic free calcium in cardiac cells is thought to result primarily from sarcoplasmic reticulum which seems to be triggered by the early peak of calcium influx (Fabiato 1983; Erne and Hermsmeyer 1988; duBell and Houser 1989). Inversely, the reduction of intracellular free calcium concentration depends mainly on cellular calcium extension and sarcoplasmic reticulum calcium uptake mechanisms. The Ca^{2+}-pumping by sarcoplasmic reticulum is mediated by a Ca^{2+}-dependent ATPase and regulated by a low molecular weight integral protein, phospholambam (James et al. 1989), which, if phosphorylated by cAMP-dependent protein kinase, increases Ca^{2+} uptake by elevating the turnover rate of the Ca^{2+}-ATPase. A reduction of Ca^{2+}-ATPase activity depends on the reduction of phosphorylation of phospholambam and dephosphorylated phospholambam may act as an Ca^{2+}-channel (Schoutsen et al. 1989). Some studies indicate a decreased Ca^{2+}-uptake by sarcoplasmic reticulum during pressure overload hypertrophy (Lamers and Stinis 1979), while increased uptake occurs in thyreotoxic hypertrophy (Limas 1978). More recently it has been postulated that hypertrophy might induce changes in membrane assembly and evidence has been provided that sarcoplasmic reticulum of the failing heart exhibits a decreased concentration of Ca^{2+}-ATPase rather than altered isoforms of this protein, which might result in a reduced calcium uptake by sarcoplasmic reticulum due to changes of transcription and/or mRNA stability (Nagai et al.1989).

Intracellular Ca^{2+}-release and reuptake are necessary for normal contraction and relaxation of the human heart. An increase in the cytosolic calcium concentration may cause transient depolarizations, and overload calcium extrusion mechanisms during diastole. It thus contributes to diastolic (Gwathmey et al. 1987) and subsequently to systolic dysfunction (Morgan et al. 1990) with markedly prolonged Ca^{2+}-transients and a diminished capacity to restore a low resting Ca^{2+}-level during diastole. These abnormalities have important functional manifestations in terms of the force-frequency relation of myopathic hearts and can largely reflect the effects of deficient intracellular production of cyclic AMP, perhaps in a compartmentalized store.

Sarcoplasmic reticulum plays a central role in calcium handling abnormalities in hypertrophic cardiomyopathy which might be explained as a result of an energy deficiency for calcium reuptake by sarcoplasmic reticulum (Lompre et al. 1988), by spontaneous recycling of calcium in the sarcoplasmic reticulum during diastole (Lakatta 1989), alterations of sarcoplasmic reticulum pumps (Whithmer et al. 1988) or a modification of proteins involved in calcium transport across sarcoplasmic membranes. Altered calcium metabolism in cardiac hypertrophy may be associated or induce changes in expression or affinities of the calcium binding proteins. Contraction and relaxation are regulated and tuned by the action of calcium binding proteins; several proteins similar to parvalbumin (postulated to be involved in skeletal muscle relaxation) have been observed in cardiac tissue, but their functions are largely unknown and as yet undefined. Generally these Ca^{2+}-binding proteins are characterized by the common structural motif, the EF-hand (Hunziker and Heizmann 1990; Heizmann and Braun 1990; chapter 1 of this Vol.). The presence of some of these low molecular weight highly conserved proteins, such as S-100 proteins, in cardiac tissue (Haimoto and Kato 1988) might be of importance with regard to its Ca^{2+}-buffer capacity and to induce calcium release from terminal cisternae (Fano et al. 1989). Since these S-100 proteins from human myocardium can be detected subsequent to cellular damage and necrosis in serum from patients during acute myocardial infarction, its estimation has recently been suggested as a diagnostic tool for the differentiation of myocardial ischemia with and without infarction (Usui et al. 1990).

CALCIUM AND HYPERTENSIVE CARDIAC HYPERTROPHY

Elevated intracellular free calcium concentrations and increased compartmentalized calcium levels have been documented in both platelets from hypertensive patients and vascular muscle cells from hypertensive animals. The altered metabolism of high calcium levels points to fundamental alteration of sarcoplasmic reticulum function and indicates a causal or subsequent involvement of intracellular calcium binding proteins in the maintenance of this equilibrium. A dysfunction of sarcoplasmic reticulum has been observed as a basic mechanism for cardiac hypertrophy. This suggests that a common pathway might be altered for both hypertension and cardiac hypertrophy. Recently, we investigated a number of proteins in platelets from hypertensive patients in the presence or absence of a hypertrophic cardiomyopathy (Crabos et al. submitted) and observed a significant elevation of intraplatelet tropomyosin levels in presence of left ventricular hypertrophy. The identification of tropomyosin was based on its binding to tropomyosin antibodies as well as on partial sequence comparison to horse platelet tropomyosin. Furthermore, the presence of tropo-myosin was identified in both human platelets and cardiac tissue. Tropomyosins from both these human sources bind calcium (after SDS-PAGE, electrophoretic transfer to nitrocellulose membranes, and incubation with $^{45}Ca^{2+}$), in contrast to tropomyosin from other tissues and animals. However, a consensus sequence typical for EF-hand calcium-binding proteins could not be found within the identified amino acid sequences of peptides from the human platelet tropomyosin isoform. Nevertheless, using a calcium sensitive electrode the tropomyosin from human platelets and cardiac tissue exhibited a calcium binding with an apparent K_D of 20 μM, suggesting that the calcium binding properties of these tropomyosin-isoforms might be important for contraction and relaxation in human platelets and cardiac cells. Although the mechanism responsible for induction of tropomyosin expression is unknown, molecular changes in contractile protein synthesis in the myocardium play a major role in the adaptation of the hypertrophied heart, and adult cardiac cells respond to overload by synthesis of fetal isoforms of tropomyosin (Izuma et al. 1988).

CONCLUSION

Intracellular calcium transients regulated by intracellular calcium release and uptake mechanisms play a pivotal role in the control of the contraction-relaxation cycle. Abnormalities of intracellular calcium metabolism have been documented and linked to the pathogenesis of hypertension and cardiac hypertrophy. Conceivably, these alterations of calcium metabolism are associated with primary or secondary alterations of the affinity or expression of calcium binding proteins. Although a number of potential calcium binding proteins might play a role in responding to chronic intracellular calcium overload, the identification of these proteins and their role in the sequence of these events need further investigations.

ACKNOWLEDGMENTS

The authors work discussed herein was supported by the Swiss Science National Foundation (32-9499.88 and 31-9409.88), the Swiss Foundation for Cardiology, Roche Research Foundation and the Helmut Horten Stiftung.

REFERENCES

Bolli P, Erne P, Hulthen LH, Ritz R, Kiowski W, Ji BH, Bühler FR (1984) Parallel reduction of calcium influx dependent vasoconstriction and platelet free calcium concentration with calcium entry and beta-adrenoceptor blockade. J Hypertension 6 (Suppl. 7):996-1001

Bruschi G, Bruschi ME, Caroppo M, Orlandini G, Spaggiari M, Cavatorta A (1985) Cytoplasmic free Ca^{2+} is increased in the platelet of spontaneously hypertensive rats and hypertensive patients. Clin Sci 66:179-184

Bühler FR, Erne P, Kiowski W (1990) Calcium antagonists for antihypertensive care. In: Bühler FR, Laragh JH (eds), Handbook of Hypertension, Vol. 13, The management of hypertension. Elsevier Science Publishers B.V., pp 255-276

Buja LM, Hagler HK, Willerson JT (1988) Altered calcium homeostasis in the pathogenesis of myocardial ischemic and hypoxic injury. Cell Calcium 9:205-217

Cirillo M, Trevisan M, Laurenz M (1989) Calcium binding capacity erythrocyte membrane in human hypertension. Hypertension 14:152-155

Crabos M, Yamakado T, Heizmann CW, Cerletti N, Bühler FR, Erne P (1990) The calcium binding protein tropomyosin in human platelets and cardiac tissue: elevation in hypertensive cardiac hypertrophy. (submitted)

Devynck MA, Pernollet MG, Nunez AM, Meyer P (1981) Analysis of Ca^{2+} handling in erythrocyte membrane of genetically hypertensive rats. Hypertension 3:397-403

duBell WH, Houser IR (1989) Voltage and beat dependence of Ca^{2+} transient in feline ventricular myocytes. Am J Physiol 257:746-759

Eberhard M, Erne P (1989) Kinetics of calcium binding to fluo-3 determined by stopped flow fluorescence. Biochem Biophys Res Commun 163:309-314

Erne P, Bolli P, Bürgisser E, Bühler FR (1984) Correlation of platelet calcium with blood pressure: effect of antihypertensive therapy. N Engl J Med 310:1084-1088

Erne P, Conen D, Kiowksi W, Bolli P, Müller FB, Bühler FR (1987) Calcium antagonist induced vasodilation in peripheral, coronary and cerebral vasculature as important factors in the treatment of elderly hypertensives. Eur Heart J 8 (Suppl. K):49-56

Erne P, Hermsmeyer K (1988) Desensitization to norepinephrine includes refractoriness of calcium release in myocardial cells. Biochem Biophys Res Commun 151:333-338

Erne P, Rusch N, Hermsmeyer K (1989a) Relation of altered calcium metabolism in platelets and vascular muscle cells to hypertension. In: Meyer P, Elghozy JL (eds) Les antagonistes calciques. Masson S.A., Paris, pp 55-64

Erne P, Hermsmeyer K (1989b) Intracellular vascular Ca^{2+} modulation in genetic hypertension. Hypertension 14:145-151

Fabiato A (1983) Calcium-induced release of calcium from the cardiac sarcoplasmic reticulum. Am J Physiol 245:H59-H66

Fano G, Marsili V, Aisa MC, Giambanco I, Donato R (1989) S-100a_0 protein stimulates Ca^{2+}-induced Ca^{2+} release from isolated sarcoplasmic reticulum vesicles. FEBS Lett 255:381-384

Folkow B (1971) The hemodynamic consequences of adaptive structural changes of resistance vessels in hypertension. Clin Sci 41:1-12

Gwathmey JK, Copelas L, MacKinonn R, Schoen FJ, Feldman MD, Grossman W, Morgan JP (1987) Abnormal intracellular calcium handling in myocardium from patients with end-stage heart failure. Circ Res 61:70-76

Haimoto H, Kato K (1988) S100a_0 protein in cardiac muscle: isolation from human cardiac muscle and the ultrastructural localization. Eur J Biochem 171:409-415

Heizmann and Braun (1990) Calcium binding proteins: molecular and functional aspects. In: The role of calcium in biological systems, L.J. Anghileri (ed), CRC Press, Vol. 5, pp 21-66, Boca Raton, Florida

Heizmann CW, Hunziker W (1990) Intracellular calcium-binding molecules. In: Bronner F (ed), Intracellular calcium regulation. Alan R. Liss Inc., pp 211-248

Hermsmeyer K (1987) Vascular muscle membrane cation mechanisms and total peripheral resistance. Hypertension 10 (Suppl. 1):20-22

Hermsmeyer K, Erne P (1990) Vascular muscle electrophysiology and platelet calcium in hypertension. In: Laragh JH, Brenner B.M. (eds) Hypertension: Pathophysiology, New York, pp 661-666

Hermsmeyer K, Erne P (1990) Subsarcolemmal increase in intracellular calcium in vascular muscle cells from spontaneously hypertensive rats. (submitted)

Huang SL, Wen YI, Kupranycz DB, Pang SC, Schlager G, Hamet P, Tremblay J (1988) Abnormality of calmodulin activity in hypertension. Evidence of the presence of an activator. J Clin Invest 82:276-281

Izumo S, Nadal-Ginard B, Makdavi V (1988) Protoneogene induction and reprogramming of cardiac gene expression produced by pressure overload. Proc Natl Acad Sci USA 85:339-343

James P, Iomi M, Tada M, Chiesi M, Carafoli E (1989) Nature and sites of phospholembran regulation of the Ca^{2+} pump of sarcoplasmic reticulum. Nature 342:90-92

Kiowski W, Erne P, Bühler FR (1989) Effects of calcium antagonists on atherogenesis. Clin and Exper Hyper Theory and Practice A11:1085-1096

Kowarski S, Cowen LA, Schachter D (1986) Decreased content of integral membrane calcium-binding protein (IMCAL) in tissues of the spontaneously hypertensive rat. Proc Natl Acad Sci USA 83:1097-1100

Kurabayashi M, Komuro I, Tsuchimochi H, Takaku F, Yazaki Y (1988) Molecular closing and characterization of human atrial and ventricular myosin alkali light chain cDNA clones. J Biol Chem 263:13930-13936

Lakatta EG (1989) Chaotic behavior of myocardial cells: possible implications regarding the pathophysiology of heart failure. Perspect Biol Med 32:421-433

Lamers JM, Stinis JT (1979) Defective calcium pump in the myoplasmic reticulum of the hypertrophied rabbit heart. Life Sci 24:2313-2320

Le Quang Sang KH, Montenay-Garestier T, Devynck MA (1985) Platelet cytosolic free calcium concentration in essential hypertension. Nouv Rev Fr Hematol 27:279-283

Limas CI (1987) Enhanced phosphorylation of myocardial sarcoplasmic reticulum in experimental hyperthyroidism. Am J Physiol 235:H745-H751

Lindner A, Kenny M, Meadam AJ (1987) Effects of a circulating factor in patients with essential hypertension on intracellular free calcium in normal platelets. N Engl J Med 316:509-513

Lompre AM, Levitzki D, de la Bastie D, Mercadier JJ, Rappaport L, Schwartz K (1988) Function of sarcoplasmic reticulum and expression of its Ca^{2+}-ATPase gene in pressure overload rat myocardium. Circulation 78 (Suppl. 2):535A

Morgan JP, Raymond EE, Allen PD, Grossman W, Gwathmey JK (1990) Abnormal intracellular calcium handling, a major cause of systolic and diastolic dysfunction in ventricular myocardium from patients with heart failure. Circulation 81 (Suppl. 3):21-32

Morgan JP, Erny RE, Allen PD, Grossman W, Gwathmey JK (1990) Abnormal intracellular calcium handling, a major cause of systolic and diastolic dysfunction in ventricular myocardium from patients with heart failure. Circulation 81 (Suppl. 3):21-32

Nagai R, Zarain-Herzberg A, Brandt CJ (1989) Regulation of myocardial Ca^{2+}-ATPase and phospholambam mRNA expression in response to pressure overload and thyroid hormone. Proc Natl Acad Sci USA 86:2966-2970

Nojima H, Kishi K, Sokabe H (1986) Organization of calmodulin genes in the spontaneously hypertensive rat. J Hypertension 4 (Suppl. 3):275-277

Orlov SN, Postnov YV (1982) Ca^{2+} binding and membrane fluidity in essential and renal hypertension. Clin Sci 63:281-284

Phair RD (1988) Cellular calcium and atherosclerosis: a brief review. Cell Calcium 9:275-284

Pontremoli S, Salamino F, Spartore B, De Tullio R, Pontremoli R, Melloni E (1988) Characterization of the calpastatin defect in erythrocytes from patients with essential hypertension. Biochem Biophys Res Commun 157:867-874

Pritchard K, Raine AEG, Ashley CC, Castell LM, Somers V, Osborn C, Ledingham IGL, Conway J (1989) Correlation of blood pressure in normotensive and hypertensive individuals with platelet but not lymphocyte intracellular free calcium concentrations. Clin Sci 76:631-635

Resink TJ, Tkachuk VA, Erne P, Bühler FR (1986) Platelet membrane calmodulin-stimulated calcium-adenosine triphosphatase. Altered activity in essential hypertension. Hypertension 8:159-166

Resink TJ, Dimitrov D, Zschauer A, Erne P, Tkachuk VA, Bühler FR (1987) Platelet calcium-linked abnormalities in essential hypertension. Ann NY Acad Sci 488:252-265

Rinaldi G, Bohr D (1988) Plasma membrane and its abnormalities in hypertension. Am J Med Sci 294:389-395

Rusch NJ, Hermsmeyer K (1988) Calcium currents are altered in the vascular muscle cell membranes of spontaneously hypertensive rats. Circ Res 63:997-1002

Schaub MC, Hirzel HO (1987) Atrial and ventricular isomyosin composition in patients with different forms of cardiac hypertrophy. Basic Research in Cardiology II:357-367

Sharma RV, Bhalla RC (1988) Calcium and abnormal reactivity of vascular smooth muscle in hypertension. Cell Calcium 9:267-274

Schoutsen B, Blom JJ, Verdouw PD, Lamers JMJ (1989) Calcium transport and phospholambam in sarcoplasmic reticulum of ischemic myocardium. J Mol Cell Cardiol 21:719-727

Sturek M, Hermsmeyer K (1986) Calcium and sodium channels in spontaneously contracting vascular muscle cells. Science 233:475-478

Swynghedauw B (1989) Remodelling of the heart in response to chronic mechanical overload. Eur Heart J 10:935-943

Usui A, Kato K, Sasa H, Minaguchi K, Abe T, Murase M, Tanaka M., Takeuchi E (1990) S-100a$_0$ protein in serum during acute myocardial infarction. Clin Chem 36, 4:639-641

Vasdev S, Thompson P, Triggle C, Fernandez P, Bolli P, Ananthanaryara VS (1988) Fura-2 used as a probe to show elevated intracellular free calcium in platelets of Dahl-sensitive rats fed a high salt diet. Biochem Biophys Res Commun 154:380-386

Whitmer JT, Kumar P, Solaro J (1988) Calcium transport properties of cardiac sarcoplasmic reticulum from cardiomyopathic Syrian hamsters (BIO 53.58 and 14.6): Evidence for quantitative defect in dilated myopathic hearts not evident in hypertrophic hearts. Circ Res 57:836-843

Young EW, Bukoski RD, McCarron DA (1990) Calcium metabolism in experimental genetic hypertension. In: Laragh JH, Brenner BM (eds) Hypertension: pathophysiology, diagnosis and management. Raven Press Ltd., New York, pp 977-987

Zemel MB, Zemel PC, Berry S, Norman G, Kowalczyk C, Sokoy RJ, Standlex PR, Walsch MF, Sowers JR (1990) Altered platelet calcium metabolism as early predictor of increased peripheral vascular resistance and preeclampsia in urban black women. New Engl J Med 323:434-438

Section VI

Calcium-Dependent and Phospholipid-Binding Proteins in Health and Disease

DIVERSITY IN THE ANNEXIN FAMILY

Stephen E. Moss, Helena C. Edwards, and Michael J. Crumpton

INTRODUCTION

The discovery in recent years of a major novel family of calcium-binding proteins, most commonly known as the lipocortin or calpactin family, has brought together several diverse areas of interest. One consequence of this, has been the generation of a cumbersome and confusing nomenclature, creating problems even for those working directly in the field. In this chapter we begin by attempting to clarify the nomenclature, adopting the generic name annexin together with the numbering system first applied to the lipocortins (Crumpton and Dedman 1990). The remainder of the chapter is divided into the following sections. Firstly, we discuss the primary structures, and the sequence homologies which lead to the identification of the relationship between the annexins. We will also consider the emergence of data relating to crystal structure and gene regulation. Secondly, the biochemical properties of the annexins are reviewed, particularly with respect to calcium and phospholipid-binding, and the implications this has for the ability of all members of the family to inhibit phospholipase A_2 and act as anticoagulants. Thirdly, the potential functional significance of phosphorylation by transforming and normal protein kinases is examined. Fourthly, we discuss the comparative patterns of tissue distribution and developmental regulation of the annexins and finally, we consider the diverse physiological roles which have been proposed for the individual annexin family proteins, and attempt to reconcile these with their known properties within a realistic conceptual framework.

NOMENCLATURE

The term lipocortin was first adopted to describe proteins which had been shown by various groups to be potent inhibitors of phospholipase A_2 (PLA_2) (Di Rosa et al. 1984). Previously, these anti-phospholipase proteins had been known as renocortin (Russo-Marie et al. 1979), lipomodulin (Hirata et al. 1980) and macrocortin (Blackwell et al. 1980). The apparent inducibility of lipocortin by glucocorticoids, prompted the suggestion that lipocortin mediated certain physiological anti-inflammatory effects by inhibiting PLA_2, thus preventing the synthesis of arachidonic acid and inflammatory downstream metabolites such as prostaglandins and leukotrienes. The controversy surrounding this hypothesis began with the discovery by molecular cloning that both lipocortin and another protein inhibitor of PLA_2, namely lipocortin II, were identical with p35/calpactin II and p36/calpactin I heavy chain respectively, which had been identified through their phosphorylation on tyrosine in growth factor stimulated and transformed cells respectively (Radke et al. 1980; Fava and Cohen 1984: reviewed by Klee 1988; Crompton et al. 1988a; Hunter 1988). The ability of both proteins to inhibit PLA_2 led to speculation that rather than having anti-inflammatory roles *in vivo*, they have functions related to stimulus-response coupling via PLA_2 and also, that PLA_2 may itself be involved in cell transformation.

The last three to four years have witnessed the expansion of the annexin family to include at least ten apparently distinct but related gene products, as judged by molecular cloning. The concomitant discovery that all members of the family have the ability (*in vitro*) to inhibit both PLA_2 and blood coagulation, has raised the important question of whether or not this property is a true reflection of the physiological roles of all or indeed any of the annexins (reviewed by Davidson and Dennis 1989). The alternative argument is that these inhibitory properties are simply a product of their uniform capacity to bind calcium and phospholipids (Davidson et al. 1987). If the annexins are neither anti-inflammatory proteins nor anticoagulants, then the question of their genuine functions remains unanswered.

Table 1. Nomenclature of the annexin family

Annexin	Synonyms	Annexin	Synonyms
I	Lipocortin I Calpactin II p35 Chromobindin 9	V	Placental anticoagulant protein I Inhibitor of blood coagulation Lipocortin V 35 kDa-Calelectrin Endonexin II
II	Calpactin I heavy chain Lipocortin II p36 Chromobindin 8 Protein I heavy chain Placental anticoagulant protein IV		Placental protein 4 Vascular anticoagulant-a 35-γ Calcimedin Calphobindin I Anchorin CII
III	Lipocortin III Placental anticoagulant protein III 35-α Calcimedin	VI	p68, p70, 73k 67 kDa-calelectrin Lipocortin VI Protein III Chromobindin 20 67 kDa-calcimedin Calphobindin II
IV	Endonexin I Protein II 32.5 kDa-Calelectrin Lipocortin IV Chromobindin 4 Placental anticoagulant protein II Placental protein 4-X 35-ß Calcimedin	VII	Synexin
		VIII	Vascular anticoagulant-ß
		IX and X	from *Drosophila*

Lipocortin I (calpactin II) and calpactin I (lipocortin II) were the first proteins of the annexin family to be described, although their relationship did not become clear until both had been cloned and sequenced (Huang et al. 1986; Saris et al. 1986; Wallner et al. 1986). The publication of these sequences was followed by those of a 32 kDa polypeptide, protein II (Weber et al. 1987), and p68, a 68 kDa polypeptide (Crompton et al. 1988b; Moss et al. 1988) also known as 67 kDa-calelectrin (Sudhof et al. 1988). The emerging family was then augmented by the addition of an inhibitor of blood coagulation (IBC, PAP, VAC, endonexin II) (Iwasaki et al. 1987; Pepinsky et al. 1988; Grundmann et al. 1988; Maurer-Fogy et al. 1988), lipocortin III (Pepinsky et al. 1988), synexin (Burns et al. 1989), VAC-ß (Hauptmann et al. 1989) and two *Drosophila* proteins, annexins IX and X (Johnston et al. 1990). Most of these proteins have now been cloned and sequenced in a variety of species by different groups using different terminologies. In an effort to alleviate growing confusion in the field, the generic term annexin which was initially suggested by Geisow

(1986), in combination with the lipocortin numbering system of Pepinsky et al. (1988) has been adopted by most (Crumpton and Dedman 1990) but not all (Browning et al. 1990) workers in the field. Thus lipocortin I becomes annexin I, calpactin I heavy chain (lipocortin II) becomes annexin II and so on. Table 1 shows the extent of the family, together with the annexin nomenclature.

A FAMILY OF RELATED PROTEINS

The first indication that a relationship might exist within this group of Ca^{2+}- dependent phospholipid binding proteins, arose from a series of biochemical and immunological observations. Several laboratories had independently described a group of polypeptides with molecular weights in the 30-40 kDa range and at 67-72 kDa, that were isolated in Ca^{2+}-dependent association with various cell membranes; e.g., those of B-lymphocytes (Davies et al. 1984), synaptic vesicles (Sudhof et al. 1984), liver and intestine (Shadle et al. 1985), smooth muscle (Raeymaekers et al. 1985), placental syncytiotrophoblasts (Edwards and Booth 1987) and chromaffin cells (Creutz et al. 1987). The Ca^{2+}-dependency of membrane association is worthy of emphasis, in that these proteins were found to remain tightly bound to cell cytoskeletons (i.e., the nonionic detergent insoluble fraction of cellular membranes) prepared in the presence of Ca^{2+}, despite exhaustive extraction with nonionic detergents. Their highly selective Ca^{2+}-dependent membrane association has formed the basis of almost all subsequent purification protocols, and (helped by their relative abundance in certain tissues) facilitated their further characterization. Using purified proteins, it was thus established that although variation existed in their affinity for Ca^{2+}, it was generally increased in the presence of exogenously added phospholipid (discussed in detail later). The ease with which the proteins could be purified, lead to the important discovery of their immunological cross-reactivity, with the implication of close structural relationships. Thus, antisera raised to calelectrin, a 34 kDa Ca^{2+}-binding protein from the electric fish *Torpedo marmorata*, cross-reacted immunologically with the 32 kDa and 67 kDa mammalian Ca^{2+}-binding proteins (Geisow et al. 1984; Sudhof et al. 1984).

PRIMARY STRUCTURE

Unequivocal confirmation that the shared biochemical and immunological properties were based on similarities in primary amino acid sequences, was provided by the molecular cloning of annexin I (lipocortin I) (Wallner et al. 1986; Huang et al. 1986) and annexin II (calpactin I) (Saris et al. 1986; Huang et al. 1986). Alignment of the primary structures of annexins I and II showed that the two proteins shared sequence identity of approximately 50% at both the nucleotide and amino acid levels. Apart from their overall sequence homology, they also shared an unusual internal structure, being comprised of four highly conserved repetitive domains, each approximately 70 amino acids in length, and each preceded by a variable region of 7-40 amino acids (see Figure 1).

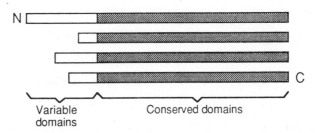

Fig. 1. Schematic representation of annexin structure. The figure shows the four-repeat structure common to the annexins (annexin VI has eight such repeats). The N- and C-terminal ends are indicated

The N-termini of annexins I and II were found to be the regions of greatest diversity, both in composition and length, although they were found to share one important characteristic, that being the presence of tyrosine and serine residues, identified as the sites of *in vivo* phosphorylation (Glenney and Tack, 1985; De et al. 1986; Isacke et al. 1986; Gould et al. 1986; Johnsson et al. 1986a; Varticovski et al. 1988). One interesting feature of the derived amino acid sequences was the apparent lack of an EF-hand helix-loop-helix structure, known to be responsible for the Ca^{2+}-binding properties of proteins such as calmodulin, parvalbumin and troponin C (Tufty and Kretsinger 1975). The possibility that Ca^{2+}-binding in the annexin family is mediated by an alternative helix-loop-helix (Taylor and Geisow, 1987; Moss and Crumpton, 1990a), will be discussed later.

The identification by molecular cloning of a further eight related annexins, has created what is clearly a major novel class of Ca^{2+}-binding proteins. Of the eight family members, all but one have structures which conform to the four-repeat model first identified in annexins I and II, the exception being annexin VI (p68, 67 kDa-calelectrin) which has eight internal repeats. The linking region between the fourth and fifth repeats of annexin VI is significantly longer than any of the other connecting sequences, and essentially divides the protein into two halves, which when aligned with one another share approximately 50% sequence identity. Examination of internal homology within the annexin VI conserved domains reveals that repeat 1 has greatest homology with repeat 5, 2 with 6, and so on. Therefore, in evolutionary terms it seems likely that the annexin VI gene arose by the duplication of a gene encoding a four-repeat protein, and that this in turn may have arisen following two successive duplications of a gene encoding a hypothetical single repeat polypeptide. Although no clear candidates have emerged for either the one or two repeat annexins, it was at one time considered that uteroglobin (a potent PLA_2 inhibitory protein which has limited homology with annexin I; Miele et al. 1988), may represent the ancestral single repeat annexin. However, on the basis of several criteria to be discussed later, this now seems improbable.

A second interesting feature, which is unique (within the family) to annexin VI, is the generation of protein diversity by alternative splicing. Thus, two forms of annexin VI exist, differing with respect to a hexapeptide which lies close to the start of the seventh internal repeat. Alternative splicing provides both a convenient and plausible explanation for the presence of the annexin VI protein as a characteristic closely spaced polypeptide doublet on denaturing gel electrophoresis (Moss and Crumpton, 1990b). Although the annexin VI gene directly encodes an eight-repeat polypeptide, it should be emphasized that annexins I, II, IV and V have all been shown to exist in dimeric forms, which effectively create similar eight-repeat structures. The biology of annexin dimerization, and the interactions between annexins and other proteins will be discussed later. Of the four-repeat proteins, annexin VII (synexin) differs from the others in having an unusually long hydrophobic N-terminus of 167 amino acids (Burns et al. 1989). Annexin VII was initially characterized as a component of and membrane fusogen in bovine chromaffin granules, and has also been shown to generate classical capacitative gating currents in isolated acidic phospholipid bi-layers (Rojas and Pollard 1987). Whether annexin VII functions as a Ca^{2+}-channel *in vivo* is not clear, and it is not obvious how this activity may be reconciled with the promotion of membrane fusion.

SECONDARY AND TERTIARY STRUCTURE

Little is presently known about the secondary and tertiary structures of the annexins, although the emergence in the literature of preliminary reports describing X-ray crystallographic studies, suggests that significant advances are being made. Secondary structure predictions for annexins I, II and IV (Taylor and Geisow 1987) were modelled on the EF-hand helix-loop-helix (Tufty and Kretsinger 1975) exemplified by calmodulin and parvalbumin. The former authors suggested that the annexin EF-hand, which lacks the ß-carboxyl groups occurring at positions 1 and 3 of the canonical EF-hand, could use main chain carbonyls to provide coordinating ligands, and that a highly conserved arginine residue at position 16 would form an additional stabilizing ionic interaction. This arginine residue is conserved in all repeats sequenced in mammalian members of the annexin family, supporting the view that it performs a crucial structural or biological function. However, the sequences of annexins IX and X (from *Drosophila*) both lack this arginine residue in at least one of their four repeats (Johnston et al. 1990) which questions the absolute functional significance of this amino-acid in Ca^{2+}/ phospholipid binding. However, it is not yet known how. many Ca^{2+}-binding sites are present in the *Drosophila* proteins, and it may be that certain repeats either do not bind Ca^{2+} or that valid interactions can occur with other amino acids. Functional studies on peptides derived from annexin V suggested that a histidine residue close to the highly conserved arginine might be involved in Ca^{2+}/phospholipid binding (Funakoshi et al. 1990), although the lack of similarly placed histidine residues in annexins IX and X (both of which bind Ca^{2+} and phospholipid) argues against this. However, in support of the alternative EF-hand model, a Ca^{2+}-binding protein from *Streptomyces erythraeus* described by Swan et al. (1987) was found to contain both calmodulin-like EF-hand's and a 17 amino acid Ca^{2+}-binding domain (Moss and Crumpton 1990a) bearing 50% homology to the annexin consensus sequence (Geisow 1986). This observation, thus raises the intriguing possibility that the annexin family and the EF- hand proteins may share common ancestry.

Recently, Newman et al. (1989) reported the 2D-crystallization of annexin VI on lipid monolayers, and demonstrated the requirement for Ca^{2+} in this model system. Interestingly, annexin VI was found to form regular two dimensional crystalline arrays, suggesting that these proteins may form paracrystalline structures *in vivo*. Recent, preliminary reports on the crystallization of annexin V (vascular anticoagulant-ß) by Lewit-Bentley et al. (1989) and Seaton et al. (1990), provide optimism for full 3D structural analysis. These authors crystallized annexin V from rat kidney and human umbilical cord arteries respectively, and both achieved a diffraction resolution of 2.2 Å Bragg spacing. Although detailed crystal structure analysis has not yet been reported, it must be hoped that such studies will eventually lead to the identification of the functionally important Ca^{2+} and phospholipid binding sites.

QUATERNARY STRUCTURE AND PROTEIN INTERACTIONS

Annexin II is unique within the annexin family of proteins in that it associates *in vivo* with another polypeptide, namely p11, to form a heterotetramer of M_r 85000, such that two molecules of annexin II bind noncovalently to a p11 dimer ($p36_2p11_2$; Gerke and Weber 1984; Johnsson et al. 1986b). This complex of proteins is often referred to as protein I (Gerke and Weber 1984) or calpactin I, on the basis of its ability to bind Ca^{2+}-dependently to actin (Glenney et al. 1987). p11 has significant amino acid sequence homology to the EF-hand related S-100 family of proteins (Hexham et al. 1986) though it does not bind Ca^{2+} itself, nor is the association with annexin II Ca^{2+}-dependent. The interaction between the two proteins occurs via a short amphipathic helix at the N-terminus of annexin II (Johnsson et al. 1988), close to the region which contains the sites for tyrosine and serine phosphorylation (see Table 2). Evidence for a role for p11 in the modulation of annexin II interaction with the cytoskeleton *in vivo*, arises from several observations. Firstly, two pools of annexin II exist within cells, an insoluble pool associated with the cytoskeleton and a soluble or cytosolic pool. Secondly, all p11 present in cells is exclusively located to the submembranous cytoskeleton, and thirdly, the K_d for the association of p11 and annexin II (estimated to be <30 nM) suggests that *in vivo*, all p11 would be present as the heterotetramer (Zokas and Glenney 1987; Osborn et al. 1988; Johnsson et al. 1988). Thus the complex, containing p11, is bound to the cytoskeleton and any excess annexin II exists mainly as monomer in the soluble pool.

Both annexin II monomer and the heterotetramer have been shown to bind to F-actin *in vitro* experiments. However, whereas monomeric annexin II was able to bundle actin filaments only at very high ratios, the heterotetramer bound to F-actin in a cooperative manner readily forming bundles. The interaction was of high affinity (Kd ~0.23 μM) with a stoichiometry of 1 heterotetramer: 2 F-actin, and half maximal binding occurred at Ca^{2+}-concentrations of 0.1-2 μM (Ikebuchi and Waisman 1990). These values, which are similar to physiological Ca^{2+} levels, provide further evidence for the *in vivo* involvement of annexin II in cytoskeleton/actin filament interactions. Interestingly, when actin polymerization is allowed to proceed in the presence of heterotetramer, shorter filaments are formed suggesting that the heterotetramer is acting either as an actin severing or actin filament capping agent, perhaps in a similar manner to that described for villin or gelsolin (Martin et al. 1988). With respect to this, it may be relevant that annexin II and gelsolin share a region of homology (Burgoyne 1987). The heterotetramer was originally described as also binding Ca^{2+}-dependently to fodrin (Gerke and Weber 1984) although more recent evidence would suggest that the association with fodrin is probably nonspecific as it is of relatively low affinity and is not readily saturable (Cheney and Willard 1989). This also seems to be the case for several of the other proteins with which the heterotetramer/annexin II is reported to be associated. Notably, p11 binding by annexin II has been reported to confer calmodulin binding activity, not seen with monomeric annexin II (Martin et al. 1988).

The ability to bind Ca^{2+}-dependently to the cytoskeleton has also been demonstrated for several other annexins, and has often been used as the basis of their isolation (eg., Owens and Crumpton 1984; ; Shadle et al. 1985; Edwards and Booth 1987; Glenney et al. 1987) such that annexins which bind to the detergent-insoluble matrix (cytoskeleton) at millimolar levels of Ca^{2+} are then selectively eluted using chelating agents such as EGTA. Where the affect of Ca^{2+}-concentration on the association of individual annexins with the cytoskeleton has been studied, large differences have been reported in their binding capacity. Annexins

I and IV only bind at relatively high Ca^{2+} (>0.1 mM, Rhoads et al. 1985; Edwards and Booth 1987; Zokas and Glenney 1987) whereas annexins II and VI require micromolar levels of Ca^{2+} (Owens and Crumpton 1984; Shadle et al. 1985; Edwards and Booth 1987). These findings, also support the proposal that different pools of annexins exist within the cell. Indeed there is evidence that annexin I largely exists in the soluble pool (Zokas and Glenney 1987). Reports of *in vitro* binding to actin by annexins I, IV, V and VI have so far failed to demonstrate specific or high affinity binding at physiological levels of Ca^{2+}, and as such, it remains to be seen whether any of these bind to actin *in vivo*. On the basis of their Ca^{2+}-dependent association with the cytoskeleton only annexin VI is a likely candidate.

In addition to the association with other proteins, the annexins have been reported to exist as covalent dimers. Thus an annexin I dimer has been identified both in placenta (Pepinsky et al. 1989) and in fibroblasts (Ando et al. 1989a), where it is also phosphorylated by tyrosine kinases (see Table 2) and its formation can be stimulated by elevation of intracellular Ca^{2+} levels. Interestingly, dimerization of annexin I does not occur through the classic disulfide bridge, but by transglutaminase mediated cross-linking of N-terminal glutamine residues. In contrast, the dimeric forms of annexins IV and V are both sensitive to reducing agents, and are therefore likely to be formed via disulfide bridges (Funakoshi et al. 1987; Ahn et al. 1988). Dimerization may be an important property of the annexins, perhaps providing large enough multi-valent complexes to form bridges between vesicles and membranes or cytoskeletal-membrane attachments, such as those described by Nakata et al. (1990).

GENE REGULATION

The elucidation of the mechanisms of gene regulation in the annexin family is likely to be a subject of particular interest over the next few years. Although little is presently known about the structural basis of annexin gene regulation, certain aspects may be inferred from biochemical data, and the distinct patterns of tissue expression for individual annexins (see later) certainly suggests that each is regulated in a unique way. The glucocorticoid induction of the annexin I (lipocortin I) gene has been most extensively studied and is also the most controversial. The induction of annexin I by glucocorticoids was believed to be the basis of the mechanism by which this protein was involved in mediating anti-inflammatory responses (Hattori et al. 1983; Wallner et al. 1986). These reports were contradicted by others who suggested that in many cell types annexin I gene expression is unaffected by glucorticoids both at the mRNA and protein levels (Bronnegard et al. 1988; Isacke et al. 1989). The confusion arising from these conflicting observations should be resolved by the cloning of the annexin I gene. With respect to this, Pepinsky et al. (1988) have reported the presence of glucocorticoid responsive elements in the regulatory upstream sequences within the annexin I gene, although it is not known whether these are functional. It seems likely therefore, that annexin I gene expression is regulated by multiple factors, perhaps with some tissue specificity and that glucocorticoid-induction may simply represent one mechanism of control (Philipps et al. 1989). This proposal is supported by the findings of Horseman, (1989) who cloned and sequenced annexin I on the basis of it's inducibility by prolactin in the pigeon crop-sac. The potential functional significance of this finding is discussed later.

The first structural analysis of an annexin family gene has recently been reported by Amiguet et al. (1990) who described the intron/exon arrangement of the murine annexin II gene. The authors made the surprising observation that the high level of conservation found in the repeated peptide domains, was not reflected in the positions of the intron/exon boundaries. This questions the proposed evolution of the annexin family as a series of simple gene duplication events (Crompton et al. 1988a). Although the annexin II gene promoter region was not studied, a recent study by Keutzer and Hirschhorn (1990) demonstrated that serum could directly stimulate expression of the annexin II gene in quiescent BALB/c 3T3 cells. Interestingly, epidermal growth factor, fibroblast growth factor and insulin all similarly induced expression of the annexin II gene, whereas platelet-derived growth factor had no effect. Furthermore, gene expression was found not to be dependent on *de novo* protein synthesis, as judged by studies with cycloheximide. These results and the previous characterization of annexin II as a substrate for phosphorylation by tyrosine kinases are consistent with it having a role in cell proliferation.

BIOCHEMICAL CHARACTERIZATION

The annexins share the common property of being able to bind phospholipids in a Ca^{2+}-dependent manner. They often show immunological cross-reactivity, have similar molecular weights (30-40 kDa or 68-72 kDa) and similar pI values (see Table 2) and are, consequently, frequently isolated under similar conditions. The marked similarities between the properties of individual members of the annexin family, emphasizes the importance of their accurate characterization. Indeed, it was not until amino acid sequence data became available that the separate identities of some of the annexins became clear. Historically, this led to some confusion. For example, annexin II has a proteolytically sensitive site at its N-terminus (Johnsson et al. 1986b), which when cleaved results in the formation of a ~33 kDa core domain that retains the Ca^{2+} and phospholipid binding characteristics of the native protein. It was therefore widely considered that polypeptides of this molecular weight which co-isolated with annexin II were in fact cleavage products, rather than novel members of the annexin family. Another area of confusion surrounds annexins with molecular weights in the region of 70 kDa. There are conflicting reports as to whether the 67 kDa-calcimedin and annexin VI are the same protein, or two different proteins which share some common properties (Moore 1988; Morse and Moore 1988; Kobayashi and Tashima 1989). This and recent evidence that annexin I also exists as a nonreducible dimer of ~70 kDa *in vivo* (Ando et al. 1989a; Pepinsky et al. 1989) further emphasize the necessity of the clear identification of proteins under investigation.

GENERAL PROPERTIES

A summary of the biochemical properties of the annexins is given in Table 2. This includes information about the oligomeric states, molecular weights, pI's, numbers of Ca^{2+}-binding sites, Kd's for Ca^{2+}-binding and both the *in vivo* and *in vitro* sites for phosphorylation with the corresponding kinases. In general the presence of phospholipid enhances the affinity of annexins for Ca^{2+}-binding, in some instances by as much as 1000 fold.

Table 2. General properties of the annexins (References are indicated by superscripted numbers)

Annexin	$M_r \times 10^{-3}$	pI	Phosphorylation	Ca^{2+}-binding
I monomer	38	7.2	EGFr Tyr-21[10,54,57,116,121] Ins r Tyr-21[75] pp60[c-src] Tyr-21[131] pp50[v-abl] Tyr-21[131] fps Tyr [57] PKA Thr-216[131] PTK Thr[1] PKC Thr-24[7,77,117] Thr-41[131] Ser-27[117,131] Ser-28[116]	1 site (-PS), 4 sites(+PS), Kd 75 μM[116] 2 sites, Kd 10 μM(+PS)[50]
*dimer[2,5,54,103]	~70	6.4	EGFr[103,121] fps Tyr[59] PKC[35]	
tetramer?[39,74]	170		EGFr Ser[74] Ins r Ser[74]	
II monomer	36	7.4	pp60[v-src] Tyr-23[33,47,49,57,106] Ins r Tyr[75] fps Tyr[53] PKA[69] CamK[69] PKC Ser-25[51,70,77]	2 sites, Kd 4.5 μM (+PS/PI)[48] 4 sites, Kd 0.5 μM (+PS)[105] Kd 0.5 mM (-PS)[105]
dimer[2]	~70			
heterotetramer p36₂p11₂	85		PKC Ser[77] Ser-25[69]	Kd 0.5 mM (-PS)[105] Kd 1.3 μM (+PS)[105] Kd 0.1 mM[120]
III monomer	34	5.9	PKC [21]	
IV monomer	33-36	5.9	PKC Thr-6[133]	Kd 0.4 μM[120]
dimer[2]	~70			
V monomer	33-35	4.8		4 sites, Kd 0.5mM (-PS)[118] Kd 0.1mM (+PS)[118]
**dimer[2,39]	~69			
VI monomer	68-73 (doublet)	5.8	PKC[21] (poorly)	1 site (-phospholipid), Kd 1.2 μM[101] 1 site(-PS)/8 sites(+PS)[33] Kd 1.0 μM[120] Kd 0.4 μM[92] Kd 20 μM[86]
dimer[2]	~160			
VII monomer[136]	51-54 human 47 bovine			
VIII monomer[56]	33-36		5.4	

Abbreviations used in Table 2:

EGFr	epidermal growth factor receptor kinase	pp50[v-abl]	retroviral protein kinase
Ins r	insulin receptor kinase	fps	protein kinase
PKA	cAMP-dependent kinase	PS	phosphatidylserine
PTK	protein threonine kinase	PI	phosphatidylinositol
PKC	protein kinase C	* glutamine bridge	
CamK	calmodulin dependent kinase	**disulfide bridge	

Similarly the number of Ca^{2+}-binding sites detected increases in the presence of phospholipid, providing some indication for a common Ca^{2+} and phospholipid binding site such as proposed by Taylor and Geisow (1987). Annexins I and II are major *in vivo* substrates for the EGF receptor and pp60[v-src] tyrosine kinases respectively, as well as *in vitro* substrates for many other kinases. However, in view of the promiscuity of protein kinase activities *in vitro*, the significance of many of these studies must be highly questionable, in the context of annexin function. Although both annexins I and II are substrates for the insulin receptor tyrosine kinase *in vitro* (Karasik et al. 1988a), only annexin I is phosphorylated *in vivo*, suggesting that only annexin I interacts with the insulin receptor *in vivo*. The sites of phosphorylation all reside in the N-termini of the proteins with the notable exception of threonine 216 in annexin I, which is phosphorylated by protein kinase A. Interestingly, the dimerization of annexin I and the interaction of annexin II with p11, which occur through their respective N-termini, apparently do not affect the tyrosine phosphorylation of these proteins. This provides evidence for distinct sites within the N-termini for phosphorylation and protein interactions respectively, and is consistent with the report by Amiguet et al. (1990) that the p11-binding and phosphorylated regions are encoded by separate exons in the murine annexin II gene.

PHOSPHOLIPID BINDING

It was originally thought that the annexins bound Ca^{2+}-dependently to hydrophobic matrices such as phenyl-Sepharose (often used in their isolation) through the exposure of hydrophobic residues, in a manner similar to that of calmodulin. In contrast, the purified proteins were found not to bind directly to phenyl-Sepharose (eg., Edwards and Booth 1987) even in the presence of millimolar Ca^{2+}. Furthermore, the degree of exposure of hydrophobic residues in response to Ca^{2+}-binding, as judged by ultraviolet difference spectroscopy, detected in proteins such as annexin II (e.g., Gerke and Weber 1984; Pigault et al. 1990) is relatively small when compared to that seen for calmodulin. These observations have led to the suggestion that the proteins bound to hydrophobic matrices via Ca^{2+}-dependent association with phospholipids present in the crude preparations from which the proteins were prepared.

Extensive studies have been conducted on the specificity and Ca^{2+}-dependency of binding to phospholipids using a variety of different techniques, but most commonly based on the co-sedimentation of annexins and phospholipid liposomes in the presence or absence of Ca^{2+}. These data are summarized in Table 3. In general, the annexins bind to a broad range of phospholipids at millimolar levels of Ca^{2+}, with the notable exceptions of phosphatidylcholine and sphingomyelin, both of which have a phosphorylcholine headgroup. The nature of the acyl chains does not significantly affect the capacity of the proteins to bind, lending further weight to the hypothesis that the proteins bind Ca^{2+}-dependently to the headgroup of the phospholipid. Different laboratories have reported large variations in the experimentally determined Ca^{2+}-concentrations required for half maximal binding ($K_{0.5}$) of the individual proteins to specific phospholipids. For example, the $K_{0.5}$ for the binding of monomeric annexin I to phosphatidylserine has been reported to be <1 µM (Genge et al. 1989) and between 80 and 120 µM (Pepinsky et al. 1989), also both Pepinsky et al. (1989) and Ando et al. (1989a) have reported that the dimeric form of annexin I requires lower Ca^{2+}-concentration for binding to phosphatidylserine vesicles than monomeric annexin I, the actual recorded $K_{0.5}$ values being 40-80 µM (dimer) and 80-120 µM (monomer), and 3 µM (dimer) and 20 µM (monomer) respectively. Much of the difficulty in obtaining reproducible data most probably stems from problems in obtaining phospholipid liposomes of constant size and density for use in the binding assays. In experiments examining the effects of Ca^{2+}-concentration on phospholipid binding by the annexins, the effect of changes in Ca^{2+}-concentration on the physical nature of the liposomes *per se* also needs to be taken into account. Deterioration of liposomes and lipids on storage (Powell and Glenney 1987) and the presence of contaminants such as free fatty acids in commercial preparations may also influence binding (Davidson et al. 1990). At least some of the artifacts arising from variations in liposome size and density, which affect the sedimentation of liposomes, can be alleviated by the use of phospholipid affinity columns, where the phospholipids are adsorbed onto hydrophobic matrices (eg., Edwards and Booth 1987; Ando et al. 1989b).

Despite the variable results which have been obtained, some general trends emerge from the data summarized in Table 3. Most striking is the range of affinities reported for Ca^{2+}-dependent phospholipid-binding for individual annexins. In particular, annexin V exhibited very low affinities for binding (high $K_{0.5}$) for all phospholipids studied, with the notable exception of phosphatidylethanolamine/phosphatidic acid mixtures, whereas annexins III and IV have intermediate affinities and annexins I, II and IV showed the highest affinities. The most variable data pertained to binding to phosphatidylethanolamine; in some cases no binding was detected even at millimolar Ca^{2+}, whereas in other cases binding was reported at micromolar Ca^{2+}. Interestingly, when Ca^{2+}-dependent binding to mixtures of phosphatidyl-ethanolamine and phosphatidic acid was investigated, very low $K_{0.5}$ values (~1 μM Ca^{2+}) were obtained for all the annexins studied, compared with low affinity (high $K_{0.5}$) binding to phosphatidic acid alone, phosphatidylethanolamine alone and mixtures of phosphatidyl-ethanolamine and phosphatidylserine (see Table 3). Studies in our laboratory, on annexins IV and VI, have revealed similar reductions in the Ca^{2+} concentrations required for half maximal binding when fatty acid/phospholipid mixtures (e.g., arachidonic acid/phosphatidyl-choline) were compared to phospholipid alone (Edwards and Crumpton unpubl. observations). These results indicate that liposome composition can have a profound affect on Ca^{2+}-dependent binding of the annexins, and that binding to phospholipids is more complex than at first supposed. The variability in the results obtained even when apparently the same protein, phospholipid and technique are used clearly indicates the importance of establishing better procedures for studying Ca^{2+}-dependent phospholipid binding of the annexins.

In a few instances, post-translational modification of annexins has been found to alter their Ca^{2+}-dependency for phospholipid-binding *in vitro*. Thus, tyrosine phosphorylation of annexin I monomer by the EGF receptor kinase resulted in a reduction in the Ca^{2+}-concentration required for phospholipid binding, whereas tyrosine phosphorylation of annexin II by pp60[v-src] kinase decreased its affinity (i.e., increased the Ca^{2+}-requirement) for phospholipid binding (Powell and Glenney 1987; Schlaepfer and Haigler 1987).

Table 3. Phospholipid binding data

Annexin	Phospholipids bound at 1mM Ca^{2+}	unbound at 1mM Ca^{2+}	$K_{0.5/\mu M}$		References
I monomer	PS,PE,PI,CL PA	PC,SM,PE	PS	22	5,6,9,27,32,
				20	43,57,103,116
				5	121
				1.3	
				<1	
				10 +tyr P^n	
				4 +tyr P^n	
				*80-120	
			PI	2	
			PS/PE	*30 (15 + P^n)	
				*1-10	
			PE/PA	0.6	
dimer	PS,PI	PC	PS	3	5,59,103
				*40-80	
			PS,PI	*50 (PI>PS)	
II monomer	PS,PE,PI,CL PA	PC,SM,PE	PS	0.65	9,27,32,47,
				*5	57,106,136
				>*10	
			PI	1.3	
			PS/PE	*1-10	
				*4.5	
			PE/PA	0.2	
p36$_2$p11$_2$	PS	PC	PS	5	57,106,136
				*<0.0	
III	PS,PE,PI,PA	PC	PS	8.8	9,34,73,127
				2	
			PE	270	
			PI	8	
			PE/PA	1.2	
IV	PS,PE,PI,CL PA	PC,SM,PI (50 μM^{58})	PS	†4.5	9,32,58,73
				2.4	81,82,127
				*100	
				*50	
			PE	5.2	
				†4.5	
				*100	
			PI	†20	
				4.4	
				*100	
			PE/PA	0.95	

Table 3. Continued

V	PS,PE,PI,CL PA	PC,SM	PS	220 53 16 800 *100	4,9,11,39,58, 82,109,118, 127
			PE	860 270 40 *100	
			PI	470 130 *100	
			CL	39	
			PA	750 *100	
			PE/PA	1.2	
VI	PS,PE,PI,CL PA,PC[80]	PC,SM	PS	2 †0.6 *100 *3	9,27,32,59, 80,98,135
			PE	33 †0.4 *100 *3	
			PI	34 †3.5 *100	
			PS/PE	*<10	
			PE/PA	0.55	
			PC	*100[80]	
VII	PS,PI		PS	*>100 15	17,111,136
			PI	15	
VIII	PS				56

$K_{0.5}$ the Ca^{2+}-concentration required for half maximal binding to occur

* Ca^{2+}-concentrations (μM) at which binding occurred when no value for $K_{0.5}$ is given

† Edwards and Crumpton, unpubl. results

Abbreviations

PS	phosphatidylserine	PC	phosphatidylcholine
PE	phosphatidylethanolamine	PA	phosphatidic acid
PI	phosphatidylinositol	CL	cardiolipin
SM	sphingomyelin	Pn	phosphorylated protein

Interestingly, dimeric annexin I (which is reported to occur in response to a transient increase in the concentration of intracellular free Ca^{2+}) was found to have enhanced phospholipid binding when compared to that of the monomeric form. Taken together, these observations provide evidence for regulation of Ca^{2+}-dependent membrane association of certain annexins *in vivo*, during stimulus response coupling.

INHIBITION OF PHOSPHOLIPASE-A$_2$ AND BLOOD COAGULATION

The anti-inflammatory actions of glucocorticoids are considered to be at least partly mediated by proteins which prevent prostaglandin production via the inhibition of phospholipase-A$_2$ (PLA$_2$). This proposal is based on several observations. Firstly, the anti-inflammatory effects of glucocorticoids are dependent on mRNA and protein synthesis; secondly, glucocorticoids inhibit the release of [^{3}H]-arachidonic acid from prelabeled cells in response to inflammatory stimuli and thirdly, glucocorticoids appear to act at some stage prior to cycloxygenase in the production of prostaglandins. Since PLA$_2$ plays a pivotal role in the synthesis of the precursor metabolites of prostaglandins, it seems reasonable to suppose that protein inhibitors of PLA$_2$ could thereby mediate the activity of glucocorticoids. One approach to the designation of the mechanisms underlying the decrease in arachidonic acid production, has focused on the characterization of proteins present in the lavage fluids and supernatants of glucocorticoid-stimulated animals and cells respectively. As has been described previously, this led to the cloning and sequencing of two proteins, which proved to be identical to the EGF-dependent tyrosine kinase substrate p35 (annexin I), and the closely related cellular substrate for pp60^{src} kinase, p36 (annexin II) [for reviews see Flower (1988) and Davidson and Dennis (1989)]. This important discovery questions the significance of PLA$_2$ inhibition in the context of anti-inflammatory responses, since there is no obvious link between anti-inflammatory responses, and phosphorylation of annexins by tyrosine kinases. The possibility remains that certain proteins of the annexin family are involved in PLA$_2$ inhibition in signal transduction pathways. It may be significant that two proteins capable of potent PLA$_2$ inhibition (of approximate molecular weight 33 kDa) have recently been characterized as proteolytically derived fragments of complement (Suwa et al. 1990). It is tempting to speculate that these proteins may have been the original lipocortins.

The assignation of the annexins as *in vivo* PLA_2-inhibitors/anti-inflammatory agents remains controversial for several reasons. Firstly, conflicting reports describe the stimulation or regulation of annexin synthesis by glucocorticoids (see Gene regulation). Secondly, there is a lack of evidence for active secretion of annexins (although they are found in certain extracellular fluids), which neither have signal sequences nor are glycosylated, and which are usually either cytosolic or situated in close apposition with intracellular membranes. Thirdly, the mechanism for the *in vitro* inhibition of PLA_2 (described below), which is an undoubted property of all annexins examined, suggests that PLA_2 inhibition is unlikely to be a specific function of these proteins *in vivo*. Finally, the recent discovery of annexins in lower eukaryotes (Gerke 1989; Johnston et al. 1990), clearly makes general roles in anti-inflammatory responses (or the inhibition of blood coagulation) untenable.

If the annexins are not involved in glucocorticoid-mediated anti-inflammatory responses, then the significance of their ability to inhibit PLA_2 must be seriously re-examined. The ability of these proteins to inhibit both PLA_2 and prothrombin-induced blood coagulation appears to be attributable to their common Ca^{2+}-dependent phospholipid binding properties, rather than to their specific interaction(s) with the enzyme or clotting factor. Observations that the dose dependent inhibition of PLA_2 by annexins can be abolished by increasing the phospholipid substrate concentration, that pre-incubation of annexins with the substrate results in greater inhibition than pre-incubation with the enzyme, and that the levels of annexin required for inhibition are relatively high, led to the substrate depletion model proposed by Davidson et al. (1987). In this model the annexins bind Ca^{2+}-dependently to the phospholipid substrate under the conditions of the assay, resulting in the sequestration of substrate or more probably coating of the surface of the liposomes, thus preventing the enzyme from physically interacting with its substrate. If substrate depletion is sufficient to explain the PLA_2 inhibitory and anti-blood coagulation activities of the annexins, then the efficiency of inhibition by individual annexins should reflect their relative affinities for phospholipid-binding. There is some evidence for this; for example, in parallel with their phospholipid-binding affinities, annexins IV and V are less efficient inhibitors of PLA_2 than annexin I (Kobayashi et al. 1990a), and annexin VIII is twice as efficient as annexin V as an inhibitor of blood coagulation (Hauptmann et al. 1989). Also, the ability of individual annexins to bind to a range of phospholipids (see Phospholipid binding), should reflect a capacity to inhibit a corresponding range of phospholipases. This proposal is corroborated by the reported inhibition of phospholipase-C by annexins I and II (Machoczek et al. 1989). Although the vast majority of biochemical evidence supports the substrate depletion hypothesis, there are findings which are inconsistent with this view. For example, Miele et al. (1988) reported that peptides derived from the regions of highest sequence homology between annexin I and uteroglobin (a potent PLA_2-inhibitor), were able to elicit a potent anti-inflammatory response *in vivo*. The authors claimed, on the basis of *in vitro* studies, that the peptides achieved their effect via the inhibition of PLA_2, with which they apparently interacted directly. The authors further claimed to have shown the specificity of the interaction with PLA_2, by demonstrating that no inhibition occurred with phospholipase-C. These observations have proved to be completely irreproducible, and have recently been strongly contradicted by van Binsbergen et al. (1989) who attempted to repeat these experiments with the same peptides.

Despite the inability of the annexins to bind phosphatidylcholine (PC) even at millimolar Ca^{2+} (see Table 3), inhibition of PLA_2 when PC-liposomes were used as a substrate has been reported (e.g., Ahn et al. 1988). Davidson et al. (1990) suggest that the inhibition observed when PC is the substrate, was due to low levels of binding and that the degree of inhibition was dependent on the nature of the fatty acid attached to the sn-2 position. Again the highest levels of inhibition were seen at low concentrations of PC, suggesting that this is a substrate dependent process. Conversely, there are reports of annexins unable to inhibit PLA_2 when PC-liposomes were used (e.g., Rothut et al. 1987). A possible explanation for these discrepancies, yet to be explored, is that the annexins are able to bind to contaminants (such as fatty acids) present in the substrate. It should be noted that the $K_{0.5}$-values for binding to phosphatidic acid/phosphatidylethanolamine mixtures (see Table 3) are amongst the lowest reported, implying that only low levels of contamination would be required for inhibition to occur. The observation that annexins were able to bind to PC-liposomes when arachidonic acid was present (Edwards and Crumpton unpubl. data), also suggests that the annexins might bind to the products of PLA_2 action on PC-liposomes, preventing further interaction between the enzyme and its substrate, in a similar manner to that described for substrate depletion.

THE ROLE OF PHOSPHORYLATION

The varied biochemical properties of the annexins have prompted considerable speculation as to the actual physiological roles of individual family members. One of the most intriguing observations is the highly reproducible and well-documented phosphorylation of annexins I and II. Tyrosine phosphorylation of annexin I (lipocortin I) was first described in EGF-treated A431 cells (Sawyer and Cohen 1985) and permeabilized human fibroblasts (Giugni et al. 1985). Annexin I is also phosphorylated on serine following EGF treatment (Sawyer and Cohen 1985). It is interesting to note that whereas annexin I is an excellent substrate for the EGF receptor kinase, it is a poor substrate for pp60^{v-src} kinase, whilst annexin II (calpactin I) shows the opposite specificity. These differences most probably reflect the fact that despite the overall homology between the two proteins, the sites for tyrosine phosphorylation are in the N-terminal domains, which are the regions of least similarity. In addition, the principal sites for phosphorylation on serine by protein kinase C also lie in the N-termini, reinforcing the idea that these domains mediate/regulate particular functions. It is not surprising to note that the region within the N-terminus of annexin II which contains the serine and tyrosine phosphorylation sites, is encoded by a distinct exon from that which encodes the p11 binding domain (Amiguet et al. 1990). If the phosphorylation of annexins I and II is a means of regulating their physiological activities, then it is important to identify the functional changes induced by phosphorylation. There is evidence that tyrosine phosphorylation of annexins I and II alters their Ca^{2+}-dependence of phospholipid binding. Thus, phosphorylation of annexin I at Tyr 20 reduces by fivefold the amount of Ca^{2+} required for half maximal association with phospholipid (Schlaepfer and Haigler 1987), whereas phosphorylation of annexin II at Tyr 23 increases the Ca^{2+}-requirement for phospholipid binding (Powell and Glenney 1987). In view of the closely related structures of annexins I and II, interpretation of the contrary effects of tyrosine phosphorylation on Ca^{2+}-dependent phospholipid binding is clearly complex.

Since Ca^{2+} mediates the association of these proteins with phospholipids *in vitro* and probably also with the cytoskeleton *in vivo*, it is possible that by changing their membrane-binding properties, phosphorylation regulates the cellular compartmentalization of annexins I and II during stimulus-response coupling. In addition, tyrosine phosphorylation of annexin I in A431 cell membranes increases the sensitivity of the 12 amino acid N-terminal domain to proteolytic cleavage (Chuah and Pallen 1989). The proteolytic cleavage of the N-terminus *in vitro* has been shown to dramatically increase the Ca^{2+}-sensitivity of phospholipid binding by annexin I (Ando et al. 1989b). However, uncertainty over the functional significance of tyrosine 20 phosphorylation of annexin I by the EGF receptor has increased, with the observation that this tyrosine phosphorylation site was not conserved in a pigeon annexin I homologue (Horseman 1989).

The discovery by molecular cloning of a further eight members of the annexin family, has inevitably been accompanied by speculation that these proteins may similarly provide targets for phosphorylation *in vivo*. However, phosphorylation of other annexins has been especially difficult to demonstrate, although several are readily phosphorylated *in vitro*. One exception is annexin VI (p68, 67 kDa-calelectrin) which we have found to be phosphorylated in murine Swiss 3T3 fibroblasts and human T-lymphoblasts, in response to stimulation with serum and interleukin-2 respectively (Moss and Crumpton unpubl. observations). Although we have shown that phosphorylation of p68 occurs at least partly on serine and threonine, we have not yet established any functional differences between the phosphorylated and unphosphorylated forms. It will be important to identify the phosphorylated residues in annexin VI, since although both serine and tyrosine residues are present in the N-terminal domain, they are not contained within sequences recognized as optimal for phosphorylation either by protein kinase C or protein-tyrosine kinases.

The controversy surrounding the various proposed functions of the annexins is largely due to the *in vitro* derivation of much data, with inevitable discrepancies between laboratories, arising from the use of purified proteins from different sources, reagents from different manufacturers and different technical approaches. Such problems emphasize the requirement for physiological systems in which to study these proteins, and in this context, the phosphorylation of annexins I and II by the epidermal growth factor receptor and pp60[v-src] tyrosine kinases respectively, remains one of the few important *in vivo* observations reported for individual members of the annexin family.

TISSUE DISTRIBUTION AND DEVELOPMENTAL REGULATION

The annexins are notable for their broad tissue distribution. Thus, all mammalian tissues examined have been found to express representatives of the family, although no single cell type apparently expresses the full complement of annexins. Recent reports based on immunological cross-reactivities, have also indicated the presence of annexins in several lower eukaryotes (Gerke 1989), and higher plants (Boustead et al. 1989). These reports have been substantiated by the molecular cloning of two novel annexins from *Drosophila melanogaster* (Johnston et al. 1990), neither of which share any preferential homology with any of the mammalian annexins, but which have the characteristic internally repetitive structure and share 50% homology with each other.

In an effort to aid assignment of function to members of the annexin family, several laboratories have examined patterns of tissue distribution in the hope that expression of individual members may correlate with a particular functional phenotype. Studies in this laboratory identified annexin VI in Ca^{2+}-dependent association with lymphocyte plasma-membrane (Davies et al. 1984; Owens and Crumpton 1984), and a similar protein was identified by other groups in the electric organ of *Torpedo* (Sudhof et al. 1984), and porcine liver (Shadle et al. 1985). However, it rapidly became clear that both annexin VI and other members of the family were widely expressed; e.g., in chromaffin cells, fibroblasts, muscle cells, placenta, lung, aorta etc (Creutz et al. 1987; Raeymaekers et al. 1985; Glenney et al. 1987; Martin et al. 1987). Annexins I, II, IV and VI have all been detected in secretory epithelial cells, although annexins IV and VI are generally restricted to breast, sweat gland, and salivary gland ductal epithelia (Fava et al. 1989; Silva et al. 1986; Clarke, Moss and Crumpton unpubl. observations). However, there does not appear to be any obvious functional characteristic shared by the annexin VI positive epithelia that distinguishes them from annexin VI negative intestinal and endometrial epithelia. Lozano et al. (1989) have recently demonstrated that annexins IV and VI are functionally down-regulated in murine mammary epithelia in culture, during progression from a nonsecretory to a secretory phenotype. These authors have indicated the potential usefulness of a manipulable cell system in which protein expression can be regulated by the induction of phenotypic changes. Since prolactin can switch on the secretory phenotype in mammary epithelia, it is possible that annexin I which is prolactin inducible (Horseman 1989), may act antagonistically to annexins IV and VI which are apparently down-regulated. Annexins are also abundant in endocrine cells, with annexin VI being highly expressed in all such cells with one notable exception, namely the Chief cells of the parathyroid. Such a correlation again suggests a functional link with secretion and it is interesting to note that the adrenal cortex, testicular Leydig cells, ovary, thyroid, islets of Langerhans and intestinal neuroendocrine cells which are all annexin VI positive, secrete in response to transient increases in the concentration of free intracellular Ca^{2+}, whereas the Chief cells of the parathyroid secrete in response to a drop in the level of free Ca^{2+} (Burgoyne 1990).

Evidence that certain of the annexins may be functionally and developmentally regulated, has added a particularly interesting and important aspect to such studies. Thus, expression of the tyrosine phosphorylation substrate annexin II (p36) is increased during avian limb bud development in skin, cartilage, connective tissue and blood vessel endothelia (Carter et al. 1986). Similarly, expression of the EGF receptor kinase substrate annexin I (p35) appears to be closely regulated during embryonic development of the floor plate of the rat central nervous system (McKanna and Cohen 1989). The two *Drosophila* annexins were found to be differentially expressed in development, with annexin X abundant in adult flies and weakly expressed in early embryos (Johnston et al. 1990). In contrast, annexin IX was expressed at all stages in development, although most abundantly in adult flies. In our investigations into the tissue distribution of annexin VI (p68), we discovered evidence for developmental regulation of it's expression during both T- and B-lymphocyte ontogeny (Gallagher, Clark, Moss and Crumpton unpubl. observations). Thus, in lymph nodes, annexin VI was largely restricted to the small lymphocytes of the cortex and paracortex, with the centroblasts, centrocytes and immunoblasts of the germinal centers appearing negative. In the thymus, immature cortical thymocytes did not express detectable amounts of annexin VI, whereas more mature medullary thymocytes appeared positive for annexin VI. Clearly, it is important to establish a correlation between annexin VI expression and the acquisition of a particular functional phenotype in B- and T-cell development. In this respect, there is some suggestion that in B-lymphocytes, expression of annexin VI coincides with the ability to secrete Ig. If substantiated, this is an important correlation, since it encourages the view that annexin VI function is related to secretion.

SUMMARY

In this chapter we have reviewed the central aspects of annexin biology, including structure, gene regulation, biochemical properties, developmental regulation and phosphorylation. It is clear that despite the wealth of anecdotal biochemical data, there is still a dearth of well defined functional data, and some of the most significant advances in the field continue to be the identification by molecular cloning of novel members of the annexin family. Whilst the primary sequences in themselves do not necessarily convey any directly useful functional data, the discoveries of new annexins, particularly those in *Drosophila* (and preliminary reports of plant annexins), have important implications for previously held functional hypotheses. Thus, general roles in anti-inflammatory responses and/or the inhibition of blood coagulation now seem especially unlikely, although one cannot preclude the possibility that certain members of the annexin family could act as PLA_2 inhibitors in other biochemical contexts, e.g., signal transduction. A second important consequence of the discovery of annexins in lower eukaryotes, is the opportunity for the use of genetic approaches to the question of function. With the use of molecular recombination, in conjunction with a steady advance towards 3D-crystallographic structural analyses and the understanding of annexin gene regulation, there is a degree of cautious optimism for the emergence of some conclusive functional data over the coming few years.

REFERENCES

1   Abdel-Ghany M, Kole HK, Saad MAE, Racker E (1989) Stimulation of phosphorylation of lipocortin at threonine residues by epidermal growth factor (EGF) and the EGF receptor. Proc Natl Acad Sci USA 86:6072-607

2   Ahn NG, Teller DC, Bienkowski MJ, McMullen BA, Lipkin EW, de Haen C (1988) Sedimentation equilibrium analysis of five lipocortin-related phospholipase A_2 inhibitors from human placenta. J Biol Chem 263:18657-18663

3   Amiguet P, D-Eustachio P, Kristensen T, Wetsel RA, Saris CJM, Hunter T, Chaplin DD, Tack BF (1990) Structure and chromosome assignment of the murine p36 calpactin I heavy chain gene. Biochemistry 29:1226-1232

4   Andree HAM, Reutelingsperger CPM, Hauptmann R, Hemker HC, Hermens WT, Willems GM (1990) Binding of vascular anticoagulant a (VAC a) to planar phospholipid bilayers. J Biol Chem 265:4923-4928

5   Ando Y, Imamura S, Owada MK, Kakunaga T, Kannagi R (1989a) Cross-linking of lipocortin I and enhancement of its Ca^{2+} sensitivity by tissue transglutaminase. Biochem Biophys Res Comm 163:944-951

6   Ando Y, Imamura S, Hong Y-M, Owada MK, Kakunaga T, Kannagi R (1989b) Enhancement of calcium sensitivity of lipocortin I in phospholipid binding induced by limited proteolysis and phosphorylation at the amino terminus as analyzed by phospholipid affinity column chromatography. J Biol Chem 264:6948-6955

7   Antonicelli F, Omri B, Breton MF, Rothut B, Russo-Marie F, Pavlovic-Hournac M, Haye B (1989) Identification of four lipocortin proteins and phosphorylation of lipocortin I by protein kinase C in cytosols of porcine thyroid cell cultures. FEBS Lett 258:346-350

8   Blackwell GJ, Carnuccio R, Di Rosa M, Flower RJ, Parente L, Persico P (1980) Macrocortin: a polypeptide causing the anti-phospholipase effect of glucocorticoids. Nature 287:147-149

9   Blackwood RA, Ernst JD (1990) Characterization of Ca^{2+}-dependent phospholipid binding, vesicle aggregation and membrane fusion by annexins. Biochem J 266:195-200

10   Blay J, Valentine-Braun KA, Northup JK, Hollenberg MD (1989) Epidermal-growth factor-stimulated phosphorylation of calpactin II in membrane vesicles shed from cultured A-431 cells. Biochem J 259:577-583

11   Boustead CM, Walker JH, Geisow MJ (1988) Isolation and characterization of two novel calcium-dependent phospholipid-binding proteins from bovine lung. FEBS Lett 233:233-238

12   Boustead CM, Smallwood M, Small H, Bowles DJ, Walker JH (1989) Identification of calcium-dependent phospholipid-binding proteins in higher plant cells. FEBS Lett 244:456-460

13   Bronnegard M, Andersson O, Edwall D, Lind J, Norsted G, Carlstedt-Duke J (1988) Human calpactin II (lipocortin I) messenger ribonucleic acid is not induced by glucocorticoids. Mol Endocrinol 2:732-739

14   Browning J, Pepinsky B, Wallner B, Flower RJ, Peers SH (1990) Too soon for consensus? Nature 346:324

15   Burgoyne RD (1987) Gelsolin and p36 share a similar domain. TIBS 12:85-86

16   Burgoyne RD (1990) Intracellular membrane fusion in cell-free systems. Trends Biochem Sci 15:123-124

17   Burns AL, Magendzo K, Shirvan A, Srivastava M, Rojas E, Alijani MR, Pollard HB (1989) Calcium channel activity of purified human synexin and structure of the human synexin gene. Proc Natl Acad Sci USA 86:3798-3802

18   Carter VC, Howlett AR, Martin GS, Bissel MJ (1986) The tyrosine phosphorylation substrate p36 is developmentally regulated in embryonic avian limb and is induced in cell culture. J Cell Biol 103:2017-2024

19   Chuah SY, Pallen CJ (1989) Calcium-dependent and phosphorylation-stimulated proteolysis of lipocortin I by an endogenous A431 cell membrane protease. J Biol Chem 264:21160-21166

20   Cheney RE, Willard MB (1989) Characterization of the interaction between calpactin I and fodrin (nonerythroid spectrin). J Biol Chem 264:18068-18075

21   Comera C, Rothut B, Cavadore J-C, Vilgrain I, Cochet C, Chambaz E, Russo-Marie F. (1989) Further
      characterization of four lipocortins from peripheral blood mononuclear cells.
      J Cell Biochem 40:361-370

22   Creutz CE, Zaks WJ, Hamman HC, Crane S, Martin WH, Gould KL, Oddie KM, Parsons SJ (1987)
      Identification of chromaffin granule-binding proteins. J Biol Chem 262:1860-1868

23   Crompton MR, Moss SE, Crumpton MJ (1988a) Diversity in the lipocortin/calpactin family. Cell 55:1-3

24   Crompton MR, Owens RJ, Totty NF, Moss SE, Waterfield MD, Crumpton MJ (1988b) Primary structure
      of the human, membrane-associated Ca^{2+}-binding protein p68: a novel member of a protein family.
      EMBO J 7:21-27

25   Crumpton MJ, Dedman JR (1990) Protein terminology tangle. Nature 345:212

26   Davidson FF, Dennis EA (1989) Biological relevance of lipocortins and related proteins as inhibitors of
      phospholipase A_2. Biochem Pharmacol 38:3645-3651

27   Davidson FF, Dennis EA, Powell M, Glenney JR (1987) Inhibition of phospholipase A_2 by "lipocortins"
      and calpactins. J Biol Chem 262:1698-1705

28   Davidson FF, Lister MD, Dennis EA (1990) Binding and inhibition studies on lipocortins using
      phosphatidylcholine vesicles and phospholipase A_2 from snake venom, pancreas, and a
      macrophage-like cell line. J Biol Chem 265:5602-5609

29   Davies AA, Wigglesworth NM, Allan D, Owens RJ, Crumpton MJ (1984) Nonidet P-40 extraction of
      lymphocyte plasma-membranes. Characterization of the insoluble residue. Biochem J 219:301-308

30   De BK, Misono KS, Lukas TJ, Mroczkowski B, Cohen S (1986) A calcium-dependent 35-kilodalton
      substrate for epidermal growth factor receptor/kinase isolated from normal tissue.
      J Biol Chem 261:13784-13792

31   Di Rosa M, Flower RJ, Hirata F, Parente L, Russo-Marie F (1984) Nomenclature announcement.
      Anti-phospholipase proteins. Prostaglandins 28:441-442

32   Edwards HC, Booth AG (1987) Calcium sensitive, lipid binding cytoskeletal proteins of the human
      placental microvillar region. J Cell Biol 105:303-311

33   Erikson E, Erikson RL (1980) Identification of a cellular protein substrate phosphorylated by the avian
      sarcoma virus-transforming gene product. Cell 21:829-836

34   Ernst JD, Hoye E, Blackwood RA, Jaye D (1990) Purification and characterization of an abundant
      cytosolic protein from human neutrophils that promotes Ca^{2+}-dependent aggregation of isolated
      specific granules. J Clin Invest 85:1065-1071

35   Fauvel J, Vicendo P, Roques V, Ragab-Thomas J, Granier C, Vilgrain I, Chambaz E, Rochat H, Chap H,
      Douste-Blazy L (1987) Isolation of two 67 kDa calcium-binding proteins from pig lung differing in
      affinity for phospholipids and in anti-phospholipase A_2 activity. FEBS Lett 221:397-402

36   Fava RA, Cohen S (1984) Isolation of a calcium-dependent 35-kilodalton substrate for the epidermal
      growth factor receptor/kinase from A431 cells. J Biol Chem 259:2636-2645

37   Fava RA, McKanna J, Cohen S (1989) Lipocortin I (p35) is abundant in a restricted number of
      differentiated cell types in adult organs. J Cell Physiol 141:284-293

38   Flower RJ (1988) Lipocortin and the mechanism of action of the glucocorticoids.
      Br J Pharmacol 94:987-1015

39   Funakoshi T, Heimark RL, Hendrickson LE, McMullen BA, Fujikawa K (1987) Human placental
      anticoagulant protein:isolation and characterization. Biochemistry 26:5572-5578

40   Funakoshi T, Abe M, Sakata M, Shoji S, Kubota Y (1990) The functional site of placental anticoagulant
      protein: essential histidine residue of placental anticoagulant protein.
      Biochem Biophys Res Comm 168:125-134

41   Geisow MJ (1986) Common domain structure of Ca^{2+} and lipid-binding proteins. FEBS Lett. 203:99-103

42   Geisow M, Childs J, Dash B, Harris A, Panayotou G, Sudhof T, Walker J (1984) Cellular distribution
      of three mammalian calcium-binding proteins related to *Torpedo* calelectrin. EMBO J 3:2969-2974

43   Genge BR, Wu LNY, Wuthier RE (1989) Identification of phospholipid-dependent calcium-binding
      proteins as constituents of matrix vesicles. J Biol Chem 264:10917-10921

44   Gerke V (1989) Consensus peptide antibodies reveal a widespread occurrence of Ca^{2+}/lipid-binding
      proteins of the annexin family. FEBS Lett 258:259-262

45   Gerke V, Weber K (1984) Identity of p36K phosphorylated upon *Rous sarcoma virus* transformation with
      a protein purified from brush borders. EMBO J 3:227-233

46   Giugni TD, James LC, Haigler HT (1985) Epidermal growth factor stimulates tyrosine phosphorylation of specific proteins in permeabilized human fibroblasts. J Biol Chem 260:15081-15090

47   Glenney JR (1985) Phosphorylation of p36 *in vitro* with pp60^{src}. FEBSLett 192:79-82

48   Glenney J (1986) Phospholipid-dependent Ca^{2+} binding by the 36-kDa tyrosine kinase substrate (calpactin) and its 33-kDa core. J Biol Chem 261:7247-7252

49   Glenney JR Jr, Tack BF (1985) Amino-terminal sequence of p36 and associated p10: identification of the site of tyrosine phosphorylation and homology with S-100. Proc Natl Acad Sci USA 82:7884-7888

50   Glenney JR, Tack B, Powell MA (1987) Calpactins: two distinct Ca^{++}-regulated phospholipid- and actin-binding proteins isolated from lung and placenta. J Cell Biol 104:503-511

51   Gould KL, Woodgett JR, Isacke CM, Hunter T (1986) The protein-tyrosine kinase substrate p36 is also a substrate for protein kinase C *in vitro* and *in vivo*. Mol Cell Biol 6:2738-2744

52   Grundmann U, Abel KJ, Dohn H, Lobermann H, Lottspeich F, Kupper H (1988) Characterization of cDNA encoding human placental anti-coagulant protein (PP4): homology with the lipocortin family. Proc Natl Acad Sci USA 85:3708-3712

53   Hagiwara M, Ochiai M, Owada K, Tanaka T, Hidaka H (1988) Modulation of tyrosine phosphorylation of p36 and other substrates by the S-100 protein. J Biol Chem 263:6438-6441

54   Haigler HT, Schlaepfer DD, Burgess WH (1987) Characterization of lipocortin I and an immunologically unrelated 33-kDa protein as epidermal growth factor receptor kinase substrates and phospholipase A$_2$ inhibitors. J Biol Chem 262:6921-6930

55   Hattori T, Hoffman T, Hirata F (1983) Differentiation of a histiocytic lymphoma cell-line by lipomodulin, a phospholipase inhibitory protein. Biochem Biophys Res Comm 111:551-558

56   Hauptmann R, Maurer-Fogy I, Krystek E, Bodo G, Andree H, Reutelingsperger CPM (1989) Vascular anticoagulant ß: a novel human Ca^{2+}/phospholipid binding protein that inhibits coagulation and phospholipase A$_2$ activity. Eur J Biochem 185:63-71

57   Hayashi H, Sonobe S, Owada MJ, Kakunaga T (1987a) Isolation and characterization of three forms of 36-kDa Ca^{2+}-dependent actin- and phospholipid-binding proteins from human placenta membrane. Biochem Biophys Res Comm 146:912-918

58   Hayashi H, Owada MK, Sonobe S, Kakunaga T, Kawakatsu H, Yano J. (1987b) A 32-kDa protein associated with phospholipase A$_2$-inhibitory activity from human placenta. FEBS Lett 223:267-272

59   Hayashi H, Owada MK, Sonobe S, Kakunaga T (1989) Characterization of two distinct Ca^{2+}-dependent phospholipid-binding proteins of 68-kDa isolated from human placenta. J Biol Chem 264:17222-17230

60   Hexham JM, Totty NF, Waterfield MD, Crumpton MJ (1986) Homology between the subunits of S-100 and a 10 kDa polypeptide associated with p36 of pig lymphocytes. Biochem Biophys Res Commun 134:248-254

61   Hirata F, Schiffmann E, Venkatasubramanian K, Salomon D, Axelrod JA (1980) Phospholipase A$_2$ inhibitory protein in rabbit neutrophils induced by glucocorticoids. Proc Natl Acad Sci USA 77:2533-2536

62   Horseman ND (1989) A prolactin-inducible gene product which is a member of the calpactin/lipocortin family. Mol Endocrinol 3:773-779

63   Huang K-S, Wallner BP, Mattaliano RJ, Tizard R, Burne C, Frey A, Hession C, McGray P, Sinclair LK, Chow EP, Browning JL, Ramachandran KL, Tang J, Smart JE, Pepinsky RB (1986) Two human 35-kDa inhibitors of phospholipase A$_2$ are related to substrates of pp60^{v-src} and of the epidermal growth factor receptor/kinase. Cell 46:191-199

64   Hunter T (1988) The Ca^{2+}/phospholipid-binding proteins of the submembranous skeleton. Adv Exp Med Biol 234:169-193

65   Ikebuchi NW, Waisman DM (1990) Calcium-dependent regulation of actin filament bundling by lipocortin-85. J Biol Chem 265:3392-3400

66   Isacke CM, Trowbridge IS, Hunter T (1986) Modulation of p36 phosphorylation in human cells:studies using anti-p36 monoclonal antibodies. Mol Cell Biol 6:2745-2751

67   Isacke CM, Lindberg RA, Hunter T (1989) Synthesis of p36 and p35 is increased when U-937 cells differentiate in culture but expression is not inducible by glucocorticoids. Mol Cell Biol 9:232-240

68 Iwasaki A, Suda M, Nakao H, Nagoya T, Arai K, Mizoguchi T, Sato F, Toshizaki H, Hirata M, Miyata T, Shidara Y, Murata M, Maki M (1987) Structure and expression of cDNA for inhibitor of blood coagulation isolated from human placenta: a new lipocortin-like protein. J Biochem 102:1261-1273

69 Johnsson N, Van PN, Soling H-D, Weber K (1986a) Functionally distinct serine phosphorylation sites of p36, the cellular substrate of retroviral protein kinase; differential inhibition of reassociation with p11. EMBO J 5:3455-3460

70 Johnsson N, Vanderkerckhove J, van Damme J, Weber K (1986b) Binding sites for calcium, lipid and p11 on p36, the substrate of retroviral tyrosine-specific protein kinases. FEBS Lett 198:361-364

71 Johnsson N, Marriott G, Weber K (1988) p36, the major cytoplasmic substrate of src tyrosine protein kinase, binds to its p11 regulatory subunit via a short amino-terminal amphiphatic helix. EMBO J 7:2435-2442

72 Johnston PA, Perin MS, Reynolds GA, Wasserman SA, Sudhof TC (1990) Two novel annexins from *Drosophila melanogaster*. Cloning, characterization and differential expression in development. J Biol Chem (in press)

73 Kaetzel MA, Hazarika P, Dedman JR (1989) Differential expression of three 35-kDa annexin calcium-dependent phospholipid binding proteins. J Biol Chem 264:14463-14470

74 Karasik A, Pepinsky RB, Kahn CR (1988a) Insulin and epidermal growth factor stimulate phosphorylation of a 170-kDa protein in intact hepatocytes immunologically related to lipocortin 1. J Biol Chem 263:18558-18562

75 Karasik A, Pepinsky RB, Shoelson SE, Kahn CR (1988b) Lipocortins 1 and 2 as substrates for the insulin receptor kinase in rat liver. J Biol Chem 263:11862-11867

76 Keutzer JC, Hirschhorn RR (1990) The growth-regulated gene 1B6 is identified as the heavy chain of calpactin I. Exp Cell Res 188:153-159

77 Khanna NC, Tokuda M, Waisman DM (1986) Phosphorylation of lipocortins *in vitro* by protein kinase C. Biochem Biophys Res Comm 141:547-554

78 Klee CB (1988) Ca^{2+}-dependent phospholipid- (and membrane-) binding proteins. Biochem 27:6645-6652

79 Kobayashi R, Tashima Y (1989) An immunological and biochemical comparison of 67 kDa calcimedin and 67 kDa calelectrin. Biochem J 262:993-996

80 Kobayashi R, Tashima Y (1990) Purification, biological properties and partial sequence analysis of 67-kDa calcimedin and its 34-kDa fragment from chicken gizzard. Eur J Biochem 188:447-453

81 Kobayashi R, Hidaka H, Tashima Y (1990a) Purification, characterization, and partial sequence analysis of 32-kDa calcimedin from chicken gizzard. Arch Biochem Biophys 277:203-210

82 Kobayashi R, Nakayama R, Ohta A, Sakai F, Sakuragi S, Tashima (1990b) Identification of the 32-kDa components of bovine lens EDTA-extractable protein as endonexins I and II. Biochem J 266:505-511

83 Lewit-Bentley A, Doublie S, Fourme R, Bodo G (1989) Crystallization and preliminary X-ray studies of human vascular anticoagulant protein. J Mol Biol 210:875-876

84 Lozano JJ, Silberstein GB, Hwang S-I, Haindl AH, Rocha V (1989) Developmental regulation of calcium-binding proteins (calelectrins and calpactin I) in mammary glands. J Cell Physiol 138:503-510

85 Machoczek K, Fischer M, Soling H-D (1989) Lipocortin I and lipocortin II inhibit phosphoinositide- and polyphosphoinositide-specific phospholipase C. FEBS Lett 251:207-212

86 Mani RS, Kay CM (1989) Purification and spectral studies on the Ca^{2+}-binding properties of 67 kDa calcimedin. Biochem J 259:799-804

87 Martin F, Derancourt J, Capony J-C, Colote S, Cavadore J-C (1987) Sequence homologies between p36, the substrate of pp60src tyrosine kinase and a 67 kDa protein isolated from bovine aorta. Biochem Biophys Res Comm 145:961-968

88 Martin F, Derancourt J, Capony J-P, Watrin A, Cavadore J-C (1988) A 36-kDa monomeric protein and its complex with a 10-kDa protein both isolated from bovine aorta are calpactin-like proteins that differ in their Ca^{2+}-dependent calmodulin-binding and actin-severing. Biochem J 251:777-785

89 Maurer-Fogy I, Reutelingsperger CPM, Pieters J, Bodo G, Stratowa C, Hauptmann R (1988) Cloning and expression of cDNA for human vascular anticoagulant, a Ca^{2+}-dependent phospholipid-binding protein. Eur J Biochem 174:585-592

90 McKanna JA, Cohen S (1989) The EGF receptor kinase substrate p35 in the floor plate of the embryonic rat CNS. Science 243:1477-1479

91   Miele L, Cordella-Miele E, Facchiano A, Mukherjee AB (1988) Novel anti-inflammatory peptides from the region of highest similarity between uteroglobin and lipocortin I. Nature 335:726-730

92   Moore PB (1986) 67 kDa calcimedin, a new Ca^{2+}-binding protein. Biochem J 238:49-54

93   Moore PB (1988) Calcimedin, calelectrin: correlation of relatedness. Biochem Biophys Res Comm 156:193-198

94   Morse SS, Moore PB (1988) 67K-calcimedin (67 kDa) is distinct from p67 calelectrin and lymphocyte 68 kDa Ca^{2+}-binding protein. Biochem J 251:171-174

95   Moss SE, Crompton MR, Crumpton MJ (1988) Molecular cloning of murine p68, a Ca^{2+}-binding protein of the lipocortin family. Eur J Biochem 177:21-27

96   Moss SE, Crumpton MJ (1990a) The lipocortins/calpactins and the EF-hand proteins: Ca^{2+}-binding sites and evolution. Trends Biochem Sci 15:11-12

97   Moss SE, Crumpton MJ (1990b) Alternative splicing gives rise to two forms of the p68 Ca^{2+}-binding protein. FEBS Lett 2:299-302

98   Nakata T, Sobue K, Hirokawa N (1990) Conformational change and localization of calpactin I complex involved in exocytosis as revealed by quick-freeze deep-etch electron microscopy and immunocytochemistry. J Cell Biol 110:13-25

99   Newman R, Tucker A, Ferguson C, Tsernoglou D, Leonard K, Crumpton MJ (1989) Crystallization of p68 on lipid monolayers and as three-dimensional single crystals. J Mol Biol 206:213-219

100 Osborn M, Johnsson N, Wehland J, Weber K (1988) The submembranous location of p11 and its interaction with the p36 substrate of pp60 src kinase in situ. Exp Cell Res 175:81-96

101 Owens RJ, Crumpton MJ (1984) Isolation and characterization of a novel 68000-M_r Ca^{2+}-binding protein of lymphocyte plasma membrane. Biochem J 219:309-316

102 Pepinsky RB, Tizard R, Mattaliano RJ, Sinclair LK, Miller GT, Browning JL, Chow EP, Burne C, Huang K-S, Pratt D, Wachter L, Hession C, Frey AZ, Wallner BP (1988) Five distinct calcium and phospholipid binding proteins share homology with lipocortin I. J Biol Chem 263:10799-10811

103 Pepinsky RB, Sinclair LK, Chow EP, O'Brine-Greco B (1989) A dimeric form of lipocortin-1 in human placenta. Biochem J 263:97-103

104 Philipps C, Rose-John S, Rincke G, Furstenberger G, Marks F (1989) cDNA-cloning, sequencing and expression in glucocorticoid-stimulated quiescent Swiss 3T3 fibroblasts of mouse lipocortin I. Biochem Biophys Res Comm 159:155-162

105 Pigault C, Follenius-Wund A, Lux B, Gerard D (1990) A fluorescence spectroscopy study of the calpactin I complex and its subunits p11 and p36. Biochem Biophys Acta 1037:106-114

106 Powell MA, Glenney JR (1987) Regulation of calpactin I phospholipid binding by light chain binding and phosphorylation by $p60^{v\text{-src}}$. Biochem J 247:321-328

107 Radke K, Gilmore T, Martin GS (1980) Transformation by Rous sarcoma virus: a cellular substrate for transformation-specific phosphorylation contains phosphotyrosine. Cell 21:821-828

108 Raeymaekers L, Waytack F, Casteels R (1985) Isolation of calelectrin-like proteins associated with smooth muscle membranes. Biochem Biophys Res Comm 132:526-532

109 Reutelingsperger CPM, Kop JMM, Hornstra G, Hemker HC (1988) Purification and characterization of a novel protein from bovine aorta that inhibits coagulation. Eur J Biochem 173:171-178

110 Rhoads AR, Lulla M, Moore PB, Jackson CE (1985) Characterization of calcium-dependent membrane binding proteins of brain cortex. Biochem J 229:587-593

111 Rojas E, Pollard HB (1987) Membrane capacity measurements suggest a calcium-dependent insertion of synexin into phosphatidyl-serine bilayers. FEBS Lett 217:25-31

112 Rothut B, Comera C, Prieur B, Errasfa M, Minassian G, Russo-Marie F (1987) Purification and characterization of a 32-kDa phospholipase A_2 inhibitory protein (lipocortin) from human peripheral blood mononuclear cells. FEBS Lett 219:169-175

113 Russo-Marie F, Paing M, Duval D (1979) Involvement of glucocorticoid receptors in steroid-induced inhibition of prostaglandin secretion. J Biol Chem 254:8498-8504

114 Saris CJM, Tack BF, Kristensen T, Glenney JR, Hunter T (1986) The cDNA sequence for the protein-tyrosine kinase substrate p36 (calpactin I heavy chain) reveals a multi-domain protein with internal repeats. Cell 46:201-212

115 Sawyer ST, Cohen, S (1985) Epidermal growth factor stimulates the phosphorylation of the calcium-dependent 35000-dalton substrate in intact A-431 cells. J Biol Chem 260:8233-8236

116 Schlaepfer DD, Haigler HT (1987) Characterization of Ca^{2+}-dependent phospholipid binding and phosphorylation of lipocortin I. J Biol Chem 262:6931-6937

117 Schlaepfer DD, Haigler HT (1988) *In vitro* protein kinase C phosphorylation sites of placental lipocortin. Biochemistry 27:4253-4258

118 Schlaepfer DD, Mehlman T, Burgess WH, Haigler HT (1987) Structural and functional characterization of endonexin II, a calcium and phospholipid binding protein. Proc Natl Acad Sci USA 84:6078-6082

119 Seaton BA, Head JF, Kaetzel MA, Dedman JR (1990) Purification, crystallization, and preliminary X-ray diffraction analysis of rat kidney annexin V, a calcium-dependent phospholipid-binding protein. J Biol Chem 265:4567-4569

120 Shadle PJ, Gerke V, Weber K (1985) Three Ca^{2+}-binding proteins from porcine liver and intestine differ immunologically and physiochemically and are distinct in Ca^{2+} affinities. J Biol Chem 260:16354-16360

121 Sheets EE, Giugni TD, Coates GC, Schlaepfer DD, Haigler HT (1987) Epidermal growth factor dependent phosphorylation of a 35-kilodalton protein in placental membranes. Biochemistry 26:1164-1172

122 Silva GF, Sherrill K, Spurgeon S, Sudhof TC, Stone DK (1986) High level expression of the 32.5-kilodalton calelectrin in ductal epithelia as revealed by immunocytochemistry. Differentiation 33:175-183

123 Sudhof TC, Ebbecke M, Walker JH, Fritsche U, Boustead C (1984) Isolation of mammalian calelectrins. Biochemistry 23:1103-1109

124 Sudhof TC, Slaughter CA, Leznicki I, Barjon P, Reynolds GA (1988) Human 67-kDa calelectrin contains a duplication of four repeats found in 35-kDa lipocortins. Proc Natl Acad Sci USA 85:664-668

125 Suwa Y, Kudo I, Imaizumi A, Okada M, Kamimura T, Suzuki Y, Chang HW, Hara S, Inoue K (1990) Proteinaceous inhibitors of phospholipase A_2 purified from inflammatory sites in rats. Proc Natl Acad Sci USA 87:2395-2399

126 Swan DG, Hale RS, Dhillon N, Leadlay PF (1987) A bacterial calcium-binding protein homologous to calmodulin. Nature 329:84-85

127 Tait JF, Sakata M, McMullen BA, Miao CH, Funakoshi T, Hendrickson LE, Fujikawa K (1988) Placental anticoagulant proteins: isolation and comparative characterization of four members of the lipocortin family. Biochemistry 27:6268-6276

128 Taylor WR, Geisow MJ (1987) Predicted structure for the calcium-dependent membrane-binding proteins p35, p36, and p32. Protein Eng 1:183-187

129 Tufty RM, Kretsinger RH (1975) Troponin and parvalbumin calcium binding regions predicted in myosin light chain and T4 lysozyme. Science 187:167-169

130 van Binsbergen J, Slotboom AJ, Aarsman AJ, de Haas GH (1989) Synthetic peptide from lipocortin I has no phospholipase A_2 inhibitory activity. FEBS Lett 247:293-287

131 Varticovski L, Chahwala SB, Whitman M, Cantley L, Schindler D, Chow EP, Sinclair LK, Pepinsky RB (1988) Location of sites in human lipocortin I that are phosphorylated by protein tyrosine kinases and protein kinases A and C. Biochemistry 27:3682-3690

132 Wallner BP, Mattaliano RJ, Hession C, Cate RL, Tizard R, Sinclair LK, Foeller C, Chow EP, Browning JL, Ramachandran KL, Pepinsky RB (1986) Cloning and expression of human lipocortin, a phospholipase A_2 inhibitor with potential anti-inflammatory activity. Nature 320:77-80

133 Weber K, Johnsson N, Plessmann U, Van PN, Soling H-D, Ampe C, Vanderkerckhove J (1987) The amino acid sequence of protein II and its phosphorylation site for protein kinase C; the domain structure Ca^{2+}-modulated lipid binding proteins. EMBO J 6:1599-1604

134 Yoshizaki H, Arai K, Mizoguchi T, Shiratsuchi M, Hattori Y, Nagoya T, Shidara Y, Maki M (1989) Isolation and characterization of an anticoagulant protein from human placenta. J Biochem 105:178-183

135 Yoshizaki H, Mizoguchi T, Arai K, Shiratsuchi M, Shidara Y, Maki M (1990) Structure and properties of calphobindin II an anti-coagulant protein from human placenta. J Biochem 107:43-50

136 Zhuang Q, Stracher A (1989) Purification and characterization of a calcium binding protein with "synexin-like" activity from human blood platelets. Biochem Biophys Res Comm 159:236-241

137 Zokas L, Glenney JR (1987) The calpactin light chain is tightly linked to the cytoskeletal form of calpactin I. J Cell Biol 105:2111-2121

ANNEXIN I AND ITS BIOCHEMICAL PROPERTIES

Eisuke F. Sato, Yoshikazu Tanaka, Keisuke Edashige, Junzo Sasaki,
Masayasu Inoue, and Kozo Utsumi

INTRODUCTION

Signal transduction triggered by ligand-receptor interaction often induces interaction of transmembrane proteins with the cytoskeletal net work at the inner surface of cell membranes. It is well known that many cytoplasmic proteins bind to phospholipid liposomes or the inner surface of plasma membranes and the interaction induces physicochemical changes in the membranes (Utsumi et al. 1981, 1982; Okimasu et al. 1982, 1986, 1987; Noboriet al. 1987). In some cases, phosphoinositides in plasma membranes are hydrolyzed and diacylglycerol (DG) is transiently produced during ligand receptor interaction (Berridge 1984). These DG and IP_3 activate proteinkinase C (PKC) (Nishizuka 1984) and increase intracellular calcium levels by releasing calcium from intracellular stores (Berridge and Irvine 1984). In this case, it has been demonstrated that DG increases the affinity of PKC for Ca^{2+} (Takai et al. 1979), and enhances the binding of the enzyme to phospholipids of plasma membranes, such as phosphatidylserine (PS), phosphatidic acid (PA), and cardiolipin (CL), by way of exposing its hydrophobic amino acid residues. No such interaction occurs with liposomes composed of uncharged phospholipids, such as phosphatidylcholine (PC) (Konig et al. 1985). Thus, the physicochemical nature of the constituent phospholipids is important for specific interaction of PKC with membranes leading to its activation: different physical forms of synthetic phospholipid vesicles or multilamellar vesicles are differentially capable of complying with the phospholipid requirement of PKC (Boni and Rando 1985). Besides PKC, other proteins such as protease (Inoue et al. 1977), actinin (Rotman et al. 1982), thrombin (Le Compte et al. 1984) and lipocortins (Sato et al. 1987) are known to be activated by interacting with phospholipids. These observations indicate that the interaction of these cytoplasmic proteins with the inner surface of plasma membranes plays a critical role in signal transduction that modulates cellular metabolism.

Based on such a concept, we systematically looked for some cytosolic proteins suitable for studying the molecular mechanism for signal transduction, and found 39 kDa protein in neutrophil cytoplasma (Utsumi et al. 1986). The protein was isolated (Sato et al. 1987), purified (Sato et al. 1988b), and cDNA sequenced (Sato et al. 1989a). In the presence of appropriate phospholipid and calcium ions, this protein reversibly binds to phospholipid membrane bilayers (Sato et al. 1987; Utsumi and Sato 1990). Physicochemical analysis revealed that this protein belongs to a new family of annexins or lipocortins (Geisow et al. 1986) including calpactins (Powell and Glenney 1987), lipocortins, protein I and II (Gerke and Weber 1984), calelectrins (Sudhof et al. 1984), calcimedins (Moore and Dedman 1987), chromobindins (Creutz et al. 1987), endonexin (Schlaepfer et al. 1987), a human lymphocyte p68 (Crompton et al. 1988), placental anticoagulant protein (Funakoshi et al. 1987), an inhibitor of blood coagulation (Iwasaki et al. 1987), and some PLA_2 inhibitory proteins (Rothhut et al. 1987). Annexins have homologous amino acid sequences (Pepinsky et al. 1988). and serve as substrates for PKC (Varticovski et al. 1988) and tyrosine kinases such as pp60^{src} and EGF receptor/kinase (De et al. 1986). They have several biochemical characteristics, such as calcium-dependent binding with acidic phospholipid, F-actin and spectrin (Hayashi et al. 1987; Sato et al. 1988a), and PLA_2 inhibitory activity (Davidson et al. 1987). Though these properties could be regulated by intracellular calcium ion concentration and their phosphorylation, their physiological role is still unknown.

Concerning the annexins of blood cells, six proteins have been identified, such as lipocortin I-VI. Among these proteins, four were studied extensively (Comera et al. 1989): lipocortin I (identical to calpactin II) is localized in polymorphonuclear leukocytes (PMN), monocytes, and epithelial cells such as spleen and lung cells; this protein can be phosphorylated by EGF receptor/kinase (Huang et al. 1986). Lipocortin III (identical to 35K protein) is localized specifically to PMN and can be phosphorylated by PKC but not by EGF receptor/kinase. Lipocortin V (identical to the coagulation inhibitor, IBC-PAP, endonexin II, p68 and 32K) is localized in lymphocytes and monocytes; this protein does not serve as a substrate for EGF receptor/kinase (Pepinsky et al. 1988). Calelectrin (identical to 73K and

35K which is 72/2 K) is localized specifically in lymphocytes, and is phosphorylated by PKC but not by EGF receptor/kinase (Comera et al. 1989).

The 39K isolated from guinea pig neutrophils is different from other annexins in their distribution and biochemical properties. The protein is distributed mainly in neutrophils. Although the amino acid sequence of 39K is quite similar to that of lipocortin I, it is localized mainly in neutrophils and serves as a substrate for PKC and EGF receptor/kinase (Utsumi and Sato 1990; Sato et al. 1988a,b). It also binds to F-actin and inhibits PLA_2 (Sato et al. 1989b). To understand the physiological significance of 39K protein in neutrophils, we constructed a plasmid for overexpressing recombinant 39K protein.

Recombinant lipocortin I has been overexpressed in *E. coli* (Wallner et al. 1986), and confirmed to inhibit the catalytic activity of PLA_2. It was also reported that the recombinant protein inhibited the production of eicosanoid synthesis such as prostacycline and thromboxane A_2 in isolated perfused lungs from the guinea pig (Cirino et al. 1987), and suppressed carrageenin-induced paw edema of the rat (Cirino et al. 1989). Since this recombinant protein is overexpressed in *E. coli*, the α-amino group of its N-terminal residue is not blocked by the acyl group. Thus, the mode of processing by glycosylation and its biological properties may not be identical with those of native lipocortin I. Therefore, we constructed a high expression vector for the recombinant 39K protein (r39K) in eukaryotic yeast, *Saccharomyces cerevisiae*, expressed it on a large scale, and analyzed the biochemical properties of r39K (Sato et al. 1989b).

CONSTRUCTION OF EXPRESSION PLASMID AND ITS OVEREXPRESSION IN YEAST

Oligonucleotides corresponding to partial amino acid sequence of guinea pig neutrophil 39K protein were chemically synthesized based on the determined nucleotide sequence corresponding to human annexin I (Wallner et al. 1986).

Using ^{32}P-labeled probes of these oligonucleotides, recombinant cDNA clones from the libraries were screened and positive plaques, such as λgpL7 and λgpL8 clones, were obtained. Each clone contained the 3'- terminal and 5'-terminal regions of the cDNA for 39K (Figure 1A).

We also used λgpL9 cDNA clone which contained the 5'-terminal region. These cDNA clones were subcloned into the EcoRI site of pUC9 and designated pGPL7, pGPL8 and pGPL9, respectively. Each of the 0.8 Kb BglII-SalI fragment from pGPL7, 0.4 kb EcoRI-HhaI fragment from pGPL8 and the 0.2 kb HhaI-BglII fragment from pGPL9 were ligated with the 8 kb EcoRI-SalI fragment of pYHCC101, a yeast expression vector, just after glyceraldehyde-3-phosphate dehydrogenase (GAP) promoter, as described by Tanaka et al. (1988), and the resultant plasmid was designated as pYGPL100 (Figure 1B).

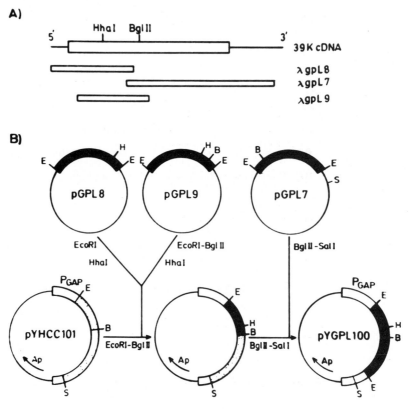

Fig. 1A,B. Construction of a vector for expression 39k cDNA in yeast A Restriction map of 39K cDNA's overlapping cDNA clones. The coding sequence is indicated by *thick box*. B pYHCC101 was described by Tanaka et al. (1988). The *solid boxes* denote 39K cDNA. P_{GAP} promoter of glyceraldehyde-3-phosphate dehydrogenase; *AP* ampicillin resistance; *E* EcoRI; *B* BglII; *H* HhaI; *S* SalI

The strain of *Saccharomyces cerevisiae* used in this experiment was HE-13-15 (MAT, α, trp 1). Cells were transformed with yeast expression vector containing pYGPL100 by the method of Ito et al. (1983) and grown at 30 °C for 40 h in Burkholder's medium.

About 40 g of cells were obtained and disrupted in 50 ml of 50 mM Tris-HCl (pH 7.5) containing 5 mM EGTA, 1 mM PMSF, and 0.01% leupeptin, using bead beaders (Biospeck products) at 4 °C (Figure 2,2). The resultant lysate was centrifuged at 5000 g for 30 min at 4 °C. The supernatant fraction (Figure 2,3) was further incubated with 6 mM $CaCl_2$ at 4 °C for 30 min. Upon centrifugation at 100000 g for 30 min at 4 °C, the Ca^{2+}-dependent phospholipid binding proteins were precipitated (Figure 2,4) after removing the supernatant fraction, and the pellet was suspended in 20 mM Hepes-KOH buffer (pH 7.4) containing 0.1 M KCl, 25 mM EGTA, 1 mM PMSF, 0.01% leupeptin and incubated for 60 min at 4 °C.

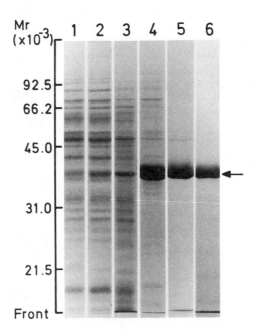

Fig. 2. SDS-PAGE analysis of recombinant 39K at different steps of purification. Samples were analyzed by 12% SDS-PAGE and stained by Coomassie brilliant blue. Lanes: *1* total, lysate of yeast (control); *2* total lysate of yeast containing pYGPL100; *3* post-nuclear fraction; *4* a protein fraction which bound to membranes Ca^{2+}-dependently; *5* Mono P fraction; *6* TSK gel G-3000SW fraction

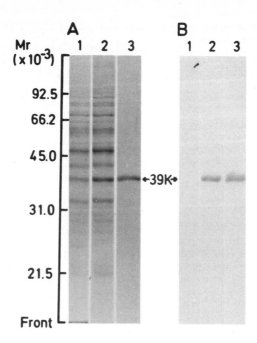

Fig. 3A,B. SDS-PAGE and Western blotting of total yeast lysate and purified recombinant 39K. **A** Coomassie brilliant blue stain. **B** electroblotted onto Durapore filter and immunostained with mouse antiserum against native 39K. Lanes: *1* total yeast lysate (control); *2* total yeast lysate containing pYGPL100; *3* purified r39K

Solid ammonium sulfate was added to the supernatant to give a final saturation level of 100%. The mixture was allowed to stand overnight at 4 °C and centrifuged again. The pellet was redissolved in a minimum volume of 25 mM triethanolamine-iminodiacetic acid (pH 8.3) (buffer A) and dialyzed overnight again to 1-1 volumes of buffer A. After dialysis, insoluble material was precipitated by centrifugation at 100000 g for 30 min. The supernatant was then loaded onto a Mono P column (200 x 5 mm) pre-equilibrated with buffer A. Proteins were eluted with elution buffer B (10% polybuffer-iminodiacetic acid, pH 5.0), at a flow rate of 1.0 ml/min and determined by 12% SDS-PAGE. The 39K-rich fractions (Figure 2,5) were loaded onto a column (300 x 5 mm) of TSK G3000SW pre-equilibrated with 0.1 M KCl, 20 mM Hepes-KOH buffer (pH 7.4), and proteins were eluted with the same buffer at a flow rate of 0.5 ml/min (Figure 2,6).

The r39K was highly enriched by these procedures; the purified protein was electrophoresed as a single band in SDS-PAGE with a molecular weight of 35000.

Table 1. Amino acid composition (in mol%) of native 39k, recombinant 39K (r39K) and that calculated from amino acid sequence analysis

	Predicted from nucleotide sequence	Analyzed[a] (39K)	Analyzed[a] (r39K)
Asx	40	38.9	40.3
Thr	20	19.9	19.7
Ser	22	23.4	22.3
Glx	36	34.7	35.6
Pro	6	3.5	7.3
Gly	19	19.2	19.5
Ala	34	35.7	31.9
Val	19	19.2	19.5
Met	8	6.2	5.2
Ile	22	21.0	19.7
Leu	37	34.5	35.5
Tyr	13	5.9	15.5
Phe	8	4.1	8.1
Lys	33	31.4	32.6
His	6	4.0	5.9
Arg	18	16.8	17.8
Cys	5	n.d.	n.d.
Trp	0	n.d.	n.d.
Total	346		

[a]Average values obtained from 24, 48, 72h hydrolysis with 5.7 M HCl.

BIOCHEMICAL PROPERTIES OF r39K
NATURE OF PROTEIN AND AMINO ACID COMPOSITION

Amino acid analysis of the purified r39K is shown in Table 1 together with the amino acid composition of native 39K. The composition of r39K was more closely similar with that of 39K calculated from the amino acid sequence analysis than that of purified native 39K (Table 1). Immunological identity of native 39K and r39K was also confirmed by immunoblot analysis with the antibody against native 39K (Figure 3). Both of the N-terminal amino acids of the native 39K and r39K were blocked. Wallner et al. (1986) suggested that human lipocortin I contains a single potential glycosylation site, [43]N-P-S, corresponding to the nucleotide sequence at base pairs 121-129. However, in 39K the [43]N(Asn) residue at the

574

site for glycosylation is replaced by D(Asp) (Figure 4), suggesting glycosylation may not occur in the protein.

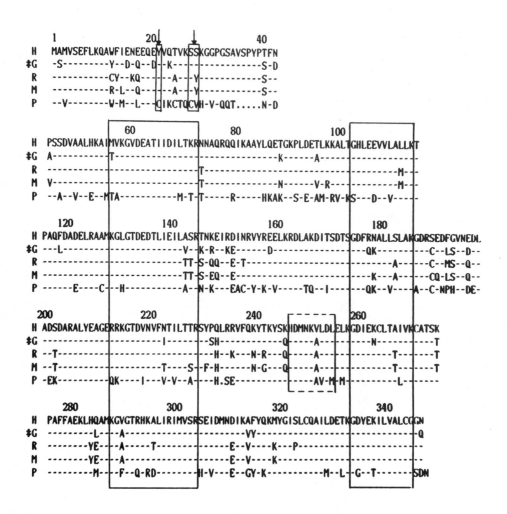

Fig. 4. Sequence alignments for 39K, rat lipocortin I, human lipocortin I, and pigeon lung lipocortin I. Amino acid residues of the proteins which are not identical with human lipocortin I, are described in the figure. *Boxed regions* correspond to the two consensus sequences of the lipocortin I(annexin I). The (---) box indicates the PLA$_2$ inhibitory site. *H* human lipocortin I (Wallner et al. 1986); *G* guinea pig 39K (Sato et al. 1989a); *R* rat lipocortin I (Tamaki et al. 1987); *M* mouse lipocortin I (Sakata et al. 1988); *P* pigeon lipocortin I (Horseman 1980)

INHIBITION OF PHOSPHOLIPASE A$_2$

As described above, both native and recombinant 39K possess a potent inhibitory action on PLA$_2$; 1 μg of r39K inhibited 50 ng of porcine PLA$_2$ by about 50% as measured by the release of (^{3}H) oleic acid from *E. coli* membranes.

When the enzyme was preincubated with r39K, the inhibitory action was slightly low (Figure 5). This result suggested that like native 39K, r39K seems to interact with substrate membranes. A similar mechanism for the inhibitory action was proposed by Davidson et al. (1987); annexin might inhibit PLA$_2$ activity by masking the surface of substrate membranes. This concept is also supported by sedimentation equilibrium analysis (Ahn et al. 1988).

Similar inhibitory action of annexin on phospholipase C was also observed. Machoczek et al. (1989) showed the inhibitory effect of annexin on phosphatidyl-inositide-specific phospholipase C.

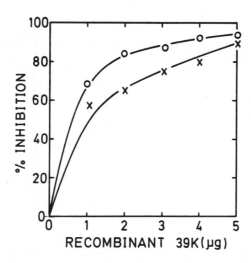

Fig.5. Inhibition of PLA$_2$ by r39K. Aliquots of the purified r39K were preincubated with 50 ng porcine pancreatic PLA$_2$ (o) or ^{3}H-labeled *E. coli* membrane (x) at 4 °C for 10 min. The reaction was initiated by adding ^{3}H-labeled *E. coli* membranes (o) or 50 ng pancreatic PLA$_2$ (x) at 4 °C and terminated 5 min later by adding 100 μl of 2 N HCl and 100 mg/ml of bovine serum albumin. After centrifugation to remove *E. coli* membrane, the radio activity in the supernatant fraction was determined in a liquid scintillation spectrometer

In addition, ultracentrifugation studies revealed the direct interaction of annexin with posphatidylinositol, and indicated that the inhibitory action of annexin is predominantly due to interaction with the substrate lipids (Sato et al. 1988). In this context, other mechanisms for PLA$_2$ inhibition have also been proposed, such as depletion of substrate of PLA$_2$ by annexin (Davidson et al. 1987).

BINDING TO F-ACTIN

Annexin is a basic protein (Cooper and Hunter 1982), and binds to F-actin, spectrin, and fodrin in the presence of high concentrations of Ca^{2+} (10^{-4}-10^{-3}M). However, the Ca^{2+} sensitivity of annexins to Ca^{2+} markedly increased by binding to phospholipids; the change in intracellular Ca^{2+} levels might possibly affect the organization of cytoskeletal systems by way of modulating the affinity of annexins to these proteins. In fact, annexin has been shown to translocate from cytoplasm to plasma membranes when intracellular Ca^{2+} is increased (Sato et al. 1987, 1988b). Presumably due to such properties of annexin, ligand-induced changes in intracellular Ca^{2+} levels might affect the movement and phagocytotic activity of neutrophils.

The native and r39K also bind to F-actin by an EGTA-inhibitable mechanism as is the case in other annexin proteins (Hayashi et al. 1987). However, its physiological role in cells is not yet well understood.

PHOSPHORYLATION OF r39K BY EGF RECEPTOR/KINASE AND PROTEIN KINASE C

Lipocortin I and II serve as substrates for EGF-and insulin-receptor-kinase, and for pp60^{src}, respectively. Because of such properties, they have been postulated to play critical roles in the signal transduction required for triggering mitosis.

Both native and r39K are also phosphorylated by EGF receptor/kinase and by protein kinase C (Utsumi and Sato 1990). It is known that the phosphorylation of lipocortin I by tyrosine phosphokinase occurs at the position of 17-21 (-[17]E-E-Q-E-Y-). In the case of 39K, this sequence is replaced by -[17]Q-E-Q-D-Y-. The 39K also contains two other domains that can be phosphorylated by threonine phosphokinase (-[212]R-R-K-G-T- at the position of 212-216) and serine phosphokinase at serine at the position of 27 (Wallner et al. 1986). The sequences around these phosphorylation sites remains unchanged in 39K.

PHOSPHORYLATION BY *EGF* RECEPTOR/KINASE

To elucidate the mechanism for phosphorylation of 39K by EGF receptor/kinase, the protein was incubated with cell membranes of A431 under various conditions. Plasma membranes of the cultured A431 cells were prepared by the method of De et al. (1986). The phosphorylation of r39K by EGF receptor/kinase was analyzed by measuring the incorporation of (γ-^{32}P)ATP into the protein at 0 °C in the presence of EGF (2 μg/ml).

Figure 6 shows the phosphorylation of r39K protein by EGF receptor/kinase. Kinetic analysis revealed that, in the presence of Ca^{2+}, the r39K protein first bound to the membrane fraction of A431 cells and then phosphorylation by the enzyme occurred. In contrast, neither phosphorylation nor binding of 39K to the membranes occurred in the absence of Ca^{2+}. The activity of r39K to bind to plasma membranes remained unchanged during the over-expression processes. The requirements of Ca^{2+} and binding to membrane surface for the activation of the enzyme were also observed by Fava et al. (1984) and Cohen and Fava (1985). These observations support the concept that the binding of these proteins to membrane/lipid bilayers is prerequired for the phosphorylation by EGF receptor/kinase. Autoradiographic analysis by SDS-PAGE also revealed two bands for phosphorylated r39K. The amount of the phosphorylated protein on the retarded position increased with incubation time. Thus, physicochemical properties and the conformation of r39K might significantly be changed by phosphorylation of its tyrosyl residues.

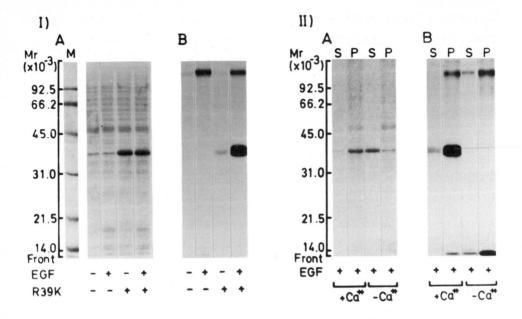

Fig. 6A,B. EGF-dependent phosphorylation of 39K. **I** A431 cells were grown at 37 °C in Dulbecco's modified Eagle's medium supplemented with 10% fetal calf serum. In vitro phosphorylation reaction using isolated A431 cell membranes was analyzed by SDS-PAGE (A) and the radioactive products were visualized by autoradiography (B). *EGF* epidermal growth factor; *R39K* recombinant 39K (200 μg/ml); *M* molecular marker. **II** After phosphorylation by EGF receptor/kinase with or without Ca^{2+}, the recombinant samples were centrifuged at 10000 g for 10 min at 4 °C. Each of the supernatant (S) or precipitant (P) fractions was analyzed by 12% SDS-PAGE (A) and the radioactive products were visualized by autoradiography (B). +Ca^{2+}, 0.5 mM $CaCl_2$; -Ca^{2+}, 1 mM EGTA

PHOSPHORYLATION BY PROTEIN KINASE C

The mode of phosphorylation of r39K by protein kinase C was also studied in vitro. The phosphorylation of r39K by PKC was analyzed by measuring the ^{32}P of $(\gamma\text{-}^{32}P)ATP$ in a medium containing dipalmitoyl-phosphatidylcholine/cholesterol/phosphatidylserine liposomes (2/1/1 in molar ratio), 100 nM PMA and $(\gamma\text{-}^{32}P)ATP$, and PKC in the presence or absence of Ca^{2+} for 3 min at 30 °C by the method of Boni and Rondo (1985). Free and membrane bound proteins were separated by Ficoll density gradient centrifugation (flotation method)

(Utsumi et al. 1982). Protein kinase C was partially purified from the brains of male Sprague Dawley rats by the method of Kikkawa et al. (1982).

Figure 7 shows the binding of phosphorylated r39K to phospholipid membrane: binding of r39K required Ca^{2+}. Binding of r39K to lipid membrane surface is essential for phosphorylation as is the case of EGF receptor/kinase. Although phosphorylation of 39K was greatly enhanced by millimolar levels of Ca^{2+} as was the case of calpactin (Glenney 1985), its sensitivity to Ca^{2+} could be decreased from mM to μM levels by the presence of phospholipids in a manner similar to that of the activation of protein kinase C.

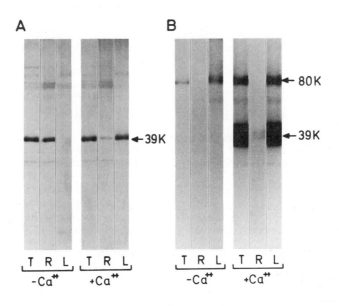

Fig. 7A,B. Autoradiograms of the phosphorylated r39K by PKC. r39K was incubated in the medium containing dipalmyitoylphosphatidylcholine /cholesterol/ phosphatidylserine liposomes (2:1:1 in molar ratio), 100 nM PMA and (γ-^{32}P) ATP in the presence or absence of Ca^{2+} for 30 min at 30 °C. Free and membrane bound proteins were separated by Ficoll density gradient centrifugation (flotation method) (Utsumi et al. 1981). Proteins were analyzed by 12% SDS-PAGE and silver staining (**A**). Phosphorylated proteins were analyzed by autoradiography (**B**). *T* total protein fraction; *R* unbound protein fraction; *L* membrane-bound protein fraction. + Ca^{2+}, 0.3 mM Ca^{2+}; - Ca^{2+}, 1 mM EGTA

In this connection, it has been reported that the phosphorylation decreases the affinity of these proteins for Ca^{2+} and their ability to bind to membranes, and neutralizes the activity to inhibit phospholipase A_2 (Haigler et al. 1987; Powel and Glenney 1987; Hirata 1981; Blanquet et al. 1988).

However, the activity of r39K to bind onto membrane surface remained unchanged after phosphorylation by protein kinase C or by EGF receptor/ kinase. These results suggest that membrane binding activity of 39K might not correlate with its inhibitory action against PLA_2.

OLIGOPEPTIDE DOMAIN OF 39K THAT INHIBITS *PLA₂*

Miele et al. (1988) have reported that synthetic oligopeptides corresponding to a highly homologous region of uteroglobin and lipocortin I had a potent activity to inhibit PLA_2 in vitro and revealed striking anti-inflammatory effects in vivo. Like lipocortin I, the third repetition of 39K (residues 246-254) has the highest similarity with those of other proteins (Figure 4). The two anti-inflammatory oligopeptides, H-D-M-N-K-V-L-D-L and H-D-M-N-K-A-L-D-L, corresponding to the homologous regions of lipocortin I and 39K respectively, were synthesized and tested for their ability to inhibit pancreatic PLA_2 by the methods reported by Clarke et al. (1986) and Davidson et al. (1987). Clark's method uses 1-stearoyl,2-(1-^{14}C)arachidonyl phosphatidylcholine as a substrate for PLA_2, while Davidson's method uses (^{3}H)oleic acid-labeled *E.coli* membrane as a substrate. However, all of these peptides failed to inhibit PLA_2 activity as measured by the two assayed methods. A similar result with our experiment was also reported by Binsbergen et al. (1989).

DISTRIBUTION OF ANTI-r39K ANTIBODY REACTIVE PROTEIN

Native 39K distributed mainly in cytoplasm of guinea pig neutrophils (Sato et al. 1987). To test the distribution of anti-r39K antibody reactive protein in the cells, similar immunofluorescence assays were performed with guinea pig bone marrow cells and other tissues. As shown in Figure 8, the anti-r39K antibody reactive protein was detected only in cytoplasma of pre- and matured neutrophils as observed in the cells which reacted with anti-39K antibody (Sato et al. 1987). The localization of anti-r39K antibody reactive protein was also studied in various tissues by immunoblotting method. The homogenate supernatant of various tissues obtained from guinea pigs was applied to SDS-PAGE and used for immunoblotting with anti-r39K antibody. The antibody reacted strongly with the 39K protein of lung and spleen as observed in native anti-39K antibody. However, the reactivity of anti-r39K antibody was stronger than that of native anti-39K antibody. Thus specificity was slightly reduced and a weak reaction was observed in the intraperitoneal macrophages.

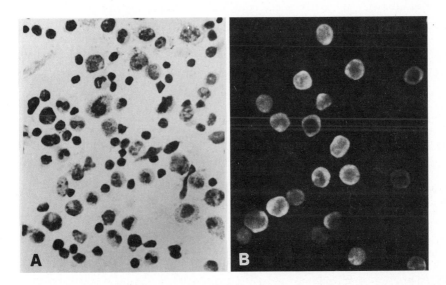

Fig.8A,B. Indirect immunofluorescent staining of guinea pig bone marrow cells by polyclonal anti-r39K antibody. Bone marrow cells of guinea pigs were stained with Giemsa solution or polyclonal antibody to the r39K. **A** Giemsa staining. **B** Distribution of the anti-r39K reacting protein visualized by FITC labeling

INHIBITION OF CARRAGEENIN-INDUCED PAW EDEMA BY r39K AND RELATED PEPTIDES

Miele et al. (1988) reported that the oligopeptides from human lipocortin I revealed a potent inhibitory effect on carrageenin-induced paw edema of the rat. To test whether r39K also has anti-inflammatory activity, we tested the effect of r39K, and its peptide corresponded to anti-inflammatory domains on carrageenin-induced paw edema. Briefly, 1 ml of saline solution containing either r39K or its oligopeptide was injected intraperitoneally 30 min before carrageenin treatment. Under pentobarbital anesthesia, animals were subcutaneously injected with 10 mg carrageenin in 0.2 ml saline solution in the right paw. For the control experiments, the same volume of saline solution was also injected into the left paw, and time-dependent changes in paw volume were measured by a plethysmohmete (Union Model TK-101). As shown in Figure 9, carrageenin markedly increased the paw volume. However, the increase in paw volume was significantly inhibited by r39K. Since the oligopeptide from human lipocortin I also possesses inhibitory activity on carrageenin-induced paw edema (Miele et al. 1988), the effect of the homologous oligopeptide from r39K was tested. As expected, the oligopeptide for 39K also showed the inhibitory effect on paw edema; the inhibitory action of the 39K peptide was slightly weaker than that of the human lipocortin I peptide. Miele et al. (1988) also reported that the core tetrapeptide K-V-L-D is the common structure found in all these active peptides.

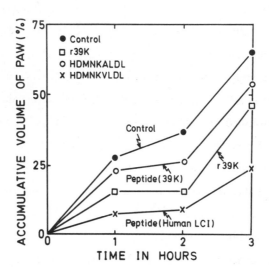

Fig. 9. Effect of r39K, oligopeptides corresponding to the PLA_2 inhibitory region of lipocortin I and 39K on carrageenin-induced inflammation. Under light ether anesthesia, animals were subcutaneously injected with 0.2 ml saline solution in the left paw. The same volume of carrageenin solution (10 mg/0.2 ml) was administrated in the right paw, and the change in paw volume was measured at the indicated times. The control group was treated with the intraperitoneal injection of 1 ml saline solution, and the experimental group was administered the same volume of saline containing oligopeptide or r39K 30 min prior to the experiment (Winter et al. 1963). The amount of added inhibitors was 0.5 nmol equivalent of r39K. Each *point* is a mean + SD of five experiments

In case of r39K, ^{251}V(Val) is replaced by ^{251}A(Ala). The reason why the 39K peptide revealed a weaker inhibitory activity against paw edema than did the human lipocortin I peptide may possibly be accounted for by such a minor difference in peptide structure. At present, the mechanism by which r39K inhibits the carrageenin-induced paw edema is not clear. Apart from the mechanism of its inhibitory action, the pharmacological effects of r39K and its active peptides permit studies on the development of novel anti-inflammatory agents for therapeutic use.

PERSPECTIVE

The present article describes the importance of the interaction of annexin in cytoplasmic proteins with the inner surface of plasma membranes for transmembrane signaling and regulation of cellular metabolism. Since neutrophils show dynamic responses to various stimuli, we studied some cytoplasmic proteins whose physicochemical properties were modulated by membrane stimulation, and found a 39K protein that reversibly bound onto the lipid membrane surface in a Ca^{2+}-dependent manner. To obtain molecular insight into 39K protein, we have established the overexpression system in yeast. In contrast to the recombinant lipocortin I previously obtained from *E.coli*, the r39K protein from yeast has properties identical with those of naturally occurring 39K protein; N-terminal amino acid is blocked. Comparative studies revealed that physicochemical and biochemical properties of the naturally occurring and recombinant 39K protein from yeast are identical. The following biochemical properties of r39K protein are described. First, r39K binds to F-actin and acidic phospholipids in a Ca^{2+}-dependent manner. Second, anti r39K antibody reactive protein is localized in cytosol and inner surface of cell membranes, and is not excreted from cells. Third, the r39K inhibits PLA_2 in vitro. Fourth, the r39K can be phosphorylated by protein kinase C and EGF receptor/kinase only when bound onto membrane surface with acidic phospholipids.

Predominantly because of various inflammatory activities and suppressed generation of various inflammatory mediators, the concept that they belong to the family of anti-inflammatory proteins has generally been accepted. However, there are several investigators who hold divergent views on such functions of lipocortins (Isacke et al. 1987; Bienkowski et al. 1989). In fact, experiments with various cell types revealed that neither lipocortin-like proteins nor their mRNA were induced by anti-inflammatory steroids. Some of the lipocortin-like proteins have no conventional signal peptides required for protein secretion and, hence, are not secreted by treatment with steroid hormones (Isacke et al. 1989).

The lipocortin-like proteins are the major substrates for tyrosine kinase, oncogene products and EGF receptor/kinase (tyrosine kinase). The proteins also serve as substrates for protein kinase C that plays crucial roles in signal transduction. Furthermore, like proteins in cytoskeletal systems, lipocortins are fairly basic in nature and, hence they interact with actin, spectrin, and fodrin in the presence of intracellular levels of Ca^{2+}. Furthermore, human lipocortin and rat gene products have similar amino acid sequences (Munn and Mueis 1988). These observations suggest that the proteins might play important roles in the regulation of not only metabolism but also cellular growth, transformation, and differentiation. Further studies on the role of interaction of 39K and other lipocortin-like proteins with the inner surface of plasma membranes with respect to the change in intracellular Ca^{2+} status will shed new light on the mechanism for transmembrane signaling.

ACKNOWLEDGMENTS

This work was supported in part by a Grant-in-Aid from the Ministry of Education, Science and Culture of Japan. The authors wish to thank Dr. K. Owada for the gift of A-431 cells and Mrs. Taeko Okanishi and Miss Yoko Okazoe for preparing the manuscript.

REFERENCES

Ahn NG, Bienkowski MJ, McMullen BA, Lipkin EW, Haen CD (1988) Sedimentation equilibrium analysis of five lipocortin-related phospholipase A_2 inhibitors from human placenta. J Biol Chem 263:18657-18663

Berridge MJ (1984) Inositol trisphosphate and diacylglycerol as second messengers. Biochem J 220:345-360

Berridge MJ and Irvine RF (1984) Inositol trisphosphate, a novel second messenger in cellular signal transduction. Nature 312:315-320

Bienkowski MJ, Petro MA, Robinson LJ (1989) Inhibition of thromboxian synthesis in U937 cells by glucocorticoids. J Biol Chem 264:6536-6544

Binsbergen JV, Slotboom AJ, Aarsnab AJ, deHass GH (1989) Synthetic peptide from lipocortin I has no phospholipase A$_2$ inhibitory activity. FEBS Lett 247:293-297

Blanquet PR, Paillard S, Courtois Y (1988) Influence of fibroblast growth factor on phosphorylation and activity of a 34 kDa lipocortin-like protein in bovine epithelial lens cells. FEBS Lett 229:183-187

Boni LY and Rando RR (1985) The nature of protein kinase C activation by physically defined phospholipid vesicles and diacylglycerols. J Biol Chem 260:10819-10825

Cirino G, Flower RJ, Browning JL, Sinclair LK, Pepinksy RB (1987) Recombinant human lipocortin I inhibits thromboxan release form guinea-pig isolated perfused lung. Nature 328:270-272

Cirino G, Peers SH, Flower RJ, Browning JL, Pepinsky RB (1989) Human recombinant lipocortin I has acute local anti-inflammatory properties in the rat paw edema test.
Proc Natl Acad Sci USA 86:3428-3432

Clark MA, Littlejohn D, Conway TM, Mong S, Steiner S, Crooke ST (1986) Leukotriene D$_4$ treatment of bovine aortic endothelial cells and murine smooth muscle cells in culture results in an increase in phospholipase A$_2$ activity. J Biol Chem 261:10713-10718

Cohen S and Fava RA (1985) Internalization of functional epidermal growth factor: receptor/kinase complexes in A-431 cells. J Biochem 260:12351-12358

Comera C, Rothhut B, Cavadore J, Vilgrain I, Cochet C, Chambaz E, Russo-Marie F (1989) Further characterization of four lipocortins from human peripheral blood mononuclear cells.
J Cell Biochem 40:361-370

Cooper JA and Hunter T (1982) Discrete primary location of tyrosine protein kinase and of three proteins that contain phosphotyrosine in virally transformed chick fibroblasts. J Cell Biol 94:287-296

Creutz CE, Zaks WJ, Hamman HC, Crane S, Martin WH, Gould KL, Oddie KM, Parsons SJ (1987) Identification of chromaffin granule-binding proteins. J Biol Chem 262:1860-1868

Crompton MR, Owens RJ, Totty NF, Moss SE, Waterfield MD, Crumpton MJ (1988) Primary structure of the human, membrane-associated Ca^{2+}-binding protein p68 a novel member of a protein family.
EMBO J 7:21-27

Davidson EF, Dennis EA, Powell M, Glenney JR (1987) Inhibition of phospholipase A$_2$ by "Lipocortins" and calpactins. An effect of binding to substrate phospholipids. J Biol Chem 262:1698-1705

De BK, Misono KS, Lukas TJ, Moroczkowski B, Cohen S (1986) A calcium-dependent 35-kilodalton substrate for epidermal growth factor receptor/kinase isolated from normal tissue.
J Biol Chem 261:13784-13792

Fava RA and Cohen S (1984) Isolation of a calcium-dependent 35-kilodalton substrate for the epidermal growth factor receptor/kinase from A-431 cells. J Biol Chem 259:2636-2645

Funakoshi T, Heirmark RL, Hendrikson LE, McMullen BA, Fujikawa K (1987) Human placental anticoagulant protein: isolation and characterization. Biochemistry 26:5572-5578

Geisow MJ, Fritsche U, Hexham JM, Dash B, Johnson T (1986) A consensus amino-acid sequence repeat in Torpedo and mammalian Ca^{2+}-dependent membrane-binding proteins. Nature 320:636-638

Gerke V and Weber K (1984) Identity of p36K phosphorylated upon Rous sarcoma virus transformation with a protein purified from brushborders; calcium-dependent binding to non-erythroid spectrin and F-action.
EMBO J 3:227-233

Glenney JR Jr (1985) Phosphorylation of p36 in vitro with pp60^{src}. Regulation by Ca^{2+} and phospholipid.
FEBS Lett 192:79-82

Haigler HT, Schlaepfer DD, Burgess WH (1987) Characterization of lipocortin I and an immunologically unrelated 33-kDa proteins as epidermal growth factor receptor/kinase substrates and phospholipase A$_2$ inhibitors. J Biol Chem 262:6921-6930

Hayashi H, Owada MK, Sonobe S, Kakunaga H, Yano J (1987) A 32-kDa protein associated with phospholipase A$_2$-inhibitory activity from human placenta. FEBS Lett 223:267-272

Hirata F (1981) The regulation of lipomodulin, a phospholipase inhibitory protein, in rabbit neutrophils by phosphorylation. J Biol Chem 256:7730-773

Horseman ND (1989) A prolactin-inducible gene product which is a member of the calpactin/lipocortin family.
Mol Endo 3:773-77

586

Huang KS, Wallner BP, Mattaliano RJ, Tizard R, Burne C, Frey A, Hession C, McGray P, Sinclair LK, Chow EP, Browning JL, Ramachandran KL, Tang J, Smart JE, Pepinsky RB (1986) Two human 35 kd inhibitors of phospholipase A_2 are related to substrates of pp60^{v-src} and of the epidermal growth factor receptor/kinase. Cell 46:191-199

Inoue M, Kishimoto A, Takai Y, Nishizuka Y (1977) Studies on a cyclic nucleotide-independent protein kinase and its proenzyme in mammalian tissues. II. Proenzyme and its activation by calcium-dependent protease from rat brain. J Biol Chem 252:7610-7616

Isack CM, Lindberg RA, Hunter T (1989) Synthesis of p36 and p35 is increased when U-937 cells differentiate in culture but expression is not inducible by glucocorticoids. Mol Cell Biol 9:232-240

Ito H, Fukuda Y, Murata K, Kimura A (1983) Transformation of intact yeast cells treated with alkali cations. J Bacteriol 153:163-168

Iwasaki A, Suda M, Nakao H, Nagoya T, Saino Y, Arai K, Mizoguchi T, Sato F, Yoshizaki H, Hirata M, Miyata T, Shidara Y, Murata M, Maki M (1987) Structure and expression of cDNA for an inhibitor of blood coagulation isolated from human placenta: a new lipocortin-like protein. J Biochem 102:1261-1273

Kikkawa U, Takai Y, Minakuchi Y, Inohara S, Nishizuka Y (1982) Calcium-activated, phospholipid-dependent protein kinase from rat brain. Subcellular distribution, purification and properties. J Biol Chem 257:13341-13348

Konig B, Di Nitto PA, Blumberg PM (1985) Phospholipid and Ca^{++} dependency of phorbol ester receptors. J Cell Biochem 27:255-265

Lecompte MF, Rosenberg I, Gitler C (1984) Membrane iserion of prothrombin. Biochem Biophys Res Comm 125:381-386

Machozaek K, Fischer M, Soling H (1989) Lipocortin I and lipocortin II inhibit phosphoinositide- and polyphosphoinositide-specific phospholipase C. The effect results from interaction with the substrates. FEBS Lett 251:207-212

Miele L, Cordella-Miele C, Facchiano A, Mukherjee AB (1988) Novel anti-inflammatory peptides from the region of highest similarity between uteroglobin and lipocortin I. Nature 335:726-730

Moore PB and Dedman JR (1987) Calcium-dependent protein binding to phenothiazine columns. J Biol Chem 257:9663-9667

Munn TZ and Mues GI (1988) Human lipocortin similar to ras gene products. Nature 322:314-315

Nishizuka Y (1984) The role of protein kinase C in cell surface signal transduction and tumor promotion. Nature 308:693-698

Nobori K, Okimasu E, Sato EF, Utsumi K (1987) Activation of proteinkinase C with cardiolipin-containing liposomes in relation to membrane-protein interaction. Cell Struct Funct 12:375-385

Okimasu E, Shiraishi N, Kobayashi S, Morimoto YM, Miyahara M, Utsumi K (1982) Permeability change of phospholipid liposomes caused by pancreatic phospholipase A_2: analysis by means of phase transition release. FEBS Lett 145:82-86

Okimasu E, Fuji Y, Utsumi T, Yamamoto M, Utsumi K (1986) Cytoplasmic proteins, association with phospholipid vesicles and its dependency on cholesterol. Cell Struct Funct 11:273-283

Okimasu E, Nobori K, Kobayashi S, Suzaki E, Terada S, Utsumi K (1987) Inhibitory effect of cholesterol on interaction between cytoplasmic actin and liposomes, and resorptive effect of high osmotic pressure. Cell Struct Funct 12:187-195

Pepinsky RB, Tizard R, Mattaliano RJ, Sinclair LK, Miller GT, Browning JL, Chow EP, Burne C, Huang KS, Pratt D, Wachter L, Hession C, Frey AZ, Wallner BP (1988) Five distinct calcium and phospholipid binding proteins share homology with lipocortin I. J Biol Chem 263:10799-10811

Powell MA and Glenney JR (1987) Regulation of calpactin I phospholipid binding by calpactin I light-chain binding and phosphorylation by pp60^{v-src}. Biochem J 247:321-328

Rothhut B, Comera C, Prieur B, Errasfa M, Minassian G, Russo-Marie F (1987) Purification and characterization of a 32-kDa phospholipase A_2 inhibitory protein (lipocortin) from human peripheral blood mononuclear cells. FEBS Lett 219:169-175

Rotman A, Heldman J, Linder S (1982) Association of membrane and cytoplasmic proteins with the cytoskeleton in blood platelets. Biochemistry 21:1713-1719

Sakata T, Iwagami S, Tsuruta Y, Suzuki R, Hojo K, Sato K, Teraoka H (1988) Mouse lipocortin I cDNA. Nucleic Acids Res 16:11818

Sato EF, Morimoto YM, Matsuno T, Miyahara M, Utsumi K (1987) Neutrophil specific 33 kDa protein: its Ca²⁺- and phospholipid-dependent intracellular translocation. FEBS Lett 214:181-186

Sato EF, Miyahara M, Utsumi K (1988a) Purification and characterization of a lipocortin-like 33 kDa protein from guinea pig neutrophils. FEBS Lett 227:131-135

Sato EF, Okimasu E, Takahashi R, Miyahara M, Matsuno T, Utsumi K (1988b) Lipocortin-like 33 kDa protein of guinea pig neutrophil. Its distribution and stimulation-dependent translocation detected by monoclonal anti-33kDa protein antibody. Cell Struct Funct 13:89-96

Sato EF, Tanaka Y, Utsumi K (1989a) cDNA cloning and nucleotide sequence of lipocortin-like 33 kDa protein in guinea pig neutrophils. FEBS Lett 244:108-112

Sato EF, Tanaka Y, Edashige K, Kobuchi H, Morishita S, Sugino YN, Inoue M, Utsumi K (1989b) Expression of the cDNA encoding lipocortin-like 39 kDa protein of guinea pig neutrophils in yeast. FEBS Lett 255:231-236

Schlaepfer DD, Mehlman T, Burgess WH, Haigler HT (1987) Structural and functional characterization of endonexin II. A calcium- and phospholipid-binding. Proc Natl Acad Sci USA 84:6078-6082

Sudhof TC, Ebbecke M, Walker JH, Fritsche U, Boustead C (1984) Isolation of mammalian calelectrins : a new class of ubiquitous Ca²⁺-regulated proteins. Biochemistry 23:1103-1109

Takai Y, Kishimoto A, Kikkawa U, Mori T, Nishizuka Y (1979) Unsaturated diacylglycerol as a possible messenger for the activation of calcium-activated, phospholipid-dependent proteinkinase system. Biochem Biophys Res Comm 91:1218-1224

Tamaki F, Nakamura E, Nishikubo C, Sakata T, Shin M, Teraoka H (1987) Rat lipocortin I cDNA. Nucleic acids Res 15:7637

Tanaka Y, Ashikari T, Shibano Y, Amachi T, Yoshizumi H, Matsubara H (1988) Construction of a human cytochrome c gene and its functional expression in *Saccharomyces cerevisiae*. J Biochem 103:954-961

Utsumi K and Sato EF (1990) Ca²⁺-dependent phospholipid binding 39 kDa protein of neutrophils with special reference to stimulation coupled lipid-protein interaction. In: Dedman JR and Smith VL (ed) Stimulus-response coupling: the role of intracellular calcium. The Telford Press, pp 417-449

Utsumi K, Okimasu E, Takehara Y, Watanabe S, Miyahara M, Moromizato Y (1981) Interaction of cytoplasmic protein with liposomes and their cell specificity. FEBS Lett 124: 257-260

Utsumi K, Okimasu E, Morimoto YM, Nishihara Y, Miyahara M (1982) Selective interaction of cytoskeletal proteins with liposomes. FEBS Lett 141:176-180

Utsumi K, Sato E, Okimasu E, Miyahara M, Takahashi R (1986) Calcium-dependent association of 33 kDa protein in polymorphonuclear leukocytes with phospholipid liposomes containing phosphatidylserine or caldiolipin. FEBS Lett 201: 277-281

Varticovski L, Chahwala SB, Whitman M, Cantley L, Schindler D, Chow EP, Sinclair LK, Pepinsky RB (1988) Location of sites in human lipocortin I that are phosphorylated by protein tyrosin kinases and protein kinase A and C. Biochemistry 27:3682-3690

Wallner BP, Mattaliano RJ, Hession C, Cate RL, Tizard R, Sinclair LK, Foeller C, Chow EP, Browning JL, Ramachandran KL, Pepinsky RB (1986) Cloning and expression of human lipocortin, a phospholipase A₂ inhibitor with potential anti-inflammatory activity. Nature 320:77-81

THE ANTI-INFLAMMATORY ACTIVITY OF HUMAN RECOMBINANT LIPOCORTIN 1

G. Cirino and R.J. Flower

INTRODUCTION

Glucocorticoids are widely used as potent anti-inflammatory drugs to treat many inflammatory and allergic diseases. The mechanism of action of these drugs has been widely researched and over the last ten years several hypotheses have evolved to account for the way in which steroids bring about their dramatic anti-inflammatory effects.

Some years ago the most popular notion (Weissmann and Thomas 1964) was that glucocorticoids stabilized lysosomal membranes by a physicochemical interaction with the bilayer, but the rapid developments which took place in the field of eicosanoid research in the early 1970's, particularly the observation that aspirin-like drugs inhibited the production of prostaglandins, focused the attention of many workers upon the ability of glucocorticoid steroids to control the production of pro-inflammatory lipids. It was found subsequently that steroids had a potent inhibitory action on prostaglandin generation but that unlike the aspirin-like drugs this was not brought about by direct inhibition of the cyclo-oxygenase enzyme but seemed to require that the target cells were intact and viable. It was also observed that the inhibitory effect could be reversed by the addition of substrate fatty acids (Gryglewski et al. 1975; Hong and Levine 1976; Nijkamp et al. 1976). Further work demonstrated that the steroid action on prostaglandin biosynthesis was confined to the glucocorticoids, that there was a latency in their action and that the block seemed to occur at the level of the substrate supply (Blackwell et al. 1978). Furthermore it was demonstrated that the blocking action of steroids was itself prevented by inhibition of macromolecule biosynthesis (Danon and Assouline 1978; Flower and Blackwell 1979) or by glucocorticoid receptor antagonists and, significantly, that a factor generated by steroid treated cells was able to reproduce the

590

prostaglandin inhibitory effect of the steroids (Flower and Blackwell 1979). This factor was shown eventually to be a protein of about 40 kDa (Blackwell et al. 1982) and was subsequently named lipocortin (Di Rosa et al. 1984). Several similar proteins were observed and all members of this family of proteins can inhibit phospholipase A_2 activity in vitro (Pepinsky et al. 1988). Partially purified preparation of these proteins also inhibit the release of eicosanoids and lyso-PAF (Parente and Flower 1985) from cells and inhibit carrageenan-induced pleurisy (Blackwell et al. 1982) and paw oedema in the rat (Parente et al. 1984), but lack of adequate amounts of highly purified protein prevented more definitive experimental work.

Lipocortin 1 was the first member of this protein family to be sequenced and cloned (Wallner et al. 1986). Subsequently an entire group of related proteins that bind to phospholipids in the presence of calcium were identified. Generically this family of proteins has continued to be called lipocortins or annexins (Crumpton and Dedman 1990) yet independent discoveries of individual members have generated over a dozen different names (reviewed by Crompton et al. 1988; Flower 1988; Klee 1988; Wallner 1989). Sequences for seven distinct but related human proteins have now been determinated by cloning. The proteins are composed of four or eight repeated homologous domains, each comprising about 70 amino acids. Each repeat contains a calcium binding site that is unrelated to previously defined calcium binding structures such as the EF-hand.

Many roles have been postulated for these proteins including signal transduction, glucocorticoid-like effects, components of exocytotic machinery, calcium channels, cytoskeletal elements, anti coagulation, modulation of the immune response and collagen binding. At the time of writing a coherent universally-accepted function for any of these proteins has not been established.

According to the original hypothesis lipocortin 1 should be anti-inflammatory because it limits arachidonic acid production, but in the last few years this aspect has become very contentious and which elements of the model are valid remain to be unequivocally established (e.g, Bienkowski et al. 1989; Davidson et al. 1987; Hullin et al. 1989; Bronnegard et al. 1988).

As a generic name, the term lipocortin implies both a lipid interaction and steroid-inducibility which may not be descriptive of all members of this family. However, lipocortin 1 does have these properties and therefore, the original nomenclature has been retained throughout this chapter. Recombinant preparations of lipocortin 1 have been shown to be anti-inflammatory in a number of systems (Cirino et al. 1989a; Errasfa and Russo Marie 1989) although Northup (1988) using naturally occurring lipocortin purified from human placental membranes did not observe anti-inflammatory activity.

In this chapter we will focus upon the general properties of lipocortin 1, the anti-inflammatory activity in carrageenan and other models of rat paw oedema as well in other *in vivo* tests of inflammation.

GENERAL PROTEIN CHEMISTRY

Lipocortin 1 is a very polar molecule, with approximately one third of its amino acids being charged. These are distributed throughout the molecule separated by short stretches of hydrophobic residues. There are four cysteines close to the C-terminal (263, 270, 324, and 343) which presumably could form disulfide bridges.

Human recombinant lipocortin 1 does not apparently contain a leader peptide and although it is clearly recovered as an extracellular protein, the mechanism whereby its release from cells is effected is unclear at the moment. Lipocortin 1 contains a single potential glycosylation site (Asn-43-Ser-45), although there is no evidence for the native species being glycosylated, and also contains the consensus sequences for both tyrosine and threonine phosphorylation sites (at Tyr 21 and Thr 212). The latter point is of special significance since it had already been demonstrated that the naturally occurring molecule could be phosphorylated, and that the phosphorylated form was inactive as a phospholipase A_2 inhibitor (Hirata 1981).

To confirm that human recombinant lipocortin 1 was a phospholipase inhibitory protein, a full length coding sequence was expressed in *E. coli* and crude extracts of these *E. coli* organisms, or highly purified fractions containing the recombinant protein were tested and found to be strongly inhibitory in a popular conventional cell-free phospholipase A_2 enzyme assay (Pepinsky et al. 1986a).

Huang et al. (1986) observed that human placental extracts contained two types of lipocortin-like proteins. One species was identical to the recombinant lipocortin 1 protein previously cloned but the other (christened lipocortin 2) was a different protein. Like lipocortin 1, lipocortin 2 does not contain a signal sequence; both proteins are very polar molecules with approximately one third of their total amino acids being charged. Overall there was approximately 50% sequence homology suggesting that the genes for the two proteins arose by gene duplication. It was within the central region of lipocortin 2 that the greatest homology with lipocortin 1 was observed. Interestingly the sequences near the N-terminus of the two proteins were substantially different although both contained a sequence which could be phosphorylated by tyrosine kinases.

Both lipocortin 1 and 2 have very comparable molecular weight and similar pI values (7.9) both are inhibitors of phospholipase A_2 activity and require the presence of calcium before association with membranes can occur. Lipocortin 2, has approximately the same anti-phospholipase activity as lipocortin 1 and purified preparations of human recombinant lipocortin 2 from yeast are active in the carrageenin paw oedema in rat when given locally (Parente et al. 1990). However, it will be some time before a complete profile of the biological activity of lipocortin 2 is available.

LIPOCORTIN 1 AND OTHER MEMBRANE ASSOCIATED PROTEINS

Kretsinger and Creutz (1986) observed the presence of three repeating consensus sequences in lipocortin 1, and Munn and Mues (quoted in Huang et al. 1986) have gone further and suggested that, except for the first 43 amino acids the primary structure of lipocortin 1 is built from 4 repeats of a single unit. The same basic structure has been observed in lipocortin 2 and similar sequences are also seen in some other calcium and lipid binding proteins (see Table 1). In fact, some recent work has demonstrated that the cytoskeletal protein previously referred to as calpactin 2, is in all probability, identical to lipocortin 1 (Glenney et al. 1987).

It is likely that lipocortin 1 and lipocortin 2 are related, or identical to, some other hitherto unsequenced proteins and may belong to a super-family of membrane-associated proteins. For example, a very important discovery made by the Biogen group (Pepinsky and

Sinclair 1986; Huang et al. 1986) was that lipocortin 1 was identical to the substrate for the EGF receptor kinase and that lipocortin 2 was identical to the pp60 src substrate.

Table 1. Lipocortin and other calcium and phospholipid binding proteins

M_r (kDa)	Alternative nomenclature/identity
37 *	Lipocortin 1; Annexin 1;P35 EGF receptor kinase substrate; Chromobindin 9; Calpactin 2
37 *	Lipocortin 2; Annexin 2; P36 PP60V-SRC substrate; Chromobindin 8; Calpactin 1; Protein 1
35	Lipocortin 3; Annexin 3; Endonexin 1; Chromobindin 4; Calcimedin p35; Calelectrin p32.5; Renocortin
35	Lipocortin 4; Annexin 4; Protein 2
35	Lipocortin 5; Annexin 5; Chromobindin 5; Endonexin 2; Placental anti-coagulant factor 2
68	Lipocortin 6; Annexin 6; Protein 3; P68; Chromobindin 20; Calcimedin p67; Calelectrin p70
47	Lipocortin 7; Annexin 7; Synexin 1

* Dimers have been reported for both species at 68 kDa

BIOLOGICAL ACTIVITY OF LIPOCORTIN 1

Cloning of the human lipocortin 1 gene was a major breakthrough because it enabled large amounts of the protein to be produced for biological testing. Because there were several, apparently distinct, anti-phospholipase components in peritoneal lavage fluid, an obvious question was whether lipocortin 1 could account for the properties observed in the early experiments with macrocortin containing extracts.The activity of lipocortin 1 has been checked in three systems known to be sensitive to macrocortin they are; the guinea-pig perfused lung system, the release of eicosanoids by endothelial cells and the carrageenin-induced paw oedema in the rat.

594

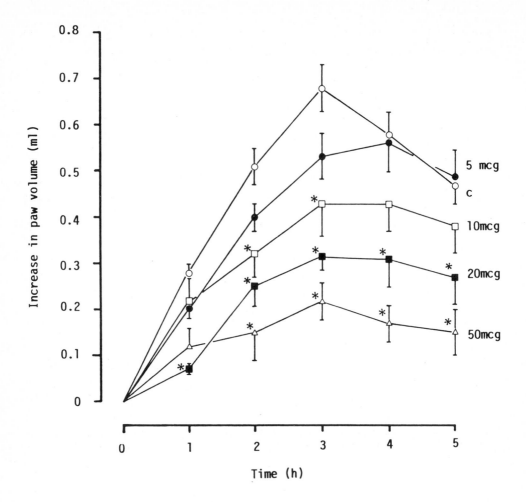

Fig. 1. A series of matched experiments demonstrating the dose dependent inhibition of carrageenin-induced rat paw oedema by 5-50 µg of lipocortin. Data are expressed as the mean oedema (ml) ± SEM at each point. Values marked with an *asterisk* are statistically different from control values (O) with a minimum level of significance of P< 0.05. Data from lipocortin doses of 5 µg (•), 10 µg (□), 20 µg (■), and 50 µg (△) are shown

Cirino et al. and Cirino and Flower (1987) demonstrated that recombinant human lipocortin 1 (ex *E. coli*) strongly inhibited the release of thromboxane A_2, induced by leukotriene C_4, from isolated perfused guinea-pig lung and of prostacyclin by human umbilical arterial endothelial cells.

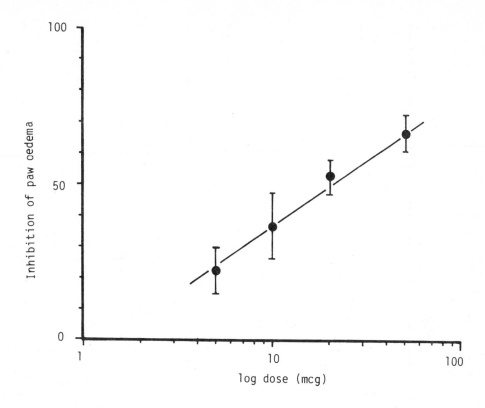

Fig. 2. Dose effect relationship between four different doses of lipocortin calculated as percent inhibition of the maximum oedema at the 3 h time point

When dealing with the anti-inflammatory activity of recombinant proteins, particularly when derived from *E. coli* (the most popular source), the perennial problem of purity assumes prominence. Even when the protein is (as in the case of lipocortin 1) greater than 99% pure, traces of *E. coli* proteins, endotoxin or DNA can cause false positive and negatives in many experiments, particularly those involving isolated leukocytes or models of local inflammation. To circumvent this type of problem a particular type of control was devised for use in our work. Sham lipocortin 1 was a preparation produced by extracting *E. coli* organisms, which did not contain the lipocortin 1 plasmid and processing the extract in an identical fashion to the preparations produced from organisms which did. In this way, one obtains a control sample containing both qualitatively and quantitatively the same contaminants as the lipocortin 1 preparations, but without the active protein.

EFFECT OF LIPOCORTIN ON CARRAGEENIN, BRADYKININ, DEXTRAN, PAF, AND COMPOUND 48/80 OEDEMA IN THE RAT

To test the ability of lipocortin 1 to inhibit carrageenin-induced oedema, the protein has been administered locally and systemically (i.v. and i.p.) to rats. Figure 1 shows a series of matched experiments in which various doses of lipocortin 1 were given locally. The inhibition seen at the time of the maximal oedema (3 h) was linearly related (Figure 2) to the logarithm of dose, and the ED_{50} fell in the range of 10-20 µg (0.27-0.54 nmol). Injection of 100 µg of lipocortin 1 (not shown in the figure), produced no greater inhibition than that of 50 µg. Thus, the maximal inhibition in this test is 60-70% at the 3 h point.

Boiled lipocortin 1 (100 °C for 15 min) was without effect at any time in this assay. Sham lipocortin 1 was also tested at a dose corresponding to the amount of *E. coli* material contained in the active lipocortin preparation. No inhibition was seen, in fact a substantial potentiation was seen relative to the saline-injected control.

Finally, to exclude the possibility that lipocortin 1 preparations contained a factor that stimulated the adrenal-cortex, thereby reflexly suppressing the inflammatory response, the protein was tested in adrenalectomized rats. An inhibitory effect indistinguishable from that observed in normal rats was seen. Subsequently tested was the ability of lipocortin 1 to inhibit carrageenan oedema once the inflammation was established.

In this test, a dose of lipocortin 1 (10 µg) was administered 1 and 2 h after carrageenin. There was a reduction in the oedema relative to controls in both treated groups at subsequent time points, which achieved statistical ($P < 0.05$) significance within 1 h of dosing and which had reached $77.8 \pm 11\%$ and $74 \pm 10.6\%$ inhibition (relative to vehicle-treated controls), respectively 1 or 2 h after giving the protein.

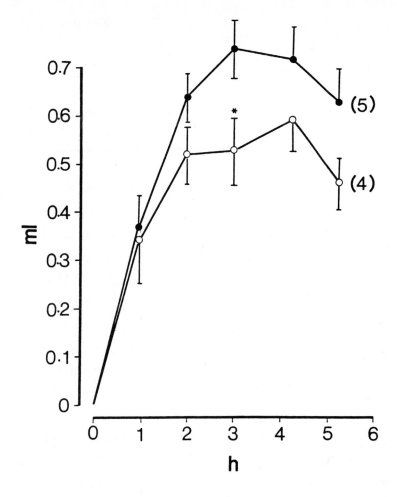

Fig. 3. Effect of an ED$_{50}$ dose (20 µg/paw) of lipocortin (o) in rats made leukopenic with methotrexate. Control rats (•) received locally the same volume of vehicle. Data are expressed as mean oedema (ml) ± SEM, and values marked with as *asterisk* are significantly different from control values at a significance level of P< 0.05

Carrageenan oedema depends upon several factors, including the release of mast-cell amines, the generation at the site of inflammation of eicosanoids, the migration of polymorphonuclear leukocytes and the action of kinins.

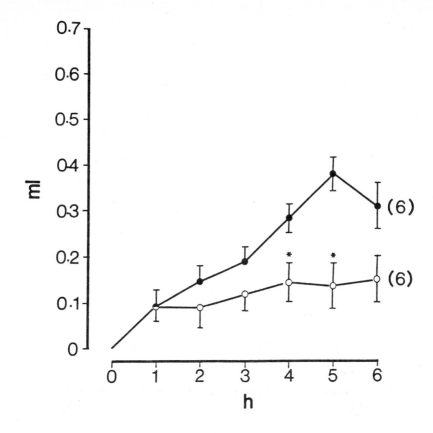

Fig. 4. Effect of ED_{50} (20 µg/paw) dose of lipocortin (o) in rats previously depleted of kininogen with cellulose sulfate. Control rats (•) received locally the same volume of vehicle. Data are expressed as mean oedema (ml) ± SEM, and values marked with an *asterisk* are significantly different from control values at a significance level of $P < 0.05$

Because lipocortin 1 generally produced a maximal inhibition of 60-70%, it was evident that, unlike the glucocorticoids themselves, lipocortin 1 could not inhibit all components of the inflammatory response. To determine which mediators were sensitive to lipocortin 1, a series of depletion experiments was performed in which an ED_{50} dose of lipocortin 1 (20 µg) was tested in rats pretreated with agents to modify various aspects of the inflammatory response.

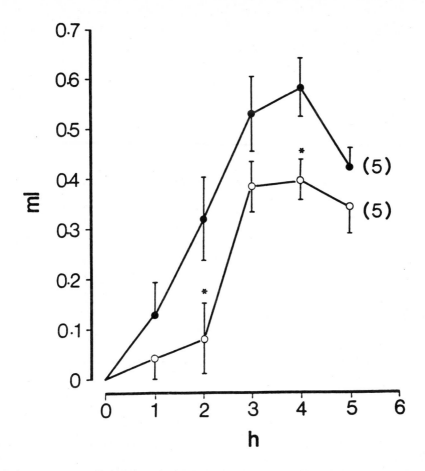

Fig. 5. Effect of an ED_{50} dose (20 μg/paw) of lipocortin (o) in rats previously depleted of mast-cell amines with 48/80. Control (•) rats received locally the same volume of vehicle. Data are expressed as mean oedema (ml) ± SEM, and values marked with an asterisk are significantly different from control values at a significance level of $P < 0.05$

Treatment of rats with agents that depleted mast-cell amines, kininogen or eicosanoids reduced the overall oedema response, indicating that one or more inflammogenic components had been removed. Treatment with methotrexate reduced the lipocortin effect somewhat (Figure 3). Lipocortin 1 reduced the residual oedema further after rats were treated with cellulose sulfate to remove kininogen (Figure 4) or compound 48/80 (Figure 5) to remove mast-cell amines (Di Rosa et al. 1970a,b). When rats received all three depleting agents, lipocortin 1 was strikingly effective in reducing oedema (Figure 6).

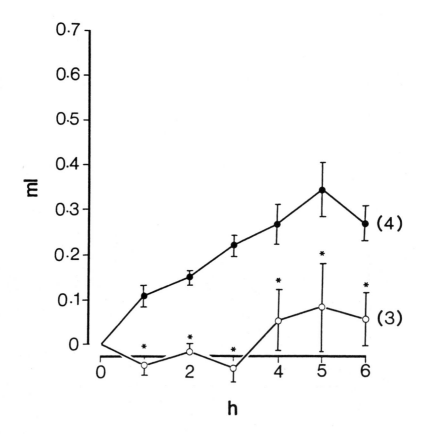

Fig. 6. Effect of an ED_{50} dose (20 µg/paw) of lipocortin (o) in rats previously depleted of mast-cell amines with compound 48/80, kininogen with cellulose sulfate, and made leukopenic with methotrexate. Control rats (•) received locally the same volume of vehicle. Data are expressed as mean oedema (ml) ± SEM, and values marked with an *asterisk* are significantly different from control values at a significance level of $P< 0.05$

In these fully depleted rats there was a small peripheral oedema in all paws, which was itself lipocortin 1 sensitive, and treatment with the latter reduced the basal-paw volume as well as the carrageenan-induced oedema (Figure 6). Lipocortin 1 could not further reduce the oedema of rats that had been depleted of eicosanoids by prior treatment with a maximal dose

(50 mg/kg) of the dual cycloxygenase and lipoxygenase inhibitor BW 755C (Figure 7). A puzzling feature observed after treatment with methotrexate was the reduction of the effect of lipocortin 1.

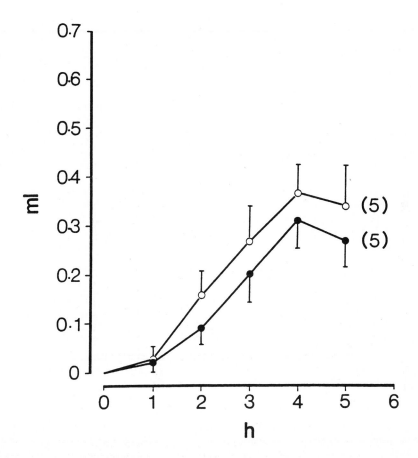

Fig. 7. Lipocortin (o) 20 μg/paw (ED$_{50}$) does not increase the inhibitory effect of a maximal dose of BW755C (•) in carrageenin-induced paw oedema. Both groups received 50 mg/kg po of BW775C. Data are expressed as mean oedema (ml) ± SEM, and values marked with an *asterisk* are significantly different from control values at a significance level of P< 0.05

This observation could mean that lipocortin 1 exerts some of its action by decreasing the migration of these cells (Figure 3).

The mechanism of action of locally administered lipocortin 1 was further studied in a series of experiments in which inflammation was induced by serotonin, bradykinin, dextran or platelet activating factor (PAF). A fixed dose of lipocortin 1 (20 µg=ED$_{50}$) was ineffective when oedema was produced by any of these agonists (Table 2).

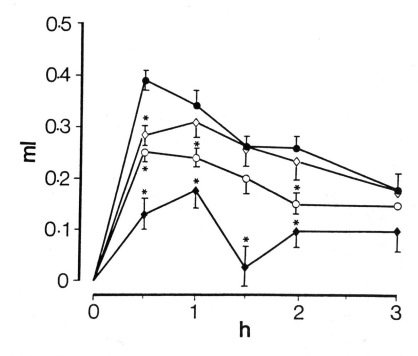

Fig. 8. Effect of lipocortin, indomethacin and dexamethasone on the oedema produced by compound 48/80 in the rat paw. The data are expressed as mean oedema (ml) ± SEM, and values marked with an *asterisk* are statistically different from controls at a minimum significance level of $P < 0.05$. Control oedema (•); 40 µg of indomethacin given locally into the paw (◊); 20 µg of lipocortin given locally into the paw (○); and dexamethasone at 1 mg/kg given 3 h before the inflammogenic stimulus (♦)

Indomethacin and BW755C in doses that completely suppress eicosanoid formation were also without effect in these tests whereas glucocorticoids invariably inhibited the inflammation.

TABLE 2. Summary of lipocortin effects in other inflammatory models

Oedema (ml+/-S E M (n)) in response to

Drug treatment	Dextran 0.1ml, 3% soln.	PAF-Acether 1µg	Bradykinin 25µg	Serotonin 10µg
Saline Only	0.24+/-0.06 (5)	0.33+/-0.003 (13)	0.52+/-0.02 (18)	0.55+/-0.03 (20)
20µg lipocortin %inhibition (p)	0.34+/-0.05 (5) 19.0% (NS)	0.28+/-0.03 (10) 15.2% (NS)	0.44+/-0.05 (10) 15.4% (NS)	0.54+/-0.08 (9) 1.8% (NS)
1mg/kg dexamethasone %inhibition (p)	NT	0.08+/-0.02 (10)* 75.8% (<0.001)	0.37+/-0.04(8) 28.9% (<0.02)	0.32+/-0.05 (10) 41.8% (<0.001)
10µg methyergide %inhibition (p)	NT	0.34+/-0.03 (12)* (NS)	NT	0.13+/-0.09 (5) 76.4% (<0.001)
5mg/kg indomethacin %inhibition (p)	NT	0.30+/-0.03 (12)* 9.1% (NS)	0.45+/-0.07 (8) 13.5% (NS)	NT
50mg/kg BW755C % inhibition (p)	NT	0.30+/-0.03 (5) 9.1% (NS)	0.51+/-0.04 (5) 1.9% (NS)	0.54+/-0.04 (10) 1.8% (NS)

* Data taken from Cirino et al. 1989a
NT= Not tested in this laboratory
NS= Not significant at the 5% level

When oedema was induced by compound 48/80 however, lipocortin 1 produced a small ($35.9 \pm 5.1\%$), but significant ($P> 0.001$), inhibition at the time of the maximal oedema as did 40 µg of indomethacin ($28 \pm 5\%$; $P< 0.001$). These data strongly suggest that eicosanoids were involved at the time of the maximal oedema as well as mast-cell amines. Once again, dexamethasone inhibited the oedema at all time points (Figure 8).

These data demonstrate that locally-administered human recombinant lipocortin 1 possesses anti-inflammatory properties in a well-characterized model of acute inflammation and support an earlier contention regarding the biological activity of this protein. Indomethacin locally administered as an inhibitor of carrageenin-induced paw oedema has an ED_{50} of 30 µg (84 nmol) while the ED_{50} for lipocortin is 150-fold more active on a molar basis at the 3 h time point than is indomethacin. In doses as high as 5 mg/kg orally (which almost completely suppress prostaglandin production) indomethacin produced a maximal inhibition of $62.8 \pm 2.7\%$ (n=30) of carrageenan oedema at the 3 h time point a maximum very similar to that of lipocortin 1.

The failure of lipocortin 1 to produce any anti-inflammatory effects when tested against many of the agents mentioned above indicates that lipocortin 1 is selective in its action. Because glucocorticoids are effective against all forms of oedema it is obvious that an additional mechanism must be invoked to explain how the glucocorticoids bring about other types of anti-inflammatory action. Carnuccio et al. (1987) have suggested that the induction of another protein vasocortin mediates the ability of steroids to reduce dextran oedema.

In addition to studies in which lipocortin 1 was administered locally, other work demonstrates that the protein inhibits carrageenin oedema when given intravenously. Browning et al. (1990) used doses of lipocortin between 50-100 µg (0.3-0.6 mg/kg) administered at the same time as the carrageenin. A substantial inhibition was observed comparable to that seen with a dose of dexamethasone of 1 mg/kg (s.c.) given 2 h prior to the carrageenin. In these experiments recombinant preparations either from *E. coli* or from CHO cells were tested. In many recombinant (and even natural preparations) there are a substantial number of inactive conformers present in the mixture. As shown in Figure 9 the refolded conformer (given i.v.) can inhibit the inflammatory response to carrageenin whereas the nonrefolded lipocortin 1 is completely without effect. The apparent lack of activity in the nonrefolded preparations may not indicate a complete loss of activity, but rather that the effective dose may have dropped into an undetectable range. Conformer presence and related dosage considerations may be an explanation for the inactivity of some natural lipocortin 1 preparations in this assay (Northup et al. 1988).

Other experiments which demonstrate a systemic action of recombinant lipocortin 1 were reported by Errasfa and Russo-Marie (1989). In these studies inflammation was induced in mice by the subcutaneous injection of polyacrylamide gel. Human recombinant lipocortin 1, as well as a natural preparation of a 36 kDa lipocortin from mouse tissues, strongly inhibited the migration of cells into the lesion apparently by reducing the local biosynthesis of leukotriene B_4.

To conclude this section it is worth noting the interesting work reported by Miele et al. (1988). A region of strong homology between lipocortin 1 and another (sex) steroid-induced protein with anti-phospholipase properties, uteroglobin, prompted the manufacture of two nonapeptides "Antiflammins" based upon the most common conserved sequence. These peptides proved to have anti-phospholipase and anti-inflammatory properties reminiscent of the intact molecule.

OEDEMA PRODUCED BY PHOSPHOLIPASE A₂

Lipocortin 1 was originally described as a protein with anti-phospholipase activity and this facet of the protein's action was tested in experiments in which oedema was induced in rat paw by a single subplantar injection of phospholipase A_2 from Naja mozambique (Cirino et al. 1989b).

Lipocortin 1, when premixed in doses of 2-20 μg with phospholipase A_2, inhibited the formation of phospholipase oedema caused by this enzyme. Maximal inhibition varied with the batch of the enzyme, but never exceeded 85% and was always achieved with equimolar ratios of lipocortin to phospholipase.

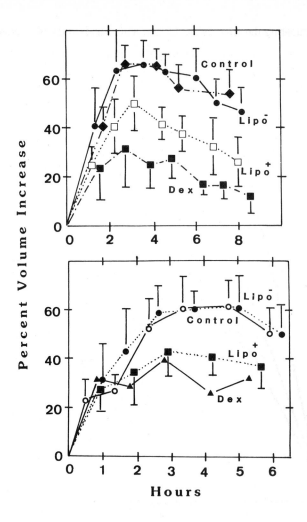

Fig. 9. Effects of protein refolding on the ability of lipocortin to block carrageenin-induced oedema in rat paw. Lipocortin was administered as a single iv dose of 0.8 mg/kg (*upper panel*) or 0.5 mg/kg (*lower panel*). *Lipo⁻* and *Lipo⁺* refer to the same preparations before and after the refolding and *Dex* refers to animals group treated with 1 mg/kg dexamethasone 2 h prior to carrageenin injection. (With permission of Browning et al. 1990)

Increasing the ratio beyond 1:1 produced no further inhibition. Figure 10 shows an experiment with a single batch of venom enzyme in which the time course of the oedema produced by injecting 5 µg of the venom phospholipase and the dose-dependent inhibition produced by pre-incubating the enzyme for 15 min with 5 and 10 μg of human recombinant lipocortin 1.

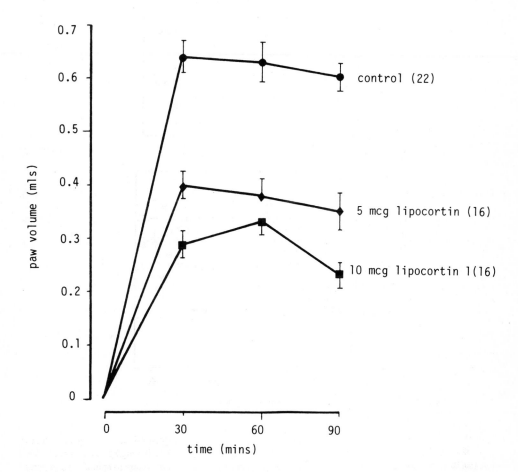

Fig. 10. Lipocortin inhibition of the oedema in rat paw caused by the injection of 5μg of venom phospholipase A_2. Data are expressed as mean oedema (ml) ± SEM, and values marked with an *asterisk* are significantly different from control values at a significance level of P< 0.01

The inhibitory activity of lipocortin was maximal when incubated for 15 min with the enzyme at 37 °C but decreased following longer incubations (e.g., 30 min) perhaps because of the instability of the protein.Phospholipase oedema is caused by mast cell degranulation which is probably secondary either to a direct lytic effect on the cells either by the phospholipase itself, or by products of its catalytic activity. The oedema can be profoundly inhibited by pretreatment of the enzyme with the active-site inhibitor p-bromophenacyl bromide. The inhibition of phospholipase-induced oedema may be biologically relevant, as lipocortin could directly neutralize the activity of extracellular phospholipases thought to be important in some forms of inflammation (Vadas 1981). This model of inflammation is the only instance in which lipocortin 1 was found to be effective against oedema caused by mast cell degranulation.

AUTO-ANTIBODIES TO LIPOCORTIN 1

Hirata and his colleagues (1981) made the important and original observation that a high proportion of patients with chronic rheumatic diseases have a detectable auto-antibodies to lipocortin in their sera, whereas disease-free people did not. The titre of antibody were particularly high in patients with systemic lupus erythematosus and rheumatoid arthritis, diseases in which auto-antibodies to a variety of cell surface proteins and frequently observed. This work led to the concept that the presence of these auto-antibodies may contribute to the aetiology of these diseases by removing protective proteins which regulate eicosanoid generation.

This observation was followed up by Goulding et al. (1989). Using recombinant human lipocortin 1 as a target these workers devised an ELISA assay for measuring anti-lipocortin 1 auto-antibodies and used this to examine the sera of selected patients attending the Royal National Hospital for Rheumatic Diseases (Bath, UK) as well as a large number of healthy controls. They confirmed the presence of both IgG and IgM anti-lipocortin 1 antibodies in both lupus and rheumatoid arthritis patients but also made two striking original observations. In patients with systemic lupus the auto-antibodies seemed present whether or not the patients had already received steroid therapy, but in the group with rheumatoid arthritis significantly elevated titre were only seen in patients who were receiving oral glucocorticoid therapy. In both groups the presence of a high titre of auto-antibodies correlated strongly with the clinical phenomenon of steroid resistance. Many other patients with other inflammatory diseases were also investigated but no auto-antibodies were detected, whether or not they were receiving steroids or were steroid resistant.

The notion that the administration of glucocorticoids may encourage the formation of auto-antibodies to lipocortin 1 in certain susceptible groups and that if present to excess these auto-antibodies may lead to steroid resistance, or at least some manifestations of it, is obviously a radical one with important implications for a correct understanding of glucocorticoid therapy.

CONCLUSIONS

Many proteins and protein extracts have been reported to possess anti-inflammatory activity, but with few exceptions (e.g., superoxide dismutase) these proteins have never been completely characterized or the activity has been found to be associated with endotoxin or reflex stimulation of the adrenal cortex. That human recombinant lipocortin 1 is anti-inflammatory is particularly significant because the glucocorticoids induce the formation of these proteins (Browning et al. 1990). Lipocortin 1 (and possibly other related proteins) seems to be a sort of second messenger of the glucocorticoids, which has the ability to control the mediators which promote development of the symptoms of the inflammatory response.

We believe that the synthesis and release of lipocortin 1 may partially explain the marked anti-inflammatory actions of steroids and that the administration of lipocortin 1 produces an anti-inflammatory action by inhibition of eicosanoid generation.

ACKNOWLEDGMENT

Some of this work was funded by grants from Biogen Research Corp. Inc. and the Wellcome Trust and the authors express their gratitude for the support. RJF is currently supported by an endowment from Lilly. Much of the data received in this chapter was reproduced with permission from Cirino et al. 1989a.

610

REFERENCES

Bienkowski MJ, Petro MA, Robinson LJ (1989) Inhibition of thromboxane A_2 synthesis in U937 cells by glucocorticoids. J Biol Chem 264:6536-6544

Blackwell GJ, Flower RJ, Nijkamp FP, Vane JR (1978) Phospholipase A_2 activity of guinea-pig isolated perfused lungs: stimulation and inhibition by anti-inflammatory steroids. Br J Pharmacol 62:78-89

Blackwell GJ, Carnuccio R, Di Rosa M, Flower RJ, Langham CSJ, Parente L, Persico P, Russell-Smith NC, Stone DF (1982) Glucocorticoids induce the formation and release of anti-inflammatory and anti-phospholipase proteins into the peritoneal cavity of the rat. Br J Pharmacol 76:185-194

Bronnegard M, Andersson O, Edwall D, Lund J, Norstedt G, Carlstedt-Duke J (1988) Human calpactin II (lipocortin 1) messenger ribonucleic acid is not induced by glucocorticoids. Mol Endocrinol 88:732-739

Browning JL, Ward MP, Wallner BP, Pepinsky RB (1990) Studies on the structural properties of lipocortin 1 and the regulation of its synthesis by steroids. In: Cytokines and lipocortin in inflammation and differentiation. Wiley-Liss Inc. New York, pp 27-45

Carnuccio R, Di Rosa M, Guerrasio B, Iuvone T, Sautebin L (1987) Vasocortin: a novel glucocorticoid-induced anti-inflammatory protein. Br J Pharmacol 90:443-445

Cirino G, Peers SH, Flower RJ, Browning JL, Sinclair LK, Pepinsky RB (1987) Recombinant human lipocortin 1 inhibits thromboxane release from guinea-pig isolated perfused lung. Nature 328:270-272

Cirino G, Flower RJ (1987). Human recombinant lipocortin 1 inhibits prostacyclin production by human umbilical artery in vitro. Prostaglandins 34:59-62

Cirino G, Peers SH, Flower RJ, Browning JL, Pepinsky RB (1989a) Human recombinant lipocortin 1 has acute local anti-inflammatory properties in the rat paw oedema test. Proc Natl Acad Sci USA 86:3428-3432

Cirino G, Peers SH, Wallace JL, Flower RJ (1989b) A study on phospholipase A_2-induced oedema in rat paw. Eur J Pharmacol 166:505-510

Crompton MR, Moss SE, Crumpton MJ (1988) Diversity in the lipocortin-calpactin family. Cell 55:1-3

Crumpton MJ, Dedman JR (1990) Protein terminology tangle. Nature 345:212

Danon A, Assouline G (1978) Inhibition of prostaglandin biosynthesis by corticosteroids requires RNA and protein synthesis. Nature 273:552-554

Davidson FF, Dennis EA, Powell M, Glenney JR (1987) Inhibition of phospholipase A_2 by "lipocortins" and calpactins. J Biol Chem 262:1698-1705

Di Rosa M, Giroud JP, Willoughby DA (1970a) Studies on the mediators of the acute inflammation response induced in rats in different sites by carrageenan and turpentine. J Pathol 104:15-29

Di Rosa M, Sorrentino L (1970b). Some pharmacodynamic properties of carrageenan in the rat.
Br J Pharmacol 38:214-220

Di Rosa M, Flower RJ, Hirata F, Parente L, Russo Marie F (1984). Nomenclature announcement. Anti-phospholipase proteins. Prostaglandins 28:441-442

Errasfa M, Russo-Marie F (1989) A purified lipocortin shares the anti-inflammatory effect of glucocorticoids in vivo in mice. Br J Pharmacol 97:1051-1058

Flower RJ (1988) Lipocortin and the mechanism of action of the glucocorticoids. Br J Pharmacol 94:987-1015

Flower RJ and Blackwell GJ (1979) Anti-inflammatory steroids induce biosynthesis of phospholipase A_2 inhibitor which prevents prostaglandin generation. Nature 278:456-459

Glenney JR, Tack B, Powell MA (1987) Calpactins: two distinct Ca^{2+}-regulated phospholipid and actin binding proteins isolated from lung and placenta. J Cell Biol 104:503-511

Goulding HJ, Podgorski MR, Hall ND, Flower RJ, Browning JL, Pepinsky RB, Maddison PJ (1989). Auto-antibodies to recombinant lipocortin 1 in rheumatoid arthritis and systemic lupus erythematosus. Ann Rheum Diseases 48:843-850

Gryglewski RJ, Panczenko B, Korbut R, Grodzinska L, Ocetkiewicz A (1975) Corticosteroids inhibit prostaglandin release from perfused mesenteric blood vessels of rabbit and from perfused lungs of sensitized guinea-pigs. Prostaglandins 10:343-355

Hirata F (1981) The regulation of lipomodulin, a phospholipase inhibitory protein, in rabbit neutrophils by phosphorylation. J Biol Chem 256:7730-7733

Hirata F, Del Carmine R, Nelson CA, Axelrod J, Schiffmann E, Warabi A, De Blas AL, Nirenberg M, Manganiello V, Vaughan M, Kumagal S, Green I, Decker JL, Steinberg AD (1981) Presence of auto-antibodies for phospholipase inhibitory protein, lipomodulin, in patients with rheumatic diseases. Proc Natl Acad Sci USA 78:3190-3194

Hong SC, Levine L (1976) Inhibition of arachidonic acid release from cells as biochemical action of anti-inflammatory steroids. Proc Natl Acad Sci USA 73:1730-1734

Huang KS, Wallner BP, Mattaliano RJ, Tizard R, Burne C, Frey A, Hession C, McGray P, Sinclair LK, Pingchong Chow E, Browning JL, Ramachandran KL, Tang J, Smart JE, Pepinsky RB (1986) Two human 35kd inhibitors of phospholipase A_2 are related to substrates of pp60 v-SRC and of epidermal growth factor receptor/kinase. Cell 46:191-199

Hullin F, Raynal P, Ragab-Thomas JMF, Fauvel J, Chap H (1989) Effect of dexamethasone on prostaglandin synthesis and on lipocortin status in human endothelial cells. J Biol Chem 264:3506-3513

Klee CB (1988) Ca^{2+}-dependent phospholipid- (and membrane) binding proteins. Biochemistry 27:6645-6653

Kretsinger RH, Creutz CE (1986) Consensus in exocytosis. Nature 320:573-575

Miele L, Cordella-Miele E, Facchiaro A and Mukherjee AB (1988) Novel anti-inflammatory peptides from the region of highest similarity between uteroglobin and lipocortin 1. Nature 335:726-730

Nijkamp FP, Flower RJ, Moncada S, Vane JR (1976) Partial purification of RCS-RF (rabbit aorta contracting substance-releasing factor) and inhibition of its activity by anti-inflammatory corticosteroids. Nature 263:479-482

Northrup JK, Valentine-Braun KA, Johnson LK, Severson DL, Hollenberg MD (1988) Evaluation of the anti-inflammatory and phospholipase-inhibitory activity of calpactin-II/lipocortin 1. J Clin Invest 82:1347-1352

Parente L, Di Rosa M, Flower RJ, Ghiara P, Meli R, Persico P, Salmon JA, Wood JN (1984) Relationship between the anti-phospholipase and anti-inflammatory effects of glucocorticoid-induced proteins. Eur J Pharmacol 99:233-239

Parente L, Flower RJ (1985) Hydrocortisone and macrocortin inhibit the zymosan-induce release of Lyso-PAF from rat peritoneal leukocytes. Life Sci 36:1225-1231

Parente L, Becherucci C, Perretti M, Solito E, Mugridge KG, Galeotti GL, Raugei G, Melli M, Sanso M (1990) Are the lipocortins the second messengers of the anti-inflammatory action of glucocorticoids? In: Cytokines and lipocortins in inflammation and differentiation. Wiley-Liss, Inc. New York, ppxx-xx

Pepinsky RB, Sinclair LK, Browning JL, Mattaliano RJ, Smart JE, Chow EP, Falber T, Ribolini A, Garwin J, Wallner BP (1986) Purification and partial sequence analysis of 37 kDa protein that inhibits phospholipase A_2 activity from rat peritoneal exudates. J Biol Chem 261:4239-4246

Pepinsky RB, Sinclair LK (1986) Epidermal growth factor-dependent phosphorylation of lipocortin. Nature 223:29-35

Pepinsky RB, Tizard R, Mattaliano RJ, Sinclair LK, Miller TG, Browning JL, Pinchang Chow E, Burne C, Huang KS, Pratt D, Watcher L, Hession C, Frey AZ, Wallner BP (1988) Five distinct calcium and phospholipid binding proteins share homology with lipocortin 1. J Biol Chem 263:10799-10811

Vadas P (1981) Extracellular phospholipase A_2: a mediator of inflammatory hyperemia. Nature 293:583-585

Wallner BP, Mattaliano RJ, Hession C, Cate RL, Tizard T, Sinclair LK, Foeller C, Pingchong Chow E, Browning JL, Ramachandran KL, Pepinsky RB (1986) Cloning and expression of human lipocortin, a phospholipase A_2 inhibitor with potential anti-inflammatory activity. Nature 320:77-81

Wallner B (1989) Lipocortins, a family of phospholipase A_2-inhibitory proteins. In: Genes and signal transduction in multistage carcinogenesis (NH Colburn, ed), Marcel Dekker, Inc. New York

Weissmann G and Thomas L (1964). The effects of corticosteroids upon connective tissue and lysosomes. Recent Prog Horm Res 20:215-239

Author Index

Subject Index

619

624